Edward J. Miers

Report on the Scientific Results of the Voyage of H.M.S. Challenger during the Years 1873-76

Edward J. Miers

Report on the Scientific Results of the Voyage of H.M.S. Challenger during the Years 1873-76

ISBN/EAN: 9783742831354

Manufactured in Europe, USA, Canada, Australia, Japa

Cover: Foto ©berggeist007 / pixelio.de

Manufactured and distributed by brebook publishing software (www.brebook.com)

Edward J. Miers

Report on the Scientific Results of the Voyage of H.M.S. Challenger during the Years 1873-76

THE

VOYAGE OF H.M.S. CHALLENGER.

ZOOLOGY.

REPORT on the BRACHYURA collected by H.M.S. Challenger during the
Years 1873-76. By EDWARD J. MIERS. F.L.S., F.Z.S.

INTRODUCTION.

THE Collection of Brachyura obtained during the Expedition of H.M.S. Challenger
contains a large number of new and very interesting types, which deserve perhaps a
fuller treatment than it has been found possible to give to their description and illustration. The groups assigned to me as Brachyura are those included in this group by
Professors H. Milne Edwards and J. D. Dana, who may perhaps still be regarded as
the leading authorities on the classification of the higher Crustacea, and the arrangement
followed is, as regards the four principal subdivisions, that of those naturalists, although
considerably modified as regards the minor groups.

In the present Report no attempt has been made to enter into the anatomical or
palæontological affinities of the Brachyura,—subjects which it seemed to me should be
left to those students of the group, who, by special training and experience, would be
better qualified than myself to discuss them—but merely to furnish a systematic account of
the numerous species collected, to amend and modify the classification where necessary, and
to give lists as complete as possible, of the recent species of each genus, so far as these
are not included in the works of MM. H. and A. Milne Edwards, Professor J. Bell, F.R.S.,
and others who have revised families or subfamilies of the Brachyura; the numerous
genera have also been redescribed on a uniform system, and it is thus hoped that the
Report will prove, in the absence of any catalogue of the group, of assistance both to
present and future systematists, and a useful supplement to the monographs of A. Milne
Edwards (the first living authority upon this special group of the Crustacea), and to the

Histoire naturelle des Crustacés ; and may thus facilitate the preparation, at some future time, of a Catalogue of the Brachyura, a work much needed by all students of the higher Crustacea.

The collection itself was well suited to the preparation of a memoir of this kind, on account of the relatively large number of littoral and shallow-water forms represented in it from nearly all parts of the globe, the sorting, determining and incorporation of which with the collection of the Museum occupied much more of the time which was originally allotted to me for the preparation of the Report than was anticipated, and although this time was generously extended by Mr. John Murray, and no limitations whatever placed upon me in the performance of my task by the authorities of the British Museum, yet in its accomplishment I am conscious of having fallen far short of what might have been effected, on account of the weakness of my health, and the necessary duties which fell upon me as the assistant (until quite recently) in charge of the collection of Crustacea in that Institution.

With regard to the collection itself, as I have elsewhere noted,[1] the groups richest in new genera and species are the Oxyrhyncha (Maioidea) and Oxystomata (Leucosoidea), and to these belong most of the new forms collected at depths exceeding 100 fathoms. No Brachyurous crab occurs in those deepest abysses of the ocean (exceeding 2000 fathoms), where the Challenger dredgings have shown the lower forms of animal life may occur, and but very few at depths exceeding 500 fathoms. The localities furnishing the greatest proportion of new and interesting types are stations at, among, or near the Islands of the Malaysian Archipelago ; e.g., at the Philippine Islands (Station 210), at the Ki Islands (Station 192), at the Admiralty Islands (Station 219), and also at the Fijis (Station 173). Some interesting types were also taken on the Agulhas Bank off the South African coast (Station 142). For further details, I may refer to the List of Stations, and that indicating the bathymetrical distribution of the species.

As regards the geographical distribution of the species, a large number of the littoral and shallow-water species are widely distributed Atlantic or Indo-Pacific forms. Of the new species collected in deeper water, many, of course, are known from but a single locality, but instances are not wanting to show that of the species inhabiting deeper water some may also prove to have a wide geographical range ; such are—*Lispognathus thomsoni* (Norman), and *Ebalia tuberculosa*, A. Milne Edwards. From a geographical point of view, the collection is interesting, also, as furnishing indications of the Crustacean fauna of localities hitherto rarely visited or unexplored by the carcinologist. I may instance the island of Fernando Noronha, where nine species were taken in shallow water, of which two are apparently undescribed. Also the range of some well-known species has been shown to be more extended than was hitherto known, as, e.g., *Geocarcinus lagostoma*, Milne Edwards.

---

[1] Narr. Chall. Exp., vol. i. part ii. p. 585, 1885.

With regard to the terminology employed in the descriptions of new genera and species, I have found the simple divisions of the "regions" of the carapace, long ago indicated by Desmarest, and modified by H. Milne Edwards in 1834–40, sufficient for all practical purposes. These regions are (besides the frontal, orbital and antennal regions, and the epistoma) the gastric, cardiac and intestinal regions, situated in the median dorsal line, the hepatic, situated ordinarily behind the orbits and near to the antero-lateral margins of the carapace, the branchial regions, covering the whole of the postero-lateral parts of the dorsal surface, and the pterygostomian regions, situated on the inferior surface, and between the antero-lateral margins and the buccal cavity. The description of the carapace and post-abdomen is followed by that of the appendages and limbs in regular sequence, a complicated terminology being avoided wherever possible.

With regard to the synonymical citations, it has been thought sufficient, as a rule, in the case of well-known species, to refer to the original authority for the species and to a recognisable figure, also to the latest or principal authority for the group to which the species belongs, but the synonyma are occasionally given more fully when there has been anything to add or elucidate with regard to them.

Had health and time allowed, I had hoped to give a complete bibliographical list of the works and memoirs relating to the Brachyura, but health, which necessitated the resignation of my post in the British (Natural History) Museum, compelled also the abandonment of this idea, and I thought it right to place the manuscript of the Report in the hands of Mr. Murray for publication. The whole of the systematic part, which had been completed as far as the description of the genera and species was concerned, though not finally revised for press, was accordingly printed off before my health again permitted me to return to the subject. I have therefore found it impossible to insert many details that would have been desirable regarding the affinities and limitation of the various families and subfamilies, &c., and a few errors and omissions occur, which are referred to in the list of errata at the end of the Report; perhaps also the lists of species are not in all cases as complete as a thorough revision of the literature would have made them. I have to thank the Editor of the Reports, and Mr. T. Wemyss Fulton, M.B., of the Editorial Staff of the Challenger Office, for the care that was taken to render the text as correct as possible while the sheets were passing through the press.

The generic descriptions of the earlier part of the Report, relating to the Oxyrhyncha, and drawn up when I was much engaged with other work, might, perhaps, have been extended with advantage to embrace further structural details, and some of the earlier plates are inferior to the later in finish; nevertheless they will in all cases, it is hoped, serve the purpose for which they were intended, namely, to illustrate and identify the species; and my best thanks are due to Mr. R. Morgan for the pains which he at all times bestowed upon this work, in which he had had previously but little experience.

SYSTEMATIC CLASSIFICATION OF THE BRACHYURA.

In place of a complete bibliographical list I subjoin the following brief notice of the classifications proposed by the leading systematists. If time and opportunity had permitted, even this short abstract of the subject could have been with advantage treated in greater detail and thereby rendered more complete.[1]

Professor H. Milne Edwards, in 1834, in the first volume of his great work,[2] separated from the Brachyura of earlier authors, and designated Anomura, those forms in which the sternum is linear, and the post-abdomen is less closely inflexed beneath the sternal surface of the body, and bears more or less well-developed appendages upon the penultimate segment. He divided the restricted Brachyura into the four great natural groups or subdivisions, Oxyrhinques (Oxyrhyncha), Cyclomètopes (Cyclometopa), Catomètopes (Catometopa), and Oxystomes (Oxystomata), which have been retained by most succeeding authors, and are adopted in the present Report, with these modifications only, that I follow Professor Dana in placing the Thelphusines (which are regarded by Milne Edwards as a tribe of the Catometopa) and the somewhat heterogeneous group Corystoidea in the Cyclometopa; the affinities of the Corystoidea, however restricted, seeming to be rather with the Oxyrhyncha and the Cyclometopa than with the Oxystomata where Milne Edwards places them.

This method of restricting the Brachyura, indicated in the Histoire naturelle des Crustacés, was apparently adhered to by Milne Edwards in 1852 in the article entitled "Observations sur les affinités zoologiques et la classification naturelle des Crustacés,"[3] where, however, the term Ocypodidæ is adopted for the group designated Catometopa in his earlier work, and an arrangement of the genera proposed, which I think to be in many particulars less natural and convenient to systematists than that of Dana, which appeared almost contemporaneously, and which in its turn has been modified by Dr. Stimpson and in the present Report.

It will be unnecessary to refer in further detail to the arrangement of the families and subfamilies indicated in the Histoire naturelle des Crustacés, a work which is in the hands of every student of the group.

De Haan, in 1835–1849, in his great work on the Crustacea of Japan,[4] which is a standard work of reference with all students of the Crustacea, divides the Brachyura into two great primary sections or groups, as follows:—(1) Brachygnatha, with the four subdivisions Cancroidea, Majacea, Dromiacea, and Trichidea, and (2) Oxystomata,

---

[1] Reference is made here only to those works which deal in a general way with the arrangement of the whole group, and not to several papers where special families or subfamilies are dealt with, nor, of course, to many memoirs relating to the faunæ of particular regions, or describing collections from special localities.

[2] Hist. Nat. des Crust., vol. i. pp. 247, 263, 1834.   [3] Ann. d. Sci. Nat., ser. 3 (Zool.), xviii. p. 126, 1852.

[4] Crust. in v. Siebold, Fauna Japonica, Introd., p. xi., 1849.

including the groups Dorippidea, Calappidea, Matutoidea, Leucosidea, and Raninoidea. In the first named division (Brachygnatha) the subdivision Cancroidea includes not only the Cyclometopa, but also the Catometopa or Grapsoid Crabs, and the Corystoidea (Corystiens) of Milne Edwards.

This system, based in large measure upon the structure of the maxillipedes (upon the study and illustration of which in the different types of the Brachyura de Haan bestowed so much time and labour) has been adopted by few other authors.[1]

W. S. MacLeay in 1849 [2] somewhat fancifully divides the Brachyura into two primary sections, the first, Tetragonostoma, including the stirpes Pinnotherina, Grapsina, Cancrina, Parthenopina, and Inachina, the second, Trigonostoma, including the stirpes Dromiina, Dorippina, Corystina, Calappina, and Leucosina.

The value of the great subdivisions proposed by de Haan, and of his minor groups or genera, was discussed at length by Professor J. D. Dana in 1852, in the introduction to his elaborate Report on the Crustacea collected during the U.S. Exploring Expedition under Captain (Commodore) Wilkes, U.S.N.,[3] and the defects of his classification are pointed out. It will be unnecessary here to reproduce in detail the system of arrangement proposed by Dana, who not only characterised anew the families and subfamilies of the Brachyura, but gave diagnoses of all of the then known genera; it will be sufficient to note, that the four great groups of the Brachyura proposed by Milne Edwards are retained nearly as they were defined by that author, and the Dromiacea and Raninoidea, included by de Haan in the Brachyura, are restored to the Anomura. Dana's classification, as regards the subfamilies and minor subdivisions, has been considerably modified by A. Milne Edwards in 1861–65 as regards the Cancroidea,[4] and by myself in 1879 [5] as regards the Oxyrhyncha.

He divides the Oxyrhyncha into the legions Maiinea, Parthenopinea, and Oncininea, the latter section restricted to the genus Oncinopus, which in my revision of the group is placed near Macrocheira in the subfamily Inachinae; the Cyclometopa or Cancroidea into the legions Cancrinea, Thelphusinea, and Cyclinea (the latter restricted to the genus Acanthocyclus); for the Catometopa and Oxystomata the division is into families only, for which I must refer to his Report; no primary sections or legions are established in these groups.

His subdivisions in the Oxyrhyncha, Cyclometopa, and Catometopa seem to me sometimes needlessly numerous, but his primary sections and his arrangement of the leading groups of the Oxystomata are followed in the present Report.

Professor Dana's system, offering as it does facilities for the classification and

---

[1] It was, however, followed by Dr. F. Krauss in his work entitled Die süd-afrikanischen Crustaceen, Stuttgart. 4to, 1843.

[2] Annulosa of South Africa in Smith's Illustr. of Zoology of South Africa, p. 54, 1849.

[3] Crust. in U.S. Explor. Exped., vol. xiii. (i.) pp. 69–75, 1852.

[4] Archives du Museum, vol. x. pp. 309–421, 1861; Nouvelles Archives du Museum, vol. i. pp. 177–308, 1865.

[5] Journ. Linn. Soc. Lond. (Zool.), vol. xiv. pp. 634–673, pls. xii., xiii., 1879.

determination of large collections beyond that of any other writer since the author of the Histoire naturelle des Crustacés, has been adopted by many later writers on the Crustacea, e.g., by Dr. Heller, in 1865, in his Report on the Crustacea of the "Novara" Expedition,[1] and by Professor Targioni-Tozzetti in 1877, in the volume dealing with the Crustacea of the Italian steam corvette "Magenta,"[2] and by myself,[3] and by Mr. W. A. Haswell,[4] and other carcinologists. This full and exhaustive report must be regarded, after the Histoire Naturelle des Crustacés, as the work by which the study of the Crustacea, at least the systematic study of the recent Crustacea, has been most advanced.[5]

Professor A. Milne Edwards in 1860, in the introductory article prefixed to his Histoire des Crustacés Podophthalmaires Fossiles,[6] separated the Decapoda into two primary sections, the Brachyura and Macrura, in the first of which, the Brachyura, he included not only the groups included by M. H. Milne Edwards under that designation, but also the various Anomurous groups referred by the elder Milne Edwards to his family Anomoures Apterures. The remainder of the Anomura are referred by A. Milne Edwards in this important memoir to the Macrura. The family Anomoures Apterures of H. Milne Edwards becomes, therefore, in the classification of A. Milne Edwards, the section des Brachyures anormaux (tom. cit., p. 181). In similar manner, the section "Brachyures proprement dits or Brachyures normaux," which includes the groups constituting the Brachyura of H. Milne Edwards and of the present Report, is subdivided into two principal groups; in the first of which, Brachyures macrocephales, are included the Oxyrhyncha, Cyclometopa, Catometopa, and the greater part of the Oxystomata. The second section, Brachyures microcephales, characterised by the very small facial region, rudimentary eyes and epistoma, and the form of the branchial chambers (which are closed at the bases of the legs, and open externally only at the antero-lateral angles of the buccal cavity), is restricted to the single abnormal family Leucosiidæ.

The Brachyures macrocephales are further subdivided into two parallel series :—

(a) Eustomés, including the Cyclometopa, Catometopa, and Oxyrhyncha.

(b) Oligorhynques, including the Oxystomata, except the Leucosiidæ, and the Corystidæ.

This classification has been adopted by M. Brocchi, in 1875,[7] and also by M. F. Mocquard in 1883,[8] but has not been generally used, so far as I know, by systematists ;

[1] Reise der Œsterreichischen Fregatte "Novara," Zoologischer Theil, Crustaceen, 1867.

[2] Crostacei Brachiuri and Anomuri in Zoologia del Viaggio intorno al globo della R. piro-corvetta, "Magenta," 8vo, Firenze, 1877.

[3] Catalogue of the stalk and sessile-eyed Crustacea of New Zealand, 8vo, London, 1876.

[4] Catalogue of the Australian stalk and sessile-eyed Crustacea, 8vo, Sydney, 1882.

[5] For some later remarks on the classification of the Crustacea and of the relationship existing between the Brachyura and Anomura see a memoir by Professor Dana, in the Amer. Journ. Sci. and Arts for 1856, p. 14.

[6] Ann. d. Sci. Nat., ser. 4 (Zool.), vol. xiv. p. 175, 1860.

[7] Recherches sur les organes génitaux males des Crustacés Decapodes, Ann. d. Sci. Nat., ser. 6 (Zool.), vol. iii. Art. 2, 1875.

[8] Recherches anatomiques sur l'estomac des Crustacés Podophthalmaires, op. cit., vol. xvi. Art. 1, 1883.

it, in several ways, however, brings into greater prominence true natural affinities
existing between the different groups; as, e.g., in the approximation of the Oxyrhyncha
to the Oxystomata, and perhaps in the definite separation of the Leucosiidæ from the
latter group, nor can it be denied that the Dromiidæ, at least, are so nearly related to the
Brachyura that they may with almost equal justice be arranged with them or with the
Anomura. Perhaps, therefore, this classification will upon further study be adopted by
systematists in preference to the older one, which is followed in the present Report.

The classification proposed in 1861 by Dr. Strahl,[1] who, basing his system upon the
modifications of the structure and the position of the exterior antennæ and especially of
the basal joint, proposed four entirely new subdivisions of the Brachyura designated (1)
Orbata, (2) Liberata, (3) Incuneata, and (4) Perfusa, needs no extended remark. The
artificiality of his arrangement, and the inconvenience resulting from the dismemberment
of the long-established groups, was exposed shortly after by the late Dr. W. Stimpson,[2]
the well-known American carcinologist, and it is to be regretted that the minor
subdivisions of the Brachyura indicated by this author in his Preliminary Synopsis
of the Crustacea collected by the U.S. Exploring Expedition to the North Pacific[3]
were never fully recharacterised, and the classification never worked out in detail.
This latter remark applies also to his memoir on the Crustacea dredged in the Florida
Straits,[4] where such definitions as are given are brief and incomplete.

Several of the families and subfamilies indicated, but not always properly defined, by
Stimpson, are adopted in the present Report.

Dr. Camil Heller in 1863[5] limited the Brachyura in the sense indicated by M. H.
Milne Edwards and Dana, and retains the four great subdivisions, Oxyrhyncha,
Cyclometopa, Catometopa, and Oxystomata. Dana's family Eriphiidæ is not sustained;
and the Corystoid genera Atelecyclus, Thia, and Corystes are classed with the Oxystomata
as in H. Milne Edwards' system.

Professor C. Claus in his Zoologie[6] divides the Brachyura into five tribes :—
(1) Notopoda (including not only the Dorippidæ, but also the groups Porcellanidæ,
Lithodidæ, and Dromiidæ, which have been generally included in the Anomura);
(2) the Oxystomata, including besides the groups referred to this tribe by Dana, the
Anomurous family Raninidæ; (3) the Oxyrhyncha (Majacea); (4) the Cyclometopa
(Arcuata) or Cancroidea; and (5) the Catometopa. The three last groups are limited as
by Professor Dana; the Corystidæ and the Thelphusidæ are included in the
Cyclometopa.

[1] Monatsber. d. k. preuss. Akad. d. Wiss. Berlin, pp. 713, 1804, 1862.
[2] Amer. Journ. Sci. and Arts, vol. xxxv. p. 139, 1863; Ann. and Mag. Nat. Hist., ser. 3, vol. xi. p. 233, 1863.
[3] Prodromus descriptionis Animalium evertebratorum, &c., Proc. Acad. Nat. Sci. Philad., 1857, p. 216; 1858,
pp. 31, 93, 159.
[4] Bull. Mus. Comp. Zool., vol. ii. p. 109, 1870.
[5] Die Crustaceen des südlichen Europa, Wien, 1863, 8vo.
[6] Grundzüge der Zoologie, 4ter Auflage, Bd. i. p. 632, 1880.

The remainder of the Anomura (Hippidæ, Paguridæ, Galatheidæ) are included by Dr. Claus in the Macrura.[1]

Dr. E. Nauck, in 1880,[2] who based his classification on the solid stomachal plates of the Brachyura, proposed to divide this group into the following sections :—

    I. Heterodonta (to include the Gelasimidæ, &c., and Pinnotheridæ).

    II. Cyclodonta ; subdivided as follows:—
        A. Cœlostylidea, including the Catometopa of Milne Edwards (except the Gelasimidæ and Pinnotheridæ), and the Oxyrhyncha, Milne Edwards.
        B. Platystylidea, including the Oxystomata, Milne Edwards; and Cyclometopa, Milne Edwards.

Trapezia, in which the structure of these parts is very peculiar, is separated from the remainder of the Cyclometopa and establishes a connecting link with the Heterodonta. This classification is not very natural or convenient, and is not followed by M. F. Mocquard in his elaborate memoir referred to above.

By J. E. V. Boas, 1880,[3] the Dromiidæ are included with the Brachyura, and the group is divided into (*a*), Brachyura genuina (=Brachygnatha, de Haan, except the Dromiacea and the Oxygnatha) and (*b*), Dromiacea. This author regards the group Oxystomata as of doubtful value.

In the classification adopted in the following pages, a synoptical view of which is given below, the arrangement followed is generally that of Professor Dana, so far as the leading subdivisions are concerned.

In subdividing the Oxyrhyncha, I have adopted the arrangement indicated by myself in a memoir submitted to the Linnean Society in 1879.[4]

As regards the Cancroid Cyclometopa, I have followed Professor A. Milne Edwards in his monograph, unfortunately never completed, of the Canceridæ,[5] in not sustaining Dana's family Eriphiidæ, but have not ventured to propose any detailed classification of the genera which are not very numerously represented in the Challenger collection. The Trapeziinæ, which are placed in a separate section, should probably rank as a distinct family intermediate between the Cyclometopa and Catometopa.

As regards the swimming crabs (Portunidæ), I follow A. Milne Edwards[6] in uniting

[1] This classification has been followed by J. V. Carus in his recently published Prodromus Faunæ Mediterraneæ (Arthropoda), Stuttgart, 1885.

[2] Das Kaugerüst der Brachyuren, Zeitschr. f. wiss. Zool., Bd. xxxiv. pp. 17, 24, 64, 1880.

[3] Studier over Dekapodernes Slaegtskabsforhold, Dansk. Vidensk. Selsk. Skrifter. (6te R.), Bd. i., ii., pp. 141, 159, 200, 202, 1880.

[4] On the classification of the Maioid Crustacea or Oxyrhyncha, Journ. Linn. Soc. Lond. (Zool.), vol. xiv. pp. 634-673, pl. xii.-xiii., 1879.

[5] Etudes Zoologiques sur les Cancériens, Nouv. Archiv. Mus. Hist. Nat., i. pp. 177-308, pls. xi.-xix., 1865.

[6] Etudes Zoologiques sur les Crustacés recents de la famille des Portuniens, Archiv. Mus. Hist. Nat., x. pp. 309-428, pls. xxviii.-xxxviii., 1861.

Dana's families Portunidæ and Platyonychidæ, and the new arrangement of the genera, in which I have reduced the number of the subfamilies, is a modification of that indicated by the eminent French carcinologist.

In the Catometopa, or Grapsoid Crabs, the increase in the number of the genera in the family Ocypodidæ necessitated a revision of their classification, and I have accordingly proposed a new arrangement, differing alike from that of Dana and of H. Milne Edwards in 1852–54.[1]

Here, also, I have somewhat reduced and recharacterised the subfamilies, regarding as sectional divisions several which have been proposed, and thereby, it is hoped, simplifying the arrangement. Dana's family, Mycteridæ, is not sustained.

In the Pinnotheridæ, I have constituted a new subfamily, Hexapodinæ, for those curious forms in which the fifth ambulatory legs are rudimentary or aborted. The arrangement of the Ocypodoidea now proposed agrees much more nearly with Dana's than with that of Professor Milne Edwards, by whom the genera *Mycteris*, *Scopimera* and *Dotilla* (*Doto*) are separated from the Pinnotheridæ. In uniting these three genera as a distinct subfamily, Mycterinæ, of the Pinnotheridæ, I have, I trust, violated no natural affinities, but rather relegated them to their true position in the system.

In the Oxystomata, a somewhat heterogeneous group, which it will perhaps be found hereafter impossible to sustain in its entirety,[2] the arrangement followed is, as already stated, that of Dana. In rearranging the genera of the Leucosiidæ, however, I have thought it advisable to establish two new subfamilies, pending a more thorough revision of the subject than time and opportunity allowed to be made in the present Report.

[1] *Cf.* Milne Edwards' Memoire sur la famille des Ocypodiens, *Ann. d. Sci. Nat.*, Zool., ser. 3, vol. xviii. pp. 128–164, pls. iii., iv., 1852 ; vol. xx. pp. 163–224, pls. vi.-xi., 1853.

[2] The diagnosis of this group, which was wanting in the MSS. when sent to Edinburgh, is supplied in the appendix to the Report.

BRACHYURA.

OXYRHYNCHA or MAIOIDEA.

Oxyrhynchi, Latreille (pt.).
Oxyrhinques et Canceriens Crytopodes, Milne Edwards.
Maioidea vel Oxyrhyncha, Dana.

Legion I. MAIINEA, Dana.

Family I. INACHIDÆ, Miers.

Subfamily 1. LEPTOPODIINÆ, Miers.

Challenger Genera—

Leptopodia, Leach.
Metoporaphis, Stimpson.
Stenorhynchus, Lamarck (pt.).

Achæus, Leach.
Podochela, Stimpson, with *Coryrhynchus*, Kingsley (= *Podonema*, Stimpson) subgenus (?).

Subfamily 2. INACHINÆ.

Eurypodiidæ, Stimpson.

Challenger Genera—

Platymaia, n. gen.
Cyrtomaia, n. gen.
Apocremnus, A. Milne Edwards.
Achæopsis, Stimpson.
Inachus, Fabricius (pt.).
Oncinopus, de Haan.
Eurypodius, Guérin-Méneville.

Gonatorhynchus, Haswell.
Anamathia, Smith (= *Amathia*, Roux).
Lispognathus, A. Milne Edwards (= *Dorhynchus*, Norman).
Ergasticus, A. Milne Edwards.
Echinoplax, n. gen.
Macrocheira, de Haan.

Subfamily 3. ACANTHONTCHINÆ.

Acanthonychidæ, Stimpson.

Challenger Genera—

Huenia, de Haan.
Menæthius, Milne Edwards.
Oxypleurodon, n. gen.

Dehaanius, MacLeay.
Pugettia, Dana.
Acanthonyx, Latreille.

Family II. MAIIDÆ, Miers.

Subfamily 1. MAIINÆ.

Raiens cryptophthalmes, Milne Edwards (pt.).

Challenger Genera—

Egeria, Leach (*Leptopus*, Lamarck, pt.).
Chorilibinia, Lockington.
Hyas, Leach.
Herbstia, Milne Edwards (= *Rhodia*, Bell; *Micropisa*, Stimpson) with subgenus *Herbstiella*, Stimpson.
Chlorinoides, Haswell.

Pisa, Leach (? = *Arctopsis*, Lamarck).
Hyastenus, White (= *Chorilia*, Dana; *Lahaina*, Dana).
Naxia, Milne Edwards (= *Naxioides*, A. Milne Edwards; *Podopisa*, Hilgendorf).
Scyra, Dana.
Notolopas, Stimpson.

Subfamily 2. Schizophrysinæ, Miers.

Challenger Genus—

Schizophrys, White (= Dione, de Haan, nom. præocc.).

Subfamily 3. Micippinæ, Miers.

Challenger Genera—

Paramicippa, Heller (? = Microhalimus, | Micippa, Leach.
Haswell). |

Family III. Periceridæ, Miers.

Maians cryptophthalmus, Milne Edwards (pt.).

Subfamily 1. Pericerinæ.

Pericerinæ, Stimpson.

Challenger Genera—

Libinia, Leach. | Macrocœloma, Miers.
Pericera, Latreille. | Microphrys, Milne Edwards (= Milnia, Stimp-
Pericceroides, n. gen. | son; Pericera, Dana; Fisheria, Lockington).

Subfamily 2. Mithracinæ.

Mithracinæ, Stimpson.

Challenger Genus—

Mithrax, Latreille, with subgenera Nemausa, A. Milne Edwards, and Mithrax, Latreille (= Mithrax
and Mithraculus, Miers, and Teleophrys, Stimpson).

Legion II. PARTHENOPINEA, Dana.

Family IV. Parthenopidæ.

Parthenopiens et Cancriens cryptophthalmus, Milne Edwards.

Subfamily 1. Parthenopinæ, Miers.

Challenger Genera—

Lambrus, Leach, with the subgenera Lambrus | Cryptopodia, Milne Edwards.
(= Platylambrus, Stimpson, pt.; Rhino- | Heterocrypta, Stimpson.
lambrus, A. Milne Edwards, pt.); Aulaco- |
lambrus, A. M. Edwards; and Parthen- |
olambrus, A. Milne Edwards (= Partheno- |
poides, Miers). |

Subfamily 2. EUMEDONINÆ, Miers.

Challenger Genus—

Ceratocarcinus, Adams and White.

CYCLOMETOPA or CANCROIDEA.

Cyclometopa, Milne Edwards (pt.).
Cancroidea, Dana (pt.).
Cyclometopa, Miers, Cat. New Zealand Crust.

Legion I. CANCRINEA, Dana.

Family I. CANCRIDÆ.

Cancridæ et Eriphidæ, Dana.
Cancrinea, Milne Edwards and A. Milne Edwards.

Section I. Cancrinæ.

Cancriens arqués, Milne Edwards.
Cancriens quadrilatères, Milne Edwards (pt.).

Challenger Genera—

Cancer, Lamarck.
Carpilius, Desmarest.
Atergatis, de Haan.
Lophactæa, A. Milne Edwards.
Lophoxanthus, A. Milne Edwards.
Medæus, Dana.
Actæa, de Haan.
Atergatopsis, A. Milne Edwards.
Xantho, Leach.
Xanthodes, Dana.
Panopeus, Milne Edwards.

Micropanope, Stimpson.
Etisus, Milne Edwards (pt.).
Carpilodes, Dana.
Zozymus, Leach.
Actæodes, Dana.
Leptodius, A. Milne Edwards.
Phymodius, A. Milne Edwards.
Eurytium, Stimpson.
Pseudozius, Dana (subgenus *Euryozius*, nov.).
Sphærozius, Stimpson.
Pilumnus, Leach.

Eriphia, Latreille.

Section II. Trapeziinæ.

Trapezidæ, A. Milne Edwards.

Challenger Genus—

Trapezia, Latreille.

Family II. PORTUNIDÆ.

Portunidæ et Platyonychidæ, Dana.
Portuniens, Milne Edwards and A. Milne Edwards.

Section I. Portuninæ.

Portuniens normaux, A. Milne Edwards.

Challenger Genera—

Neptunus, de Haan (subgenera *Amphitrite*, de Haan; *Achelous*, de Haan; *Hellenus*, A. Milne Edwards).
Xiphonectes, A. Milne Edwards.
Scylla, de Haan.
Cronius, Stimpson.
Goniosoma, A. Milne Edwards.

Thalamonyx, A. Milne Edwards.
Thalamita, Latreille.
Lupocyclus, Adams and White (subgenus *Pseudothranites*, nov.).
Platyonychus, Latreille (pt.).
Portunus, Fabricius.
Lissocarcinus, Adams and White.

Section II. Podophthalminæ.

Portuniens anormaux, A. Milne Edwards.

Challenger Genus—
Podophthalmus, Lamarck.

Legion II. CYCLINEA, Dana.

Challenger Genus—
Acanthocyclus, Milne Edwards and Lucas.

Legion III. CORYSTOIDEA.

Corystoidea, Dana.
Corystiens, Milne Edwards.

Challenger Genera—

Hypopeltarium, nov. (= *Peltarion*, Jacquinot and Lucas).

Gomeza, Gray.

Legion IV. THELPHUSINEA.

Thelphusiens, Milne Edwards.

Challenger Genus—
Thelphusa, Latreille (subgenera *Potamonautes*, MacLeay; *Geothelphusa*, Stimpson).

CATOMETOPA or OCYPODIIDEA.

Catométopes, Milne Edwards (pt.).
Grapsoidea, Dana.
Ocypodiens, M. Edwards, 1852-53.
Catometopa, Miers, Cat. New Zealand Crust., p. 32.

Family I. GEOCARCINIDÆ.

Géocarciniens, Milne Edwards.
Geocarcinidæ, Dana.

Challenger Genera—

Geocarcinus, Leach.   *Cardiosoma,* Latreille.

Family II. OCYPODIDÆ.

Ocypodiens, Milne Edwards.
Macrophthalmidæ (pt.), Dana.

Subfamily 1. CARCINOPLACINÆ.

Challenger Genera—

Geryon, Kroyer.   *Litocheira,* Kinahan.
Pilumnoplax, Stimpson.   *Ceratoplax,* Stimpson.
Bathyplax, A. Milne Edwards.   *Notonyx,* A. Milne Edwards.

Subfamily 2. OCYPODINÆ.

Challenger Genera—

Ocypoda, Fabricius.   *Macrophthalmus,* Latreille.
Gelasimus, Latreille.   *Hemiplax,* Heller.
Gonoplax, Leach.   *Euplax,* Milne Edwards (subgenus *Chæno-*
Ommatocarcinus, White.   *stoma,* Stimpson).

Family III. GRAPSIDÆ.

Grapsoidiens, Milne Edwards.
Grapsidæ, Dana.

Subfamily 1. GRAPSINÆ.

Challenger Genera—

Nautilograpsus, Milne Edwards.   *Platygrapsus,* Stimpson.
Grapsus, Lamarck.   *Brachynotus,* de Haan.
Leptograpsus, Milne Edwards.   *Varuna,* Milne Edwards.
Metopograpsus, Milne Edwards.   *Epigrapsus,* Heller.
Pachygrapsus, Randall.   *Goniopsis,* de Haan.
Geograpsus, Stimpson.   *Helice,* de Haan.
Pseudograpsus, Milne Edwards.   *Sesarma,* Say.

Subfamily 2. PLAGUSIINÆ.

Challenger Genus—
Plagusia, Latreille.

Family IV. PINNOTHERIDÆ.

Pinnotheriens, Milne Edwards.

Subfamily 1. PINNOTHERINÆ.

Challenger Genus—
Pinnotheres, Latreille.

Subfamily 2. MYCTERINÆ.

Challenger Genus—
Mycteris, Latreille.

Subfamily 3. HYMENOSOMINÆ.

Hymenosomiens, Milne Edwards.
Hymeciclæ, Dana.

Challenger Genera—

Hymenosoma, Desmarest. | *Halicarcinus*, White (= *Lévinpea*, Nicolet).

OXYSTOMATA or LEUCOSIIDEA.

Orbiculata, Latreille (pt.).
Oxystomes, Milne Edwards (pt.).
Leucosidea vel *Oxystomata*, Dana.

Family I. CALAPPIDÆ.

Calappiens (pt.), Milne Edwards.
Calappidæ, Dana.

Subfamily CALAPPINÆ, Dana.

Challenger Genera—

Calappa, Fabricius (= *Lophos*, *Gallus*, de Haan, subgenera). | *Mursia*, Desmarest (= *Thealia*, Lucas).
Paracyclois, n. gen. | *Cryptosoma*, Brullé (= *Cyclïes*, de Haan).

Family II. MATUTIDÆ, Dana.

Subfamily MATUTINÆ.

Challenger Genus—
Matuta, Fabricius.

Family III. Leucosiidæ.

Leucosiens, Milne Edwards.
Leucosiidæ, Dana.

Subfamily 1. Iliinæ.

Iliæ, Stimpson.
Ebaliinæ, Stimpson.

Challenger Genera—

Myrodes, Bell.
Arcania, Leach (= *Iphis*, Leach).
Ixa, Leach.
Iliacantha, Stimpson.
Ebalia, Leach (= *Bellidilia*, Kinahan; *Philyra*, Leach).

Persephona, Leach (= *Guia*, Milne Edwards).
Myra, Leach.
Randallia, Stimpson.
Lithadia, Bell.
Merocryptus, A. Milne Edwards.
Philyra, Leach.

Subfamily 2. Leucosiinæ.

Challenger Genus—

Leucosia, Fabricius.

Family IV. Dorippidæ.

Dorippiens, Milne Edwards (pt.).
Dorippidæ, Dana.

Challenger Genera—

Dorippe, Fabricius.
Ethusa, Roux (with *Ethusina*, Smith, subgenus).

Cymopolia, Roux.

GEOGRAPHICAL DISTRIBUTION.

STATION LIST,

Showing the Localities at which Brachyura were Collected during the Expedition, with the Names of the Species Obtained at each.

As in my former Memoirs, four principal regions under which the higher Crustacea (or at least the littoral and shallow-water forms) may be distributed are distinguished. They are (1) the Arctic or Boreal Circumpolar Region (not represented by any species in the collection of H.M.S. Challenger); (2) the Atlantic Region; (3) the Oriental or Indo-Pacific Region; and (4) the Antarctic or Austral Circumpolar Region. In the present Report I have found it convenient to include in the latter region the species obtained south of 40° S. lat.; those obtained at the Cape of Good Hope and on the Agulhas Bank are included in the Atlantic Region, to which their affinities show they decidedly belong. The only Brachyurous Crab distributed throughout the Antarctic Region is *Halicarcinus planatus*, Fabricius. Of these regions, Nos. 1, 3 and 4 were established by Dana.[1]

The Atlantic Region includes not only the European Kingdom of Dana, but also in parts the Occidental Kingdom of that author, since it will embrace the Crustacean fauna of the West Indian Seas and of the eastern shores of the American Continent. Should this designation be generally adopted, it may be found convenient to restrict Dana's Occidental Region to the west coast of America and Islands adjacent, for the Crustacean fauna of these coasts must be regarded as upon the whole distinct from that of the Indo-Pacific Region, although the researches of naturalists are always adding to the number of species common to that Region and the Indo-Pacific and Atlantic Regions.

ATLANTIC REGION.

Tenerife, 78 fathoms.

    Pisa (Arctopsis) tribulus (Linné).

    Neptunus (Amphitrite) hastatus (Linné). This species was perhaps obtained at the next-mentioned locality.

[1] See the Appendix to his Report on the Crustacea of the U.S. Exploring Expedition, p. 1554, 1853.

Off Gomera, Canary Islands, 75 fathoms.

Xanthodes melanodactylus, A. Milne Edwards.

Madeira.

Leptopodia sagittaria (Fabricius).

Azores, near Fayal, 50 to 90 fathoms.

Inachus leptochirus, Leach.
Pisa (Arctopsis) tribulus (Linné).
Lambrus (Parthenolambrus) massena, Roux.
Lambrus (Parthenolambrus) expansus, Miers.

Heterocrypta maltzani, Miers.
Xanthodes melanodactylus, A. Milne Edwards.
Pilumnus spinifer, Milne Edwards.
Portunus corrugatus (Pennant).
Calappa granulata (Linné).

STATION 73.—Azores, 1000 fathoms; June 30, 1873; lat. 38° 30′ 0″ N., long. 31° 14′ 0″ W.; bottom, Pteropod ooze; bottom temperature, 39°·4. Dredged.

Ethusa microphthalma, Smith.

STATION 75.—Azores, off Fayal, 450 fathoms; July 2, 1873; lat. 38° 38′ 0″ N., long. 28° 28′ 30″ W.; bottom, volcanic mud. Dredged.

Heterocrypta maltzani, Miers.

St. Vincent, Cape Verde Islands.

Leptopodia sagittaria (Fabricius).
Acanthonyx lunulatus (Risso).
Herbstia rubra, A. Milne Edwards.
Herbstia violacea (A. Milne Edwards).
Herbstia ovata, Stimpson.
Lambrus (Parthenolambrus) massena, Roux.
Lambrus (Parthenolambrus) massena, Roux, var. *atlanticus*, Miers.
Xanthodes melanodactylus, A. Milne Edwards.

Leptodius punctatus, Miers.
Pilumnus africanus, A. Milne Edwards.
Portunus corrugatus (Pennant).
Ocypoda cursor (Linné).
Grapsus maculatus (Catesby).
Pachygrapsus transversus, Gibbes.
Plagusia depressa, Fabricius.
Calappa gallus (Herbst).
Cryptosoma cristatum, Brullé.
Cymopolia caronii, Roux.

Cape Verde Islands, Porto Praya, St. Iago.

> *Cardiosoma armatum,* Herklots. | *Ocypoda cursor* (Linné).

Ascension Island.

> *Pseudozius bouvieri,* A. Milne | *Geocarcinus lagostoma,* Milne
> Edwards, var. *mellissii,* Miers. | Edwards.
> *Grapsus maculatus* (Catesby).

West North Atlantic (Gulf weed).

> *Neptunus sayi,* A. Milne Edwards.

STATION 33.—Bermudas, 435 fathoms; April 4, 1873; lat. 32° 21′ 30″ N., long.
64° 35′ 55″ W.; bottom, coral mud. Dredged.

> *Geryon incertus,* n. sp.

STATION 48.—South of Nova Scotia, 51 fathoms; May 8, 1873; lat. 43° 4′ 0″ N., long.
64° 5′ 0″ W.; bottom, rock. Dredged.

> *Hyas aranea* (Linné). | *Hyas coarctata,* Leach.

STATION 49.—South of Nova Scotia, 85 fathoms; May 20, 1873; lat. 43° 3′ 0″ N., long.
63° 39′ 0″ W.; bottom, gravel, stones; bottom temperature, 35°·0. Dredged.

> *Hyas coarctata,* Leach. | *Neptunus sayi,* A. Milne Edwards.

Bermuda, shallow water and shore.

> *Podochela riisei* (Stimpson).   *Neptunus (Achelous) depressi-*
> *Macrocœloma trispinosa* (Latreille).   *frons* (Stimpson).
> *Microphrys bicornutus* (Latreille).   *Geocarcinus lagostoma,* Milne
> *Mithrax (Nemausa) rostrata,* A.   Edwards.
> Milne Edwards.   *Cardiosoma guanhumi,* Latreille.
> *Mithrax forceps,* A. Milne Edwards.   *Ocypoda arenaria* (Catesby).
> *Lophactæa lobata* (Milne Edwards).   *Grapsus maculatus* (Catesby).
> *Panopeus herbstii,* var. *serratus*   *Pachygrapsus transversus,* Gibbes.
> (de Saussure).   *Calappa flammea* (Herbst).
> *Eriphia gonagra* (Fabricius).   *Calappa gallus* (Herbst).

Bermuda, Hungry Bay (swamps).

Eurytium limosum (Say).   *Goniopsis cruentatus* (Latreille).

Bahia, shallow water (7 to 20 fathoms).

Leptopodia sagittaria (Fabricius).

Metoporaphis forficulatus, A. Milne Edwards.

Notolopas brasiliensis, n. sp.

Picroceroides tubularis, n. gen. and sp.

Macrocæloma trispinosa (Latreille).

Macrocæloma septemspinosa (Stimpson).

Macrocæloma concava, n. sp.

Mithrax cornutus, de Saussure.

Mithrax forceps, A. Milne Edwards.

Lambrus guérinii, F. de B. Capello.

Lambrus serratus, Milne Edwards.

Heterocrypta granulata (Gibbes).

Actæa rufopunctata, Milne Edwards, var. *nodosa*, Stimpson.

Micropanope spinipes, A. Milne Edwards.

Pilumnus brasiliensis, n. sp.

Pilumnus floridanus, Stimpson.

Pilumnus fragosus, A. Milne Edwards.

Cronius bispinosus, n. sp.

Ocypoda arenaria (Catesby).

Sesarma mülleri, A. Milne Edwards.

Persephona punctata (Browne).

Lithadia cariosa, Stimpson, var.

Iliacantha intermedia, n. sp.

STATIONS 122–122c.—Off Barra Grande, 30 to 400 fathoms ; lat. 9° 5′ 0″ to 9° 10′ 0″ S., long. 34° 49′ 0″ to 34° 53′ 0″ W. ; bottom, red mud. Trawled.

Leptopodia sagittaria (Fabricius).

Metoporaphis forficulatus, A. Milne Edwards.

Podochela riisei (Stimpson).

Herbstia (Herbstiella) depressa, Stimpson.

Mithrax hispidus (Herbst), var. *pleuracanthus*, Stimpson.

Pilumnus floridanus, Stimpson.

Neptunus (Hellenus) spinicarpus, Stimpson.

Bathyplax typhlus, var. *oculiferus*, nov.

Fernando Noronha Island, shallow water (7 to 20 fathoms).

Apocremnus septemspinosus, A. Milne Edwards.

Picroceroides tubularis, n. gen. and sp.

Macrocæloma concava, n. sp.

Mithrax coronatus (Herbst).

Mithrax sp.

Mithrax (Mithraculus) sculptus (Lamarck).

Mithrax forceps, A. Milne Edwards.

Grapsus maculatus (Catesby).

Calappa gallus (Herbst).

St. Paul's Rocks, 10 to 80 fathoms (and shore ?).

Stenorhynchus (?) spinifer, n. sp. | *Grapsus maculatus* (Catesby).

Tristan da Cunha Group, Nightingale Island, 100 fathoms.

Pilumnoplax heterochir, Studer.

Cape of Good Hope, Simon's Bay, shallow water (5 to 18 fathoms).

Stenorhynchus falcifer, Stimpson. | *Pericera cornuta,* Milne Edwards.
Achæopsis spinulosus, Stimpson. | *Plagusia chabrus* (Linné).
Dehaanius dentatus (Milne Ed- | *Hymenosoma orbiculare* (Desmarest).
wards). | *Calappa flammea* (Herbst).

Mursia cristimana (Desmarest).

South Africa, Wellington and Cape Town (rivers).

Thelphusa (Potamonautes) perlata, Milne Edwards.

Sea Point.

Mursia cristimana (Desmarest).

Station 142.—Agulhas Bank, 150 fathoms ; December 18, 1873 ; lat. 35° 4′ 0″ S., long.
18° 37′ 0″ E.; bottom, green sand ; bottom temperature, 47°·0. Dredged.

Lispognathus thomsoni (Norman). | *Litocheira kingsleyi,* n. sp.
Pilumnoplax heterochir, Studer. | *Mursia cristimana* (Desmarest).

Ebalia tuberculosa (A. Milne Edwards) (yg. ?).

INDO-PACIFIC OR ORIENTAL REGION.

Station 161.—Port Philip, 33 fathoms ; April 1, 1874 ; lat. 38° 22′ 30″ S., long.
144° 36′ 30″ E. ; bottom, sand. Trawled.

Achæus tenuicollis, n. sp. | *Portunus corrugatus* (Pennant).
Pilumnus tomentosus, Latreille. | *Halicarcinus ovatus,* Stimpson.

Ebalia (Phlyxia) intermedia, n. sp.

STATION 162.—Bass Strait, East Moncœur Island, 38 fathoms; April 2, 1874; lat. 39° 10′ 30″ S., long. 146° 37′ 0″ E.; bottom, sand and shells. Dredged.

Achæus tenuicollis, n. sp.

Chlorinoides coppingeri, Haswell.

Micippa spinosa, var. affinis, Miers.

Actæa peronii (Milne Edwards).

Pilumnus tomentosus, Latreille.

Portunus corrugatus (Pennant).

Ebalia tuberculosa (A. Milne Edwards).

Ebalia (Phlyxia) crassipes, Bell.

Merocryptus lambriformis, A. Milne Edwards.

South Australian Coast, shallow water (2 to 10 fathoms), April 1874.

Oncinopus aranea, de Haan.

Gonatorhynchus tumidus, Haswell.

Micippa spinosa, Stimpson.

Actæa peronii (Milne Edwards).

Pilumnus rufopunctatus, Stimpson.

Neptunus (Neptunus) sanguinolentus (Herbst).

Thalamita sima, Milne Edwards.

Platyonychus bipustulatus, Milne Edwards.

Lissocarcinus polybioides, Adams and White.

Halicarcinus ovatus, Stimpson.

Calappa depressa, n. sp.

Leucosia australiensis, n. sp.

Philyra platycheira, de Haan.

Ebalia (Phlyxia) crassipes, Bell.

Ebalia (Phlyxia) undecimspinosa (Kinahan), var. orbicularis, Haswell.

Ebalia (Phlyxia) dentifrons, n. sp.

STATION 163A.—Off Twofold Bay, 150 fathoms; April 4, 1874; lat. 36° 59′ 0″ S., long. 150° 20′ 0″ E.; bottom, green mud. Dredged.

Medæus haswelli, n. sp.

Ebalia tuberculosa (A. Milne Edwards.

Merocryptus lambriformis, A. Milne Edwards.

New South Wales, Botany Bay (beach).

Mycteris platycheles, Milne Edwards.

New South Wales, Hawkesbury (brackish water).

Sesarma schüttei (Hess).

STATION 164B.—Off Sydney, 410 fathoms; June 13, 1874; lat. 34° 13′ 0″ S., long. 151° 38′ 0″ E.; bottom, green mud. Trawled.

Lispognathus thomsoni (Norman).

Sydney.

Pinnotheres villosulus, Guérin-Méneville.

Port Jackson, shore, 3 to 15 fathoms.

Micippa spinosa, Stimpson.
Pilumnus vestitus, Haswell.
Pachygrapsus transversus, Gibbes.
Helice crassa, Dana.
Matuta læviductyla, Miers.
Leucosia australiensis, n. sp.

Ebalia (Phlyxia) crassipes, Bell.
Ebalia (Phlyxia) undeciuspinosa (Kinahan), var. *orbicularis*, Haswell.
Ebalia (Phlyxia) quadridentata (Gray), var. *spinifera*, nov.

Port Jackson, 30 to 35 fathoms.

Calappa lophos (Herbst).
Ebalia tuberculosa (A. Milne Edwards).

Ebalia (Phlyxia) crassipes, Bell.

STATION 167.—New Zealand, 150 fathoms; June 24, 1874; lat. 39° 32′ 0″ S., long. 171° 48′ 0″ E.; bottom, blue mud.

Ebalia lævis, Bell.

Ebalia tuberculosa (A. Milne Edwards).

STATION 167A.—New Zealand, Queen Charlotte Sound, near Long Island, 10 fathoms; June 27, 1874; lat. 41° 4′ 0″ S., long. 174° 19′ 0″ W.; bottom, mud. Dredged.

Hemiplax hirtipes (Heller).
Ommatocarcinus macgillivrayi, White.

Ebalia lævis (Bell).

STATION 172.—Tongatabu, off Nukalofa, 18 fathoms; July 22, 1874; lat. 20° 58′ 0″ S., long. 175° 9′ 0″ W.; bottom, coral mud. Dredged.

Menæthius monoceros, Latreille, var. *angusta*, Dana.
Lambrus (Parthenolambrus) calappoides, Adams and White.
Trapezia cymodoce (Herbst).
Pilumnus longicornis, Hilgendorf, var.

Xiphonectes longispinosus, Dana.
Thalamonyx danæ, var. *gracilipes*, nov.
Thalamita sima, Milne Edwards, var. *granulata*, nov.
Thalamita sexlobata, n. sp.
Thalamita stimpsoni, A. Milne Edwards.

xxivTHE VOYAGE OF H.M.S. CHALLENGER.

STATION 172A.—Tongatabu, off Nukalofa, 240 fathoms; July 22, 1874; lat. 20° 56′ 0″ S., long. 175° 11′ 0″ W.; bottom, coral mud. Dredged.

Randallia granulata, n. sp.

Tongatabu, reefs.

Etisus lævimanus, Randall.
Thalamita admete (Herbst).

Thalamita prymna (Herbst).
Calappa hepatica (Linné).

Fiji Islands, Matuku.

Gelasimus annulipes, Milne Edwards.

Fiji Islands, Kandavu (reefs).

Zozymus æneus (Linné).
Trapezia guttata, Rüppell, var.
Trapezia areolata, var. *inermis*, A. Milne Edwards.
Cardiosoma carnifex (Herbst).
Euplax (*Chænostoma*) *boscii*, Audouin.

Gelasimus vocans, Linné.
Gelasimus annulipes, Milne Edwards.
Ocypoda ceratophthalma (Pallas).
Metopograpsus messor (Forskål).
Geograpsus grayi (Milne Edwards).
Pseudograpsus albus, Stimpson.

Helice latreillei, Milne Edwards.

Fiji Islands, Levuka (reefs).

Thalamita admete (Herbst).

STATION 173.—Fiji Islands, off Matuku, 315 fathoms; July 24, 1874; lat. 19° 9′ 35″ S., long. 179° 41′ 50″ E.; bottom, coral mud. Dredged.

Pilumnoplax abyssicola, n. sp.
Mursia curtispina, n. sp.
Randallia granulata, n. sp.

STATION 173A.—Fiji Islands, 310 fathoms; July 24, 1874; lat. 19° 9′ 32″ S., long. 179° 41′ 55″ E.; bottom, coral mud. Dredged.

Ethusa orientalis, n. sp.

New Hebrides, Api.

Epigrapsus politus, Heller.

Sesarma (Holometopus) aubryi, A. Milne Edwards.

Torres Strait (some 3 to 11 fathoms).

Chorilibinia gracilipes (Miers).
Schizophrys aspera (Milne Edwards).
Lambrus affinis, A. Milne Edwards.
Lambrus hoplonotus, var. granulosus, Miers.
Actæa granulata (Audouin).
Pilumnus pulcher, Miers.
Pilumnus seminudus, Miers.

Neptunus (Amphitrite) rugosus, A. Milne Edwards.
Macrophthalmus podophthalmus, Eydoux and Souleyet.
Ceratoplax ciliata, Stimpson.
Pinnotheres villosulus, Guérin Méneville.
Myra affinis, Bell.

North Australia, Raine Island.

Ocypoda ceratophthalma (Pallas).

STATION 186.—Off Cape York, 8 fathoms; September 8, 1874; lat. 10° 30′ 0″ S., long. 142° 18′ 0″ E.; bottom, coral mud. Dredged.

Huenia proteus, de Haan.
Menæthius monoceros, Latreille, var. angusta (Dana).
Pseudomicippa varians, Miers.
Lambrus (Aulacolambrus) hoplonotus, Adams and White.
Actæa hirsutissima, Rüppell.

Actæa hystrix, n. sp.
Pilumnus pulcher, Miers.
Pilumnus labyrinthicus, Miers.
Neptunus (Achelous) unispinosus (Miers).
Thalamita sima, Milne Edwards.
Thalamita intermedia, n. sp.

STATION 187.—Torres Strait, 6 fathoms; September 9, 1874; lat. 10° 36′ 0″ S., long. 141° 55′ 0″ E.; bottom, coral mud. Dredged.

Lambrus intermedius, Miers.   Matuta inermis, Miers.
Myra australis, Haswell.

Albany Island.

Thalamita crenata, Rüppell.   Mycteris longicarpus (Latreille).
(ZOOL. CHALL. EXP.—PART XLIX.—1886.)   Ccc d

STATION 188.—South of New Guinea, 28 fathoms; September 10, 1874; lat. 9° 59′ 0″ S.,
long. 139° 42′ 0″ E.; bottom, green mud.   Dredged and trawled.

Egeria arachnoides (Rumph).
Chorilibinia gracilipes (Miers).
Hyastenus planasius (Adams and White).
Hyastenus (Chorilia) oryx, A. Milne Edwards.
Lambrus longimanus (Linné).
Lambrus hoplonotus, var. *longioculis*, Miers.
Cryptopodia fornicata (Fabricius).
Atergatopsis granulatus, A. Milne Edwards.
Neptunus (Amphitrite) hastatoides, (Fabricius).
Neptunus (Amphitrite) gladiator, (Fabricius).
Neptunus (Achelous) whitei, A. Milne Edwards.
Ceratoplax arcuata, Miers.
Notonyx nitidus, A. Milne Edwards.
Leucosia haswelli, n. sp.
Leucosia ocellata, Bell.
Leucosia whitei, Bell.
Leucosia craniolaris (Linné).
Myra fugax (Fabricius).
Ebalia lambriformis, Bell.
Myrodes eudactylus, Bell.

Arafura Sea.

Oncinopus aranea, de Haan.   *Huenia proteus*, de Haan.
Macrophthalmus podophthalmus, Eydoux and Souleyet (yg. ?).

STATION 190.—Arafura Sea, 49 fathoms; September 12, 1874; lat. 8° 56′ 0″ S.,
long. 136° 5′ 0″ E.; bottom, green mud.   Trawled.

Egeria arachnoides (Rumph.).
Hyastenus diacanthus (de Haan).
Goniosoma ornatum, A. Milne Edwards.
Leucosia craniolaris (Linné).
Myra fugax (Fabricius).

STATION 191.—Banda Sea, 800 fathoms; September 23, 1874; lat. 5° 41′ 0″ S.,
long. 134° 4′ 30″ E.; bottom, green mud; bottom temperature, 39°·5.   Trawled.
Ethusa (Ethusina) gracilipes, var. *robusta*, nov.

STATION 192.—Ki Islands, 140 fathoms; September 26, 1874; lat. 5° 49′ 15″ S.,
long. 132° 14′ 15″ E.; bottom, blue mud.   Trawled.

Cyrtomaia murrayi, n. gen. and sp.
Oxypleurodon stimpsoni, n. gen. and sp.
Pugettia velutina, n. sp.
Hyastenus elegans, n. sp.
Pilumnus normani, n. sp.
Lupocyclus (Parathranites) orientalis, n. subgen. and sp.
Platyonychus iridescens, n. sp.

Moluccas, Amboina, 100 fathoms.

Nuxia hystrix, n. sp. | *Lambrus longimanus* (Linné).

Amboina, 10 to 25 fathoms.

Lambrus turriger, White. | *Calappa hepatica* (Linné).
Gonoplax sinuatifrons, n. sp. | *Calappa gallus* (Herbst).

STATION 195.—Banda Sea, 1425 fathoms; October 3, 1874; lat. 4° 21′ 0″ S.,
long. 129° 7′ 0″ E.; bottom, blue mud; bottom temperature, 38°. Trawled.
Ethusa (Ethusina) gracilipes, var. *robusta*, nov.

Banda (fresh water).
Varuna litterata (Fabricius).

STATION 196.—Moluccas Passage, 825 fathoms; October 13, 1874; lat. 0° 48′ 30″ S.,
long. 126° 58′ 30″ E.; bottom, hard ground; bottom temperature, 36°·9. Trawled.
Oncinopus aranea, de Haan.

Ternate.

Atergatis floridus (Linné). | *Thalamita prymna* (Herbst).
Neptunus (Neptunus) sanguino- | *Matuta victrix* (Fabricius), var.
lentus (Herbst). | *crebrépunctata*, Miers.

Banda (fresh water).
Varuna litterata (Fabricius).

Arrou (Aröe) Islands.

Ceratocarcinus longimanus, Adams | *Ocypoda ceratophthalma* (Pallas)
and White. | (yg.).
Neptunus (Amphitrite) tubercu- | *Grapsus strigosus* (Herbst).
losus, A. Milne Edwards. | *Sesarma (Holometopus) aubryi*,
Gelasimus vocans (Linné). | A. Milne Edwards.
Gelasimus tetragonon (Herbst).

Hong Kong, 7 to 10 fathoms.

Neptunus (Amphitrite) hasta- | *Leucosia craniolaris* (Linné).
toides (Fabricius). | *Arcania septemspinosa* (Fabricius).
Goniosoma cruciferum (Fabricius). | *Dorippe facchino* (Herbst).

STATION 203.—Philippine Islands, near Masbate, 20 fathoms; October 31, 1874; lat. 11° 6′ 0″ N., long. 123° 9′ 0″ E.; bottom, mud. Trawled.

> *Neptunus (Neptunus) pelagicus* (Linné).

Manila, 1 fathom.

> *Ixa cylindrus* (Fabricius), var. *megaspis*, Adams and White.

STATION 207.—Philippine Islands, 700 fathoms; January 16, 1875; lat. 12° 21′ 0″ N., long. 122° 15′ 0″ E.; bottom, blue mud; bottom temperature, 51°·6. Trawled.

> *Ethusa (Ethusina) gracilipes*, n. sp.

STATION 208.—Philippine Islands, 18 fathoms; January 17, 1875; lat. 11° 37′ 0″ N., long. 123° 31′ 0″ E.; bottom, blue mud. Trawled.

> *Hyastenus diacanthus* (de Haan).
> *Hyastenus oryx*, A. Milne Edwards.
> *Pilumnus dehaani*, Miers.
>
> *Neptunus (Amphitrite) rugosus*, A. Milne Edwards.
> *Neptunus (Amphitrite) spinipes*, n. sp.

Samboangan, 10 to 20 fathoms, and beach.

> *Naxia hirta*, A. Milne Edwards.
> *Lambrus contrarius* (Herbst).
> *Xantho bidentatus*, A. Milne Edwards.
> *Carpilodes bellus* (Dana).
> *Actæodes tomentosus* (Milne Edwards).
> *Leptodius exaratus*, var. *sanguineus* (Milne Edwards).
> *Phymodius monticulosus* (Dana).
> *Pilumnus scabriusculus*, Adams and White.
> *Trapezia cymodoce* (Herbst).
> *Trapezia rufopunctata* (Herbst).
>
> *Neptunus (Achelous) granulatus* (Milne Edwards).
> *Thalamita prymna* (Herbst).
> *Thalamita stimpsonii*, A. Milne Edwards.
> *Gelasimus rubripes*, Jacquinot and Lucas.
> *Gelasimus annulipes*, Milne Edwards.
> *Ocypoda ceratophthalma* (Pallas), (yg. ?)
> *Helice latreillei*, Milne Edwards.
> *Matuta victrix* (Fabricius).
> *Matuta banksii*, Leach.

Philippine Islands, Zebu (reefs).

> *Atergatis floridus* (Linné).

Philippine Islands, Mindanao, Pasananea (rivers).

Thelphusa sinuatifrons, Milne Edwards.

Varuna litterata, Fabricius.

STATION 210.—Philippine Islands, 375 fathoms; January 25, 1875; lat. 9° 26′ 0″ N., long. 123° 45′ 0″ E.; bottom, blue mud; bottom temperature, 54°·1. Trawled and dredged.

Anamathia pulchra, n. sp.
Echinoplax moseleyi, n. gen. and sp.

Oxypleurodon stimpsoni, n. gen. and sp.

STATION 212.—Celebes Sea, 10 fathoms; January 30, 1875; lat. 6° 54′ 0″ N., long. 122° 18′ 0″ E.; bottom, sand. Trawled and dredged.

Egeria arachnoides (Rumph).
Neptunus (Amphitrite) rugosus, A. Milne Edwards.
Neptunus (Amphitrite) gladiator (Fabricius), var. *argentatus*, White.

Lissocarcinus lævis, n. sp.
Leucosia haswelli, n. sp.
Myra darnleyensis, Haswell.
Cymopolia jukesii, White.
Gomeza bicornis, Gray (10 to 20 fathoms).

STATION 214.—Tulur Islands, 500 fathoms; February 10, 1875; lat. 4° 33′ 0″ N., long. 127° 6′ 0″ E.; bottom, blue mud; bottom temperature, 41°·8. Dredged.

Cyrtomaia suhmi, n. sp.

Off North Coast of New Guinea (drift wood).

Varuna litterata (Fabricius).

Plagusia immaculata, Lamarck.

Admiralty Islands, Nares Harbour.

Cardiosoma carnifex (Herbst).

STATION 219.—North of the Admiralty Islands, 150 fathoms; March 10, 1875; lat. 1° 54′ 0″ S., long. 146° 39′ 40″ E.; bottom, coral mud. Trawled.

Platymaia wyville-thomsoni, n. gen. and sp.
Ergasticus naresii, n. sp.

Lupocyclus (Parathranites) orientalis, n. subgen. and sp.
Paraeyclois milne-edwardsi, n. gen. and sp.

Admiralty Islands, 16 to 25 fathoms.

    Neptunus (Achelous) granulatus (Milne Edwards).

Admiralty Islands, Wild Island (beach).

    Sesarma (Holometopus) aubryi, A. Milne Edwards.

STATION 232.—Japan, off Inosima, 345 fathoms; May 12, 1875; lat. 35° 11′ 0″ N., long. 139° 28′ 0″ E.; bottom, green mud; bottom temperature, 41°·1. Trawled and dredged.

    Macrocheira kämpferi, de Haan.

Japan, off Yokoska, 10 fathoms.

    Pugettia incisa (de Haan).

    Pilumnoplax vestita (de Haan), var. *sexdentata*, Haswell.

Japan, Kobé (some 8 to 10 fathoms).

    Thelphusa (Geothelphusa) dehaanii, White.

    Macrophthalmus serratus, Adams and White.

    Arcania septemspinosa (Fabricius).

STATION 233A.—Japan, near Kobé, 50 fathoms; May 19, 1875; lat. 34° 38′ 0″ N., long. 135° 1′ 0″ E.; bottom, sand. Dredged.

Achæus japonicus, de Haan.
Scyra compressipes, Stimpson.
Lophozozymus (Lophoxanthus) bellus, Stimpson, var. *leucomanus*, Lockington.
Sphærozius nitidus, Stimpson.
Pilumnus minutus, var. *hirsutus*, Stimpson.
Pilumnus globosus, Dana.

Neptunus (Amphitrite) hastatoides, (Fabricius).
Goniosoma variegatum (Fabricius) var. *bimaculatum*, nov.
Platygrapsus depressus, de Haan.
Leucosia craniolaris (Linné).
Dorippe japonica, v. Siebold (8 to 50 fathoms).

Japan, Oosima (pools).

    Brachynotus (Heterograpsus) penicillatus (de Haan).

STATION 233B.—Japan, 15 fathoms; May 26, 1875; lat. 34° 18′ 0″ N., long. 133° 35′ 0″ E.; bottom, blue mud. Trawled.

Neptunus (Amphitrite) hastatoides, (Fabricius.)

Pilumnoplax restita (de Haan), var. *sexdentata,* Haswell.

Arcania septemspinosa (Fabricius).

Japan, near Lake Biva.

Thelphusa (Geothelphusa) dehaanii, White.

Japan, Hakouni (2500 feet).

Thelphusa (Geothelphusa) dehaanii, White.

STATION 237.—Japan, near Yokohama, 1875 fathoms; June 17, 1875; lat. 34° 37′ 0″ N., long. 140° 32′ 0″ E.; bottom, blue mud; bottom temperature, 35°·3. Trawled.

Ethusa (Ethusina) challengeri, n. sp.

Sandwich Islands, Hawaii.

Neptunus (Neptunus) sanguinolentus (Herbst).

Sandwich Islands, Hilo (beach).

Metopograpsus messor (Forskål).

Sandwich Islands, Honolulu (reefs), some 16, 18 to 20 fathoms.

Carpilius maculatus (Linné).
Lophactæa granulosa (Rüppell).
Actæa nodulosa, White.
Etisus lævimanus, Randall.
Trapezia rufopunctata, var. *guttata,* Rüppell.

Neptunus (Neptunus) sanguinolentus (Herbst).
Thalamita integra, Dana.
Podophthalmus vigil (Fabricius).
Calappa hepatica (Linné).

Society Islands, Tahiti, Papiete.

Scylla serrata (Forskål).
Cardiosoma carnifex (Herbst).

Gelasimus tetragonon (Herbst).
Metopograpsus messor (Forskål).

Chili, Valparaiso (shore).

    Cancer longipes, Bell.   *Leptograpsus variegatus* (Fabricius).
    Acanthocyclus gayi, Milne Edwards
      and Lucas.

Messier Channel (fresh water).

        Acanthocyclus gayi, Milne Edwards and Lucas.

ANTARCTIC OR AUSTRAL CIRCUMPOLAR REGION.

STATION 304.—Chiloe, off Cape Tres Montes, 45 fathoms; December 31, 1875; lat. 46° 53′ 15″ S., long. 75° 12′ 0″ W.; bottom, green sand. Dredged.

    Eurypodius   latreillei, Guérin-   *Libinia gracilipes*, n. sp.
      Méneville.   *Hypopeltarium spinulosum*, White.

STATION 308.—Chiloe, 175 fathoms; January 5, 1876; lat. 50° 8′ 30″ S., long. 74° 41′ 0″ W.; bottom, blue mud. Trawled.

                    Eurypodius longirostris, n. sp.

STATION 311.—Coast of Chiloe, 245 fathoms; January 11, 1876; lat. 52° 45′ 30″ S., long. 73° 46′ 0″ W.; bottom, blue mud; bottom temperature, 46°. Trawled.

                      Libinia smithii, n. sp.

Magellan Strait, Port William.

                    Eurypodius latreillei, Guérin-Méneville.

STATION 312.—Magellan Strait, Port Famine, 9 fathoms; January 13, 1876; lat. 53° 37′ 30″ S., long. 70° 56′ 0″ W.; bottom, blue mud. Dredged.

                    Eurypodius latreillei, Guérin-Méneville.

STATIONS 313, 314.—Magellan Strait, 55 and 70 fathoms; January 20, 21, 1876; lat. 52° 20′ 0″ S. to 51° 35′ 0″ S., long. 67° 39′ 0″ W. to 65° 39′ 0″ W.; bottom, sand. Trawled.

                    Eurypodius latreillei, Guérin-Méneville.

STATIONS 315, 316.—Falkland Islands, 12 and 4 fathoms; January 26 and February 3, 1876; lat. 51° 40′ 0″ S. to 51° 32′ 0″ S., long. 57° 50′ 0″ W. to 58° 6′ 0″ W.; bottom, sand, gravel (Station 315), and mud (Station 316).   Dredged.

Eurypodius latreillei, Guérin-Méneville.	*Hypopeltarium spinulosum*, White.
	Halicarcinus planatus (Fabricius).

Port Stanley.

Hypopeltarium spinulosum, White.

Port William.

Halicarcinus planatus (Fabricius).

Near Marion Island, 50 to 75 fathoms.

Halicarcinus planatus (Fabricius).

Off Prince Edward Island, 85 to 150 fathoms.

Halicarcinus planatus (Fabricius).

Off Kerguelen Island (rock pools).

Halicarcinus planatus (Fabricius).

New Zealand, South Island, Cape Campbell.

Halicarcinus planatus (Fabricius).

BATHYMETRICAL DISTRIBUTION.[1]

The greater number of the species collected by the Expedition are littoral or shallow-water forms; and this observation applies especially to the families Cancridæ in the Cyclometopa, and Grapsidæ in the Catometopa; in these groups, also, the new species obtained are comparatively few and unimportant. About 280 species and well-marked varieties are in the collection; of these, between 180 and 190 were taken on the shore, or in 1 to 20 fathoms, and perhaps between 70 to 80 in 20 to 100 fathoms; of which, however, not a few (as, e.g., the Brazilian species, dredged between 30 to 350 fathoms in Station 122) may have been taken at greater depths. The new forms obtained, as I have already noted in the abstract of my Report, published in the introductory volume (pt. ii.) of the Challenger Reports, are chiefly in the groups Oxyrhyncha and Oxystomata, and were dredged at depths greater than 200 fathoms; thus of the littoral and shallow-water species and varieties, scarcely 20 are new to science, and of those obtained in 20 to 100 fathoms, scarcely 15; of 21 species obtained in 200 to 500 fathoms, about half are undescribed forms; between 500 to 1000 fathoms, but 3 species (2 previously described) were obtained; and but two species, *Ethusa* (*Ethusina*) *challengeri*, n. sp., dredged in 1875 fathoms, and *Ethusa* (*Ethusina*) *gracilipes*, nov., var. *robusta*, nov., taken in 1425 fathoms, occurred beyond this depth.

Shore and 1 to 20 fathoms.

(Littoral and shallow-water species).

Leptopodia sagittaria (Fabricius).

Metoporaphis forficulatus, A. Milne Edwards.

Stenorhynchus falcifer, Stimpson.

? *Stenorhynchus spinifer*, n. sp.

Podochela riisei, Stimpson.

? *Apocremnus septemspinosus*, A. Milne Edwards.

Achæopsis spinulosus, Stimpson.

Oncinopus aranea, de Haan.

Eurypodius latreillei, Guérin-Ménéville.

Gonatorhynchus tumidus, Haswell.

Huenia proteus, de Haan.

Menæthius monoceros (Latreille), var. *angusta*, Dana.

Dehaanius dentatus (Milne Edwards).

Pugettia incisa (de Haan).
Acanthonyx lunulatus, Risso.
Egeria arachnoides, Rumph.
Chorilibinia gracilipes, Miers.
Herbstia rubra, A. Milne Edwards.
Herbstia violacea (A. Milne Edwards).
Herbstia ovata, Stimpson.
Hyastenus diacanthus (de Haan).
Hyastenus oryx, A. Milne Edwards.
Naxia hirta, A. Milne Edwards.
Notolopas brasiliensis, n. sp.
Schizophrys aspera, Milne Edwards.
Pseudomicippa varians, Miers.
Micippa spinosa, Stimpson.
Pericera cornuta, Milne Edwards.
Picroceroides tubularis, n. gen. and sp.
Macrocœloma trispinosa, Latreille.
Macrocœloma septemspinosa (Stimpson).
Macrocœloma concava, n. sp.
Microphrys bicornutus (Latreille).
Mithrax (Nemausa) rostrata, A. Milne
    Edwards.
Mithrax cornutus, de Saussure.
Mithrax forceps, A. Milne Edwards.
Mithrax coronatus (Herbst).
Lambrus contrarius (Herbst).
Lambrus affinis, A. Milne Edwards.
Lambrus intermedius, Miers.
? Lambrus turriger, White.
Lambrus guérini, F. de B. Capello.
Lambrus serratus (Milne Edwards).
Lambrus (Aulacolambrus) hoplonotus,
    Adams and White, var. granu-
    losus, Miers.
Lambrus (Parthenolambrus) massena,
    Roux, with var. atlanticus, Miers.
Lambrus (Parthenolambrus) calap-
    poides, Adams and White.
Heterocrypta granulata, Gibbes.

Ceratocarcinus longimanus, Adams and
    White.
Cancer longipes, Bell.
Carpilius maculatus (Linné).
Atergatis floridus (Linné).
Lophactæa lobata (Milne Edwards).
Lophactæa granulosa, Rüppell.
Actæa granulata, Audouin.
Actæa hirsutissima (Rüppell).
Actæa nodulosa, White.
Actæa hystrix, n. sp.
Actæa peronii, Milne Edwards.
Actæa rufopunctata, Milne Edwards,
    var. nodosa, Stimpson.
Xantho bidentatus, A. Milne Edwards.
Xanthodes melanodactylus, A. Milne
    Edwards.
Panopeus herbstii,   var.   serratus
    (Saussure).
Micropanope spinipes, A. Milne Edwards.
Etisus lævimanus, Randall.
Carpilodes bellus (Dana).
Zozymus æneus, Linné.
Actæodes tomentosus, Milne Edwards.
Leptodius exaratus, var. sanguineus,
    Milne Edwards.
Leptodius punctatus, Miers.
Phymodius monticulosus, Dana.
Eurytium limosum, Say.
Pseudozius bouvieri, A. Milne Edwards,
    var. mellissii, Miers.
Pilumnus africanus, A. Milne Edwards.
Pilumnus brasiliensis, n. sp.
Pilumnus fragosus, A. Milne Edwards.
Pilumnus dehaani, Miers.
Pilumnus scabriusculus, Adams and
    White.
Pilumnus longicornis, Hilgendorf, var.
Pilumnus vestitus, Haswell.

Pilumnus rufopunctatus, Stimpson.
Pilumnus pulcher, Miers.
Pilumnus seminudus, Miers.
Pilumnus labyrinthicus, Miers.
Eriphia gonagra, Fabricius.
Trapezia cymodoce (Herbst).
Trapezia guttata, Rüppell, var.
Trapezia areolata, var. *inermis*, A.
   Milne Edwards.
Trapezia rufopunctata (Herbst), with
   var. *intermedia*, nov.
Neptunus pelagicus (Linné).
Neptunus sanguinolentus (Herbst).
Neptunus (*Amphitrite*) *hastatoides*
   (Fabricius).
Neptunus (*Amphitrite*) *tuberculosus*, A.
   Milne Edwards.
Neptunus (*Amphitrite*) *rugosus*, A.
   Milne Edwards.
Neptunus (*Amphitrite*) *gladiator*
   (Fabricius), var. *argentatus*, White.
Neptunus (*Amphitrite*) *spinipes*, n. sp.
Neptunus (*Achelous*) *granulatus*, Milne
   Edwards.
Neptunus (*Achelous*) *unispinosus*, Miers.
Neptunus (*Achelous*) *depressifrons*
   (Stimpson).
Xiphonectes longispinosus, Dana.
Scylla serrata, Forskål.
Cronius bispinosus, n. sp.
Goniosoma cruciferum (Fabricius).
Thalamonyx danæ, var. *gracilipes*, A.
   Milne Edwards.
Thalamita admete (Herbst).
Thalamita integra, Dana.
Thalamita sima, Milne Edwards.
Thalamita intermedia, n. sp.
Thalamita sexlobata, n. sp.
Thalamita prymna (Herbst).

Thalamita stimpsoni, A. Milne Edwards.
Thalamita crenata, Rüppell.
Portunus corrugatus (Pennant).
Platyonychus bipustulatus, Milne Edwards.
Lissocarcinus polybioides, Adams and
   White.
Lissocarcinus lævis, n. sp.
Podophthalmus vigil, Fabricius.
Acanthocyclus gayi, Milne Edwards
   and Lucas.
Hypopeltarium spinulosum, White.
Gomeza bicornis, Gray.
Pilumnoplax vestita, de Haan, var.
   sexdentata, Haswell.
Ceratoplax ciliata, Stimpson.
Ocypoda ceratophthalma, Pallas.
Ocypoda cursor (Linné).
Ocypoda arenaria (Catesby).
Gelasimus vocans (Linné).
Gelasimus rubripes, Jacquinot and Lucas.
Gelasimus tetragonon (Herbst).
Gelasimus annulipes, Milne Edwards.
Gonoplax sinuatifrons, n. sp.
Ommatocarcinus macgillivrayi, White.
Macrophthalmus podophthalmus, Eydoux
   and Souleyet.
Macrophthalmus serratus, Adams and
   White.
Hemiplax hirtipes (Heller).
Euplax (*Chænostoma*) *boscii* (Audouin).
Grapsus maculatus (Catesby).
Grapsus strigosus (Herbst).
Leptograpsus variegatus (Fabricius).
Metopograpsus messor (Forskål).
Pachygrapsus transversus, Gibbes.
Geograpsus grayi (Milne Edwards).
Pseudograpsus albus, Stimpson.
Brachynotus (*Heterograpsus*) *peni-*
   cillatus, de Haan.

Varuna litterata (Fabricius).
Epigrapsus politus, Heller.
Goniopsis cruentatus, Latreille.
Helice latreillei, Milne Edwards, var.
Helice crassa, Dana.
Sesarma müllerii, A. Milne Edwards.
Sesarma schüttei, Hess.
Sesarma (Holometopus) aubryi, A. Milne Edwards.
Plagusia depressa (Fabricius).
Plagusia immaculata, Lamarck.
Plagusia chabrus (Linné).
Pinnotheres villosulus, Guérin-Ménéville.
Mycteris longicarpus (Latreille).
Mycteris platycheles, Milne Edwards.
Hymenosoma orbiculare (Desmarest).
Halicarcinus planatus (Fabricius).
Halicarcinus ovatus, Stimpson.
Calappa flammea (Herbst).
Calappa hepatica (Linné).
Calappa gallus (Herbst).
Calappa depressa, n. sp.
Mursia cristimana (Desmarest).
Cryptosoma cristatum, Leach (*ined.*), Brullé.
Matuta victrix (Fabricius), with var. *crebrépunctata*, Miers.

Matuta banskii, Leach.
Matuta lævidactyla, Miers.
Matuta inermis, Miers.
Arcania septemspinosa (Fabricius).
Ixa cylindrus, Fabricius, var. *megaspis*, Adams and White.
Iliacantha intermedia, n. sp.
Ebalia lævis, Bell.
Ebalia (Phlyxia) crassipes, Bell.
Ebalia (Phlyxia) undecimspinosa (Kinahan), var. *orbicularis*, Haswell.
Ebalia (Phlyxia) quadridentata, Gray, var. *spinifera*, nov.
Ebalia (Phlyxia) dentifrons, n. sp.
Persephona punctata (Browne).
Myra australis, Haswell.
Myra affinis, Bell.
Myra darnleyensis, Haswell.
Lithadia cariosa, Stimpson, var.
Philyra platycheira, de Haan.
? *Leucosia australiensis*, n. sp.
Leucosia haswelli, n. sp.
Leucosia craniolaris (Linné).
Dorippe facchino (Herbst).
Cymopolia caronii, Roux.
Cymopolia jukesii, White.

20 to 100 fathoms.

Leptopodia sagittaria, Fabricius.
Metoporaphis forficulatus, A. Milne Edwards.
Stenorhynchus spinifer, n. sp.
Achæus japonicus, de Haan.
Achæus tenuicollis, n. sp.
? *Podochela riisei*, Stimpson.
Inachus leptochirus, Leach.
Eurypodius latreillei, Guérin-Ménéville.

Egeria arachnoides (Rumph).
Chorilibinia gracilipes, Miers.
Hyas aranea (Linné).
Hyas coarctata, Leach.
? *Herbstia (Herbstiella) depressa*, Stimpson.
Chlorinoides coppingeri, Haswell.
Pisa (Arctopsis) tribulus (Linné).
Hyastenus diacanthus, de Haan.
Hyastenus planasius, Adams and White.

Scyra compressipes, Stimpson.
Micippa spinosa, var. *affinis*, Miers.
Libinia gracilipes, n. sp.
? *Mithrax hispidus* (Herbst), var. *pleura-
canthus*, Stimpson (?).
Lambrus longimanus (Linné).
? *Lambrus turriger*, White.
Lambrus (Aulacolambrus) hoplonotus,
var. *longioculis*, Miers.
Lambrus (Parthenolambrus) massena,
Roux.
Lambrus (Parthenolambrus) expansus,
Miers.
Cryptopodia fornicata (Fabricius).
Heterocrypta maltzani, Miers.
Lophozozymus (Lophoxanthus) bellus,
Stimpson, var. *leucomanus*,
Lockington.
Actæa peronii (Milne Edwards).
Atergatopsis granulatus, Stimpson.
Xanthodes melanodactylus, A. Milne
Edwards.
Sphærozius nitidus, Stimpson.
Pilumnus spinifer, Milne Edwards.
? *Pilumnus floridanus*, Stimpson.
Pilumnus minutus, de Haan, var.
hirsutus, nov.
Pilumnus globosus, Dana.
Pilumnus tomentosus, Latreille.
Neptunus sayi, A. Milne Edwards.
Neptunus (Amphitrite) hastatus (Linné).
Neptunus (Amphitrite) hastatoides
(Fabricius).
Neptunus (Amphitrite) gladiator
(Fabricius).

Neptunus (Achelous) whitei, A. Milne
Edwards.
? *Neptunus (Achelous) granulatus*, Milne
Edwards.
Goniosoma variegatum, Fabricius, var.
bimaculatum, nov.
Goniosoma ornatum, A. Milne Ed-
wards.
Portunus corrugatus, Pennant.
Hypopeltarium spinulosum, White.
? *Bathyplax typhlus*, Milne Edwards, var.
oculiferus, nov.
Ceratoplax arcuata, Miers.
Notonyx nitidus, A. Milne Edwards.
? *Gonoplax sinuatifrons*, n. sp.
Platygrapsus depressus (de Haan).
Halicarcinus planatus (Fabricius)
Halicarcinus ovatus, Stimpson.
Calappa granulata (Linné).
Calappa lophos (Herbst).
? *Calappa gallus* (Herbst).
? *Myrodes eudactylus*, Bell.
Ebalia lambriformis (Bell).
Ebalia tuberculosa, A. Milne Edwards.
Ebalia (Phlyxia) crassipes, Bell.
Ebalia (Phlyxia) intermedia, n. sp.
Myra fugax (Fabricius).
Merocryptus lambriformis, A. Milne
Edwards.
Leucosia haswelli, n. sp.
Leucosia ocellata (Bell).
Leucosia whitei, Bell.
Leucosia craniolaris (Linné).
Dorippe japonica, v. Siebold.

100 to 200 fathoms.

Leptopodia sagittaria (Fabricius) (?).
Metoporaphis forficulatus, A. Milne
  Edwards.
Podochela riisei, Stimpson.
Platymaia wyville-thomsoni, n. gen.
  and sp.
Cyrtomaia murrayi, n. gen. and sp.
Eurypodius longirostris, n. sp.
Lispognathus thomsoni, Norman.
Ergasticus naresii, n. sp.
Oxypleurodon stimpsoni, n. gen. and sp.
Pugettia velutina, n. sp.
Herbstia (*Herbstiella*) *depressa*,
  Stimpson (?).
Hyastenus elegans, n. sp.
Naxia hystrix, n. sp.
Mithrax hispidus (Herbst), var. *pleura-
  canthus*, Stimpson (?).

Medæus haswelli, n. sp.
Pilumnus floridanus, Stimpson (?).
Pilumnus normani, n. sp.
Lupocyclus (*Parathranites*) *orientalis*,
  n. subgen. and sp.
Platyonychus iridescens, n. sp.
Pilumnoplax heterochir, Studer.
Bathyplax typhlus, A. Milne Edwards,
  var. *oculiferus*, nov (?).
Litocheira kingsleyi, n. sp.
Halicarcinus planatus (Fabricius).
Paracyclois milne-edwardsii, n. gen.
  and sp.
Mursia cristimana, Desmarest.
Ebalia lævis, Bell.
Ebalia tuberculosa (A. Milne Edwards).
Merocryptus lambriformis, A. Milne
  Edwards.

200 to 500 fathoms.

Leptopodia sagittaria, Fabricius (?).
Metoporaphis forficulatus, A. Milne
  Edwards (?).
Podochela riisei, Stimpson.
Cyrtomaia suhmii, n. sp.
Anamathia pulchra, n. sp.
Lispognathus thomsoni (Norman).
Echinoplax moseleyi, n. gen. and sp.
Macrocheira kämpferi, de Haan.
Oxypleurodon stimpsoni, n. gen. and sp.
Herbstia (*Herbstiella*) *depressa*,
  Stimpson (?).
Libinia smithii, n. sp.

Mithrax hispidus (Herbst), var. *pleura-
  canthus*, Stimpson (?).
Heterocrypta maltzani, Miers.
Pilumnus floridanus, Stimpson (?).
Neptunus (*Hellenus*) *spinicarpus*
  (Stimpson).
Geryon (?) *incertus*, n. sp.
Pilumnoplax abyssicola, n. sp.
Bathyplax typhlus, A. Milne Edwards,
  var. *oculiferus*, nov. (?).
Mursia curtispina, n. sp.
Randallia granulata, n. sp.
Ethusa orientalis, n. sp.

500 to 1000 fathoms.

Oncinopus araneus, de Haan.
Ethusa microphthalma, Smith.

Ethusa gracilipes, n. sp.; with var. robusta, nov.

1000 to 2000 fathoms.

Ethusa (Ethusina) challengeri, n. sp.

Ethusa (Ethusina) gracilipes, n. sp., var. robusta, nov.

Gulf-weed and Pelagic Species.

Neptunus sayi, A. Milne Edwards.

Nautilograpsus minutus (Linné).

Land and Fresh-water Species.

Acanthocyclus gayi, Milne Edwards and Lucas.
Thelphusa sinuatifrons, Milne Edwards.
Thelphusa (Potamonautes) perlata, Milne Edwards.
Thelphusa (Geothelphusa) dehaanii, White.

Geocarcinus lagostoma (Milne Edwards).
Cardisoma guanhumi, Latreille.
Cardisoma armatum, Herklots.
Cardisoma carnifex, Herbst.
Varuna litterata (Fabricius).

BRACHYURA.

OXYRHYNCHA or MAIOIDEA.

INACHIDÆ.

MAIIDÆ.

Egeria arachnoides (Rumph), p. 44.
Chorilibinia gracilipes, Miers, p. 46.
Hyas aranea (Linné), p. 47.
Hyas coarctata, Leach, p. 48.
Herbstia rubra, A. Milne Edwards, p. 49, Pl. VII. fig. 1.
Herbstia violacea (A. Milne Edwards), p. 50.
Herbstia ovata (Stimpson), p. 50.
Herbstia(Herbstiella)depressa,Stimpson (?), p. 51, Pl. VII. fig. 2.
Chlorinoides coppingeri (Haswell), p. 53, Pl. VII. fig. 3.
Pisa (Arctopsis) tribulus (Linné), p. 55.
Hyastenus diacanthus (de Haan), p. 57.
Hyastenus planasius (Adams and White), p. 57.

Hyastenus oryx, A. Milne Edwards, p. 58.
Hyastenus elegans, n. sp., p. 58, Pl. VI. fig. 3.
Naxia hystrix, n. sp., p. 60, Pl. VI. fig. 4.
Naxia hirta (A. Milne Edwards), p. 61.
Scyra compressipes, Stimpson, p. 63, Pl. VII. fig. 4.
Notolopas brasiliensis, n. sp., p. 64, Pl. VIII. fig. 1.
Schizophrys aspera (Milne Edwards), p. 67.
Pseudomicippa (?) varians, Miers, p. 68.
Micippa spinosa (Stimpson), p. 70, Pl. VIII. fig. 2.
Micippa spinosa, var. affinis, Miers, p. 70, Pl. VIII. fig. 3.

PERICERIDÆ.

Libinia smithii, n. sp., p. 73, Pl. IX. fig. 1.
Libinia gracilipes, n. sp., p. 74, Pl. IX. fig. 2.
Pericera cornuta, Milne Edwards. p. 76.
Picroceroides tubularis,n. gen. and sp.,p. 77, Pl. X. fig. 1.
Macrocœloma trispinosa (Latreille), p. 80.
Macrocœloma septemspinosa (Stimpson), p. 80.
Macrocœloma concava, n. sp., p. 81, Pl. X. fig. 2.

Microphrys bicornutus (Latreille), p. 83.
Mithrax (Nemausa) rostrata, A. Milne Edwards, p. 85.
Mithrax cornutus, de Saussure, p. 87.
Mithrax hispidus (Herbst), var. pleuracanthus, Stimpson, p. 88.
Mithrax forceps (A. Milne Edwards), p. 88.
Mithrax coronatus (Herbst), p. 89.
Mithrax sp., p. 89, Pl. X. fig. 3.

PARTHENOPIDÆ.

Lambrus contrarius (Herbst), p. 94.
Lambrus longimanus (Linné), p. 95.
Lambrus affinis, A. Milne Edwards, p. 95.
Lambrus intermedius, Miers, p. 96, Pl. X. fig. 4.

Lambrus turriger, White, p. 96.
Lambrus guérinii, Capello, p. 96.
Lambrus serratus (Milne Edwards), p. 97.
Lambrus (Aulacolambrus) hoplonotus, Adams and White, p. 98.

Lambrus (Aulacolambrus) hoplonotus, Adams and White, p. 98.
Lambrus (Aulacolambrus) hoplonotus, var. granulosus, Miers, p. 98, Pl. X. fig. 5.
Lambrus (Aulacolambrus) hoplonotus, var. longioculis, Miers, p. 99.
Lambrus (Parthenolambrus) massena, Roux, p. 100.
Lambrus (Parthenolambrus) massena, var. atlanticus, Miers, p. 100.

Lambrus (Parthenolambrus) expansus, Miers, p. 100.
Lambrus (Parthenolambrus) calappoides, Adams and White, p. 101.
Cryptopodia fornicata (Fabricius), p. 102.
Heterocrypta granulata (Gibbes), p. 103.
Heterocrypta maltzani, Miers, p. 103.
Ceratocarcinus longimanus, Adams and White, p. 105.

CYCLOMETOPA or CANCROIDEA.

CANCRIDÆ.

Cancer longipes, Bell, p. 110.
Carpilius maculatus (Linné), p. 111.
Atergatis floridus (Linné), p. 112.
Lophactæa lobata (Milne Edwards), p. 113.
Lophactæa granulosa (Rüppell), p. 114.
Lophozozymus (Lophoxanthus) bellus, Stimpson, var. leucomanus, Lockington, p. 115, Pl. XI. fig. 1.
Medæus haswelli, n. sp., p. 117, Pl. XI. fig. 2.
Actæa granulata (Audouin), p. 120.
? Actæa hirsutissima (Rüppell), p. 120.
Actæa nodulosa, White, p. 120.
Actæa hystrix, n. sp., p. 121, Pl. XI. fig. 3.
Actæa peronii (Milne Edwards), p. 122.
Actæa rufopunctata (Milne Edwards), var. nodosa, Stimpson, p. 122.
Atergatopsis granulatus, A. Milne Edwards, p. 123.
Xantho bidentatus, A. Milne Edwards, p. 126.
Xanthodes melanodactylus, A. Milne Edwards, p. 128.
Panopeus herbstii, var. serratus, (de Saussure), p. 129.

Micropanope spinipes, A. Milne Edwards (?), p. 130.
Etisus lævimanus, Randall, p. 132.
Carpilodes bellus (Dana), p. 134.
Zozymus æneus (Linné), p. 134.
Actæodes tomentosus (Milne Edwards), p. 135.
Leptodius exaratus, var. sanguineus (Milne Edwards), p. 138.
Leptodius punctatus, Miers, p. 138.
Phymodius monticulosus (Dana), p. 139.
Eurytium limosum (Say), p. 141.
Pseudozius bouvieri, A. Milne Edwards, var. mellissii, Miers, p. 143, Pl. XII. fig. 3.
Sphærozius nitidus, Stimpson, p. 144, Pl. XII. fig. 4.
Pilumnus africanus, A. Milne Edwards, p. 150, Pl. XIII. fig. 1.
Pilumnus spinifer, Milne Edwards (?), p. 150.
Pilumnus brasiliensis, n. sp., p. 151, Pl. XIII. fig. 2.
Pilumnus floridanus, Stimpson, p. 152, Pl. XIII. fig. 3.

Pilumnus fragosus, A. Milne Edwards, var., p. 153.

Pilumnus minutus, de Haan, var. *hirsutus*, p. 154.

Pilumnus globosus, Dana, p. 155.

Pilumnus dehaani, Miers, p. 155, Pl. XIV. fig. 1.

Pilumnus scabriusculus, Adams and White (?), p. 155.

Pilumnus normani, n. sp., p. 156, Pl. XIV. fig. 2.

Pilumnus longicornis, Hilgendorf, var., p. 157.

Pilumnus vestitus, Haswell, p. 159, Pl. XIV. fig. 3.

Pilumnus tomentosus, Latreille (?), p. 160, Pl. XIV. fig. 4.

Pilumnus rufopunctatus, Stimpson, p. 160, Pl. XIV. fig. 5.

Pilumnus pulcher, Miers, p. 161.

Pilumnus seminudus, Miers, p. 161.

Pilumnus labyrinthicus, Miers, p. 161.

Eriphia gonagra (Fabricius), p. 163.

Trapezia cymodoce (Herbst) (?), p. 166.

Trapezia guttata, Rüppell, var., p. 166, Pl. XII. fig. 1.

Trapezia areolata, var. *inermis*, A. Milne Edwards, p. 167.

Trapezia rufopunctata (Herbst), p. 167.

Trapezia rufopunctata, var. *intermedia*, n. sp., p. 168, Pl. XII. fig. 2.

PORTUNIDÆ.

Neptunus (*Neptunus*) *pelagicus* (Linné), p. 173.

Neptunus (*Neptunus*) *sayi*, A. Milne Edwards, p. 173.

Neptunus (*Neptunus*) *sanguinolentus* (Herbst), p. 174.

Neptunus (*Amphitrite*) *hastatus* (Linné), p. 175.

Neptunus (*Amphitrite*) *hastatoides* (Fabricius), p. 175.

Neptunus (*Amphitrite*) *tuberculosus*, A. Milne Edwards, p. 176.

Neptunus (*Amphitrite*) *rugosus*, A. Milne Edwards, p. 176.

Neptunus (*Amphitrite*) *gladiator* (Fabricius), p. 177.

Neptunus (*Amphitrite*) *gladiator*, var. *argentatus*, White, p. 177.

Neptunus (*Amphitrite*) *spinipes*, n. sp., p. 178, Pl. XV. fig. 1.

Neptunus (*Achelous*) *whitei*, (A. Milne Edwards), p. 179.

Neptunus (*Achelous*) *granulatus* (Milne Edwards), p. 180.

Neptunus (*Achelous*) *unispinosus* (Miers), p. 180.

Neptunus (*Achelous*) *depressifrons* (Stimpson), p. 181.

Neptunus (*Hellenus*) *spinicarpus* (Stimpson), p. 182.

Xiphonectes longispinosus, Dana, p. 183.

Scylla serrata (Forskål), p. 185.

Lupocyclus (*Parathranites*) *orientalis*, n. subgen. and sp., p. 186, Pl. XVII. fig. 1.

Cronius bispinosus, n. sp., p. 188, Pl. XV. fig. 2.

Goniosoma variegatum (Fabricius), p. 190.

Goniosoma variegatum, var. *bimaculatum*, nov., p. 191, Pl. XV. fig. 3.

CYCLINEA.

CORYSTOIDEA.

THELPHUSINEA.

CATOMETOPA or OCYPODIIDEA.

GEOCARCINIDÆ.

OCYPODIDÆ.

Geryon (?) *incertus*, n. sp., p. 224, Pl. XVI.
fig. 3.
Pilumnoplax heterochir, Studer, p. 227,
Pl. XIX. fig. 1.
Pilumnoplax abyssicola, n. sp., p. 228,
Pl. XIX. fig. 2.
Pilumnoplax vestita, de Haan, var. *sexden-
tata*, Haswell, p. 229.
Bathyplax typhlus, A. Milne Edwards, var.
oculiferus, nov., p. 230, Pl. XX. fig. 3.
Litocheira kingsleyi, n. sp., p. 232, Pl.
XXI. fig. 1.
Ceratoplax ciliata, Stimpson, p. 234, Pl.
XIX. fig. 3.
Ceratoplax arcuata, Miers, p. 235.
Notonyx nitidus, A. Milne Edwards, p. 236.
Ocypoda ceratophthalma (Pallas), p. 238.
Ocypoda cursor (Linné), p. 240.

Ocpoda arenaria (Catesby), p. 240.
Gelasimus vocans (Linné), p. 242.
Gelasimus rubripes, Jacquinot and Lucas,
p. 243.
Gelasimus tetragonon (Herbst), p. 243.
Gelasimus annulipes, Milne Edwards,
p. 244.
Gonoplax sinuatifrons, n. sp., p. 246, Pl.
XX. fig. 2.
Ommatocarcinus macgillivrayi, White,
p. 247.
Macrophthalmus podophthalmus, Eydoux
and Souleyet, p. 249.
Macrophthalmus serratus, Adams and
White, p. 250, Pl. XX. fig. 1.
Hemiplax hirtipes, Heller, p. 251.
Euplax (*Chænostoma*) *boscii* (Audouin),
p. 252.

GRAPSIDÆ.

Nautilograpsus minutus (Linné), p. 254.
Grapsus maculatus (Catesby), p. 255.
Grapsus strigosus (Herbst), p. 256.
Leptograpsus variegatus (Fabricius), p. 257.
Metopograpsus messor (Forskål), p. 258.
Metopograpsus messor, var. *frontalis*, Miers,
p. 258.
Pachygrapsus transversus, Gibbes, p. 259.
Geograpsus grayi (Milne Edwards), p. 261.
Pseudograpsus albus, Stimpson, p. 262.
Platygrapsus depressus (de Haan), p. 263.
Brachynotus (*Heterograpsus*) *penicillatus*
(de Haan), p. 264.
Varuna litterata (Fabricius), p. 265.
Epigrapsus politus, Heller, p. 266.

Goniopsis cruentatus, Latreille, p. 267.
Helice latreillei, Milne Edwards, var., p. 268,
Pl. XXI. fig. 2.
Helice crassa, Dana, p. 269.
Sesarma müllerii, A. Milne Edwards, p. 270,
Pl. XXI. fig. 3.
Sesarma schüttei, Hess, p. 271.
Sesarma (*Holometopus*) *aubryi*, A. Milne
Edwards, p. 271.
Plagusia depressa (Fabricius), p. 272.
Plagusia immaculata, Lamarck, p. 273,
Pl. XXII. fig. 1.
Plagusia chabrus (Linné), p. 273, Pl. XXII.
fig. 1d.

PINNOTHERIDÆ.

Pinnotheres villosulus, Guérin-Méneville, p. 277, Pl. XXII. fig. 2.

Mycteris longicarpus (Latreille), p. 278.

Mycteris platycheles, Milne Edwards, p. 279.

Hymenosoma orbiculare (Desmarest), p. 280.

Halicarcinus planatus (Fabricius), p. 281.

Halicarcinus ovatus, Stimpson, p. 282.

OXYSTOMATA or LEUCOSIIDEA.

CALAPPIDÆ.

Calappa flammea (Herbst), p. 284, Pl. XXIII. fig. 1.

Calappa hepatica (Linné), p. 285.

Calappa granulata (Linné), p. 285.

Calappa gallus (Herbst), p. 286.

Calappa lophos (Herbst), p. 286.

Calappa depressa, n. sp., p. 287, Pl. XXIII. fig. 2.

Paracyclois milne-edwardsii, n. gen. and sp., p. 289, Pl. XXIV. fig. 1.

Mursia cristimana (Desmarest), p. 291.

Mursia curtispina, n. sp., p. 291, Pl. XXIV. fig. 2.

Cryptosoma cristatum, Leach (ined.), Brullé, p. 293.

MATUTIDÆ.

Matuta victrix (Fabricius), p. 295.

Matuta victrix, var. crebrépunctata, Miers, p. 295.

Matuta banksii, Leach, p. 295.

Matuta lævidactyla, Miers, p. 296.

Matuta inermis, Miers, p. 296.

LEUCOSIIDÆ.

Myrodes eudactylus, Bell, p. 298.

Arcania septemspinosa (Fabricius), p. 300.

Ixa cylindrus (Fabricius), var. megaspis, Adams and White, p. 301.

Iliacantha intermedia, n. sp., p. 302, Pl. XXVI. fig. 3.

Ebalia lambriformis (Bell), p. 306.

Ebalia lævis (Bell), p. 306.

Ebalia tuberculosa (A. Milne Edwards), p. 306, Pl. XXV. fig. 1.

Ebalia (Phlyxia) crassipes, Bell, p. 307.

(ZOOL. CHALL. EXP.—PART XLIX.—1886.)

Ebalia (Phlyxia) intermedia, n. sp., p. 308, Pl. XXV. fig. 2.

Ebalia (Phlyxia) undecimspinosa (Kinahan), var. orbicularis, Haswell, p. 309.

Ebalia (Phlyxia) quadridentata (Gray), var. spinifera, nov., p. 309, Pl. XXV. fig. 3.

Ebalia (Phlyxia) dentifrons, n. sp. p. 310, Pl. XXV. fig. 4.

Persephona punctata (Browne), p. 312. Pl. XXV. fig. 5.

Myra fugax (Fabricius), p. 313.

Ccc g

Myra affinis, Bell, p. 315.
Myra australis, Haswell, p. 315.
Myra darnleyensis, Haswell, p. 315.
Randallia granulata, n. sp., p. 317, Pl. XXVI. fig. 1.
Lithadia cariosa, Stimpson, var., p. 319, Pl. XXVI. fig. 2.
Merocryptus lambriformis, A. Milne Edwards, p. 320.

Philyra platycheira, de Haan (?), p. 321.
Leucosia australiensis, n. sp., p. 322, Pl. XXVII. fig. 1.
Leucosia haswelli, n. sp., p. 324, Pl. XXVII. fig. 2.
Leucosia ocellata, Bell, p. 325.
Leucosia whitei, Bell, p. 325.
Leucosia craniolaris (Linné), p. 325, Pl. XXVII. fig. 3.

DORIPPIDÆ.

Dorippe facchino (Herbst), p. 328.
Dorippe japonica, v. Siebold, p. 328.
Ethusa microphthalma, Smith, p. 329.
Ethusa orientalis, n. sp., p. 330, Pl. XXVIII. fig. 1.
Ethusa (Ethusina) challengeri, n. sp., p. 331, Pl. XXVIII. fig. 2.

Ethusa (Ethusina) gracilipes, n. sp., p. 332, Pl. XXIX. fig. 1.
Ethusa (Ethusina) gracilipes, var. *robusta*, nov., p. 333, Pl. XXIX. fig. 2.
Cymopolia caronii, Roux, p. 334.
Cymopolia jukesii, White, p. 335.

BRACHYURA.

Cancri Brachyuri, Lamarck (pt.), Syst. des Anim. sans Vert., p. 148, 1801.

Brachyures, Milne Edwards, Hist. Nat. Crust., i. p. 247 (1834).

Brachyura, Leach (pt.), Trans. Linn. Soc. Lond., vol. xi. p. 307, 1815.

,, Latreille (pt.), Fam. Nat. du Règne Anim., p. 267, 1825.

,, Dana, U. S. Expl. Exped., vol. xiii. Crust. 1, p. 58, 1852.

,, Miers, Cat. New Zeal. Crust., p. 1, 1876.

,, Claus (pt.), Grundzüge der Zoologie, ed. 4, vol. i. p. 632, 1880.

,, Haswell, Cat. Australian Stalk and Sessile-Eyed Crust., p. 1, 1882.

The cervical and thoracic regions of the body are covered by the carapace, which is greatly developed. The antennulary fossettes and orbits are usually well defined. The buccal cavity is distinctly defined in front. The sternum is never linear, the vulvæ are situated upon the sternum. The post-abdomen, in the male, is short; inflexed beneath the cephalothorax, and usually closely applied to it. The flagella of the antennules and antennæ are usually short or rudimentary. The exterior maxillipedes are operculiform and usually completely close the buccal cavity. The anterior legs are developed as perfectly chelate limbs (chelipedes). The ambulatory legs (i.e., those of the second to the fifth pairs) are never perfectly chelate; but rarely, in those forms which establish the passage to the Anomura (e.g., the Dorippidæ) are feeble and imperfectly prehensile.

The Brachyura, or short-tailed, i.e., true crabs, are, as stated in the introductory portion of this Report, here limited in the sense long ago indicated by Profs. H. Milne Edwards and J. D. Dana; who may perhaps still be regarded as the leading authorities on the systematic classification of the Crustacea; and are subdivided into the four secondary divisions or subtribes, Oxyrhyncha, Cyclometopa, Catometopa, and Oxystomata.

The recent species, as is well known, are extremely numerous, for the most part marine, and inhabitants of the shores or shallower waters of the temperate, subtropical, or tropical regions of the globe; but few species occurring in the Arctic or Antarctic circumpolar areas of distribution. Others (Geocarcinidæ, Thelphusidæ) are terrestrial or fluviatile, and a few (Pinnotheridæ) inhabit the shells, &c., of animals of other groups. Until recently, little was known of the bathymetrical distribution of these animals, but the Challenger and other recent expeditions have shown that species, especially of the somewhat aberrant family Dorippidæ, may occur in the abysses of the ocean to a depth exceeding 1500 fathoms.

OXYRHYNCHA or MAIOIDEA.

Oxyrhynchi, Latreille (pt.), Hist. Nat. Crust., vol. vi. p. 85, 1803.
Oxyrhinques, Milne Edwards, Hist. Nat. Crust., vol. i. pp. 263, 266, 1834.
Canceriens Cryptopodes, Milne Edwards, *tom. cit.*, pp. 368, 369, 1834.
Maioidea vel *Oxyrhyncha*, Dana, U.S. Expl. Exped., vol. xiii. Crust. 1, pp. 66, 75, 1852.
Oxyrhyncha, Miers, Cat. New Zeal. Crust., p. 2, 1876; Journ. Linn. Soc. Lond. (Zool.), vol. xiv. p. 634, 1879.

Carapace more or less narrowed in front, and usually rostrated; with the branchial regions considerably developed. Hepatic regions small. Epistome usually large. Buccal cavity quadrate, with the anterior margin straight. Branchiæ nine in number; their efferent channels terminating at the sides of the endostome or palate. Antennules longitudinally plicated. The carpal joint of the endognath of the exterior maxillipedes is articulated at the summit or at the antero-internal angle of the merus. The verges of the male are inserted at the bases of the fifth ambulatory legs.

The Oxyrhyncha, thus defined, constitute as a whole a very natural group; which however is connected with the Cyclometopa by almost insensible gradations, and no single character can be mentioned which will suffice to distinguish them universally from the other Brachyura.

Legion I. MAIINEA.

Maiinea, Dana, U.S. Expl. Exped., vol. xiii., Crust. i. pp. 76, 77, 1852.
„   Miers, Journ. Linn. Soc. Lond. (Zool.), vol. xiv. p. 640, 1879.

Basal antennal joint well developed, inserted beneath the eyes, and usually occupying a great part of the infraocular space.

Family I. INACHIDÆ.

Inachidæ, Miers, Journ. Linn. Soc. Lond. (Zool.), vol. xiv. pp. 640–642, 1879.

Eyes non-retractile, or retractile against the sides of the carapace. Usually the orbits are not defined, but there is often a well-developed præocular and postocular spine. Basal joint of the antennæ generally slender, sometimes moderately enlarged.

The carapace varies in shape, being subtriangulate or oblong-triangulate or subpyriform, rarely suborbiculate. Rostrum simple or bifid, sometimes very short. Chelipedes with the fingers never excavated at the tips. Ambulatory legs sometimes very long. Post-abdomen of male and female four to seven-jointed, two or three of the joints often coalescent.

Subfamily 1. LEPTOPODIINÆ.

Leptopodiinæ, Miers, tom. cit., p. 642, 1879, *et synonym.*

Eyes slender, non-retractile, and laterally projecting. Præocular and postocular spines minute or wanting. Basal antennal joint very slender throughout its length. The genera have been enumerated in my memoir referred to above.

Leptopodia, Leach.

Leptopodia, Leach, Zool. Miscell., vol. ii. p. 15, 1815.
„　　Milne Edwards, Hist. Nat. Crust., vol. i. p. 275, 1834.
„　　Miers, Journ. Linn. Soc. Lond. (Zool.), vol. xiv. p. 643, 1879.

Carapace triangulate, with the posterior margin straight, and not prolonged over the posterior segment of the thorax ; there are no defined orbits, but a distinct, although small, postocular spine. The rostrum is very long, simple, and horizontal, with the lateral margin serrate. The endostome or palate is without longitudinal ridges. The epistome is very large. The post-abdomen is six-jointed, with the penultimate and terminal segments coalescent. The eyes are short and non-retractile. The antennules are longitudinally plicated. The basal joint of the antennæ is elongated and very slender, and the elongated flagellum is in great part concealed beneath the rostrum. The ischium of the exterior maxillipedes is somewhat produced at its antero-internal angle ; the merus is obversely triangulate, truncated at the distal extremity, and bears the next joint at its antero-external angle. The chelipedes (in the adult male) are rather slender and greatly elongated, with the merus, carpus, and palm subcylindrical ; fingers much shorter than the palm, distally acute, and dentated on the inner margins. The ambulatory legs are very slender and extremely elongated, dactyli styliform ; the merus-joints of all the legs are spinuliferous.

The single species (*Leptopodia sagittaria*) occurs, as A. Milne Edwards has shown, in the Gulf of Mexico and West Indian Seas, at the Canary and Cape Verde Islands, and Madeira, and also on the coast of Brazil and on the west coast of Central and South America, southward to Chili.[1]

Leptopodia sagittaria has already been recorded by A. Milne Edwards from considerable depths at several localities in the West Indian Seas, e.g., at Barbados (94 fathoms)

---

[1] There are in the British Museum Collection specimens from Angola (Dr. Welwitsch). I have seen no examples of the Chilian variety designated by A. Milne Edwards, *L. ocolota*, in which the cardiac and branchial regions of the carapace are more swollen and the rostrum shorter and raised at the apex.

and Santa Cruz (115 fathoms), but these by no means indicate the extreme limit of its bathymetrical range; specimens having been recently received by the British Museum from Captain E. Cole, taken off Jamaica in 600 fathoms, and in the Mona Channel, West Indies, in 814 fathoms.

Leptopodia sagittaria (Fabricius).

Cancer sagittarius, Fabricius, Ent. Syst., ii. p. 442, 1793.
Leptopodia sagittaria, Leach, Zool. Miscell., ii. p. 16, pl. lxvii., 1815.
,,   ,,   Milne Edwards, Hist. Nat. Crust., vol. i. p. 276, 1834; Atlas, in Cuvier, Règne Animal, Crust., ed. 3, pl. xxvi. fig. 1.
,,   ,,   A. Milne Edwards, Études sur les Crust. Podophthalmaires in Miss. Sci. au Mexique et dans l' Amérique Centrale, pt. v. p. 172, 1878; and references to literature.

Fully grown specimens were obtained at Madeira, and St. Vincent, Cape Verde Islands, also smaller examples at Bahia in shallow water, 7 to 20 fathoms. Another young female was dredged at Barra Grande, south of Pernambuco, in 30 to 350 fathoms (Station 122), in lat. 9° 5′ to 9° 10′ S., long. 34° 49′ to 34° 53′ W.

An adult male from Madeira measures as follows :—

Adult ♂.		Linea.	Millims.
Length of carapace and rostrum,	.	29½	62·5
Breadth of carapace,	. .	9	19

Metoporaphis, Stimpson.

Metoporaphis, Stimpson, Ann. Lyc. Nat. Hist., New York, vol. vii. p. 198, 1860.
,,   Miers, Journ. Linn. Soc. Lond. (Zool.), vol. xiv. p. 643, 1879.

This genus, which is nearly allied to Leptopodia, is distinguished by its uneven and tuberculated carapace, the exposed flagella of the antennæ which are visible at the sides of the rostrum in a dorsal view, and by the very considerable development of the median distal spine of the merus of the ambulatory legs.

In Metoporaphis forficulatus, which is the only species I have examined, the chelipede differs markedly from that of Leptopodia, the palm (in the adult) being much shorter than the dactyle and pollex, which are slender, gaping, incurved, and meet only at the tips.

The ascertained range of this genus is from Charleston Harbour (whence the type was obtained) southward to Barra Grande on the Brazilian coast; the species are apparently of very rare and local occurrence. But two have been described, which must be regarded as doubtfully distinct one from the other; Metoporaphis calcarata (Say), and Metoporaphis forficulatus, A. Milne Edwards.

The description of the former species (from a very imperfect type) is, however, too brief for certain identification, and therefore Milne Edwards' designation for the Challenger specimens is retained.

Metoporaphis forficulatus, A. Milne Edwards.

Metoporaphis forficulatus, A. Milne Edwards, Crust. in Miss. Sci. au Mexique, pt. v. p. 174, pl. xxxi. fig. 3, 1878.

A male in somewhat mutilated condition was dredged off Bahia in 7 to 20 fathoms. A female in very fragmentary condition dredged at Barra Grande, south of Pernambuco, at Station 122, in 30 to 350 fathoms, also probably belongs to this species, but the carapace is broader, and the tubercles of its dorsal surface are less distinct.

In both specimens there is a spinule upon the outer surface of the basal joint of the antennæ, and another at the distal extremity of the same joint which are not shown in Milne Edwards' figures of this appendage (fig. 3a), but I have little doubt of the correct identification of the specimens. The type was from Guiana.

The dimensions of the male are as follows :—

Adult ♂.	Lines.	Millims.
Length of carapace and rostrum,	6½	13·5
Breadth of carapace, about	3	6·5
Length of chelipede, about	8	17

Stenorhynchus, Lamarck.

Stenorhynchus, Lamarck (pt.), Hist. des Anim. sans Vert., v., p. 236, 1818.
„ Milne Edwards, Hist. Nat. Crust., vol. i. p. 278, 1834.
„ Miers, Journ. Linn. Soc. Lond. (Zool.), vol. xiv. p. 643, 1879.

The carapace is subtriangulate, with a straight posterior margin, and is more or less distinctly spinose above; there is usually no postocular spine. The rostrum is composed of two slender, straight, contiguous spines. The post-abdomen has but six distinct segments. The eyes are slender and project laterally. The merus of the outer maxillipedes is somewhat elongated and distally rounded, and articulates at its summit with the next joint. The chelipedes have the palms somewhat inflated, the ambulatory legs are slender and much elongated, with the dactyli nearly straight or those of the fifth pair only slightly falcated.

The species, which are small and not numerous, occur commonly in the temperate waters of the northern and southern hemisphere (Australian and South African Seas) at moderate depths, and more rarely in deep water [1] and in the Tropical Seas.

[1] A. Milne Edwards mentions the occurrence of *Stenorhynchus longirostris* in the Mediterranean at 420 metres. (Rapport sur la faune marine dans les grandes profondeurs de la Méditerranée et de l'Océan Atlantique, p. 18, 1882.)

The following are the forms with which I am acquainted :—

> *Stenorhynchus rostratus* (Linné) = *Cancer phalangium*, Pennant, *Stenorhynchus inermis*, Heller. Seas of Europe; Mediterranean (to 40 fathoms, Heller); Shetlands (to 70 fathoms, Norman), &c.; Cape Verde Islands (to 38 fathoms, Studer).
>
> *Stenorhynchus rostratus*, var. *spinulosus*, Miers. Senegambia, Goree Island (9 to 15 fathoms); coast of Portugal; Ireland (20 fathoms).
>
> *Stenorhynchus longirostris* (Fabricius) = *Leptopodia tenuirostris*, Leach. Seas of Europe; Mediterranean (to 420 fathoms, A. Milne Edwards).
>
> *Stenorhynchus ægyptius*, Milne Edwards. Mediterranean; South British Seas.
>
> *Stenorhynchus czernjawskii*, Brandt. Black Sea. (Perhaps not distinct from *Stenorhynchus longirostris*).
>
> *Stenorhynchus falcifer*, Stimpson. Cape of Good Hope (12 to 18 fathoms).
>
> *Stenorhynchus spinifer*, n. sp. St. Paul's Rocks (10 to 80 fathoms).[2]

Stenorhynchus falcifer, Stimpson (Pl. I. fig. 1).

Stenorhynchus falcifer, Stimpson, Proc. Acad. Nat. Sci. Philad., p. 219, 1857.

Two males and a female were dredged in Simon's Bay, 5 to 18 fathoms, November 19, 1873; where also the types of the species were collected.

Carapace moderately convex, triangulate, with a straight or slightly concave posterior margin; with a long median spine on the gastric region, in front of which are usually two smaller spines, the three forming a triangle; a prominent median spine on the cardiac region and behind this usually a smaller intestinal spine or tubercle; two small dorsal spines or tubercles on each branchial region; a lateral spine on each hepatic and branchial region, and commonly a few small lateral spinules beneath these spines. Rostrum one-fourth to one-half the length of the carapace, composed of two straight slender contiguous spines, which are usually directed obliquely upwards. Post-abdomen in both sexes six-jointed, a median prominence on each segment, eyes of moderate length, a small distal tubercle on the upper surface of the cornea. Basal antennal joint slender, with a small distal spine (there is also usually a small tubercle near to its base on the epistome and another near to the antero-lateral angles of the buccal cavity, and a small postocular tubercle, which may represent the postocular spine), the antennal flagella are exposed and visible in a dorsal view at the sides of the rostrum. Chelipedes (in the adult male) about twice as long as the carapace; merus with some small spines on its upper and lower margins and with a strong spine at the distal extremity of its upper

---

[2] As has been noted below, *Stenorhynchus curvirostris*, A. Milne Edwards, from Bass Strait, and *Stenorhynchus falcifrons*, Haswell, from Port Jackson and New Zealand, may belong to the genus *Achæopsis*; *Stenorhynchus brevirostris*, Haswell, is, I think, a species of *Achæus*.

margin. Carpus and palm with some small spines on their superior margins, and with the lower margins minutely tuberculate or spinuliferous; palm compressed and very slightly turgid in the middle, and slightly longer than the wrist, fingers about as long as palm, compressed, incurved toward the apices and minutely dentate on the inner margins, which are nearly in contact at the base. Ambulatory legs elongate and very slender; merus-joints with a strong spine (which is sometimes bifid) at the distal extremity of the upper margin; dactyli slender, in the fourth and fifth legs slightly falciform. Colour yellowish-brown or reddish. Carapace and chelipedes minutely pubescent, and there are some longer hairs on the carapace near the bases of the spines and on the margins of the chelipedes and ambulatory legs.

Adult ♂.			Lines.	Millims.
Length of carapace and rostrum,	.	. .	7½	15·5
Breadth of carapace, nearly	.	. . . .	4½	9
Length of a chelipede, nearly .	.	. . . .	11	23
Length of a penultimate ambulatory leg (the only one remaining attached),	.	. . . . .	18	38

Stenorhynchus spinifer, n. sp. (Pl. I. fig. 2).

The carapace is subtriangulate, longer than broad, and not constricted behind the eyes, and bears a median spine upon the gastric region (not shown in the figure) and another on each side, placed midway between this and the postocular spine; there existed also apparently a rounded tubercle or prominence upon the cardiac region, and possibly others near the front of the branchial region. The spines of the rostrum being broken, their exact length cannot be given, but they exceed half the length of the carapace. There is a strong spine on the upper margin of the orbit in front of the eye, and also a strong postocular spine.

The eye-peduncles are moderately robust; the corneæ considerably thicker than the peduncles, with a small tubercle on the anterior and upper surface, the bases of the antennules are lodged in very deep longitudinal fossettes; the very slender basal joints of the antennæ, which are not joined with the front at the distal extremities, are armed with a longitudinal series of five spinules, whereof the second from the distal extremity is the longest; the following joint is short, the next considerably elongated, slightly longer than the basal joint; the flagella filiform; the merus-joint of the outer maxillipedes has its antero-internal angle slightly produced and rounded, and has two or three spinules on its exterior face, near to the outer margin; the inner margin is fringed with long hairs. The carapace is scantily clothed with fine hairs. Colour (in spirit) yellowish-brown.

Adult ♂.	Lines.	Millims.
Length of carapace to base of rostrum,	4½	9·5
Breadth of carapace, about	4	8·5

This species is distinguished from all of the genus with which I am acquainted by the spine on the upper margin of the orbit in front of the eye, and the strong postocular spine.

The single mutilated specimen (a male) was dredged off St. Paul's Rocks in 10 to 80 fathoms.

It is described from a single crushed and mutilated specimen, with broken rostrum, and wanting all the legs, and the description and figures are necessarily very imperfect.

Achæus, Leach.

Achæus, Leach, Malacostraca Podophthalmata Britanniæ, pl. xxii. fig. c., 1815.
»   Milne Edwards, Hist. Nat. Crust., vol. i. p. 281, 1834.
»   Miers, Journ. Linn. Soc. Lond. (Zool.), vol. xiv. p. 643, 1879.

This genus is very nearly allied to *Stenorhynchus*, and it is in fact only distinguished from it by the absence of rostral spines ; the rostrum in *Achæus* being composed merely of two small acute or subacute lobes. The dactyli of the fourth and fifth ambulatory legs are usually but not invariably very distinctly falcated. The merus of the outer maxillipedes in certain species (*e.g.*, the typical *Achæus cranchii*) is shorter than in *Stenorhynchus*, and distally truncated.

The species, which are small, are distributed nearly as those of *Stenorhynchus*, and two or three are found in the Seas of Japan ; whence, I believe, *Stenorhynchus* has not been as yet recorded. On the other hand, I am not aware that *Achæus* has been recorded from any intertropical locality.

The following are the species known ; some of them may prove on careful revision to be mere varieties :—

Achæus cranchii, Leach = *Macropodia gracilis*, Costa. European Seas (Mediterranean, 70 to 75 métres, A. Milne Edwards).

Achæus lacertosus, Stimpson = *Achæus breviceps*, Haswell. East and North Australia.

Achæus japonicus, de Haan. Japan (to 50 fathoms).

Achæus tuberculatus, Miers. Corean Channel (to 36 fathoms).

Achæus lorina, Adams and White. Mindanao and Borneo.

Achæus spinosus, Miers. Corean Channel (to 30 fathoms).

Achæus affinis, Miers. North-East and West Australia.

Achæus brevirostris (Haswell). East and North-East Australia.

Achæus lævioculis, Miers. Seychelles (4 to 12 fathoms).

Achæus tenuicollis, n. sp. Port Phillip and Bass Strait (33 to 38 fathoms).

Achæus japonicus, de Haan.

Achæus japonicus, de Haan, Crust. in Siebold, Fauna Japonica, p. 99, pl. xxix. fig. 3 and pl. H., 1839.

A single small female dredged near Kobi, Japan, in 50 fathoms (Station 233A), lat. 34° 38′ N., long. 135° 1′ E., is referred with doubt to this species, as the ocular peduncles are unispinose, not quadrispinose, as in the description (but not the figure) of de Haan. The fourth and fifth legs, also, are more distinctly falcated than in his figure.

<table>
<tr><td>Adult ♀.</td><td>Lines.</td><td>Millims.</td></tr>
<tr><td>Length of carapace and rostrum,</td><td>5½</td><td>11·5</td></tr>
<tr><td>Breadth of carapace,</td><td>4</td><td>8·5</td></tr>
</table>

Achæus tenuicollis, n. sp. (Pl. I. fig. 3).

The body is thinly clothed with short curled hairs; the limbs with similar hairs, interspersed among which are some longer ones. The carapace is subtriangulate, little longer than broad, with a neck-like constriction behind the orbits, and armed with spines as follows:—Three conical spines upon the gastric and another upon the cardiac region, two shorter conical spines or tubercles, whereof the anterior is the smallest, on each branchial region, behind these one very small on the posterior margin of the carapace, and another on the sides of the branchial regions above the bases of the chelipedes; also a small spine upon the rounded, lateral, hepatic protuberance, and another behind this, on the pterygostomian region; there is also a strong spinule on the upper margin of the orbit, above the eye-peduncles. The lobes of the rostrum are short, and terminate each in a spine. The sternal surface of the body bears a few spinules. The post-abdomen of the male is, as usual, six-jointed (the two last joints having coalesced). The eye-peduncles are robust, with the corneæ protuberant; a small spinule exists on the inferior margin of the eye-peduncle, and another on the upper margin of the eye, near the distal extremity. The antennules are lodged in deep longitudinal fossettes; the very slender basal joint of the antennæ is joined with the front at its distal extremity, and bears several small spinules on its inferior surface, the following joint is short, the next about as long as the basal joint, flagella slender; the ischium-joint of the outer maxillipedes is produced at its inner and distal angle which is rounded, and bears several spinules on its outer surface, as does also the merus-joint which is rounded, not truncated, at the distal extremity where it bears the next joint. The chelipedes (in the male) are rather slender, and longer than the body; with the joints clothed with rather long hairs; ischium and merus-joints with a series of spinules on their antero- and postero-inferior faces, wrist about as long as palm, with a few spinules hardly discernible amid the hairs which clothe this joint, palm slightly compressed, not dilated, armed with spinules on its upper and lower margins, fingers about as long as palm, and slightly incurved at the apices which are nearly destitute of hair; the ambulatory legs are very slender and elongated; the dactyli of the first three pairs are short and nearly straight, in the last pair only are they slightly falciform.   Colour (in spirit), light yellowish-brown.

Adult ♂.	Lines.	Millims.
Length of carapace to base of rostrum,	3	6·5
Greatest breadth of carapace, about	2½	5
Length of a chelipede, about	5½	12
Length of third ambulatory leg, about	13	28

The description is wholly taken from an adult male, but applies also in nearly every particular to the adult female.

Of *Achæus tenuicollis* a male and female were dredged off the entrance to Port Phillip, in 33 fathoms, lat. 38° 22′ 30″ S., long. 144° 36′ 30″ E. (Station 161); also an adult and smaller female, off East Moncœur Island, Bass Straits, 38 fathoms, lat. 39° 10′ 30″ S., long. 146° 37″ E. (Station 162).

This species is apparently most nearly allied to *Achæus lorina*, Adams and White,[1] from Mindanao, and to *Achæus spinosus*, Miers,[2] from the Japanese Seas; and the three indeed may possibly prove to be varieties of one and the same form, but with the material at present available for comparison I have not ventured to unite them.   From *Achæus lorina*, *Achæus tenuicollis* is distinguished by the more numerous spines of the carapace, and relatively shorter, though still elongated legs; the specimens designated *Achæus lorina* in the British Museum collection are, I may add, apparently not those from which the species was described and figured.   *Achæus spinosus* differs in the bilobated spine of the cardiac region, the much more strongly-developed chelipedes of the male, and the proportionately shorter ambulatory legs.

Podochela, Stimpson.

Podochela, Stimpson, Ann. Lyc. Nat. Hist. New York, vol. vii. p. 194, 1860.
   „   Miers, Journ. Linn. Soc. Lond. (Zool.), vol. xiv. p. 643, 1879.
   „   A. Milne Edwards, Crust. in Miss. Sci. au Mexique, pt. v. p. 189, 1879, and synonyma.
Coryrhynchus, Kingsley, Proc. Acad. Nat. Sci. Philad., p. 384, 1879.

The carapace is triangulate and somewhat depressed, with the gastric region prominent and narrow.   Rostrum short and simple; either acute (subgenus *Podochela*) or hood-shaped and rounded at the distal extremity (subgenus *Coryrhynchus*).   The post-abdomen (in the male) has the sixth and seventh segments coalescent.   The eyes are non-retractile and project laterally.   The basal antennal joint is narrow and longitudinally carinated,

[1] Zoology of H.M.S. "Samarang," Crust., p. 3, pl. xi. fig. 2. 1848.   [2] *Proc. Zool. Soc. Lond.*, p. 25, 1879.

but is without distal spines or teeth. The chelipedes (in the male) are moderately robust, and the palm is slightly turgid (as in *Stenorhynchus*). The ambulatory legs are slender and somewhat elongated; the dactyli short and retractile.

M. Milne Edwards has united *Coryrhynchus*, Kingsley (= *Podonema*, Stimpson, nom. præoc.) with *Podochela*, but it may be convenient to retain this term as a subgeneric designation for the species with a distally-rounded and (usually) hood-shaped rostrum.

The species which have been enumerated by Milne Edwards, with one exception (*Podochela vestita*, Stimpson, from California, Cape St. Lucas) inhabit the Gulf of Mexico and Caribbean Sea, where they are found in shallow water and in depths not exceeding 60 fathoms (as at present recorded).

By the Challenger researches, however, the range of *Podochela riisei* is extended southward to the Brazilian coast, and its bathymetrical range to (possibly) 350 fathoms.

Podochela riisei (Stimpson).

Podonema riisei, Stimpson, Ann. Lyc. Nat. Hist. New York, vol. vii. p. 196, pl. ii. fig. 1, 1860; Bull. Mus. Comp. Zool. Cambridge, vol. ii. p. 126, 1870.
Dryope falcipoda, Desbonne and Schramm, Crust. de la Guadeloupe, p. 2.
Podochela riisei, A. Milne Edwards, Crust. in Miss. Sci. au Mexique, pt. v. p. 193, pl. xxxiv. fig. 1, 1879.
Coryrhynchus riisei, Kingsley, American Naturalist, vol. xiii. p. 585, 1879; Proc. Acad. Nat. Sci. Philad., p. 384, 1879.

Two males and two females were collected at Bermuda in shallow water, and a male in mutilated condition south of Pernambuco, Brazil (Station 122, 30 to 350 fathoms).

These specimens have the tuberculiform prominences of the gastric region of the carapace less distinctly defined than in the specimen figured by Milne Edwards. In all of them the ridges upon the pterygostomian regions, defining the afferent channels to the branchiæ are very distinctly developed.

The largest male has the following dimensions:—

Adult ♂.	Lines.	Millims.
Length of carapace and rostrum, about	11½	24·5
Breadth of carapace,	9½	20
Length of a chelipede,	13	27·5
Length of first ambulatory leg,	31	65·5

Subfamily II. INACHINÆ.

Inachinæ, Miers, Journ. Linn. Soc. Lond. (Zool.), vol. xiv. p. 644, 1879; et synonyma.

Eyes slender and retractile. Præocular spine usually wanting, postocular distinct. Basal antennal joint usually very slender throughout its length, not narrowing distally. In the typical species the margin of the carapace often forms a slight rim over the bases of the eye-peduncles (but there are no distinct orbits). The merus of the exterior maxillipedes is sometimes distally rounded, the palms of the chelipedes are often somewhat inflated. The ambulatory legs are slender and often very long.

To the genera enumerated by me in 1879, as belonging to this group, the following are probably to be added :—

Anisonotus, A. Milne Edwards.	*Scyramathia*, A Milne Edwards.
Anasimus, A. Milne Edwards.	*Trachymaia*, A. Milne Edwards.
Apocremnus, A. Milne Edwards.	*Gonatorhynchus*, Haswell.
Ergasticus, A. Milne Edwards.	*Platymaia*, n. gen.
Lispognathus, A. Milne Edwards.	*Cyrtomaia*, n. gen.
Echinoplax, n. gen.	

Anisonotus apparently establishes the transition between *Inachoides* in this group, and the Leptopodiinæ.[1]

Platymaia, n. gen.

Carapace depressed ; suborbiculate.   Rostrum short, apparently tridentate, the median lobe arising from the distal end of the interantennulary septum.   No præocular but a supra and postocular spine.   Epistoma very small, transverse.   Post-abdomen (in the female) narrowest at base and broadening distally, thus obovate and subtruncated at the distal extremity, with all the segments distinct.   Eyes short, with a distal tubercle, corneæ somewhat dilated.   Antennæ with a short and slender basal peduncular joint, which does not reach the front ; the flagellum is well developed.   Exterior maxillipedes with the ischium-joint rather broad, with a spine at the antero-internal angle ; the merus is slenderer than the ischium, articulating with the next joint at its antero-internal angle, which is not emarginate.   Chelipedes (in the female) rather slender and short, spinuliferous ; spines of merus long.   Ambulatory legs very considerably elongated ; those of the first pair having the fourth to last joints (merus to dactyl) armed with strong spines, which are longest on the anterior margin of the penultimate joints ; the second to last legs almost devoid of spines, with the penultimate joints dilated and compressed (as in *Eurypodius*), and ciliated on the anterior margins ; dactyli elongated, slender and slightly arcuated.

Platymaia is not very nearly allied to any genus of this family known to me. Its nearest affinities are perhaps with *Euprognatha*, Stimpson,[2] but it is at once distinguished by the depressed and suborbiculate form of the carapace and the dilatation of the

[1] Professor S. I. Smith has recently proposed the designation *Anamathia* in place of *Amathia*, Roux, which name was previously used.   *Proc. U.S. Nat. Mus.*, vol. vii. p. 493, 1884.
[2] *Bull. Mus. Comp. Zool.*, vol. ii. p. 122, 1870.

penultimate joint of the ambulatory legs. As in *Euprognatha rastellifera*, the rostrum, on account of the prominence of the median spine of the interantennulary septum, is apparently trifid ; but the spines on the anterior margins of the joints of the ambulatory legs, which in *Euprognatha rastellifera* are very small, are in this species enormously developed on the ambulatory legs of the first pair. The basal antennal joint does not, as in most genera of the Inachidæ, attain to the frontal margin.

This genus and the one which follows are among the most remarkable of the Challenger Brachyura, and are of especial interest as being Malayasian representatives of a section of the subfamily Inachinæ hitherto represented only by types from Eastern America, and which Stimpson [1] separated as a distinct subfamily, Collodiinæ, on, as I think, insufficient grounds.[2]

Platymaia wyville-thomsoni, n. sp. (Pl. II. fig. 1).

Carapace depressed, suborbiculate, with the cervical suture very strongly defined, closely granulated over the whole of its upper surface ; some of the granules (e.g.) in the median line of the gastric region and on the cardiac and front of the branchial regions being of larger size. The median (or interantennulary) lobe of the short rostrum is longest and acute ; the lateral lobes very short and obtuse ; the supraocular spines small, the postocular longer ; behind these there are spinules on the hepatic and ptery-gostomian regions and two or three on the sides of the branchial regions in front ; there are also lobes, armed with one to four spinules, between the bases of the third and fourth and fourth and fifth legs ; the anterior margin of the buccal cavity has a truncated and denticulated lobe at its antero-external angle ; the basal segment of the post-abdomen (in the female) is granulated and spinuliferous ; the others are granulated only ; the first, fifth, and sixth segments longer than the others, the last segment widely transverse ; the short basal antennal joint bears two inferior spinules placed near to the distal extremity, the two following joints of the peduncle are short, slender, and nearly equal ; both the ischium and merus-joints of the outer maxillipedes are spinuliferous on their outer surface, and the merus bears also several spinules at and near to its antero-external angle ; the chelipedes are about as long as the carapace (with rostrum); merus, carpus, and palm thinly setose and spinuliferous ; merus with some longer spines on its antero- and postero-inferior surfaces ; carpus short, palm about as long as fingers, slightly compressed, nearly smooth on its inner surface, fingers slender, straight, acute, and minutely denticulated on their inner margins ; the first pair of ambulatory legs are (roughly) three and a half times the length of the carapace to base of rostrum, with all the joints spinuliferous, the spines on the anterior margins of the merus and penultimate joints (usually) alternately longer and shorter, the penultimate joint with the spines greatly elongate, rastelliform,

[1] *Bull. Mus. Comp. Zool.,* vol. ii. p. 119, 1870.   [2] *Vide Journ. Linn. Soc. Lond.,* vol. xiv. p. 644, 1879.

several of the spines of the dactyl are also very long ; the second to fourth ambulatory legs, which have been already described, are devoid of spines, except a series of spinules on the anterior margin of the merus and posterior margin of the dactylus of the second pair and a spinule at the distal ends of the merus-joints of all the legs ; the anterior surface of the merus of the third leg is granulated.   Colour (in spirit) yellowish, inclining to red in some places, as, for instance, on the gastric region of the carapace.

Adult ♀.		Lines.	Millims.
Length of carapace to base of rostrum,	.	15½	33
Breadth of carapace, .	.	17	36
Length of chelipede, .	.	18	38
Length of first ambulatory leg,	.	57	121
Length of second ambulatory leg,	.	70	148

The unique example (an adult female) was dredged north of the Admiralty Islands, lat. 1° 54′ 0″ S., long. 146° 39′ 40″ E., in 150 fathoms (Station 219).

It is one of the finest of the Brachyura dredged in the Challenger Expedition, and it is therefore very appropriately associated with the name of the late Director of the Civilian Scientific Staff of the Expedition.

Cyrtomaia, n. gen.

Carapace transverse and very convex, broadly rounded on the sides of the branchial regions ; spinose ; the spines of the gastric region greatly developed.   Spines of rostrum short, not contiguous.   Epistoma transverse.   Post-abdominal segments (in the male) distinct.   Eyes rather long, slender, or rather robust, with a spinule at the distal extremity.   Antennæ long, the basal joint slender, in contact with the front at the distal extremity.   Outer maxillipedes with the merus and ischium-joints strongly spinulose, the merus narrower than the ischium, longer than broad, and subtruncated at the distal extremity, with its antero-external angle produced, spiniferous ; the next joint articulated with it at its antero-external angle ; exognath slender.   Chelipedes (in the male) very slender and spinulose, the palm not at all dilated, the fingers slender, nearly straight, and meeting along their inner margins when closed.   Ambulatory legs very long and slender, with the joints more or less spinulose, the dactyli long, slender, and nearly straight.

This genus is distinguished from all others of the same section of the family by the remarkable convexity of the carapace, which is almost vertically deflexed at the gastric region, and the great development of its gastric spines, and by the elongated and slender spinuliferous chelipedes.

Cyrtomaia murrayi, n. sp. (Pl. III. fig. 1).

The carapace is rather broader than long, convex, with the gastric and cardiac regions much elevated, so that the gastric region is situated above rather than behind the hepatic regions; the gastric region is armed with six long spines, of which two are placed at some distance behind the eyes, three posterior to them in a transverse series, and one at the back of the gastric region; on its sides are also two or three smaller spinules, there are two spines on the cardiac region, three on each hepatic region, and three or four on each branchial region, besides some very small spinules; also two spinules on the upper margin of the orbits, and one (median) on the posterior margin of the carapace, three or four are placed in an oblique series on the pterygostomian region, and two on the sternal surface of the body, at the bases of the first four ambulatory legs. The segments of the post-abdomen, also, are spinuliferous, the second segment is transverse and very short, but broader than the following segment, the fifth and sixth longer than broad, the seventh narrowing to its distal extremity, which is subacute. The slender basal joint of the antennæ is not much elongated, and bears two spinules on its outer surface; the merus-joint of the outer maxillipedes has two longitudinal series of spinules on its outer surface and is denticulated on its inner margin, the merus-joint is also spinuliferous and has, in particular, a strong bifid spinule at its antero-external angle; the chelipedes are very slender and about two and a half times the length of the body, with the merus-joint longer than palm and fingers taken together, and, as well as the wrist and palm, armed with spinules, which are generally alternately long and short; palm about twice as long as the wrist and not more dilated than that joint, fingers about half as long as the palm and (as already stated) nearly straight, but slightly decussate at the tips. Ambulatory legs extremely long and slender, the merus and ischium-joints are armed with spinules, which are often alternately longer and shorter; a stronger spinule exists at the distal ends of the merus-joints; the dactyli of the first and second pairs are very minutely spinuliferous, the penultimate joints and dactyli of the fourth and fifth legs are devoid of spines, slightly compressed, and fringed with hairs. Colour (in spirit) light brownish-yellow.

Adult ♂.		Lines.	Millims.
Length of carapace to base of rostrum,	.	8½	18
Greatest breadth of carapace, nearly	.	10	20·5
Length of a chelipede, about	.	23	49
Length of first ambulatory leg, about	.	41	87

The description and dimensions are taken from the smaller but more perfect specimen, the larger having lost all the legs, except a chelipede.

Two imperfect examples (adult males) were collected near the Ki Islands (Station 192), in 140 fathoms, lat. 5° 49′ 15″ S., long. 132° 14′ 15″ E.

Cyrtomaia suhmii, n. sp. (Pl. III. fig. 2).

The carapace, as in the preceding species, is broader than long, very convex, especially at the gastric region, and spinose; the spines of the rostrum are much longer than in the preceding species, straight, and divergent from the base (but broken in the unique example); the spines of the carapace are disposed as follows:—two very strong straight and extremely long spines on the gastric region on each side, (one or perhaps orginally two) spines on the cardiac, two rather long and about half a dozen smaller spinules on each branchial region; a strong postocular spine, and behind this a smaller spine on the hepatic region, also a series of very small spinules on the sides of the carapace, above the bases of the ambulatory legs; there are no præocular and supraocular spines; the sternum bears several spines and spinules on its anterior surface at the bases of the first three pairs of legs, the post-abdomen (in the male) is narrow, with the terminal segment slightly constricted at base, and rounded at its distal extremity; there are two spinules on the fourth and fifth segments, one (longer) on the sixth and two (small) on the seventh segment. The eyes are shorter and much more robust than in the preceding species, and have a small tubercle at the distal extremity; the basal antennal joint is slender, rather short, and is armed with three or four spinules (of which two long) on its outer margin and one (small) near its antero-internal angle; the second and third joints are slender, the third a little longer than the second; the flagellum very long and slender and fringed with long setæ. The ischium and merus-joints of the outer maxillipedes are (as already stated) strongly spinulose, the spinules of the merus arranged in longitudinal series, the spines of the antero-external lobe of the merus prominent; the exognath straight, and narrowing slightly toward its distal extremity; the chelipedes (in the male) are rather long and slender; the merus, carpus, and palm armed with long and slender spines, the longest of which, upon the upper surface of the joints, are somewhat curved; the palm is more than twice the length of the carpus, but shorter than the merus, the fingers rather more than half as long as the palm, straight and unarmed, acute at the tips and very minutely denticulated on their inner margins. The ambulatory legs are very imperfect; the first and second pairs were apparently more robust than the following and armed with long spines and setæ; the spines longest on the last three joints; the legs of the last two pairs are rather more slender than those of the two first pairs, and nearly devoid of spines (except a strong spine at the distal end of the upper margin of the merus-joints; the dactyli nearly straight. Colour (in spirit) yellowish-white.

Adult ♂.				Lines.	Millims.
Length of carapace to base of rostrum, nearly	.		.	12	25
Greatest breadth of carapace, about	.	.	.	13	28
Length of a chelipede, about	.	.	.	28	59

The unique specimen is an adult male, and is unfortunately in very crushed and imperfect state ; it was dredged at Station 214, in 500 fathoms, in lat. 4° 33′ 0″ N., long., 127° 6′ 0″ E., near the Tulur Islands.

It is at once distinguished from the preceding species by the great development of the gastric spines, and by the much longer rostrum and the shorter, thicker eyes, not to mention other characters.

Apocremnus, A. Milne Edwards.

Apocremnus, A. Milne Edwards, Crust. in Miss. Sci. au Méxique, pt. 5, p. 184, 1879.

The carapace is subtriangulate and shaped nearly as in *Inachus,* with strong spines upon the gastric, cardiac and branchial regions ; there is a distinct supraocular spine, but no postocular spine. The epistoma is large. The post-abdomen in the male is six-jointed, the sixth and seventh joints coalescent. The eyes project laterally and are imperfectly retractile. The basal antennal joint is somewhat broader at the base than at the distal extremity, which bears a spine and is longitudinally carinated ; the flagella are visible from above at the sides of the front. The merus of the outer maxillipedes is rounded and somewhat produced at the antero-external angle and slightly notched at the antero-internal angle, where it is articulated with the next joint. The chelipedes in the male are not robust, and the palm is but very slightly turgid ; the dactyli have between them (when closed) an interspace at the base. The ambulatory legs are small and very slender, and the dactyli are partially retractile against the penultimate joint.

The single species, *Apocremnus septemspinosus,* A. Milne Edwards, has hitherto been recorded only from the Florida Straits (37 fathoms).

Apocremnus septemspinosus, A. Milne Edwards.

Apocremnus septemspinosus, A. Milne Edwards, tom. cit., p. 185, pl. xxxv. fig. 5, 1879.

A small male obtained at the Island of Fernando Noronha, in 7 to 20 fathoms, is somewhat doubtfully referred to this species.

In nearly all particulars the detailed description of M. A. Milne Edwards applies to this specimen ; the sternum, however, is less distinctly ridged, and more irregularly granulated.

Adult ♂ .	Lines.	Millims.
Length of carapace and rostrum, about	3½	7·5
Breadth of carapace, .	3	6·5
Length of a chelipede, .	5	11

Achæopsis, Stimpson.

Achæopsis, Stimpson, Proc. Acad. Nat. Sci. Philad., p. 218, 1857.
      „   Miers, Journ. Linn. Soc. Lond (Zool.), vol. xiv. p. 645, 1879.

This genus, which apparently represents the northern genus *Inachus* in the southern hemisphere, is distinguished from it merely by having the postocular as well as the præocular spine distinctly developed, and by the more or less falciform dactyli of the three posterior ambulatory legs.

Two species of the genus have been described, which both occur at the Cape of Good Hope, in shallow water, *Achæopsis spinulosus*, Stimpson, and *Achæopsis güntheri*, Miers, the latter distinguished from the type by having but a single very long perpendicular spine on the gastric region.[1]

Achæopsis spinulosus, Stimpson (Pl. I. fig. 4).

Achæopsis spinulosus, Stimpson, Proc. Acad. Nat. Sci. Philad., p. 218, 1857.

Three males were collected in Simon's Bay in 10 to 20 fathoms, and one in 5 to 18 fathoms; here also the types of the species were obtained.

In the specimens examined, the ambulatory legs, although granulated, can scarcely be described as spinuliferous, as in Stimpson's description.

The description (except measurements) which follows, and the figure are from adult dried examples in the collection of the British Museum, which are in more perfect condition than the Challenger specimens.

Carapace ovate-triangulate, moderately convex, and rounded on the sides of the branchial regions, with three spines disposed in a triangle on the gastric region, one on the cardiac, and two on each branchial region, on the dorsal surface (there are sometimes some smaller branchial spinules). There is one (or sometimes two) spinules above or in front of the eye-peduncles, and a well-developed postocular spine. The hepatic region is prominent and rounded and bears several small spinules, and there are several on the sides of the branchial regions. The rostrum is composed of two small acute lobes or spines which do not exceed in length the interocular portion of the front. The post-abdomen in both sexes is six-jointed. The eyes are of moderate size, and the corneæ have usually a small distal tubercle. The basal joint of the antennæ is slender, longitudinally sulcated, and bears several small spinules; the flagella are exposed and visible in a dorsal view at the sides of the rostrum. The merus of the exterior maxillipedes is narrower than the penultimate joint, is not distally truncated, and bears the next joint near to its antero-external angle. The chelipedes (in the adult male) are

---

[1] Possibly the Australian species *Stenorhynchus curvirostris*, A. Milne Edwards, and *Stenorhynchus fissifrons*, Haswell, which have not been figured, may be referable to this genus. *Stenorhynchus brevirostris*, Haswell, is, I think, a species of *Achæus*.

rather more than twice as long as the carapace; merus and carpus covered with unequal spinules, the merus with a stronger spine near its distal extremity; palm longer than the carpus, turgid, granulated or spinuliferous on the outer surface and on the upper margin near the base; dactyli nearly as long as the palm, smooth, nearly straight, with acute apices and dentated on the inner margins near the base. Ambulatory legs slender, of moderate length, with a small spine or tubercle at the distal extremity of the merus-joints; the dactyli of the fourth and fifth legs distinctly falcated. The body and limbs are pubescent. Colour light yellowish-brown.

A male,[1] preserved in spirit, in the Challenger collection presents the following dimensions :—

Adult ♂.					Lines.	Millims.
Length of carapace and rostrum, nearly	.	.	.	.	5	10
Breadth of carapace, .	.	.	.	.	4	8·5
Length of a chelipede, about .	.	.	.	.	9	19·5
Length of first ambulatory leg,	.	.	.	.	11	23

Inachus, Fabricius.

Inachus, Fabricius (*partim*), Entom. Syst. Suppl., p. 355, 1798.
      „   Milne Edwards, Hist. Nat. Crust., vol. i. p. 286, 1834.
      „   Miers, Journ. Linn. Soc. Lond. (Zool.), vol. xiv. p. 645, 1879.

Carapace triangulate, with the regions well defined. Rostrum very short, bilobated; no præocular spine, postocular spine large. Post-abdomen composed of six distinct segments. Eyes moderately developed. Basal antennal joint slender, and reaching the front. Outer maxillipedes with the merus-joint somewhat elongated and rounded at the distal extremity. Chelipedes with the palm more or less inflated. Ambulatory legs elongated, with the dactyli nearly straight.

The well-established species of this genus are apparently restricted to the European Seas and Western North Atlantic, and occur commonly in water of moderate depth. M. A. Milne Edwards mentions that several species were taken in the Mediterranean during the recent expedition of the "Travailleur" at 445 to 455 metres.

The following species have been described :—(1) *Inachus dorsettensis* (Pennant) = *Cancer scorpio*, Fabricius, *Inachus mauritanicus*, Lucas, and var. *latus*, Brandt ; (2) *Inachus dorhynchus*, Leach ; (3) *Inachus leptochirus*, Leach = *Inachus leptorhynchus*, Desmarest ; (4) *Inachus thoracicus*, Roux ; (5) *Inachus aguiarii*, B. Capello.[2]

[1] The figure, drawn when the author was much engaged on other work, is inadvertently taken from the adult female, which differs from the male in the rather broader carapace and less turgid chelipedes.

[2] The Australian species described by Gray as *Inachus australis*, perhaps belongs to the genus *Chlorinoides*. Ruppell's *Inachus arabicus* from the Red Sea is *Micrthium manocerus*. *Inachus tuberculatus*, Lockington, from California, does not belong to this genus, as the rostrum is entire.

Inachus leptochirus, Leach.

Inachus leptochirus, Leach, Malacostraca Podophthalmata Britanniæ, pl. xxii. B.
,,   ,,   Bell, British Crust., p. 18, 1853.
,,   ,,   Heller, Crust. des südlichen Europa, p. 32, pl. i. figs. 12, 13, 1863.
,, *leptorinchus*, Milne Edwards (after Desmarest), Hist. Nat. Crust., vol. i. p. 289, 1834.

A small male and two small females were dredged in 50 to 90 fathoms, near Fayal, Azores.

Adult ♂.						Lines.	Millims.
Length of carapace, .	.	.	.	.	.	3½	7·5
Breadth of carapace, nearly	.	.	.			2½	6·5
Length of chelipede, nearly .		.	.	.		4½	9
Length of first ambulatory leg,		.	.	.		13	27·5

Oncinopus, de Haan.

Oncinopus, de Haan, Crust. in v. Siebold, Fauna Japonica, p. 87, pl. H., 1839.
,,   Miers, Journ. Linn. Soc. Lond. (Zool.), vol. xiv. p. 645, 1879.

Carapace semi-membranaceous, elongate, narrow-triangulate, and depressed. Rostrum very short, composed of two vertically compressed laminiform lobes, no præ- or postocular spines. Post-abdomen in both sexes distinctly seven-jointed. Eyes slender and projecting laterally. Antennæ with the basal joint very short and slender, and not attaining the front, the flagella exposed and visible at the sides of the rostrum ; merus of the exterior maxillipedes elongated and articulated with the next at its summit.

Chelipedes in the male rather small, with the palm turgid, and the fingers having between them, when closed, an interspace at the base. Ambulatory legs slender and somewhat elongated, with the penultimate joints of the first and second pairs dilated, compressed and ciliated on the posterior margin ; the dactyli in all slightly arcuated and retractile against the penultimate joints.

The described species, as I have already shown, are probably all to be regarded as synonymous with *Oncinopus aranea*, de Haan, the type of the genus, which ranges from the seas of Japan through the Malaysian Seas eastward to the New Hebrides, and in bathymetrical range, from shallow water, to a depth, as the Challenger collections show, of 825 fathoms.

Oncinopus aranea, de Haan.

Oncinopus aranea, de Haan, Crust. in v. Siebold, Fauna Japonica, p. 100, pl. xxix. fig. 2, ♂, ♀, and pl. H, 1839.
,,   ,,   Miers, Crust. in Report on Zoological Collections of H.M.S. "Alert," p. 190, 1884, et synonyma.

Specimens are referred to this species from the following localities :—An adult female from a bottle labelled "April, 74, 2 to 10 fathoms," and therefore obtained on the

Southern Australian coast, but without precise indication of locality ; an adult but small male from the Arafura Sea (depth not stated), and an adult female dredged in the Moluccas Passage, in 825 fathoms (?), in lat. 0° 48′ 30″ S., long. 126° 58′ 30″ E. (Station 196).

In the latter (deep-water) specimen the lobes of the front are prominent and widely separated, and the penultimate joints of the ambulatory legs are rather slender, as in the one from the Arafura Sea.

The South Australian specimen has a more robust and hairy body, and the single remaining ambulatory leg (of the first or second pair) has stouter, thicker joints.

The relative dimensions of the three specimens are appended :—

	Lines.	Millims.
Adult ♀, from South Australia (shallow water).		
Length of carapace and rostrum,	6½	13·5
Breadth of carapace,	4½	9·5
Length of (second ?) ambulatory leg,	14	30
Adult ♂, from the Arafura Sea.	Lines	Millims.
Length of carapace and rostrum, about,	4	8·5
Breadth of carapace,	3	6·5
Length of second ambulatory leg,	11	23·5
Adult ♀, from the Moluccas Passage (825 fathoms).	Lines.	Millims.
Length of carapace and rostrum,	5	11
Breadth of carapace,	4	8·5
Length of second ambulatory leg, nearly	13	27

Eurypodius, Guérin-Méneville.

Eurypodius, Guérin-Méneville, Mém. Mus. Hist. Nat. Paris, vol. xvi. p. 345, 1828.
     ,,   Milne Edwards, Hist. Nat. Crust., vol. i. p. 283, 1834.
     ,,   Miers, Journ. Linn. Soc. Lond. (Zool.), vol. xiv. p. 646, 1879.

Carapace elongate-triangulate, moderately convex and spinose above ; a distinct post-ocular but no præocular spine. Spines of the rostrum contiguous, at least in their basal half. Post-abdomen (in both sexes) distinctly seven-jointed. Eyes retractile. Antennæ exposed and visible in a dorsal view at the sides of the rostrum ; the basal joints slender and attaining the front. Merus of the outer maxillipedes distally truncated and bearing the next joint at its antero-internal angle. Chelipedes in the male well developed, with the palm compressed or turgid and the fingers distally acute. Ambulatory legs considerably elongated, with the penultimate joints more or less dilated and compressed ; the dactyli slightly arcuate, shorter than the penultimate joints and reflexible against their inferior margins.

The described species are of large size, and they are probably varieties of the type *Eurypodius latreillei*, which is especially abundant in the Straits of Magellan and at the

Falklands, ranges northward along the western shores of South America to Peru, and occurs also, according to Professor Bell, on the coast of Brazil. It is most abundant in shallow water, but occurs at greater depths, 30 to 70 fathoms.

I have described below a species, *Eurypodius longirostris*, from Chiloe, which must be regarded as provisionally distinct, and was obtained in much deeper water (175 fathoms).

Eurypodius latreillei, Guérin-Méneville (Pl. IV. fig. 3).

> *Eurypodius latreillei*, Guérin-Méneville, Mém. Mus. Hist. Nat. Paris, vol. xvi. p. 354, pl. xiv., 1828.
> „ „ Milne Edwards, Hist. Nat. Crust., vol. i. p. 284, 1834 ; Crust. in Cuvier, Règne Animal (ed. 3), Atlas, pl. xxxiv. *bis*, fig. 1.
> „ „ Miers, Proc. Zool. Soc. Lond., p. 64, 1881, et synonyma.

Among the males in the extensive series brought home by the Challenger two very distinct varieties, if not species, may be distinguished by the form of the chelipedes. In the first (var. *a*), which is to be regarded as the typical form, as shown by Guérin-Méneville's original figure, the palm is moderately compressed, more slender, considerably longer than broad, not turgid, and the fingers are without strong tubercles on the inner margins (see Pl. IV. fig. 3).

In the second variety (var. *β*, fig. 3*a*) the palm is shorter and much more turgid, the fingers have between them (when closed) a much larger hiatus, and the dactylus and sometimes the lower finger (pollex) is armed, on its inner margin, near the base, with a strong rounded tubercle. This form has usually, but not invariably, a more slender carapace and longer rostrum. I cannot refer it to any of the so-called "species" figured and described by authors, reference to which is made in my paper above cited, where also mention is made of a specimen of this variety presented to the British Museum by Dr. R. O. Cunningham. In the Challenger series are specimens of comparatively small size presenting this peculiarity in the form of the chelipede, e.g., one from Magellan Strait, dredged in 55 fathoms (Station 313), whose principal dimensions are as follows :—

Adult ♂.			Lines.	Millims.
Length of carapace to base of rostrum,	.	.	8½	18
Length of rostrum, about	.	.	3	6·5
Greatest breadth of carapace, about	.	.	6½	14
Length of a chelipede, about	.	.	14½	30·5
Length of second ambulatory leg, nearly	.	.	29	61

It is, therefore, not a character peculiar to the old and large males ; nevertheless, as some specimens of the typical form occur presenting some approach to this variety in the greater convexity of the palm, and as, moreover, it is impossible to say to which of the two varieties the females in the collection are to be referred, I refrain from applying to them distinctive appelations (see figs. 3, 3a).

It is not necessary to repeat here what has been elsewhere noted of the variability of *Eurypodius latreillei*, and the impossibility of distinguishing the various forms described as specifically distinct by the characters usually assigned to them.

Specimens of *Eurypodius latreillei* were obtained by the Challenger at the following localities :—

CHILOE.—Off Cape Tres Montes, in 45 fathoms, lat. 46° 53′ 15″ S., long. 75° 12′ 0″ W. (Station 304), a young male (var. α); also at Port William, the east shell of a young male (var. α), with very small and slender chelipedes.

MAGELLAN STRAIT.—Station 312, Port Famine, 9 fathoms; three adult and two younger males (var. α) the young with longer rostral spines, which are slightly divergent at the apex, also three adult and one smaller male (var. β), and three females.

Station 313, lat. 52° 20′ 0″ S., long. 67° 39′ 0″ W., in 55 fathoms; three small and young males (var. α), one having the apices of the rostral spines rather remote one from another; three males, adult, small and young (var β), and seven females.

Station 314 (between Magellan Strait and Falklands), lat. 51° 35′ 0″ S., long. 65° 39′ 0″ W., in 70 fathoms; a small female.

FALKLAND ISLANDS.—Station 315, lat. 51° 40′ 0″ S., long. 57° 50′ 0″ W., in 12 fathoms; an adult male (var. β), and seven females (one a cast shell).

Station 316, lat. 51° 32′ 0″ S., long. 58° 6′ 0″ W., in 4 fathoms; an adult male (var. α). Its dimensions are :—

Adult ♂.					Lines.	Millims.
Length of carapace and rostrum,	.	.	.	.	27	57·5
Breadth of carapace, about	.	.	.	.	19	40

I may note in conclusion that the specimens collected in shallower water (4 to 15 fathoms), are generally much larger and more robust than the specimens dredged at greater depths (45 to 70 fathoms), with a much more hairy carapace and legs. An adult male, however, dredged in 55 fathoms (Station 313), resembles the shallow-water specimens in these particulars, which I am unable to regard as of specific importance.

Eurypodius longirostris, n. sp. (Pl. V. fig. 1).

This form is distinguished by the remarkably reflexed rostrum of the male, which is bent upward at an angle of nearly 45° to the front, with the spines toward their apices laterally divergent from one another.

The body and limbs are thinly pubescent, the carapace narrow in proportion to its length, with the spines disposed as commonly in specimens of *Eurypodius latreillei* of the same size and sex, *e.g.*, two upon the gastric, one upon the cardiac, one on each branchial region and one on the posterior margin, besides some smaller granules on the sides of the hepatic and branchial regions. There is a small spine on the upper margin of the orbit, besides the postocular spine. The spines of the rostrum considerably exceed half the length of the carapace, the spines are contiguous at the base, but in the distal third of their length they curve laterally and outward (as shown in the figure); there is a strong spine on the interantennulary septum. The eyes, antennæ, and maxillipedes are of the same form as in *Eurypodius latreillei.* The chelipedes have, as in young males of the typical variety (*a*) of that species, the palms not turgid, but compressed, the fingers acute, without teeth, and without any intramarginal hiatus when closed ; the merus and carpus-joints have a few distant granules on their upper margins. The ambulatory legs are very imperfect and their dimensions cannot be given, but they were evidently slender and considerably elongated, with the penultimate joints larger than the preceding and very little dilated.

Adult ♂.						Línea.	Millims.
Length of carapace to base of rostrum, about	.	.	.	.	.	9	19
Length of rostrum, about	.	.	.	.	.	5½	11
Breadth of carapace, a little over	.	.	.	.	.	7	15
Length of a chelipede, nearly	.	.	.	.	.	15½	32

The example described above (a small male) was dredged off the coast of Chiloe in 175 fathoms (a much greater depth than any recorded for the true *Eurypodius latreillei*) in lat. 50° 8′ 30″ S., long. 74° 41′ 0″ W. (Station 308). It is in a much broken condition.

An immature female, dredged at the same locality and depth, has the rostrum scarcely at all reflexed and somewhat shorter, with the spines less divaricate at the apex; the chelipedes and ambulatory legs clothed with a denser pubescence ; the latter much less elongated. This specimen, although distinguished by the narrower carapace and more elongated rostrum, much more nearly resembles the typical *Eurypodius latreillei* than does the male.

Gonatorhynchus, Haswell.

Gonatorhynchus, Haswell, Cat. Australian Stalk and Sessile-eyed Crust., p. 10, 1882.

Carapace subpyriform, moderately convex, its dorsal surface without spines ; no præocular spine, but distinct supraocular and postocular spines ; rostrum composed of two short, acute, slightly divergent spines. Post-abdomen (in the male) distinctly seven-jointed. Eyes small and partially retractile. The basal antennal joint is about twice as long as broad and reaches the front, and has a small tooth at its antero-internal angle; the base only of the flagellum is covered by the rostral spines.

The merus of the outer maxillipedes is produced and rounded at its antero-external angle and slightly notched at the antero-internal angle, where it is articulated with the next joint. Chelipedes (in the male) of moderate size; palm somewhat turgid, smooth; fingers distally acute and having between them, when closed, an interspace at the base. Ambulatory legs of moderate length, unarmed.

The single species is restricted in its range to the southern and eastern coasts of Australia.

Gonatorhynchus tumidus, Haswell.

Gonatorhynchus tumidus, Haswell, Proc. Linn. Soc. N.S.W., vol. iv. p. 437, pl. xxv. fig. 4, 1880; Cat. Australian Crust., p. 10, 1882.

An adult male, collected on the southern Australian coast in 2 to 10 fathoms, April 1874, is referred, but somewhat doubtfully, to this species.

In this example, the close curled pubescence of the carapace is so dense that the tubercles cannot be seen; the chelipedes (imperfect) are more slender than in Haswell's description and figure, the palms of the chelipedes somewhat more slender and more elongated, and the fingers less strongly dentated on their inner margins.

This species is, I think, better referred to the subfamily Inachinæ than to the Acanthonychidæ.

The Challenger specimen has the following dimensions:—

Adult ♂.					Linea.	Millims.
Length of carapace and rostrum, about	.	.		.	10½	22·5
Breadth of carapace,	.	.	.	.	9¼	16

Anamathia, S. I. Smith.

Anamathia, S. I. Smith, Proc. U.S. Nat. Mus., vol. vii. p. 493, 1884.
Amathia, Roux, Crust. de la Méditerranée, pl. iii., 1828, with accompanying description.
„   Milne Edwards, Hist. Nat. Crust., vol. i. p. 285, 1834; name previously used.

Carapace subtriangulate, rounded behind and armed with long spines on the dorsal surface; a præocular spine present or absent; postocular spine distinct; spines of the rostrum well developed, slender and divergent. Post-abdomen in the male (in the species I have examined) distinctly seven-jointed. Eyes small. The basal antennal joint is slender and usually armed with a tooth at its antero-external angle (which is absent in the typical species, Anamathia rissoana). The merus of the outer maxillipedes is truncated distally and is slightly produced at the antero-external angle; the next joint is articulated with it at its antero-internal angle; the chelipedes (in the male) are usually small, with slender palms, but are more developed in the Oriental species described below; the ambulatory legs are slender and elongated.

Of the half-dozen species described as belonging to this genus, and which are enumerated by Professor Smith (*tom. cit.*), one, *Anamathia rissoana*, inhabits the Mediterranean and the Eastern North Atlantic region; three, *Anamathia hystrix* (Stimpson), *Anamathia modesta* (Stimpson), and *Anamathia crassa* (A. Milne Edwards), the Caribbean Sea or Florida Straits; and two, *Anamathia agassizii*, Smith, and *Anamathia tanneri*, Smith, the east coast of the United States.

The fine species described below is from the far distant Philippine Islands. The species of this genus generally inhabit deep water, though not the greatest depths at which Brachyura may occur. The Mediterranean species has been recorded from comparatively shallow water (20 fathoms); the others are found at various depths of between 80 and 400 fathoms.[1]

Anamathia pulchra, n. sp. (Pl. IV. fig. 1).

The body is everywhere clothed with very close-set, short, knobbed hairs, interspersed among which are finer, longer setæ; the limbs also are closely pubescent. The carapace is much longer than broad, subpyriform rather than subtriangulate, moderately convex and armed with long spines, disposed as follows: five disposed in two transverse series, and behind these one, longer, on the gastric region; one on each hepatic region; one on the cardiac and one on the intestinal region; also four on each branchial region, the lateral one being very long; there exists also a præocular but no postocular spine, merely the rounded lobe against which the eye folds back; the spines of the rostrum are nearly straight and divergent from their bases; both are broken, but the longer (broken) spine considerably exceeds half the length of the carapace; the epistoma is about as broad as long; the pterygostomian region is armed with an oblique row of small tubercles, the sternum is without spines or tubercles. The segments of the post-abdomen (in the male) are all of them distinct and unarmed; this part of the body has the lateral margins, between the fourth and sixth segments, slightly concave; the terminal segment is slightly longer and narrower than the penultimate segment, very little longer than broad, and distally rounded. The eyes are small, with the corneæ rounded and terminal, and are completely retractile. The basal antennal joint is moderately slender, but much more robust than the two following joints (as in *Anamathia rissoana*), it is concave on its inferior surface, and has a small spine at its antero-external angle; the next joint is somewhat longer and very slender. The outer maxillipedes are smooth on their outer surface; the

---

[1] This genus is incorrectly designated *Amathia* on the plate, which was drawn and lettered before I had the opportunity of consulting Professor Smith's memoir.

ischium slightly concave, the merus-joint also externally concave, distally truncated, with its antero-external angle somewhat produced and its antero-internal angle (where it is articulated with the next joint) obliquely truncated and scarcely at all emarginate. The legs are rather elongated ; the chelipedes (in the male) are more than twice as long as the carapace to base of rostrum ; the merus rather longer than the palm, with a few granules on its upper margin near the base and on its postero-inferior margin, and with a spine at the distal end of its upper margin ; carpus short, carinated on its inner and outer and (obscurely) on its upper surface ; palm smooth, slightly compressed, carinated on its superior and more obscurely on its inferior margin, which is produced and rounded at its proximal angle ; the fingers are shorter than the palm and are regularly denticulated on the inner margins, toward the distal extremities, which are acute ; they have between them, at the base, a vacant interspace when closed, and the dactyl has on its inner margin near the base, a strong blunt tooth ; the ambulatory legs are slender and elongated, closely pubescent and without spines except a small tooth or spine at the distal ends of the merus-joints ; the dactyli nearly straight. Colour (in spirit) brownish-yellow.

Adult ♂.			Lines.	Millims.
Length of carapace to base of rostrum,	.	. .	8½	18
Greatest breadth of carapace,	.	. .	6½	13·5
Length of chelipede, nearly	. .	. .	18	37·5
Length of right ambulatory leg of first pair, .		. .	30½	64

The unique example (an adult male) was dredged at the Philippines in 375 fathoms, in lat. 9° 26′ 0″ N., long. 123° 45′ 0″ E. (Station 210).

This handsome species is distinguished from the type of the genus (*Anamathia rissoana*) by the distinct præocular spine, which exists also in the West Indian deep-water species, *Anamathia hystrix* (Stimpson), which, however, has but four gastric spines on the carapace (in *Anamathia rissoana* there are only three).

Lispognathus, A. Milne Edwards.

Lispognathus, A. Milne Edwards, Crust. in Miss. Sci. au Mexique, pt. 5, p. 349, 1880.

The carapace is subpyriform and moderately convex, with well-developed supraocular and postocular spines ; the spines of the rostrum are straight, slender, and slightly divergent. The post-abdomen (in the male) is shaped nearly as in *Inachus*, and is six-jointed in both sexes (on account of the coalescence of the penultimate with the terminal joint). The eyes are short and retractile. The basal antennal joint is slender and attains the front, and is armed on its inferior surface with small spinules and with a longer spinule at the antero-external angle. The merus of the outer maxillipedes is somewhat elongated and rounded at the distal extremity, where it is articulated with the next joint, as in *Inachus*. As in that genus, the chelipedes (in the male) are well developed, with somewhat turgid palms and with the fingers incurved, and acute at the apices. The ambulatory legs are slender and moderately elongated.

But two species of this genus have been described ; *Lispognathus thomsoni* (Norman), whose remarkable geographical and bathymetrical distribution are referred to below, and *Lispognathus furcillatus*, A. Milne Edwards, dredged at the Island of Grenada, in 291 fathoms.

This genus is allied to *Anamathia* (among the forms with well-developed rostrum), and, in my classification of the group, must be placed in the vicinity of that genus, from which, however, it is at once distinguished by the six-jointed post-abdomen and the form of the merus of the outer maxillipedes. It is also structurally very nearly allied to *Inachus* and *Achæopsis*, but from both these types it is distinguished by the well-developed rostral spines.

Lispognathus thomsoni (Norman) (Pl. V. fig. 2).

> *Dorynchus thomsoni*, Norman, in Wyville-Thomson's Depths of the Sea, p. 174, fig. 34, woodcut, 1873.
>
> *Lispognathus thomsoni*, A. Milne Edwards, Comptes rendus, vol. xciii. pp. 878, 932, 1881 ; transl. in Ann. and Mag. Nat. Hist., ser. 5, vol. ix. pp. 38, 42, 1882 ; Rapport sur la Faune sous-marine dans les grandes profondeurs de la Méditerranée and de l'Ocean Atlantique, in Archives des Missions Scientifiques and Littéraires, ser. 3, ix. pp. 16, 39, 1882 ; Recueil des Planches de l'Expédition du "Travailleur," pl. iii. (*inédita*).

Agulhas Bank, off Cape Agulhas, South Africa, 150 fathoms ; lat. 35° 4′ 0″ S., long. 18° 37′ 0″ E. (Station 142) ; four adult females and a small male. The females are very robust.

An adult female from the Agulhas Bank has the following dimensions :—

	Lines.	Millims.
Adult ♀.		
Length of carapace and rostrum, nearly . . .	7	14·5
Breadth of carapace, ...	4½	9·5
Length of a chelipede, nearly . . .	9½	19·5
Length of first ambulatory leg, nearly . . .	20	42

The male, with imperfect ambulatory leg, is much smaller, and measures as follows :—

	Lines.	Millims.
Adult ♂.		
Length of carapace and rostrum, nearly	4	8
Breadth of carapace, ...	2½	5·5
Length of a chelipede, ...	5	11

I can observe no distinctions of note between North Atlantic specimens of this species and the Challenger specimens from Agulhas Bank, except that the former (male) examples are somewhat less robust, the spinules on the sides of the carapace are more distinctly developed, and the palms of the chelipedes somewhat more dilated.

There is in the Challenger collection a mutilated male example, dredged near Sydney in 410 fathoms, in lat. 34° 13′ 0″ S., long. 151° 38′ 0″ E. (Station 164B), which I cannot venture to separate specifically from this species. It is distinguished from specimens of the typical form only by having one, not two spines, on each branchial region, and by having two tubercles (of which one is extremely small) upon the ocular peduncles, the distal one situated upon the cornea. In this specimen the rostrum is unfortunately broken, and the legs are deficient, except a chelipede and part of an ambulatory leg. Its dimensions are as follows :—

Adult ♂.		Lines.	Millims.
Length of carapace to base of rostrum, about	.	3½	7·5
Breadth of carapace, ...	.	3	6·5
Length of chelipede, ...	.	8	17

The specimen from Sydney closely approaches the nearly allied species or variety *Lispognathus furcillatus*, A. Milne Edwards, dredged by M. A. Agassiz at Grenada, in the West Indies, in 291 fathoms, in those points wherein it differs from the typical [1] *Lispognathus thomsoni*.

I may add that in the typical form of this species, there are occasionally traces of a subdistal tubercle on the cornea of the ocular peduncle, and the anterior branchial spine is often but slightly developed.

Ergasticus, A. Milne Edwards.

Ergasticus, A. Milne Edwards, Rapport sur la Faune Sous-Marine dans les grandes profondeurs de la Méditerranée and de l'Ocean Atlantique, p. 17, 1882.

,,   Studer, Verzeich. währ. d. Reise S. M. S. "Gazelle," ges. Crustaceen, Abhandl. k. Akad. Wissensch. Berlin, Abb. ii. p. 7, 1884.

Carapace subpyriform, moderately convex and tuberculated and spinose above, and on the lateral margins, and with several small spines on the upper and lower margins of the orbits. Spines of rostrum slender, straight and divergent from the base, where they are armed with one and two accessory spines. Post-abdomen (in *Ergasticus clouei*, ♀) six-jointed. Eyes rather small and retractile. The basal joint is spinuliferous

---

[1] This species is one of the very few Brachyurous Decapods inhabiting the deeper abysses of the ocean which has been already ascertained to occur commonly and abundantly over a somewhat wide area of distribution in the North Atlantic Ocean. The Rev. A. M. Norman describes it (tom. cit.) as very widely diffused over the "warm area" of the North Atlantic, extending continuously from the Faröes to the Straits of Gibraltar. M. A. Milne Edwards (Rapport, loc. cit.) found it to occur abundantly in the Gulf of Gascony, and in the numerous dredgings of the "Travailleur" off the coasts of Spain and Portugal in the years 1880–81, at depths varying from 806 to 1225 mètres, and also in the Mediterranean at 500 to 700 mètres. I have determined the Challenger specimens by actual comparison with specimens from the North Atlantic "warm area," collected during the cruise of H. M. S. "Porcupine," and presented to the British (Natural History) Museum by P. Herbert Carpenter, Esq., and named by Mr Norman.

and slender, and reaches the frontal margin ; the flagellum is not covered by the rostral spines. The merus of the outer maxillipedes is distally truncated, with a rather prominent antero-external angle, and the next joint is articulated at the antero-internal angle of the merus. The chelipedes are moderately developed and spinuliferous, palm smooth, not turgid, fingers unarmed and acute, ambulatory legs moderately elongated, with the merus-joints, at least, spinuliferous ; dactyli nearly straight.

But two species of the genus are known, the type *Ergasticus clouei*, A. Milne Edwards, dredged off Toulon in 445 (or 455) métres, and at the Cape Verde Islands in 38 fathoms, (Studer) ; and *Ergasticus naresii*, described below.

This genus is nearly allied to the following (*Echinoplax*), but the spines of the rostrum are armed only with one or two small accessory spinules near the base ; the post-abdomen (in the female specimen in the Challenger collection) is six, not seven-jointed, with the ultimate and penultimate joints coalescent, and the merus-joint of the outer maxillipedes is somewhat produced and rounded at its antero-external angle.

The orbit also, on account of the spinules which surround it, is better defined than in most genera of Inachidæ, to which group *Ergasticus* must, I think, be referred on account of the slenderness of the basal antennal joint.

Ergasticus naresii, n. sp. (Pl. V. fig. 3).

The carapace is scantily clothed with fine curled setæ, and is everywhere closely and rather finely granulated, and bears several longer spines ; of these three are placed in a transverse series on the front of the gastric region, and one behind them, two on the cardiac region, two on the intestinal region, close to the posterior margin, and one or two on each branchial region ; the sides of the carapace and the posterior margin, also, are spinuliferous ; about five spines, including the strong postocular spine, define the upper margin of the orbit (of these one is bifid or trifid), also three or four spinules define its inferior margin ; the spines of the rostrum are slender, straight, and considerably less than half the length of the carapace ; beneath the base of each is a small spine, besides the median spine of the interantennulary septum. The eyes are moderately robust, setigerous. The very slender basal antennal joint bears several small spinules on its inferior and outer margin ; the merus and ischium-joints of the outer maxillipedes are granulated. The chelipedes (in the female) are about as long as the carapace and rostrum, very slender ; the spinules of their merus-joints increase in length toward the distal extremity of the joint ; carpus shorter than the palm ; palm about as long as fingers, very slender, not more dilated than the carpus, and spinu-liferous both on its inner and outer faces ; fingers slender, smooth, very minutely denticulated along their straight inner margins and with acute apices ; the ambulatory legs are granulated, spinuliferous and setose, with none of the joints compressed and dilated; the spinules are almost entirely absent from the penultimate (as well as from the terminal) joints of the ambulatory legs of the two last pairs. Colour (in spirit) light yellowish or pinkish.

Adult ♀.			Lines.	Millims.
Length of carapace to base of rostrum, nearly			3½	8
Greatest breadth of carapace, nearly			3	7
Length of spine of rostrum, about			1	2
Length of a chelipede, about			4½	10
Length of first ambulatory leg, about			10½	22

The single specimen, although of small size, is an adult female, bearing ova. It was dredged in 150 fathoms, north of the Admiralty Islands, in lat. 1° 54′ 0″ S., long. 146° 39′ 40″ E. (Station 219).

From the type of the genus, *Ergasticus clouei*, A. Milne Edwards,[1] this species, as I learn from an unpublished figure kindly sent me by the author, is at once distinguished by the different spinulation of the carapace, and the existence of small spines on the inferior wall of the orbit; in this latter character it resembles an American type belonging to a genus apparently very closely allied to, if distinct from, *Ergasticus*, the West Indian *Trachymaia cornuta*, A. Milne Edwards,[2] which, to judge from the description and figure, is distinguishable from *Ergasticus* only by the absence of accessory spinules from the rostral spines, and by the non-spinuliferous palms of the chelipedes and first ambulatory legs.

Echinoplax, n. gen.

Carapace subpyriform, longer than broad, and covered with very numerous closely set spines and spinules; orbital margin spinose; spines of rostrum straight, acute, divergent from their bases, and bearing several (six) accessory spinules. Post-abdomen (in the female) seven-jointed. Basal antennal joint slender, spinuliferous, and in contact with the front at the distal extremity; flagellum visible from above at the sides of the rostrum. Maxillipedes of normal shape; merus truncated and not notched at the distal extremity, the antero-lateral angle not produced. Legs spinuliferous. Chelipedes (in the female) slender and feeble, with the palms not dilated. Ambulatory legs considerably elongated, with the penultimate joints not dilated; the dactyli nearly straight.

Echinoplax is distinguished from all the genera of the subfamily Inachinae, except *Ergasticus*, with which I am acquainted, by the very numerous spinules of the carapace, the accessory spinules of the rostrum, and the spinuliferous orbital margins. It is very nearly allied to *Anamathia*, and were it not that the genera of this family are generally

---

[1] *Comptes rendus*, vol. xciii. p. 879, 1881.
[2] Crust. in Miss. Sci. au Mexique, pt. 5, p. 352, pl. xxxiA. fig. 2, 1880.

separated by external characters of small importance, I should have preferred to regard it as a subgenus of that genus. It is also nearly related to *Plistacantha*, Miers, but is distinguished by the rostral spines, which are widely divergent from their base. But its nearest ally is apparently the preceding genus (*Ergasticus*), from which it is only distinguished by the more numerous accessory spinules of the rostrum, the seven-jointed post-abdomen in the female, and the somewhat different form of the merus of the exterior maxillipedes.

Echinoplax moseleyi, n. sp. (Pl. IV. fig. 2).

The body and limbs are covered with very numerous, closely set, slender spinules, interspersed among which are some rather longer spines and fine hairs. The carapace is subpyriform, rather convex, longer than broad, with the depressions between the regions almost obliterated; several spines are placed upon the upper orbital margin. The spines of the rostrum are slender, strongly divergent from their bases, nearly straight, acute, and more than half as long as the carapace; they bear three or four strong spinules upon their exterior margins and inferior surface. The pterygostomian regions and the sternal surface at the base of the ambulatory legs are spinuliferous, but the epistoma is smooth except at its margins. The post-abdomen in the female is narrow, with all its segments distinct and spinulose; the spinules arranged in transverse series; the terminal segment is distally rounded. The eye-peduncles are retractile and armed with one or two spinules; two strong spinules are placed upon the interantennulary septum; the basal joints of the antennules are spinuliferous, the basal joint of the antennæ very slender, and armed with about four longer spinules, and the two following joints each with one or two smaller ones; the ischium and merus-joints and the exognath of the outer maxillipedes are spinuliferous; the ischium-joint has its antero-internal angle somewhat produced and rounded; the merus is truncated at its distal extremity, with the antero-internal angle scarcely at all emarginated.

The left chelipede (in the female) is slender, and nearly as long as the carapace and rostrum together; all the joints (except the dactyli) are spinuliferous, the spinules longest on the merus-joint, which is longer than the palm and fingers together; the wrist is short; palm very slender, not thicker than the wrist and less than twice its length; fingers naked, slightly incurved, scarcely denticulated on their inner margins and distally acute. The ambulatory legs are very long and slender, those of the first pair about three times the length of the carapace to base of rostrum; the joints all slender, the penultimate not more dilated than the others; the dactyli straight; the joints (except the dactyli) are armed with spinules and clothed with long hairs, the dactyli are hairy, but not spinuliferous. Colour (in spirit) light brownish-yellow.

Adult ♀.			Lines.	Millims,
Length of carapace to base of rostrum, .	.	.	7	15
Breadth of carapace, nearly .	.	.	5½	12
Length of rostral spine, .	.	.	3½	7·5
Length of chelipede, .	.	.	10	21
Length of first ambulatory leg, about .	.	.	21	44

A single adult female was dredged at the Philippines, in 375 fathoms, in lat. 9° 26′ 0″ N., long. 123° 45′ 0″ E. (Station 210).

The right chelipede is unfortunately broken in this specimen.

Macrocheira,[1] De Haan.

Macrocheira, De Haan, Crust. in. v. Siebold, Fauna Japonica, dec. 4, p. 89, 1839.
,,   Miers, Journ. Linn. Soc. Lond. (Zool)., vol. xiv. p. 647, 1879.

Carapace subpyriform, and rather convex, covered with tubercles, which tend to become spiniform; a supraocular and postocular but no praeocular spine. Spines of rostrum short and divergent. Post-abdomen (in both sexes) seven-jointed. Eyes retractile. The basal antennal joint is short and slender and does not nearly attain the front, which is barely reached by the next joint; the flagella are visible from above at the sides of the rostrum. The merus of the exterior maxillipedes is elongated (somewhat as in *Inachus*) and is articulated with the next joint at the middle of its distal margin, which is not truncated. The chelipedes in the adult males are very considerably elongated; the palm subcylindrical, not turgid, and often as long as the merus; fingers straight and acute, and without an intermarginal hiatus at the base. The ambulatory legs are considerably elongated, subcylindrical and unarmed, with the dactyli straight and shorter than the penultimate joints.

The unique species of the genus, *Macrocheira kämpferi*, De Haan, is, as is well known, the largest of the Brachyura, and occurs in the Seas of Japan, to a depth, as the Challenger collections show, of 345 fathoms.

Macrocheira kämpferi, De Haan.

Inachus (Macrocheira) kämpferi, De Haan, Crust. in v. Siebold, Fauna Japonica, p. 160, pls. xxv., xxvi., ♂; pls. xxvii., xxviii. ♀, 1839.

Japan, off Inosima, in 345 fathoms, lat. 35° 11′ 0″ N., long. 139° 28′ 0″ E. (Station 232).

[1] This generic name was used in 1838, by Schœnherr, as *Macrocheirus*, for a genus of Coleoptera. I am loth to alter the designation by which this species has been so long known to carcinologists, but, should it be necessary to do so, the generic name *Kämpferia*, suggested by my friend Mr. A. G. More, F.L.S., of the Museum of Science and Art, Dublin, might be conveniently adopted for it.

Two small males of this well-known species, the "giant crab" of the Japanese Seas, were obtained at this Station; the dimensions of the larger specimen are (roughly) as follows :[1]—

Adult ♂.				Inches.	Centms.
Length of carapace and rostrum, about	.	.	.	9	23
Width over back at the branchial regions, nearly	.	.	7	718
Length of a chelipede, rather over	.	.	.	11	28
Length of first ambulatory leg, rather over	.	.	.	20	51

The tubercles and spines of the carapace are disposed much as in the large adult males, but are somewhat more acute, especially on the sides of the branchial regions; the chelipedes are comparatively small and slender, as in De Haan's figure. In both specimens the boss or tubercle upon the fifth segment of the post-abdomen is large and prominent. Both specimens are infested with numerous examples of a pedunculated Cirripede.

Subfamily 3. ACANTHONYCHINÆ.

Acanthonychinæ, Miers, Journ. Linn. Soc. Lond. (Zool.), vol. xiv. p. 647, 1879, et synonyma.

Eyes small and immobile or partially retractile, and usually concealed beneath the prominent præocular spine. Postocular spine small or absent. Basal antennal joint usually enlarged at the base and narrowing distally; rarely so slender as in *Inachus*.

The carapace is somewhat oblong or subtriangulate, rarely elongated and narrow. Rostrum simple or bifid. Merus of the exterior maxillipedes distally truncated. The chelipedes usually have the palms compressed. The ambulatory legs are of moderate length.

Besides the new genus *Oxypleurodon*, described below; *Sphenocarcinus*, A. Milne Edwards, which in 1879 I referred to the Pericerinæ, should perhaps be placed in this group.

Huenia, De Haan.

Huenia, De Haan, Crust. in v. Siebold, Fauna Japonica, dec. 4, p. 83, 1839.
,, Miers, Journ. Linn. Soc. Lond. (Zool.), vol. xiv. p. 643, 1879.

Carapace depressed, flat above and dissimilar in the two sexes; in the male it is elongate-triangulate, usually with one pair only of lateral lobes (the lateral epibranchial lobes); in the female there are, besides these lobes, which are largely developed, always an anterior pair situated upon the sides of the hepatic regions; the carapace is thus quadrilobated; there is a small præocular, but no postocular spine. Rostrum simple, acute, vertically deep and laterally compressed. Post-abdomen in the male seven-jointed;

[1] These measurements were taken with a tape over the convexities of the carapace and limbs, not as in other species, with compasses.

in the female five-jointed, with the fourth to the sixth joints coalescent. Eyes very small and almost immobile. The basal antennal joint is somewhat enlarged and coalescent at its distal extremity with the front; beneath which the flagella are inserted and thus not visible in a dorsal view. The exterior maxillipedes are small, the merus distally truncated and bearing the next joint at its antero-internal angle. The chelipedes (in the male) are moderately developed, with the palms compressed and cristate above; the fingers sub-excavated at the apices and having between them at the base, when closed, a more or less distinct interspace. Ambulatory legs of moderate length, with the penultimate joints, at least, more or less compressed; the dactyli retractile, and shorter than the penultimate joints.

In the carapace of the males a pair of antero-lateral lobes are occasionally, but rarely, developed, and a specimen presenting this peculiarity has been figured by De Haan (*tom. cit.*, pl. xxiii. fig. 4), and a male from West Island, Torres Strait, is in the collection of the British Museum.

The typical species of the genus, the very variable *Huenia proteus*, De Haan, elsewhere noted (*tom. cit.*, *infra*, p. 191) occurs in shallow water on the coasts of Japan and China, and ranges southward among the Philippines and to the coast of Queensland and islands adjacent. Two other nearly allied species belong to this genus; *Huenia grandidieri*, A. Milne Edwards, founded on a female type from Zanzibar, and *Huenia pacifica*, Miers, occurring at the Fiji Islands and Seychelles, in 4 to 12 fathoms.[1]

The following species, which have been referred to *Huenia*, do not, I think, belong to this genus; *Huenia simplex* and *Huenia brevirostrata*, Dana, which, as has been elsewhere shown, are to be regarded as male and female of a single species, for which I have established the genus *Simocarcinus*; and *Huenia depressa*, A. Milne Edwards, also probably a female *Simocarcinus*. *Huenia bifurcata*, Streets, may also belong to *Simocarcinus*; it is described as having the rostrum bifurcated at the tip; by this may be intended the slight notch, which, in some specimens of *Simocarcinus simplex*, is observable in a lateral view of the distal extremity.

Huenia proteus, De Haan.

Huenia proteus, De Haan, Crust. in v. Siebold, Fauna Japonica, dec. 4, p. 95, pl. xxiii. figs. 4, 5, ♂ (*elongata*), fig. 6, ♀ (*heraldica*), and pl. G, 1839.
  „   Miers, Crust. in Report on Zool. Collections of H.M.S. "Alert," p. 191, 1884, et synonyma.

Two adult males and two females were collected near Cape York, Australia, in 8 fathoms, in lat. 10° 30′ 0″ S., long. 142° 18′ 0″ E. (Station 186). The males are

[1] As I have observed in the Report on the Crustacea of H.M.S. "Alert," a sufficient series of specimens might demonstrate the identity of these forms with *Huenia proteus*. There is in the collection of the British (Natural History) Museum, a female example of *Huenia proteus* from Norfolk Island, 23 fathoms (H.M.S. "Herald"). This is, I believe, the greatest depth at which the occurrence of *Huenia* has been recorded.

intermediate between the two varieties of this sex figured by De Haan, having the rostrum slender and elongated as in the one, and the antero-lateral lobes of the carapace obsolete as in the other form. In the females, the rounded antero-lateral lobes scarcely exceed in width the obliquely truncated postero-lateral lobes of the carapace.

The largest male and female have the following dimensions :—

Adult ♂.						Lines.	Millims.
Length of carapace and rostrum,	.	.	.	.	.	14	29·5
Breadth of carapace, including the epibranchial lobes,		.	.	.	.	8½	18
Length of chelipede,	.	.	.	.	.	14½	30·5
Length of first ambulatory leg,	.	.	.	.	.	18½	39
Adult ♀.							
Length of carapace and rostrum,	.	.	.	.		10	21·5
Breadth of carapace at the lateral epibranchial lobes,		.	.			7½	15·5
Length of chelipede,	.	.	.	.		7½	15·5
Length of first ambulatory leg,	.	.	.	.		11½	24

Another small male example (with all the legs deficient) was obtained in the Arafura Sea (depth and other particulars not stated) ; the dorsal surface of the carapace is flat and nearly smooth ; the rostrum, in a dorsal view, short and triangulate, the antero-lateral lobes of the carapace wholly obsolete, and the postero-lateral lobes small and acute.

Adult ♂.				Lines.	Millims.
Length of carapace and rostrum,	.		.	5	10·5
Breadth of carapace, nearly	.	.	.	3	6

Menæthius, Milne Edwards.

Menæthius, Milne Edwards, Hist. Nat. Crust., vol. i. p. 338, 1834.
    „   Miers, Journ. Linn. Soc. Lond. (Zool.), vol. xiv. p. 649, 1879.

Carapace subpyriform, moderately convex, and tuberculated on the dorsal surface, with a large triangulate præocular spine, but no postocular spine. Rostrum simple, slender, acute, or emarginate at apex. Post-abdomen in the male seven-jointed, in the female usually five-jointed, the penultimate joint formed by the coalescence of three segments. Eyes small, mobile, but not perfectly retractile. Basal antennal joint slightly wider at the base than at the distal extremity, which is unarmed ; flagellum exposed and visible from above at the side of the rostrum. Merus of the exterior maxillipedes truncated at the distal extremity and with a prominent antero-external angle, and slightly notched at the antero-internal angle, where it is articulated with the next joint. Chelipedes (in the male) well developed, with the palm slightly compressed ; fingers acute, and having between them, when closed, an interspace at the base. Ambulatory legs of moderate length ; the joints subcylindrical, not dilated or compressed ; dactyli slightly curved and partially retractile.

As Professor A. Milne Edwards has shown, it is probable that the described species of this genus, eleven in number, are all to be regarded as synonymous with the type species, *Menæthius monoceros* (Latr.).[1]

This species is very widely distributed throughout the Oriental region, from the Red Sea and East African coast to the Fiji Islands, and has not, I believe, been recorded except from shallow water (depth not exceeding 18 fathoms).

Menæthius monoceros (Latreille), var. *augusta*, Dana.

Menæthius angustus, Dana, Amer. Journ. Sci. and Arts, ser. 2, vol. xi. p. 272, 1851; U.S. Explor. Exped., Crust., vol. xiii. p. 120, pl. iv. fig. 5, 1852.

Three small male specimens, obtained off Nukalofa, Tongatabu, in 18 fathoms, lat. 20° 58′ 0″ S., long. 175° 9′ 0″ E. (Station 172), are referred to the variety designated by Dana, *Menæthius angustus*, of this very common and variable species; they agree with his description and figure in the narrowness of the carapace and the slenderness of the elongated and slightly emarginate rostrum, although presenting some differences with regard to the number and arrangement of the tubercles of the carapace, *e.g.*, the smaller tubercles of the gastric, cardiac, and branchial regions, which are much less distinct than in his figure.

Another specimen, an adult male, obtained near Cape York, Australia, in 8 fathoms, lat. 10° 30′ 0″ S., long. 142° 18′ 0″ E. (Station 186), is of very different form, with a broad, very uneven, tuberculated carapace, flattened triangulate supraocular spines, and short rostrum, which scarcely exceeds in length the width of the carapace at the frontal region, and nearly resembles the variety figured by Dana (*tom. cit.*, p. 122, pl. iv. fig. 7), as *Menæthius subserratus*, Adams and White. Dana's specimens were from the Fiji and Samoan Islands.

The dimensions of these specimens are as follows:—

Adult ♂, from Tongatabu (the largest).					Lines.	Millims.
Length of carapace, rather over	.	.	.	.	3	7
Length of rostrum, nearly	.	.	.	.	2	5
Breadth of carapace, nearly	.	.	.	.	3	5·5
Length of a cheliped, about	.	.	.	.	5	11
Length of first ambulatory leg, nearly	.	.	.	.	6	12·5
Adult ♂, from Cape York.						
Length of carapace, nearly	.	.	.	.	.	5	10
Length of rostrum, nearly	.	.	.	.	.	2	4
Breadth of carapace,	.	.	.	.	.	4	8·5
Length of a cheliped, about	.	.	.	.	.	5	11
Length of first ambulatory leg, nearly	.	.	.	.	.	6½	13·5

[1] Exception may, however, perhaps be made provisionally for *Menæthius tuberculatus*, Adams and White, from the Mauritius, in which the tubercles of the carapace are more developed than in any other variety known to me, and take the form of blunt spines on the branchial, cardiac and intestinal regions.

the body are, as elsewhere, covered with short spines, which are here as on the posterior
segment of the thorax hooked; the posterior region of the abdominal shield, from the
articulation of the uropoda onwards, is smooth and entirely devoid of spines, with the
exception of the four terminal spines.

The *antennules* are displayed in fig. 10 of Pl. V.; they consist of a two-jointed
peduncle and a five or six-jointed flagellum; in the peduncle the proximal joint is
broader as well as shorter than the succeeding joint.

The *antennæ* (fig. 9) are very much longer than the antennules, but not so long as
the body; the proximal joints are short and subequal; the two distal joints of the
peduncle are of great length, the last being slightly the longest; the flagellum is shorter
than either of the two terminal joints of the peduncle; it is composed of twenty or
more joints, of which the first is the longest.

The *mandibles* terminate in a bifid masticatory process, each division of which is
again divided into two or three teeth; the masticatory edge is also furnished with several
denticulated spines; there is a stout molar process; the palp is long and three-jointed,
the middle joint is rather the longest; the terminal joint and the distal half of the
middle joint are beset with a single row of fine spines; at the extremity of the distal
joint, which is somewhat curved, are four or five longish stiff hairs, which decrease
gradually in length from before backwards.

One of the *maxillipedes* is represented in fig. 12; the palp is five-jointed, the joints
gradually decreasing in width towards the extremity; the inner margin of the stipes is
furnished with two processes shown more highly magnified in fig. 13; they evidently
correspond to similar structures in other Isopods, especially in the Munnopsidæ.

The first pair of *thoracic appendages* are modified into prehensile limbs; one of these
is displayed in fig. 14 of Pl. V.; the proximal joint is long and rather stouter than the
succeeding joint, one margin is fringed with a row of hooked spines; the following
joints are short, the second rather longer than the third and fourth, which are subequal;
the fifth joint is oval and rather swollen, the inner margin, against which the narrow
sixth joint rests, has a few slender spines.

The remaining *thoracic appendages*[1] are elongate, particularly the three posterior
pairs; the proximal joints are furnished with several rows of spines; the terminal joint
of each limb is short and bears a long, curved, slender spine and a short slender hair on the
inner side of the former; this arrangement is, however, very different from the two subequal
terminal claws that are found in the thoracic appendages of *Munna* and other genera.

[1] In the interior of several of the thoracic appendages, probably lodged in the vascular channels, were occasionally a number of green bodies of varying form, which I take to be parasitic Algæ. I am not aware that the occurrence of parasites of this class have been noted in the Isopods, though parasitic Infusorians (*Anoplophrya circulans*, Balbiani, *Recueil zool. suisse*, ii, 1885, p. 277), are known from the appendages of *Asellus*. The presence of green bodies presumably coloured by chlorophyll might be useful in determining, in disputed cases, whether a given specimen really came from the bottom or had been caught up by the dredge in the surface waters.

The *uropoda* are rudimentary ; each consists of a somewhat conical piece articulated to the abdominal shield by its broad end; at the free extremity there are a number of long hairs, and a minute articulated scale which appears to me to represent the endopodite; in figs. 16, 17 of Pl. V. the terminal portion of this appendage is shown more highly magnified.

Station 149K, off Christmas Harbour, Kerguelen, January 29, 1874 ; depth, 120 fathoms.

Ischnosoma, G. O. Sars.

Ischnosoma, G. O. Sars. Beretning om i Sommeren, 1863, foretagen Zoologisk Reise ved Kysterne af Christiania og Christiansands Stifter. Christiania, 1866, p. 34.

The Challenger collection contains two fragments of Isopods, which appear to belong to this genus and to represent two new species of it, as well as a single complete specimen of another species.

This genus was formerly regarded by Sars as belonging to the Munnopsidæ, but it has since been removed by him to the family Asellidæ. In his account of the Isopoda collected by the Norwegian Expedition, Dr. Sars has pointed out that although the general aspect of the Crustaceans belonging to this genus is not unlike that of the Munnopsidæ, the absence of any modification into natatory organs of the three posterior pairs of thoracic appendages is opposed to their being placed in this family. The same is also the case with *Mecrostylis*.

The genus is defined by Sars as follows :—

Body elongate and narrow, broadest at first segment of thorax ; hinder part of segment four and anterior part of segment five firmly connected, and forming an hourglass shaped portion which is nearly equal in length to half the body. Head small and rounded, eyes absent. The abdominal segment much longer than broad, constricted at the base, rather dilated towards the apex and obtusely rounded. Upper antennæ projecting a little beyond the first joint of the peduncle of the lower antennæ, six-jointed, the second joint narrow and elongated ; lower antennæ slender, longer than the body, the flagellum about equalling the peduncle in length and composed of about nineteen joints. Feet of the first pair short and robust, subprehensile, the antepenultimate joint strongly dilated. Following feet slender and elongated, similar to each other, six-jointed, the last joint forming a claw very elongate in the posterior three pairs of limbs. Caudal appendages simple.

Sars has described two species of this genus, viz., *Ischnosoma bispinosum* and *Ischnosoma quadrispinosum* ; the former is distinguished by the presence of a spinous prolongation of the lateral margin of the first thoracic segment ; in *Ischnosoma quadri-spinosum*, the first and fourth thoracic segments are thus provided, and the uropoda are rudimentary. Two of the species to be described in the present Report are distinguished in the first place by their very large size as compared with the two northern representa-

Adult ♂.		Lines.	Millims.
Length of carapace and rostrum,		7½	16
Breadth of carapace to base of lateral branchial spines,		5	10·5
Length of a chelipede, about .		5½	12
Length of first ambulatory leg, .		7½	16

Pugettia, Dana.

Pugettia, Dana, Amer. Journ. Sci. and Arts, ser. 2, vol. xi, p. 268, 1851.
  „   U.S. Explor. Exped., Crust., vol. xiii. p. 84, 1852.
  „   Miers, Journ. Linn. Soc. Lond. (Zool.), vol. xiv. p. 650, 1879, and synonym.

This genus is very closely allied to *Acanthonyx*, and it may suffice to indicate the distinctive characters, which are as follows :—

The carapace is usually more convex and tuberculated than in *Acanthonyx*, and the lateral branchial lobes as well as the antero-lateral lobes are large, prominent, and acute, and separated by a concave interspace. The post-abdomen (in the male) is seven-jointed. The penultimate joints of the ambulatory legs are subcylindrical, without the prominent posterior lobe or tooth which characterises *Acanthonyx*.

Of the species of this genus, two, *Pugettia gracilis*, Dana, and *Pugettia richii*, Dana, inhabit the Californian coast, the former extending its range northward to Esquimalt Harbour, British Columbia (whence it has been recorded by Mr. Spence Bate as *Pugettia lordi*). Two species, *Pugettia quadridens* (De Haan) and *Pugettia incisa* (De Haan), inhabit the seas of Japan.

I have noted the occurrence of a female of this genus, doubtfully distinct from the Californian species (*Pugettia australis*, sp. ?) from Rio de la Plata, in 28 fathoms.

None of the above species have, I believe, been recorded from very deep water; I have, however, noted the occurrence of *Pugettia incisa* (De Haan) in the Corean Seas, at a depth of 36 fathoms.

The very distinct species described below, *Pugettia velutina*, was dredged off the Ki Islands in 140 fathoms.[1]

Pugettia incisa (De Haan).

Menæthius incisus, De Haan, Crust. in v. Siebold, Fauna Japonica, p. 98, pl. xxiv. fig. 3, ♀
        (*Halmæus*), and pl. G., 1839.
Pugettia incisa, Stimpson, Proc. Acad. Nat. Sci. Philad., p. 219, 1857.
    „   „   Miers, Proc. Zool. Soc. Lond., p. 23, 1879.

Two adult males were collected off Yokoska, Japan, in 10 fathoms. In these specimens the lateral (postocular) expansions of the carapace have the anterior and

The genus *Peltinia*, Dana, is probably synonymous with *Pugettia* ; of the recorded species, one, *Pugettia scuti-formis*, Dana, from Rio de Janeiro, was probably founded on an immature type ; the other, *Pugettia nodulosa*, Dana, inhabits the Fiji Islands (Vanua Levu), and presents all the characters of *Pugettia*.

posterior angles somewhat produced and spiniform, wherein they exhibit some approach to the allied *Pugettia quadridens* (De Haan), also from the Japanese Seas, in which the lateral expansions of the carapace are so far excavated that their anterior and posterior angles present the appearance of distinct spiniform lobes or teeth (whence the specific name).

The largest male has the following dimensions :—

Adult ♂.	Lines.	Millims.
Length of carapace and rostrum,	12½	26·5
Breadth of carapace to base of lateral branchial spines, nearly	8	16·5
Length of a chelipede,	15	32
Length of first ambulatory leg,	16	34

Pugettia velutina, n. sp. (Pl. VI. fig. 2).

The body and legs are everywhere covered with a close-set, short, felty pubescence ; the carapace is subpyriform, moderately convex, its dorsal surface with several elevated rounded bosses or prominences ; the spines of the rostrum are rather less than half as long as the carapace, acute and divaricate from their bases ; of the dorsal tubercles or bosses, three are placed upon the gastric, one on the cardiac, two on each branchial region, and there are also three small median tubercles on the elevated posterior margin of the carapace, and two conical, acute, lateral spines on each side of the body, the anterior and smaller of which is situated upon the sides of the hepatic region and the posterior on the branchial regions ; the upper orbital margin is well defined, with a small præocular tooth and narrow hiatus near the postocular lobe ; on the pterygostomian region is a flattened oval tubercle or prominence, and another above the bases of the chelipedes. The basal joint of the antennæ is moderately dilated, longitudinally concave ; the following joints are slender and are not concealed by the spines of the rostrum in a dorsal view ; the ischium-joint of the outer maxillipedes is slightly concave longitudinally ; the merus-joint small, not notched at the antero-internal angle, where the following joint is articulated with it ; the chelipedes (in the female) are slender and rather longer than the carapace, the merus or arm slender and subcylindrical, with three small tubercles at the distal extremity ; carpus indistinctly carinated on its outer surface, palm somewhat compressed, fingers naked, regularly serrato-denticulated on their inner margins ; the ambulatory legs decrease successively in length from the first to the last, and are slender, with subcylindrical joints ; the dactyli not denticulated on their inferior margins ; besides the close pubescence, with which the whole animal is covered, the merus-joints of the legs bear tufts of longer hairs, placed at intervals along the anterior and posterior margins, and a few such hairs exist at the distal extremities of the two following joints and on other parts of the body. Colour (in spirit) light yellowish-brown.

Adult ♀.					Lines.	Millims.
Length of carapace to base of rostrum, about	.	.	.	.	6½	13·5
Length of rostrum, about	.	.	.	.	3	6·5
Breadth of carapace, nearly,	.	.	.	.	4	10
Length of a chelipede (in the female), about	.		.	.	7½	16
Length of first ambulatory leg, about			.	.	12	25·5

The single specimen I have seen (an adult female) is in somewhat imperfect condition, and was dredged at Station 192, near the Ki Islands, lat. 5° 49′ 15″ S., long. 132° 14′ 15″ E., in 140 fathoms.

It is at once distinguished from all the recorded species of the genus by the closely pubescent and strongly tuberculated carapace, and longer rostral spines.

Acanthonyx, Latreille.

Acanthonyx, Latreille, Crust. in Cuvier, Règne Animal, ed. 2, vol. iv. p. 58, 1829.

" Milne Edwards (part), Hist. Nat. Crust., vol. i. p. 342, 1834.

" Miers, Journ. Linn. Soc. Lond. (Zool.), vol. xiv. p. 650, 1879.

Carapace suboblong, rounded behind, and with the dorsal surface usually depressed, not markedly constricted behind the prominent antero-lateral angles, the lateral branchial spines small and not prominent. Præocular spine prominent, acute. Spines of the rostrum united at the base, acute and but little divergent. Post-abdomen in the male, in the species I have examined, six-jointed. Eyes small, mobile, but not completely retractile. Basal antennal joint narrowing slightly from the base to the distal extremity, which is unarmed; flagellum exposed and visible from above at the side of the rostrum. Merus of the exterior maxillipedes truncated at the distal extremity and but slightly notched at the antero-internal angle, where it is articulated with the next joint. Chelipedes (in the adult male) well developed; palm compressed, but slightly turgid in the middle, and often slightly carinated above; fingers acute, and having between them, when closed, an interspace at the base. Ambulatory legs short, with the penultimate joints more or less dilated and compressed and armed with a tooth or lobe on its inferior margin, against which the small acute dactylus closes.

The species of this genus are small and not numerous, and occur both in the Atlantic and Indo-Pacific regions in shallow water. The following is a list of them :—

> *Acanthonyx lunulatus*, Guérin-Méneville (= *Maia glabra*, Latreille; *Acanthonyx viridis*, Costa; *Acanthonyx brevifrons*, A. Milne Edwards). Mediterranean and Cape Verde Islands.
>
> *Acanthonyx petiverii*, Milne Edwards (= *Acanthonyx emarginatus*, Milne Edwards and Lucas; *Acanthonyx debilis*, Dana, and *Acanthonyx concameratus*, Kinahan, vars.?). West Indies to Brazil, and California to Chili; Galapagos (A. Milne Edwards).

Acanthonyx mac-leayii, Krauss. Natal ; Ceylon (Coll. Brit. Mus.).
Acanthonyx quadridentatus, Krauss. Natal.
Acanthonyx simplex, Dana. Sandwich Islands.
Acanthonyx consobrinus, A. Milne Edwards. Réunion.
Acanthonyx limbatus, A. Milne Edwards. Réunion.
Acanthonyx elongatus, White (ined.) Miers. Red Sea.[1]

Acanthonyx lunulatus (Risso).

> *Maia lunulata*, Risso, Crust. de Nice, p. 49, pl. i. fig. 4, 1816.
> *Acanthonyx lunulatus*, Milne Edwards, Hist. Nat. Crust., vol. i. p. 342, pl. xv. figs. 6–8, 1834 ;
> Atlas in Cuvier, Règne Animal, ed. 2, pl. xxvii. fig. 2.

Of this common Mediterranean species a single small immature female was obtained at St. Vincent, Cape Verde Islands. It approaches the species or variety *Acanthonyx brevifrons*, A. Milne Edwards, in the form of the front, but there are indications of three antero-lateral teeth, and the carapace, as in *Acanthonyx lunulatus*, bears several tufts of setæ.

Adult ♀.		Lines.	Millims.
Length of carapace and rostrum, nearly .	.	3	6
Breadth of carapace, . . .	.	2	4·5

Family II. MAIIDÆ.

> *Maiidæ*, Miers, Journ. Linn. Soc. Lond. (Zool.), vol. xiv. pp. 640, 653, 1879.

Eyes retractile within the orbits, which are distinctly defined, but often more or less incomplete below, or marked with open fissures in their superior and inferior margins. Basal antennal joint always more or less enlarged.

Subfamily 1. MAIINÆ.

> *Maiinæ*, Miers, Journ. Linn. Soc. Lond. (Zool.), vol. xiv. p. 653, 1879.

Carapace usually subtriangulate. Rostrum well developed. Chelipedes (in the male) enlarged ; fingers not excavate at the distal extremity.

The following genera must be added to this subfamily :—

Sisyphus, A. Milne Edwards (if the orbits in this genus are completely closed it will be better placed in the family Periceridæ). *Chlorinoides*, Haswell ; also *Scyra*, Dana, and perhaps *Rochinia*, A. Milne Edwards, formerly placed in the Periceridæ.

---

[1] *Acanthonyx scutellatus*, MacLeay, from the Cape of Good Hope, is perhaps a species of *Epialtus*.

In the numerous genera of this group the carapace is usually subtriangulate; the rostrum emarginate or two-spined; the orbits well defined and yet incomplete; eyes completely retractile; chelipedes with fingers acute, and ambulatory legs of moderate length.

Egeria, Leach.

Egeria, Leach, Zool. Miscell., vol. ii. p. 39, 1815.
" Milne Edwards, Hist. Nat. Crust., vol. i. p. 290, 1834.
" Miers, Journ. Linn. Soc. Lond. (Zool.), vol. xiv. p. 654, 1879.

The carapace is subpyriform and nearly as broad as long, moderately convex and tuberculated; rostrum prominent, vertically compressed, and more or less deeply emarginate at the distal extremity. Orbits shallow and very open above, and with two wide fissures in the superior and two in the inferior margin. Epistoma transverse. Post-abdomen in the male seven-jointed, in the female five-jointed (the fourth to sixth segments coalescent). Eyes short, retractile. Merus of the exterior maxillipedes truncated at the distal extremity, the antero-external angle not produced, the antero-internal angle (where the next joint is articulated) slightly notched. Chelipedes (in the adult male) moderately elongated; palm about as long as the merus, and but slightly compressed; fingers acute, and having between them, when closed, but a small interspace at the base. Ambulatory legs slender, subcylindrical and extremely elongated, but diminishing successively in length.

Several species of the genus have been described, all of which are, however, probably to be regarded as synonymous with or varieties of *Egeria arachnoides* (Rumph), which occurs commonly in the Australian, Indian, Malaysian and Chinese Seas, in water of moderate depth (to 49 fathoms).

Egeria arachnoides (Rumph).

Cancer arachnoides, Rumph, D'Amboinsche Rariteit-Kamer, pl. viii. fig. 4, 1741.
?? Cancer longipes, Linné, Mus. Ludovici Ulrici, p. 446, 1764; Syst. Naturæ, ed. 12, p. 1047, 1766.
Egeria arachnoides, Milne Edwards, Hist. Nat. Crust., vol. i. p. 291, 1834.
" " Miers, Crust. in Report on Zool. Coll. H.M.S. "Alert," p. 191, 1884.
Egeria herbstii, Milne Edwards, tom. cit., p. 292, 1834.
Egeria indica, Leach, Zool. Miscell., vol. ii., pl. lxxiii., 1815.
" " Milne Edwards, tom. cit., p. 292, 1834.
Leptopus longipes, Latreille, Milne Edwards, Atlas in Cuvier, Règne Animal, Crust., ed. 3, pl. xxxiv. fig. 1.

Specimens were dredged at the following localities—

Station 188, South of New Guinea, in 28 fathoms; lat. 9° 59′ 0″ S., long. 139° 42′ 0″ E. (young female). Station 190, Arafura Sea, in 49 fathoms; lat. 8° 56′ 0″ S., long.

136° 5′ 0″ E. (adult male, having lost nearly all the legs). Station 212, Celebes Sea, 10 fathoms ; lat. 6° 54′ 0″ N., long. 122° 18′ 0″ E. (adult males and females).

The males are small or imperfect. An adult female (Station 212) presents the following dimensions :—

Adult ♀.						Lines.	Millims.
Length of carapace and rostrum, nearly	.	.	.	.	.	11	22·5
Breadth of carapace, nearly	.	.	.	.	.	9	18·5
Length of a chelipede,	.	.	.	.	.	16½	35
Length of first ambulatory leg,	.	.	.	.	.	61½	130·5

The specimen dredged in the Arafura Sea (Station 190) differs slightly in its somewhat shorter and broader rostrum.

Chorilibinia, Lockington.

Chorilibinia, Lockington, Proc. Calif. Acad. Nat. Sci., vol. vii. p. 69, 1876.
  „   Miers, Journ. Linn. Soc. Lond. (Zool.), vol. xiv. p. 654, 1879.

The generic diagnosis of Lockington is in few words, and is as follows :—" Rostrum long, broad, and emarginate at tip as in *Libinia*, but the eyes concealed beneath it as in *Chorinus* and its allies. Præ- and postorbital teeth acute, separated above and below by an acute fissure, and together constituting the orbit. Carapace triangular."

The description which follows is taken partly from Mr. Lockington's specific description, partly from the Australian species, *Chorilibinia gracilipes*, which I have referred to this genus.

Carapace subpyriform, rounded behind, and spinose on the dorsal surface ; orbit divided by a hiatus or fissure above and below, upper orbital margin prominent, the supraocular lobe terminates usually, but not invariably, in a tooth or spine. Rostrum well developed, with its spines coalescent at base, and separated and divergent in their distal half or third. Post-abdomen (in the male) in *Chorilibinia gracilipes*, seven-jointed. Eyes (in *Chorilibinia gracilipes*) small, retractile. Basal antennal joint somewhat dilated, with a spine at the extero-distal angle, and (in *Chorilibinia gracilipes*) on the outer margin at the base ; merus of the exterior maxillipedes (in *Chorilibinia gracilipes*) distally truncated, and considerably produced and rounded at the antero-external angle, emarginate at the antero-internal angle, where the next joint articulates with it. Chelipedes (in the male) small, palm slender, fingers small, in contact through the greater part of their length. Ambulatory legs subcylindrical, slender, elongated, the anterior pair (or second pair in *Chorilibinia angusta* ?) much the longest ; dactyli nearly straight, acute.

Of the two described species of this genus, one, *Chorilibinia angusta*, Lockington, inhabits the Gulf of California (depth not stated); the other, *Chorilibinia gracilipes*, Miers, the shores and islands of North and North Eastern Australia and New Guinea, in water (as recorded) of moderate depth (3 to 28 fathoms).

This genus is nearly allied to *Egeria*, but is distinguished by having but a single hiatus in the margins of the orbit, by the spinose carapace, and more deeply divided rostrum, &c.

Chorilibinia gracilipes, Miers.

Chorilibinia gracilipes, Miers, Ann. and Mag. Nat. Hist., ser. 5, vol. xix. p. 7, pl. iv. fig. 4, 1879; Crust. in Report on Zool. Coll. H.M.S. "Alert," p. 192, 1884.
Chorilibinia gracilipes, Haswell, Proc. Linn. Soc. N.S.W., vol. iv. p. 439, 1880; Cat. Australian Crust., p. 17, 1882.

An adult male is in the collection from "Torres Strait, August 1874," and an adult female dredged at Station 188, south of New Guinea, in 28 fathoms, lat. 9° 59′ 0″ S., long. 139° 42′ 0″ E.

If, as I think is hardly possible, this species should prove to be not merely specifically but generically distinct from the type of the genus described by Mr. Lockington from the Gulf of California,[1] Mr. Haswell's slightly different spelling of the generic name may be conveniently adopted for it.

The dimensions of the male are as follows :—

Adult ♂.				Lines.	Millims.
Length of carapace and rostrum,	.	.	.	6	12·5
Breadth of carapace,	. .	.	.	3½	7·5
Length of a chelipede,	.	.	.	6	12·5
Length of first ambulatory leg,	.	.	.	16	34

Hyas, Leach.

Hyas, Leach, Trans. Linn. Soc. Lond., vol. xi. p. 328, 1815.
„ Milne Edwards, Hist. Nat. Crust., vol. i. p. 311, 1834.
„ Miers, Journ. Linn. Soc. Lond. (Zool.), vol. xiv. p. 654, 1879.

Carapace depressed, broadly pyriform or lyrate, not spinose on the dorsal surface. Spines of the rostrum dilated, vertically compressed, acute, and nearly in contact along their inner margins. No præocular spine. The orbits are shallow and somewhat open above, with a hiatus or fissure in the upper and lower margins. The post-abdomen distinct, seven-jointed both in the male and female. The eyes are short and partially visible in a dorsal view when retracted. The basal antennal joint is not greatly dilated, and is unarmed, the next joint is slightly dilated, and the third is slender ; these, with the flagellum, are visible from above at the sides of the rostrum. The merus of the exterior maxillipedes is distally truncated, with the antero-external angle rounded and

[1] *Chorilibinia angusta*, Lockington, Proc. Calif. Acad. Nat. Sci., vol. vii. p. 69, 1876.

not prominent, the antero-internal angle, where the next joint articulates, scarcely
emarginate. The chelipedes (in the male) are of moderate length ; palm slightly compressed but not carinated; fingers nearly straight, acute, and scarcely toothed on the
inner margins, with scarcely any intermarginal hiatus when closed. Ambulatory legs
subcylindrical, of moderate length ; dactyli nearly straight, acute.

The species are not numerous. The two which follow, *Hyas aranea* (Linné) and
Hyas coarctata, Leach, are extensively distributed throughout the North temperate seas
of the Atlantic region, occurring commonly on the more northerly European coasts at
Iceland, Spitzbergen, and on the eastern coast of North America northward to Greenland,
and through the Arctic Seas westward to Behrings Strait and the Sea of Ochotsk.[1]

A third species, *Hyas latifrons*, Stimpson, very doubtfully distinct from *Hyas
coarctata*, has been described from Behrings Strait[2] and Alaska (Lockington) ; and a
fourth and very distinct one, *Hyas lyrata*, Dana, from California, Pugets Sound, Oregon,
"deep water" (Lockington) and Vancouver Island, and British Columbia, whence there
are specimens in the collection of the British Museum.

Details are wanting as regards the bathymetrical distribution of the two last-
mentioned species ; *Hyas aranea* has been recorded from 60 fathoms by Professor
S. I. Smith (*tom. cit.*), and the same author has recently noted the occurrence of *Hyas
coarctata* at very considerable depths, e.g., 373 and 906 fathoms, on the American
coast.[3]

Hyas aranea (Linné).

Cancer araneus, Linné, Syst. Nat., ed. 12, p. 1044, 1766.
    "   "   Pennant, Brit. Zool., vol. iv. p. 6, pl. ix. fig. 16, 1777.
    "   info, Herbst, Naturg. Krabben, &c., vol. i. p. 242, pl. xvii. fig. 95, 1782.
Hyas araneus (us), Leach, Malacostraca Podophthalmata Britanniæ, pl. xxiA., 1815.
    "   "   Milne Edwards, Hist. Nat. Crust., vol. i. p. 312, 1834 ; Crust. in Cuvier
                 Règne Animal, ed. 3, Atlas, pl. xxxii. fig. 2.
    "   "   Bell, British Crustacea, p. 31, woodcut, 1853.

Station 48, South of Nova Scotia, 51 fathoms ; in lat. 43° 4′ 0″ N., long.
64° 5′ 0″ W. (a small male).

This specimen has the antero-lateral angles of the carapace (postocular lobes)
unusually prominent.

Adult ♂.		Lines.	Millims.
Length of carapace and rostrum,	.	6	12·5
Breadth of carapace, nearly	.	4	8

[1] *Vide*, S. I. Smith, Trans. Connect. Acad., vol. v. pt. 1, p. 43, 1879.
[2] With this species or variety, characterised by the shorter, broader, less acute rostrum, and the narrower or closed
fissure of the upper orbital margin, *Hyas bufonius*, White, is identical.
[3] Report on the Decapod Crustacea of the "Albatross" dredgings on the East coast of the United States, in the
Annual Report of the Commissioner of Fish and Fisheries, 1884, p. 347.

Hyas coarctata, Leach.

Hyas coarctata (sz), Leach, Malacostraca Podophthalmata Britanniæ, pl. xxiB. figs. 1 (♂) 2 (♀), 1815.

"   "   Milne Edwards, Hist. Nat. Crust., vol. i. p. 312, 1834 ; Crust. in Cuvier, Règne Animal, ed. 3, pl. xxxii. fig. 3 (outline).

"   "   Bell, British Crustacea, p. 35, woodcut, 1853.

"   "   Hoek, Niederländ. Archiv f. Zool. Suppl., Bd. i. p. 3, pl. i. fig. 1, 1882, var.

Lissa fissirostra, Say, Journ. Acad. Nat. Sci. Philad., vol. i. p. 79, 1817.

Hyas serratus, Hailstone, Ann. and Mag. Nat. Hist., vol. viii. p. 262, woodcut, 1835.

"   *coarctata*, var. *alutacea*, Brandt, in Middendorf's Sibirische Reise, vol. ii. p. 79, 1851, var. ?

"   *bufonius*, White, List Crust. Brit. Mus., p. 6, 1847.

"   *latifrons*, Stimpson, Proc. Acad. Nat. Sci. Philad., p. 217, 1857.

An adult but small male was collected at Station 48, south of Nova Scotia, in 51 fathoms, with the preceding species, and a smaller male at Station 49, near the same locality, in lat. 43° 3′ 0″ N., long. 63° 39′ 0″ W., in 85 fathoms.

These specimens belong to the typical form of the species.

The larger male measures as follows :—

Adult ♂.					Lines.	Millims.
Length of carapace and rostrum, nearly	.	.	.		15	31·5
Breadth of carapace,	.	.	.	.	10	21

Herbstia, Milne Edwards.

Herbstia, Milne Edwards, Hist. Nat. Crust., vol. i. p. 301, 1834.

"   Miers, Journ. Linn. Soc. Lond. (Zool.), vol. xiv. p. 654, 1879, et synonyma.

Carapace depressed, broadly pyriform, with the dorsal surface usually spinose or tuberculated ; the rostrum short, more or less deeply divided by a median notch or fissure into two lobes, which are vertically compressed and dilated at the base, and distally acute. Orbits shallow, with or without a præocular spine. Post-abdomen (in the male) seven-jointed. Eyes short, and when retracted usually partially visible in a dorsal view. Antennæ with the basal joint moderately dilated and armed with a spine at the antero-external angle, and often with another on the outer margin at the base ; flagellum exposed and visible from above at the sides of the rostrum. Exterior maxillipedes with the merus-joint distally truncated and not produced at the antero-external angle ; the antero-internal angle, where the next joint articulates, slightly emarginate. Chelipedes (in the adult male) moderately developed, with the palm (in the typical species) considerably enlarged ; fingers arcuate, strongly toothed, and having between them, when closed, a wide interspace ; in others, the palm is more slender, and the fingers nearly meet along the inner margins and are very obscurely dentated.

Ambulatory legs of moderate length, with the joints subcylindrical; dactyli nearly straight, acute.

The genus may be conveniently divided into two subgenera :—

1. *Herbstia*, in which the inferior margin of the orbit is not dentated, and the merus-joints of the ambulatory legs are not spinose.

2. *Herbstiella*, Stimpson,[1] in which the inferior margin of the orbit is usually armed with a tooth or spine, and the merus-joints of the ambulatory legs have a series of small spinules.

The species of the subgenus *Herbstia* are distributed as follows :—The typical species, *Herbstia condyliata* (Herbst), is common in the Mediterranean, in 20 to 40 fathoms (Heller), and has been recorded from the Canaries by Brullé ; two species, *Herbstia ovata*, Stimpson, and *Herbstia rubra*, A. Milne Edwards, occur at the Cape Verde Islands, the former, which Stimpson regarded as belonging to a distinct genus *Micropisa*,[2] in 20 fathoms ; a fourth species, *Herbstia violacea*, (A. Milne Edwards), has been recorded from Cape St. Vincent, Angola, the Cape Verde Islands, Goree Island, Senegambia (9 to 15 fathoms), and West Africa, the Gaboon ; a fifth, *Herbstia eryophora*, Rochebrune, from Senegambia ; a sixth, *Herbstia crassipes*, A. Milne Edwards, from Australia, Bass Strait ; a seventh, *Herbstia pubescens*, Stimpson, from the west coast of Mexico, Manzanillo ; and an eighth, *Herbstia pyriformis*, Bell, type of his genus *Rhodia*,[3] from the Galapagos.

The species of the subgenus *Herbstiella*, except the type *Herbstia depressa*, Stimpson, found at St. Thomas, and by the Challenger Expedition off the Brazilian Coast, in 30 to 350 fathoms, are all West American ; *Herbstia camptacantha*, Stimpson, occurs at Acapulco, on the west coast of Mexico, and at Cape St. Lucas, California ; *Herbstia tumida*, Stimpson, at Manzanillo, and *Herbstia edwardsii* (Bell), at the Galapagos. The habitat of the insufficiently known *Herbstia parviformis* (Randall) is not particularly stated.

Herbstia rubra, A. Milne Edwards (Pl. VII. fig. 1).

Herbstia rubra, A. Milne Edwards, Rev. Mag. Zool., ser. 2, vol. xxi. p. 354, 1869.

St. Vincent, Cape Verde Islands (a small male) :—

Adult ♂.	Lines.	Millims.
Length of carapace and rostrum, .	9	11
Breadth of carapace, . .	3½	7·5
Length of a chelipede, nearly .	5	10·5
Length of first ambulatory leg, .	6½	14

[1] *Ann. Lyc. Nat. Hist. New York*, vol. x. p. 93, 1871.
[2] *Proc. Acad. Nat. Sci. Philad.*, p. 217, 1857.
[3] *Proc. Zool. Soc. Lond.*, p. 169, 1835 ; *Trans. Zool. Soc. Lond.*, vol. ii. p. 43, 1841.

This specimen, in coloration and in nearly all structural characters, bears a most remarkable resemblance to the *Herbstia pyriformis* (Bell);[1] it differs, indeed, from the description and figure of that species only in having the gastric and branchial regions more distinctly granulated. The type of *Herbstia pyriformis* was from the far distant Galapagos Islands.

From *Herbstia rubra*, as described by A. Milne Edwards, the Challenger specimen differs in the coloration, in the absence of the anterior transverse series of gastric tubercles and of the spiniform median tubercles of the cardiac region, and in the more slender legs. I should not have suspected its identity with this species, had not Professor A. Milne Edwards himself remarked upon the near affinity of *Herbstia pyriformis* and *Herbstia rubra*,[2] and to facilitate the identification of the Challenger specimen with the type of the species, I think it useful to figure it.

The margins and merus-joints of the ambulatory legs are indistinctly granulated, and the granules are spinuliform on the first pair of legs, toward the distal extremity; hence this species establishes a transition to the subgenus *Herbstiella*.

Herbstia violacea (A. Milne Edwards).

Micropisa violacea, A. Milne Edwards, Nouv. Archiv. Mus. Hist. Nat., vol. iv. p. 50, pl. xvi. figs. 3-6, 1868.
Herbstia violacea, Miers, Ann. and Mag. Nat. Hist., ser. 5, vol. viii. p. 206, 1881.

Here is referred an adult female, obtained at St. Vincent, Cape Verde Islands, with the preceding species. Its dimensions are as follows :—

Adult ♀ .					Lines.	Millims.
Length of carapace and rostrum,	.	.	.	.	14	29.5
Breadth of carapace,	.	.	.	.	11	23
Length of a chelipede, rather over	.	.	.	..	12	26
Length of first ambulatory leg,	.	.	.	.	14½	30.5

This form differs from the typical species of *Herbstia* in its more slender eye-peduncles, and in the chelipedes, whose dactyli, not only in the females, but also in the males I have examined, are nearly straight, acute, and very obscurely dentated or entire on the inner margins.

Herbstia ovata (Stimpson).

Micropisa ovata, Stimpson, Proc. Acad. Nat. Sci. Philad., p. 217, 1857.
     ,,   ,,   A. Milne Edwards, Nouv. Archiv. Mus. Hist. Nat., vol. iv. p. 51, pl. xvi. figs. 1, 2, 1868.

Several specimens of both sexes and of different sizes were collected at St Vincent, Cape Verde Islands, in July 1873, with the preceding species.

[1] *Trans. Zool. Soc. Lond.*, vol. ii. p. 44, pl. ix. fig. 1, 1841.
[2] *Crust. Podophthalm. in Miss. Sci. au Mexique*, p. 77, 1875.

In the adult females the carapace is more convex and tuberculated than in the males. This species is always of a very small size; an adult male in the Challenger series has the following dimensions :—

Adult ♂.					Lines.	Millims.
Length of carapace and rostrum,	.	.	.	.	4½	9·5
Breadth of carapace, nearly	.	.	.	.	4	8
Length of a chelipede,	.	.	.	.	5	11
Length of first ambulatory leg, nearly .	.	.	.	7	14·5

The chelipedes, as in *Herbstia condyliata*, have the fingers in adult males arcuated and dentated near the apex, and with a wide intermarginal hiatus when closed.

Herbstia (Herbstiella) depressa? (Stimpson) (Pl. VII. fig. 2).

? *Herbstia depressa*, Stimpson, Ann. Lyc. Nat. Hist. New York, vol. vii. p. 185, 1860.
    ,,   ,, A. Milne Edwards, Crust. Miss. Sci. au Mexique, pt. 5, p. 77, 1875.
? *Herbstiella depressa*, Stimpson, tom. cit., vol. x. p. 93, 1871.

Three small specimens (one male and two females) dredged off the coast of Brazil at Barra Grande, in 30 to 350 fathoms, lat. 9° 5′ 0″ to 9° 10′ 0″ S., long. 34° 50′ 0″ W. to 34° 53′ 0″ W. (Station 122), are referred very doubtfully to this species.

They agree with the description in the structure of the antennæ, maxillipedes, chelipedes, and ambulatory legs, but the rostrum is divided nearly to its base, and the gastric and cardiac regions are very slightly convex, and are granulated (more distinctly in the male than in the females).

Adult ♂.				Lines.	Millims.
Length of carapace and rostrum, nearly .				4	8
Breadth of carapace,	.	.	.	3	6·5
Length of a chelipede, .	.	.	.	4½	9·5
Length of first ambulatory leg, .	.	.	.	6	12·5

I think it not improbable that a comparison of these specimens with authentic examples of Stimpson's species, which I have never seen, would demonstrate their specific distinctness.

Chlorinoides, Haswell.

Chlorinoides, Haswell, Proc. Linn. Soc. N.S.W., vol. iv. p. 442, 1880; Cat. Australian Stalk and Sessile-eyed Crust., p. 17, 1882.

Carapace subpyriform, moderately convex, and armed with strong spines on the dorsal surface. Spines of rostrum long and divergent. A well-developed præocular spine. Orbits deep, well defined above, and with two deep fissures in the upper margin; incomplete below. Post-abdomen in both sexes distinctly seven-jointed. Eyes, when retracted, concealed beneath the projecting superior orbital margin. Antennæ with the basal joint moderately enlarged, and armed (in the species I have examined) with a laterally projecting spine at the antero-external angle, and sometimes with another on the outer margin at the base ; the flagellum is shut off from the orbital cavity by the union of the basal joint with the front, and is partially visible from above at the sides of the rostrum. Merus of the exterior maxillipedes with the antero-external angle rounded and little produced ; the antero-internal angle, where the next joint articulates, slightly emarginate. Chelipedes (in the male) moderately developed, with the wrist carinated above and on the outer surface ; palm compressed and often carinated above ; fingers acute, denticulated on the inner margins, with a small interspace at base when closed. Ambulatory legs slender and of moderate length, or somewhat elongated ; merus-joints with one (or two) spines at the distal extremity ; dactyli nearly straight.

The genus thus characterised includes not only the type, *Chlorinoides tenuirostris*, Haswell, but also certain species which have been referred to *Paramithrax*, to *Chorinus*, and to *Acanthophrys*. It may indeed almost be regarded, as I have considered it elsewhere,[1] as a subgenus of *Paramithrax*, from which it is only distinguished by the well-developed præocular spine, the laterally projecting spine at the antero-external angle of the basal antennal joint, and the spinose merus-joints of the ambulatory legs. From *Acanthophrys*, A. Milne Edwards, to which it is also very nearly related, it is apparently distinguished, if *Acanthophrys cristimanus* be regarded as the type of *Acanthophrys*, by the emarginate antero-internal angle of the merus of the exterior maxillipedes, and by the spinose merus-joints of the ambulatory legs. If, however, *Acanthophrys aculeatus* be regarded as the type, *Chlorinoides* must be regarded as synonymous with *Acanthophrys*.

The following species are apparently referable to this genus ; all occurring in the Indian, Malaysian, Japanese, or Chinese Seas, and in rather shallow water (to 38 fathoms) except *Chlorinoides longispinus*, de Haan ; of this species I have described a variety, *bituberculatus*, from the Providence and Amirante group, in the Mascarene subregion (in 19 to 22 fathoms).

> *Chlorinoides tenuirostris*, Haswell (type). Torres Strait, Darnley Island.
> *Chlorinoides spatulifer* (*Paramithrax spatulifer*, Haswell). Port Stephens (5 fathoms).
> *Chlorinoides coppingeri* (*Paramithrax coppingeri*, Haswell). North Australia (12 to 17 fathoms) ; East Australia, Moreton Bay, Bass Strait (38 fathoms) ; Indian Ocean (A. Milne Edwards as *Acanthophrys aculeatus*). This species is possibly a variety of *Chlorinoides longispinus*, de Haan.
> ? *Chlorinoides filholi* (*Acanthophrys filholi*, A. Milne Edwards). Stewart Island. (The basal antennal joint and ambulatory legs are not described in this species.)

[1] Crust. Rep. Zool. Coll. H.M.S. "Alert," p. 192, 1884.

Chlorinoides longispinus (*Maia* (*Chorinus*) *longispina*, de Haan). Japanese Seas.

Chlorinoides longispinus, var. *bituberculatus* (Miers). Amirante and Providence groups (19 to 22 fathoms).

Chlorinoides halimoides (*Paramithrax halimoides* (White ined.) Miers). Oriental Seas.

Chlorinoides acanthonotus (*Chorinus acanthonotus*, Adams and White). Borneo.

Chlorinoides aculeatus (*Chorinus aculeatus*, Milne Edwards). Seas of Asia.

Chlorinoides aculeatus, var. *armatus* (Miers). North and North East Australia (3 to 11 fathoms).

Chlorinoides coppingeri (Haswell) (Pl. VII. fig. 3).

> *Paramithrax coppingeri*, Haswell, Proc. Linn. Soc. N.S.W., vol. iv. p. 443, pl. xxvi. fig. 1, 1880 ; Cat. Australian Stalk and Sessile-eyed Crust., p. 15, 1882.
>
> *Paramithrax* (*Chlorinoides*) *coppingeri*, Miers, Crust. Rep. Zool. Coll. H.M.S. "Alert," p. 192, 1884.
>
> ? *Acanthophrys aculeatus*, A. Milne Edwards, Ann. Soc. Entom. France, ser. 4, vol. v. p. 140, pl. iv. fig. 4, 1865.

Bass Strait, off East Moncœur Island, 38 fathoms. Station 162, lat. 39° 10′ 30″ S., long. 146° 37′ 0″ E. An adult male and female of small size.

In these specimens the transverse pair of spines of the cardiac region are confluent from their bases half way to their apices, which are divergent. The palms of the chelipedes (in the male) are smooth, slender, and not carinated ; the fingers, when closed, have between them but a very small interspace.

Adults ♂.					Lines.	Millims.
Length of carapace to base of rostrum,				.	6½	13·5
Length of rostrum, about	.	.	.	.	3	7
Breadth of carapace, nearly	.	.	.	.	5	10
Length of a chelipede,	.	.	.	.	8½	18
Length of first ambulatory leg,	.	.	.	.	10	21

To facilitate the proper identification of this species, and to illustrate the generic characters, I have thought it useful to figure the Challenger male.

Pisa, Leach.

> ?? *Arctopsis*, Lamarck, Syst. Anim. sans Vert., p. 155, 1801, description inaccurate.
>
> *Pisa*, Leach (subgenus?) Trans. Linn. Soc. Lond., vol. xi. p. 327, 1815.
>
> „ Milne Edwards (part), Hist. Nat. Crust., vol. i. p. 303, 1834.
>
> „ Miers, Journ. Linn. Soc. Lond. (Zool.), vol. xiv. p. 657, 1879.
>
> „ Brandt, Bull. Acad. Sci. Petersb., vol. xxvi. p. 410, 1880.

Carapace subpyriform or ovate-pyriform, moderately convex and armed with spines or tubercles. Præocular spine usually well developed. Spines of rostrum, long, simple parallel, or in contact to near their extremities, which are usually divergent. Orbits small, with a lateral aspect, with a hiatus above and below (sometimes with two hiatuses in the upper margin). Post-abdomen in both sexes distinctly seven-jointed. Eyes rather small, retractile, and when retracted, concealed within the orbits. Antennæ with the basal joint considerably enlarged and usually armed with a tooth or spine at the antero-external angle, the flagellum exposed and visible in a dorsal view at the sides of the rostrum. Exterior maxillipedes with the merus-joint distally truncated, the antero-external angle considerably produced and rounded or subacute, the antero-internal angle slightly notched. Chelipedes (in the adult male) well developed, with the palms either turgid, fingers arcuate and meeting only at the distal extremity, or more slender, with the fingers nearly straight. Ambulatory legs of moderate length (the first much the longest) with the joints subcylindrical, without spines ; the dactyli short and nearly straight.

In 1879[1] I proposed, in order to retain Leach's designation *Pisa* for some of the species with which it had been generally associated, to use this term as a *subgeneric* designation for those species in which the carapace is more ovate in shape and the palms of the chelipedes (in the adult male) turgid, with the fingers strongly arcuated and meeting only at the tips ; but I doubt whether the characters are sufficiently constant and well defined even for this purpose. Of the described species of this genus, three, viz., *Pisa tribulus* (Linn.) ; *Pisa hirticornis* (Herbst) = *Pisa corallina*, Risso, *Pisa quadricornis*, Brandt, var. ? ; and *Pisa tetraodon* (Pennant) = *Pisa convexa*, Brandt, inhabit the Mediterranean in water of moderate depth (70 to 75 mm., *Pisa tribulus*, A. Milne Edwards). Of these species, two, *Pisa tribulus* and *Pisa tetraodon*, range northward to the English and Irish coasts, and southward to the Azores and Tenerife (50 to 90 fathoms) ; *Pisa tribulus* to the Cape Verde Islands, 38 fathoms (Studer). There are also specimens of *Pisa tetraodon* and *Pisa corallina* from Aden in the British Museum collection. Another species, *Pisa carinimanus*, Miers, has been described from the Canaries and Goree Island, Senegambia (9 to 15 fathoms). A species, *Pisa brevicornis*, A. Milne Edwards, occurs at Madagascar ; and another, *Pisa acutifrons*, A. Milne Edwards, at Zanzibar. Three species, *Pisa antilocapra*, *Pisa prælonga*, Stimpson, and *Pisa erinacea*, A. Milne Edwards, occur in the Florida Straits or Caribbean Sea at depths of 37 to 118 fathoms (the last-mentioned species should perhaps be referred to the genus *Notolopas*).

[1] *Journ. Linn. Soc. Lond. (Zool.), vol. xiv. p. 657, 1879.*

Pisa (Arctopsis) tribulus (Linné).[1]

? *Cancer tribulus*, Linné, Syst. Nat., ed. 12, p. 1045, 1766.
 „ *longirostris*, Herbst, Naturg. Krabben and Krebse, i. p. 230, pl. xvi. fig. 92, 1782 ; var. *nec*, Fabricius.
? *Arctopsis lanata*, Lamarck, Syst. Anim. sans Vert., p. 155, 1801.
? *Maia rostrata*, Bosc, Hist. Nat. Crust., vol. i. p. 255, 1802 ; var.
 „ *armata*, Latreille, Hist. Nat. Crust., vol. vi. p. 98, 1803 ; var.
Cancer biaculeatus, Montagu, Trans. Linn. Soc. Lond., vol. xi. p. 2, pl. i. fig. 2, 1815.
Pisa gibbsii, Leach, Trans. Linn. Soc., vol. xi. p. 327, 1815 ; Malacostraca Podophthalmata Britanniæ, pl. xix. figs. 1–4.
 „ „ Milne Edwards, Hist. Nat. Crust., vol. i. p. 307, 1834.
 „ „ Bell, Brit. Crust., p. 27, 1853.
 „ „ Heller, Crust. südlich. Europa, p. 41, 1863.
 „ „ Brandt, Bull. Acad. Sci. Petersb., vol. xxvi. p. 410, 1880.
 „ *nodipes*, Leach, Zool. Miscell., vol. ii. p. 50, pl. lxxviii., 1815.
 „ „ Brandt, *tom. cit.*, p. 412, 1880.
Inachus muricus, Otto, Nova Acta Acad. Cæs. Leop. Carol, p. 334, pl. xx. figs. 2, 3, 1828.
Pisa armata, Milne Edwards, Hist. Nat. Crust., vol. i. p. 308, 1834 ; Atlas des Crust. in Cuvier, Règne Animal, ed. 3, pl. xxviii. fig. 1.
 „ „ Heller, Crust. südlich. Europa, p. 43, 1863 ; var.
 „ „ Brandt, Bull. Acad. Sci. Petersb., vol. xxvi. p. 410, fig., 1880.

Tenerife, 78 fathoms (a small male) ; off Fayal, 50 to 90 fathoms (a male).

The larger (but still small) male from Fayal measures as follows :—

Adult ♂.			Lines.	Millims.
Length of carapace and rostrum,	.	.	10½	22·5
Breadth of carapace, .	.	.	5	10·5[3]

Hyastenus, White.

Hyastenus, White, Proc. Zool. Soc. Lond., p. 56, 1847.
 „ Miers, Journ. Linn. Soc. Lond. (Zool.), vol. xvi. p. 658, 1879, and synonyms.

Carapace subpyriform, rounded behind, moderately convex, smooth, tuberculated or spinose on the dorsal surface. Præocular spine small or obsolete, spines of rostrum long

[1] I am in great uncertainty as to the proper designation (both generic and specific) of this common species. Linnæus' description of *Cancer tribulus* well applies to it except as regards the legs, which are described as filiform, and except that he mentions two (not one) posterior tubercles on the carapace. *Arctopsis lanata* of Lamarck has been referred to this form by A. White, and Lamarck's generic name has priority over the almost universally used *Pisa* of Leach, but Lamarck's description is so brief, vague, and obviously incorrect, that I do not think myself justified in using his name in preference to one about which there is no uncertainty, and which has been generally adopted. It may, however, as noted above, perhaps be used for a subgeneric designation.

Pisa armata (Latreille) is, I think, a mere variety. The distinctive characters attributed by Milne Edwards to this form are, however, much better marked in a specimen in the British Museum collection than in Milne Edward's figure in the Atlas of the Règne Animal and Hist. Nat. des Crustacés.

straight, slender, simple and divergent from the base. Post-abdomen (in the male) distinctly seven-jointed. Eyes completely retractile within the small orbits, which have a lateral aspect and a fissure or hiatus in the upper and lower margins. Basal antennal joint moderately enlarged, with or without a spine at the antero-external angle, the flagellum usually exposed and visible at the sides of the rostrum in a dorsal view, but sometimes partially concealed. Exterior maxillipedes with the merus-joints distally truncated; the antero-external angle slightly produced and rounded, the antero-internal angle emarginate. Chelipedes in the adult male with the palms either slender or moderately enlarged; fingers, when closed, with or without a slight intermarginal hiatus. Ambulatory legs moderately elongated (the anterior pair usually much the longest), with the joints subcylindrical and unarmed; the dactyli nearly straight.

The species, most of which have been enumerated by A. Milne Edwards,[1] are rather numerous, and occur commonly in the shallower waters of the Indo-Pacific region, to which this genus is apparently restricted. They may be conveniently arranged under the following sections, which are connected by insensible gradations.

 1. Carapace smooth and even above, with none or with a few long spines. Basal antennal joint usually without a spine at its antero-external angle. Chelipedes (in the male) with the palms small and slender. *Hyastenus diacanthus*, (de Haan); *Hyastenus aries* (Latreille); *Hyastenus spinosus* (A. Milne Edwards) (*Hyastenus verreauxii*, A. Milne Edwards, is probably a variety of *Hyastenus diacanthus*).

 2. Carapace with the dorsal surface uneven and tuberculated, but without long spines. Basal antennal joint usually with a spine or tubercle at its antero-external angle. Chelipedes in the male with the palms sometimes enlarged. (*Hyastenus*, White; *Chorilia*, Dana; *Lahaina*, Dana; *Lepidonaxia*, Targioni Tozzetti).

(The name *Hyastenus*, which was used for the first section in my revision of the group in 1879, cannot be retained in that sense, since in the type, *Hyastenus sebæ*, White, the carapace is somewhat tuberculated.)

Species:—*Hyastenus sebæ*, White; *Hyastenus planasius*, Adams and White; *Hyastenus pleione* (Herbst); *Hyastenus oryx*, A. Milne Edwards; *Hyastenus longipes* (Dana); *Hyastenus gracilirostris*, Miers; *Hyastenus japonicus*, Miers; *Hyastenus ovatus* (Dana); *Hyastenus sinope*; *Hyastenus elegans*, described below; *Hyastenus convexus*, Miers; (this species and *Hyastenus planasius* have a nearly smooth carapace, as the species of the first section). *Pisa fascicularis*, Krauss, from Natal, may be referable here, but the orbits are shown in the figure as entire above.

    [1] *Nouv. Archiv. Mus. Hist. Nat.*, vol. viii. p. 249, 1872.

As regards the distribution of the species, *Hyastenus oryx*, *Hyastenus spinosus*, and *Hyastenus ovatus* are, I believe, the only species which have as yet been recorded from the western division of the Oriental region, and *Hyastenus longipes* from its eastern limits (coast of Oregon).

Hyastenus elegans, Miers, and *Hyastenus japonicus*, Miers, are the only deep-water species as yet described; the former occurred, as mentioned below, near the Ki Islands in 140 fathoms (Station 192), the latter in the Japanese Seas, in 100 fathoms; the other species occur in the Chinese, Japanese, Indian, Malaysian, Australian, and Polynesian Seas, at depths not exceeding 50 fathoms.

Hyastenus diacanthus (de Haan).

Naxia diacantha, de Haan, Crust. in Siebold, Fauna Japonica, p. 96, pl. xxi. fig. 1, and pl. G, 1839.
Hyastenus diacanthus, A. Milne Edwards, Nouv. Archiv. Mus. Hist. Nat., vol. viii. p. 250, 1872.

Arafura Sea, in 49 fathoms, lat. 8° 56′ 0″ S., long. 136° 5′ 0″ E. (Station 190), a female in soft and very imperfect condition; a small and young female dredged at the Philippines in 18 fathoms, lat. 11° 37′ 0″ N., long. 123° 31′ 0″ E. (Station 208), with *Hyastenus oryx*, also probably belongs to this species; it has short and but slightly divergent rostral spines and the lateral branchial teeth are obsolete.

Young specimens of this species bear a very close resemblance to *Hyastenus planasius*, but may apparently be distinguished by the more hairy carapace, the generally longer and more divergent rostral spines, and the absence of the well-developed spine at the antero-external angle of the basal antennal joint, which exists in *Hyastenus planasius*.

The dimensions of the small female are—

Adult ♀.	Lines.	Millims.
Length of carapace and rostrum, rather over	5	11
Breadth of carapace,	2½	5·5
Length of a chelipede,	3½	7·5
Length of first ambulatory leg,	6	12·5

The rostral spines in this example are nearly as in *Hyastenus planasius*.

Hyastenus planasius (Adams and White).

Pisa planasia, Adams and White, Crust. Zool. H.M.S. "Samarang," p. 9, pl. xi. figs. 4, 5, 1848.
Hyastenus planasius, A. Milne Edwards, Nouv. Archiv. Mus. Hist. Nat., vol. viii. p. 250, 1872.

An adult male was dredged south of New Guinea in 28 fathoms, lat. 9° 59′ 0″ S., long. 139° 42′ 0″ E. (Station 188).

The carapace (as in the type) is nearly devoid of hair; the rostral spines are very slightly divergent; there is a well-developed spine at the antero-external angle of the basal antennal joint.

Adult ♂.	Lines.	Millims.
Length of carapace and rostrum, .	6	12·5
Breadth of carapace, rather over	3	7
Length of a chelipede, nearly .	5	10
Length of first ambulatory leg, .	8·5	18

Hyastenus oryx, A. Milne Edwards.

> *Hyastenus oryx*, A. Milne Edwards, Nouv. Archiv. Mus. Hist. Nat., vol. viii. p. 250, pl. xiv. fig. 1, 1872.
> „ „ Miers, Crust. Rep. Zool. Coll. H.M.S. "Alert," pp. 195, 522, 1884.

Philippine Islands, 18 fathoms; lat. 11° 37′ 0″ N., long. 123° 31′ 0″ E. (Station 208). Two adult but small females are referred to this variable species. As is usual in smaller examples of this species, the spines of the rostrum are considerably reduced in length; they are much shorter than in A. Milne Edwards' figure, which is an excellent delineation of the adult male.

The larger female has only the following dimensions :—

Adult ♀.	Lines.	Millims.
Length of carapace and rostrum, .	5½	11
Breadth of carapace, nearly	3	6

Hyastenus elegans, n. sp. (Pl. VI. fig. 3).

The body is covered with a close short pubescence, interspersed amid which are some longer hairs. The carapace is shaped nearly as in *Hyastenus oryx*, to which this species is nearly allied, and is armed with tubercles and spines which are much larger and more prominent than in *Hyastenus oryx*. The gastric region is tuberculated nearly as in *Hyastenus oryx*, three of the tubercles being disposed in a longitudinal median series; there is a prominent, conical tubercle on the cardiac region and a yet longer spine on the intestinal region; the hepatic, pterygostomian and branchial regions are tuberculated; there are about five small spines on each branchial region besides some smaller tubercles; the spines of the rostrum are very long (nearly as long as the carapace), slender, and but slightly divergent. The orbits, antennæ, and outer maxillipedes are nearly as in *Hyastenus* (*Chorilia*) *oryx*; the tooth at the antero-external angle of the basal antennal joint is prominent and spiniform; the chelipedes and ambulatory legs differ in nothing from the same limbs in *Hyastenus oryx*, except that there exists a small spine at the distal end of the merus-joints in all the legs, this being represented in *Hyastenus oryx* by a small blunt tooth.

Adult ♀.	Lines.	Millims.
Length of carapace to base of rostrum, nearly	8½	17·5
Length of rostrum, . . .	7½	15·5
Greatest breadth of carapace, nearly .	5½	11
Length of ambulatory leg of first pair, .	21	44

The unique female specimen in the collection was dredged near the Ki Islands in 140 fathoms, lat. 5° 49′ 15″ S., long. 132° 14′ 15″ E. (Station 192), and although of comparatively large size, is apparently not fully adult.

As this form in so many of its characters resembles *Hyastenus oryx*, I have not thought it necessary to do more than indicate the points in which it differs from that species, of which it is possible that a sufficient series would show it to be merely a well-marked deep-water variety; I cannot, however, venture to unite it with *Hyastenus oryx* on the authority of the series at present existing in the British Museum collections.

Naxia, Milne Edwards.

Naxia, Milne Edwards, Hist. Nat. Crust., vol. i. p. 313, 1834.
   „   Miers, Journ. Linn. Soc. Lond. (Zool.), vol. xiv. p. 658, 1879, et synonyma.

Carapace subpyriform, moderately convex, rounded behind, and armed with spines or tubercles on the dorsal surface. A præocular spine usually present, and when present well developed. Spines of the rostrum well developed, subcylindrical, parallel or divergent, and bearing on the inner margin, near to the extremities, a small accessory spine or spinule. Post-abdomen (in the male) distinctly seven-jointed (in the female some of the segments may be coalescent). Eyes small, retractile within the small orbits, which may have a single or a double hiatus in the superior margin, and a wider hiatus in the inferior margin. Antennæ with the basal joint enlarged, with a spine or tubercle at the antero-lateral angle, and sometimes with another on the exterior margin ; the flagellum exposed, or partially concealed in a dorsal view by the rostral spines. Merus of the exterior maxillipedes distally truncated, with the antero-external angle little, if at all, produced, and the antero-internal angle emarginate. Chelipedes (in the male) slender and moderately developed, palm usually somewhat elongated, fingers denticulated near the distal extremity, and having between them when closed a small hiatus at the base. Ambulatory legs slender and somewhat elongated, the first pair much the longest, with the joints subcylindrical ; dactyli nearly straight.

The following are, I believe, the only recorded species of this genus, which can only be distinguished from *Hyastenus* and *Pisa* by the accessory spinules of the rostrum :—

Nazia serpulifera, Milne Edwards, not uncommon on the Australian Coasts; *Nazia hirta*, A. Milne Edwards, from the East African Coast and Mascarene Islands; *Nazia robillardi*, Miers, from the Mauritius (30 fathoms); and the very distinct species described below as *Nazia hystrix*, which was dredged at Amboina in 100 fathoms.

Nazia hystrix, n. sp. (Pl. VI. fig. 4).

The body is everywhere clothed with very close set knobbed hairs, interspersed among which are longer setæ, which are most abundant upon the rostral spines; the legs, also, are rather thinly clothed with a very short pubescence, and with some longer hairs placed at wide intervals along the margins of the joints. The carapace is subpyriform, longer than broad, moderately convex, and covered with long spines, disposed as follows:—three in a transverse series, and one (median and posterior) upon the gastric region, one (long) upon the cardiac, and one long on the intestinal region, one on each hepatic region, three (besides some smaller tubercles) on each branchial region, also three small spines on each lateral margin above the bases of the ambulatory legs. There is a strong spine on the upper margin of the orbit near the front, which bears a small subbasal, posterior, accessory tubercle; behind this the upper orbital margin is widely and deeply emarginate, the notch with a median tooth or lobe; the postocular lobe is rounded, little prominent; the rostral spines are not half as long as the carapace, nearly straight, and divergent from their bases; they bear a small accessory tooth or spine on their inner margins, not far from the distal extremity; there is a spine at the antero-lateral angles of the buccal cavity, and a strong pterygostomian spine, also a strong spine on the sternum, between the bases of the chelipedes and first ambulatory legs; the segments of the post-abdomen (in the female) are spiniferous, except the small terminal segment. The eyes have short peduncles and large rounded terminal corneæ; the basal antennal joints are longer than broad, and have a strong spine or tooth on the outer margin near to the base, and another at the antero-external angle; the second and third joints are slender and visible from above at the sides of the rostral spines, the third rather longer than the second; the outer maxillipedes are without spines on their outer surface; the merus-joint is somewhat produced and rounded at its antero-external angle, and distally truncated. Chelipedes (in the female) very slender and slightly longer than the carapace and rostrum, the joints without spines, except a small spine at the distal extremity of the merus, which joint is nearly as long as the palm and fingers; the palm is not more thickened than the preceding joints, the fingers straight, acute, and closing along their inner margins. The ambulatory legs are very slender and decrease successively in length; the first pair extremely long; they are unarmed except for a small spine at the distal extremity of the merus-joints; the dactyli are slender, elongated, and nearly straight. Colour (in spirit) light yellowish or reddish-brown.

Adult ♀.	Lines.	Millims.
Length of carapace to base of rostrum, about .	7	14·5
Length of rostral spines,	2½	5·5
Greatest breadth of carapace, rather over	5	11
Length of a chelipede, nearly .	10	21
Length of ambulatory leg of first pair, over	29	62

The unique example (a female nearly adult) was obtained at the Moluccas, Amboina, in 100 fathoms.

Perhaps the nearest ally to this interesting species is the *Naxia* (*Naxioides*) *robillardi*, Miers,[1] described from a specimen dredged in 30 fathoms at the Mauritius; on account of the strongly divergent rostral spines this species may be regarded as intermediate between *Naxia* and the genus *Hyastenus*.

Naxia hirta (A. Milne Edwards).

? *Naxioides hirta*, A. Milne Edwards, Ann. Soc. Entom. France, ser. 4, vol. v. p. 143, pl. iv. fig. 1, 1865.

Podopisa petersii, Hilgendorf, Monatsber. d. k. preuss. Akad. d. Wiss. Berlin, p. 785, pl. i. fig. 5, 1878.

An adult male was dredged in 15 fathoms, at Samboangan. In the disposition of the spines and tubercles of the carapace, structure of the orbits, antennæ, and legs, it very closely resembles the type of *Podopisa petersii*, as figured by Hilgendorf,[2] from Mozambique; the chelipedes are, however, somewhat more slender. The rostrum, in the adult (which in Hilgendorf's specimen was broken), is of moderate length, with the spines at first subparallel, but slightly divergent from the accessory spines, which are placed at some distance from the distal extremity; the rostral spines are themselves not much more than one-third the length of the carapace :—

Adult ♂.	Lines.	Millims.
Length of carapace,	14½	31
Length of rostrum,	5½	11·5
Breadth of carapace,	11½	24·5
Length of a chelipede,	17	36·5
Length of first ambulatory leg,	30	64

[1] *Proc. Zool. Soc. Lond.*, p. 339, pl. xx. fig. 1, 1882.

[2] Dr. F. Hilgendorf has, according to Dr. E. von Martens (Crust. in Zool. Record, 1878, p. 17), himself recognised the identity of the genus *Podopisa* with *Naxioides*, which is not to be distinguished from *Naxia*, Milne Edwards. But it is possible, I think, that Hilgendorf's species may yet prove to be distinguishable from *Naxia hirta*. In all the specimens I have examined, the carapace (as in Dr. Hilgendorf's figure), is much more tuberculated than in *Naxia hirta*, as figured by A. Milne Edwards, the rostral spines are longer, and the posterior spine of the carapace shorter than in that figure.

Scyra, Dana.

Scyra, Dana, Amer. Journ. Sci. and Arts, ser. 2, vol. xi. p. 269, 1851; Crust. in U.S. Explor.
Exped., vol. xiii. (i), p. 80, 1852.
   „ Miers, Journ. Linn. Soc. Lond. (Zool.), vol. xiv. p. 663, 1879.

The carapace is somewhat depressed, tuberculated, but not spinose on the dorsal surface, and with well-developed lateral epibranchial and præocular spines. The rostral spines (in the two species I have examined), are vertically compressed and laterally dilated at base, acute at the distal extremity. The orbits are small, with a lateral aspect, with a narrow hiatus or a nearly closed fissure in the upper margin, and a wider hiatus below. The post-abdomen, in the male, is distinctly seven-jointed. The eyes are very small and retractile. The basal antennal joint is moderately dilated, with a very small spine or tooth at the antero-external angle; the two following joints are slightly dilated and compressed, and these, with the flagellum, are scarcely, if at all, concealed by the spines of the rostrum. The merus of the exterior maxillipedes is distally truncated, with the antero-internal angle not produced, and the antero-internal angle notched. The chelipedes (in the adult male) are well-developed, with the palm slightly compressed and carinated above, and the fingers acute, with scarcely any basal intermarginal hiatus. The ambulatory legs are of moderate length, the joints subcylindrical, without spines; the first pair does not much exceed the rest in length; the dactyli are short, acute.

The examination of a larger series of specimens has shown that the fissure in the upper margin of the orbit is sometimes open, and constitutes a distinct hiatus in this genus, which will, therefore, be better placed among the Maiinæ, near *Hyastenus*, than with the Pericerinæ, where I formerly classed it. In the form of the carapace, orbits, rostrum and antennæ it also presents obvious affinities with the genus *Hyas* and its allies.

Two well-marked species of this genus have been described, *Scyra acutifrons*, Dana, from the coasts of Oregon, California, British Columbia, and Vancouver Island,[1] and *Scyra compressipes*, Stimpson, from Japan (6 to 50 fathoms).

Scyra umbonata, Stimpson,[2] from the Gulf of Florida (143 fathoms), cannot be included in this genus, since it differs in the cylindrical spines of the rostrum and non-dilated mobile joints of the antennæ, and the merus of the exterior maxillipedes is not notched at the antero-internal angle. In the flattened protuberances of the carapace and in other characters this form somewhat resembles the Oriental *Oxypleurodon stimpsonii* (Miers).

[1] Specimens from the two last mentioned localities are in the collection of the British Museum.
[2] Bull. Mus. Comp. Zool., vol. ii. p. 115, 1870.

Seyra compressipes, Stimpson (Pl. VII. fig. 4).

Seyra compressipes, Stimpson, Proc. Acad. Nat. Sci. Philad., p. 218, 1857.

Japan, in lat. 34° 38' 0" N., long. 135° 1' 0" E. (in 50 fathoms), Station 233A (an adult male).

As this species is only known by Stimpson's short diagnosis, it may be useful to supplement the figure now given with the following description :—

The carapace is depressed, with the dorsal surface uneven and tuberculated; the gastric region large, rounded on the dorsal surface, and bearing one or two small tubercles; the depressions separating the gastric from the cardiac and branchial regions are wide and shallow; the branchial regions bear some small tubercles, and a small, slightly recurved, lateral spine. There is a spiniform tubercle on the sides of the hepatic regions; the præocular spine is well developed and acute. The rostrum is divided half way to its base by a triangular notch and the lobes thus formed are subtriangulate in shape, flattened and expanded at the base, and distally acute. The pterygostomian regions are tuberculated, the tubercles being disposed in an oblique series; the sternum is transversely sulcated between the segments of which it is composed. The segments of the post-abdomen are transverse except the last, which is subtriangulate and distally subacute. The eyes are very small. The basal joint of the antennæ has a small tubercle at the base of the exterior margin and a very small tooth at the antero-external angle. The ischium of the exterior maxillipedes is longitudinally concave on its exterior surface; the merus is nearly quadrate, with the antero-internal angle obliquely truncated, scarcely notched; exognath subacute. The chelipedes (in the adult male) are moderately developed, with the merus obscurely quadricarinated; the superior carina irregularly dentated and armed with a strong subterminal spine; carpus roughened externally, and carinated on its inner surface; palm smooth, compressed and carinated above; fingers minutely and regularly denticulated on the inner margins, and distally acute. The ambulatory legs are very slender and of moderate length, with the penultimate and anti-penultimate joints longitudinally sulcated; dactyli little, if at all shorter than the penultimate joints. The margins of the carapace and ambulatory legs are hairy, the setæ clavate; the gastric region of the carapace and rostrum also are pubescent. Colour (in spirit) light brown.

Adult ♂ .	Lines.	Millims.
Length of carapace and rostrum, nearly	11	23
Breadth of carapace,	7	14·5
Length of a chelipede,	13	27·5
Length of first ambulatory leg,	14½	31

Notolopas, Stimpson.

Notolopas, Stimpson, Ann. Lyc. Nat. Hist. New York, vol. x. p. 96, 1871.
„   Miers, Journ. Linn. Soc. Lond. (Zool.), vol. xiv. p. 657, 1879.

Carapace subpyriform, moderately convex, and rounded posteriorly; the posterior margin more or less distinctly carinated, the dorsal surface bearing a few spines; the orbits have a single hiatus, or a hiatus and notch in the superior margin, and a wider hiatus below, and bear a præocular spine or tooth. Rostrum well developed, with the spines coalescent at the base and afterwards divergent. Post-abdomen (in the male) distinctly seven-jointed. Eyes short, retractile. Antennæ with the basal joint considerably enlarged with a spine or tooth at the antero-external angle, and often another on the exterior margin; the following joints are slender and are not concealed by the rostral spines. The merus of the exterior maxillipedes is distally truncated, with the antero-external angle rounded, a little, if at all, produced, and the antero-internal angle very slightly emarginate. The chelipedes in the adult male are slender; palm somewhat elongated and slightly compressed; fingers with but a small intermarginal hiatus. The ambulatory legs are very slender, with the joints subcylindrical, the first pair considerably the longest; the dactyli slightly curved and nearly as long as the penultimate joints.

Species :—*Notolopas lamellatus*, Stimpson, from Panama and Manzanillo (depth not stated), and *Notolopas brasiliensis*, Miers, described below.[1]

Notolopas brasiliensis, n. sp. (Pl. VIII. fig. 1).

The carapace is subpyriform, considerably longer than broad; the carina of its posterior margin is less prominent than in *Notolopas lamellatus* and terminates below the branchial spines. As in that species, there are two tubercles and a spine upon the gastric region; the cardiac region is slightly convex, there is a strong spine upon each of the branchial regions and a median spine upon the crest which defines the posterior margin of the carapace. The rostrum is rather shorter than the carapace, the spines of which it is composed are at first coalescent, but divergent from a point situated a short distance above the base, and they are slender and straight. There is a strong, triangular, supraocular spine, and a blunt postocular lobe or tooth. On the pterygostomian regions near to the buccal cavity, is an oblique ridge, armed with three strong tubercles. The

---

[1] The genus *Rochinia*, noticed and figured but not described by Milne Edwards in his Études sur les Crustacés Podophthalmaires in the Mission Scientifique du Mexique, p. 86 (footnote) and pl. xviii. fig. 1, is evidently very nearly allied to *Notolopas*, and may be identical with this genus, if the type species, *Rochinia gracilipes*, A. Milne Edwards, has the posterior margin of the carapace distinctly carinated. This species differs, however, from the two species of *Notolopas* referred to above, in having but a single hiatus in the superior orbital margin, in the more numerous spines of the carapace, and the more robust palms of the chelipedes. It was taken at Cape Corrientes, and also near the mouth of the Rio Negro in 30 fathoms, and near the Patagonian coast in 44 fathoms.

eyes are small ; the basal antennal joint is somewhat dilated, and is armed on its outer margin with a laminiform tooth near the base and another at the distal extremity ; below the basal antennal joint there is another small tooth ; the antennal flagella are exposed at the sides of the rostrum and visible in a dorsal view. The epistoma is but slightly broader than long. The chelipedes are slender and somewhat elongated, merus, carpus, and palm subcylindrical, without spines or tubercles, the palm about as long as the merus, the fingers small, not half as long as the palm, and armed with small teeth on the inner margins. The ambulatory legs are slender, with the joints subcylindrical and unarmed, and they decrease successively in length ; the first pair are longer than the chelipedes ; the dactyli in all are but slightly curved. Colour (in spirit) yellowish-brown. The carapace and the rostrum are clothed with curled hairs, which are absent from parts of the dorsal surface, the inferior surface of the body and the limbs with a very short, close pubescence.

Adult ♂.	Lines.	Millims.
Length of carapace to base of rostrum, nearly	5	10
Breadth of carapace, about	3½	7·5
Length of a chelipede,	9	19
Length of first ambulatory leg,	11	23

Two males and a female were obtained at Bahia in shallow water (7 to 20 fathoms). The description is wholly taken from the adult male. In the smaller male and female the gastric spines are not developed and the branchial spines are quite small, the carina of the posterior margin of the carapace, also, is less prominent, and the chelipedes are smaller.

In an adult male of large size (see Pl. VIII. fig. 1, b) from the same locality and taken with the preceding specimens, the rostral spines are more strongly divergent, the carapace is somewhat more broadly pyriform and much more convex over the branchial regions, the spines of the gastric region are absent, those of the branchial regions very small ; the posterior carina of the carapace on either side of the median lobe is nearly obsolete, and the chelipedes are very considerably elongated. If, as is probable, this is a mere variety of this species above described, the generic character will require amendment in that which has been hitherto regarded as a most important particular, the prominence of the carina of the posterior margin of the carapace.

The dimensions of this specimen are as follows :—

Adult ♂.	Lines.	Millims.
Length of carapace to base of rostrum,	7½	16
Breadth of carapace,	6	12·5
Length of a chelipede, rather over	16	34
Length of first ambulatory leg, about	15	32

In a specimen, without any indication of locality, in the collection of the British Museum, which I suppose to belong to *Notolopas lamellatus*, Stimpson, not only is the carina of the posterior margin of the carapace well developed, but the rostrum is shorter than in *Notolopas brasiliensis* (about half as long as the first frontal portion of the carapace), and its spines are coalescent for nearly half their length and then strongly divergent (see Pl. VIII. fig. 1c).

Subfamily 2. SCHIZOPHRYSINÆ.

Schizophrysinæ, Miers, Journ. Linn. Soc. Lond. (Zool.), vol. xiv. p. 659, 1879.

Carapace very broadly triangular, or oval, or nearly circular. Rostrum very short or obsolete. Chelipedes (in the male) small, slender; the fingers usually excavated at the tips.

In the genera referred to this subfamily the carapace is broadly subtriangulate or nearly circular; the epistoma short, the basal antennal joint largely developed, and the chelipedes have the fingers more or less excavated.

Schizophrys, White.

Schizophrys, White, Ann. and Mag. Nat. Hist., ser. 2, vol. ii. p. 282, 1848.
,, Miers, Journ. Linn. Soc. Lond. (Zool.), vol. xiv. p. 660, 1879, et synonyma.

Carapace broadly subpyriform (nearly orbiculate) or narrower and more elongated, depressed or moderately convex, and armed with a series of lateral marginal spines, and with tubercles, which tend to become spinuliferous, upon the dorsal surface. Orbits large, with fissures or notches both in the upper and lower margins, and with no præocular spine. Spines of rostrum short and armed with one or more accessory spines upon their outer margins. Post-abdomen (in the male) distinctly seven-jointed. Eyes rather large, retractile. Antennæ with the basal joint moderately enlarged and armed with a spine both at the antero-external and antero-internal angles, the mobile joints and flagellum not concealed by the rostral spines. Exterior maxillipedes with the merus joints distally truncated, the antero-external angle rounded and not much produced, the antero-internal angle emarginate. Chelipedes (in the adult male) with the merus and carpus spinuliferous; palm smooth, elongated and somewhat compressed; fingers excavated or rarely acute at the distal extremity, and with an intermarginal hiatus when closed. Ambulatory legs moderately elongated, with the joints subcylindrical; dactyli shorter than the penultimate joints.

The following species are referable to this genus: *Schizophrys dichotoma*, Milne Edwards, from the Mediterranean (Balearic Islands);[1] the very variable *Schizophrys aspera* (Milne Edwards), common throughout the shallower waters of the Indo-Pacific

[1] I have seen no specimens of this species, and am not aware that its occurrence in any other locality has been recorded.

region, from the Red Sea and Mauritius to the Fijis and New Caledonia ; *Schizophrys dama* (Herbst), to which species are doubtfully referred specimens from West Australia (3 to 5 fathoms), in the collection of the British Museum, and which is apparently distinguished by the evenly granulated and narrower and more elongated carapace.[1]

Schizophrys aspera (Milne Edwards).

> *Mithrax asper*, Milne Edwards, Hist. Nat. Crust., vol. i. p. 320, 1834.
>   ,,   ,,   Dana, Crust. in U.S. Explor. Exped., vol. xiii. p. 97, pl. ii. fig. 4, 1852.
> *Dione affinis*, de Haan, Fauna Japonica, Crust., p. 94, pl. xxii. fig. 4 (*dichotomus*), 1839, adult ♂ .
> ?! *Mithrax quadridentatus*, MacLeay, in Smith, Annulosa Zool. South Africa, p. 58, 1849.
> *Mithrax spinifrons*, A. Milne Edwards, Ann. Soc. Entom. France, ser. 4, vol. vii. p. 263, 1867.
> *Schizophrys serratus*, White, Proc. Zool. Soc. Lond., p. 223, 1847, woodcut ; Crust. Zool. H.M.S. "Samarang," p. 16, 1848.
> *Schizophrys spinifer*, White, Proc. Zool. Soc. Lond., p. 223, 1847 ; Crust. Zool. H.M.S. "Samarang," p. 17, 1848.
> *Mithrax affinis*, F. de B. Capello, Jorn. Sci. Math. Phys. Nat. Lisboa, p. 264, pl. iiia. fig. 4, 1871.
> *Schizophrys aspera*, A. Milne Edwards, Nouv. Archiv. Mus. Hist. Nat., vol. viii. p. 231, pl. x. figs. 1–1f, 1872.
>   ,,   ,,   Miers, Crust. Rep. Zool. Coll. H.M.S. " Alert " p. 197, 1884.
> *Mithrax (Schizophrys) triangularis*, Kossmann, Crust. in Zool. Ergebnisse einer Reise in Küstengebiete des Rothen Meeres, pp. 11, 13, 1877.
>   ,,   ,,   ,,   var. *africana*, Kossmann, tom. cit., pp. 11, 14, 1877.
>   ,,   ,,   ,,   var. *indica*, Kossmann, tom. cit., pp. 11, 14, 1877.

An adult male was obtained in the Torres Straits, August 1874.

It belongs to what is regarded as the typical condition of this very variable species, having but a single accessory spinule on each of the cornea of the rostrum (Crust. H.M.S. " Alert," p. 197).

Adult ♂ .		Lines.	Millims.
Length of carapace and rostrum,		16½	35·5
Breadth of carapace, rather over	.	11	23·5
Length of a chelipede,	.	21·3	46
Length of first ambulatory leg,	.	18¼	39·5

This example, as the measurements show (although adult) is not of very large size. The chelipedes are not so greatly developed as in very large males, and the dactyli are much less strongly arcuated.

Subfamily 3. MICIPPINÆ.

> *Micippinæ*, Miers, Journ. Linn. Soc. Lond. (Zool.), vol. xiv. p. 660, 1879.

Carapace somewhat oblong-oval or sub-oblong. Rostrum vertically or nearly vertically deflexed ; usually broad, lamellate. Chelipedes with the fingers acute at the

[1] Herbst suspected the habitat of his type might be American. In the small specimens in the British Museum collection the fingers are distally acute, and scarcely, if at all, excavated.

distal extremity. Basal antennal joint very much enlarged. Eye-peduncles very long, geniculated and laterally projecting.

The carapace, in the genera referred to this subfamily, is commonly somewhat constricted behind the orbits. I have already, in the memoir above cited, referred to the curious modification in the structure of the rostrum.

Pseudomicippa, Heller.

Pseudomicippe, Heller, Sitzungsb. math.-nat. Cl. k. Akad. Wiss. Wien, vol. xliii. p. 301, 1861.

„   Miers, Journ. Linn. Soc. Lond. (Zool.), vol. xiv. p. 661, 1879.

? *Microhalimus*, Haswell, Proc. Linn. Soc. N.S.W., vol. iv. p. 435, 1880.

The carapace is elongate-pyriform, moderately convex and tuberculated on the dorsal surface, somewhat constricted at the hepatic regions; the orbits are very incomplete, with a hiatus in the upper margin, scarcely defined below; the praeocular spine sometimes developed, sometimes absent; the rostral spines are divergent from, or nearly from, the base, and are horizontal or deflexed. The eyes are slender and somewhat elongated; the basal antennal joint is moderately enlarged, and has usually a small tooth at its antero-external angle; the flagellum is not concealed by the rostral spines. The merus of the exterior maxillipedes is distally truncated or slightly concave; its antero-external angle rounded, and its antero-internal angle emarginate. The chelipedes (in the male) are moderately developed; palm compressed, and fingers acute, with or without an intermarginal hiatus. The ambulatory legs are slender and moderately elongated; dactyli nearly straight, acute, and usually shorter than the penultimate joints.

The following are the described species of this genus :—

Pseudomicippe nodosa, Heller, from the Red Sea;[1] *Pseudomicippe tenuipes*, A. Milne Edwards, probably from the Indian Ocean; *Pseudomicippe varians*, Miers, rather widely distributed in the shallower waters of Australia, with which, as has elsewhere been noted, *Microhalimus deflexifrons*, Haswell, may possibly be identical.

Pseudomicippa (?) varians, Miers.

Pseudomicippe (?) *varians*, Miers, Ann. and Mag. Nat. Hist., ser. 5, vol. iv. p. 12, pl. iv. fig. 8, 1879; Crust. Rep. Zool. Coll. H.M.S. "Alert," p. 197, 1884.

A nearly adult female was obtained in the Torres Straits in 8 fathoms, lat. 10° 30′ 0″ S., long. 142° 18′ 0″ E. (Station 186).

The spines of the rostrum in this specimen are very short, little more than one-fifth

---

[1] *Maia rôsellii*, Audouin, in Savigny, Descr. de l'Egypte, Crust. Atlas, pl. vi. fig. 1, may be an earlier designation for this form, but presents certain distinctions (*cf.*, Heller, *tom. cit.*, p. 304).

the length of the carapace, and strongly deflexed; the gastric region is very distinctly tuberculated, and there is a distinct spine at the antero-external angle of the basal antennal joint.

	♀.	Lines.	Millims.
Length of carapace and rostrum, nearly		9½	19·5
Breadth of carapace, ...		5½	11·5

Micippa, Leach.

Micippa, Leach, Zool. Miscell., vol. iii. p. 15, 1817.
,,   Milne Edwards, Hist. Nat. Crust., vol. i. p. 329, 1834.
,,   Miers, Journ. Linn. Soc. Lond., *tom. cit.*, p. 661, 1879 ; Ann. and Mag. Nat. Hist., ser. 5, vol. xv. p. 3, 1885.

Carapace nearly oblong, depressed, and rounded behind, with the dorsal surface spinose, granulated or tuberculated, and often with lateral marginal spinules or spines. Interorbital space broad; the orbits deep, with two fissures or notches in the superior and (usually) in the inferior margins, which are sometimes very incomplete. Rostrum broad, lamellate, and vertically or nearly vertically deflexed, more or less distinctly bilobated, and sometimes armed with lateral marginal spines. Eyes (in the species I have examined) moderately elongated and capable of being retracted within the orbital cavity. Antennæ with the basal joints usually very much enlarged, and sometimes armed with one or two small distal spines or tubercles; it occupies the space between the base of the rostrum and the orbit, and generally constitutes a part of the inferior wall of the orbit; the following joint is usually compressed and sometimes slightly dilated, and is not concealed by the rostrum. The merus of the exterior maxillipedes is distally truncated, with the antero-external angle more or less rounded, and the antero-internal angle emarginate. The chelipedes (in the male) are moderately developed or short; palm somewhat compressed or subcylindrical; fingers acute, meeting along the inner margins, or with a large intermarginal hiatus when closed. Ambulatory legs moderately elongated, with the joints subcylindrical, and the dactyli nearly straight, little shorter than the penultimate joints.

The species, so far as is known at present, are restricted to the shallower waters of the Indo-Pacific or Oriental region.

The following are the species referable to this genus :—

1. Rostrum with eight marginal spines :—
   Micippa cristata (Linn.).   Indo-Malaysian subregion.
2. Rostrum terminating in four spines :—
   Micippa mascarenica, Kossmann.   Oriental region.
   Micippa philyra (Herbst).   Oriental region.

3. Rostrum terminating in two lobes which may be distally acuminated or tuberculated :—

> *Micippa spinosa*, Stimpson. South and East Australia ; New Zealand, to 38
> fathoms (var. *affinis*).
> *Micippa curtispina*, Haswell. North and North-East Australia.

4. Rostrum terminating in two long, narrow, acute lobes or spines :—
> *Micippa thalia* (Herbst). Oriental region.

None of the species have, I believe, been recorded from deep water.

Micippa spinosa, Stimpson (Pl. VIII. fig. 2).

Micippa spinosa, Stimpson, Proc. Acad. Nat. Sci. Philad., p. 217, 1857.
   ,,   ,,   Haswell, Cat. Australian Crust., p. 26, 1882.
   ,,   ,,   Miers, Ann. and Mag. Nat. Hist., ser. 5, vol. xv. p. 8, 1885.
Paramicippa spinosa, Miers, Cat. New Zealand Crust., p. 9, 1876.
   ,,   ,,   Crust. Rep. Zool. Coll. H.M.S. "Alert," p. 199, 1884.

South Australian coast, 2 to 10 fathoms, April, 1874 (an adult male) ; Port Jackson,
6 fathoms ; on the Sow and Pigs Bank (an adult male and three females) ; Port Jackson,
8 to 15 fathoms (three females).

The adult male from 2 to 10 fathoms measures :—

		Lines.	Millims.
Adult ♂ .			
Length of carapace and (deflexed) rostrum, rather over	.	9	19·5
Breadth of carapace, .	.	7½	15·5

Micippa spinosa, var. *affinis*, Miers (Pl. VIII. fig. 3).

Paramicippa affinis, Miers, Ann. and Mag. Nat. Hist., ser. 5, vol. iv. p. 13, 1879.
Micippa spinosa, var. *affinis*, Miers, op. cit., p. 9, 1885.

In this well-marked variety the carapace is suboblong, depressed, deeply concave on
the hepatic regions, with the dorsal surface somewhat uneven and closely granulated ; on
the gastric regions are usually two somewhat larger granules, placed one behind the
other and followed by one on the cardiac region ; the lateral margins bear some larger
tubercles which tend to become short spines ; of these, three or four are on the sides of
the hepatic regions, and one or two on the sides of the branchial regions ; the fissures of
the upper orbital margins and the postocular spine are small ; the inferior margin of the
orbit is deeply concave ; the front is obliquely (nearly vertically) deflexed, subspatulate,
with the lateral margins parallel, slightly indented at the base, the antero-lateral angles
rounded, not toothed, the distal extremity with a very small triangular notch. The
basal antennal joint is nearly smooth, and is without spines at its distal extremity, and

the next (mobile) joint is not dilated; the merus of the exterior maxillipedes is broadly rounded at its antero-external angle; the chelipedes (in the adult male) are moderately developed; merus and arm granulated, but without spines; palm rather short and moderately enlarged, granulated; fingers smooth, acute, and, when closed, meeting only at and near the tips; the ambulatory legs are densely hairy. An adult male has the following dimensions :—

	Lines.	Millims.
Adult ♂.		
Length of carapace to base of rostrum,	6	12·5
Breadth of carapace, .	5½	11·5
Length of a chelipede, about .	7	15
Length of first ambulatory leg, about .	7½	16

Habitat.—Bass Strait; a female was found among the fishes of H.M.S. Challenger's collection, and two males are in the present collection from East Moncœur Island; lat. 39° 10′ 30″ S., long. 146° 37′ 0″ E. (Station 162).

This variety is distinguished by the absence of spines from the dorsal surface of the carapace and by the form of the front, which has subparallel sides, and is much less deeply notched at the distal extremity.

Family III. PERICERIDÆ.

Periceridæ, Miers, Journ. Linn. Soc. Lond. (Zool.), vol. xiv. pp. 640–662, 1879.

Eyes retractile within the small circular and well-defined orbits, which are not incomplete as in the Maiidæ. Basal antennal joint well developed, and constituting a great part of the inferior wall of the orbit; this joint is usually very considerably enlarged.

Subfamily 1. PERICERINÆ.

Pericerinæ, Miers, Journ. Linn. Soc. Lond. (Zool.), vol. xiv. p. 662, 1879.

Carapace more or less subtriangulate in shape. Rostrum well developed. Second joint of antennæ not dilated. Chelipedes with the fingers acute at tips.

In this subfamily are included what may be regarded as the typical Periceridæ, in which the rostral spines are well developed and often in contact one with another. The interorbital space is usually broad, and the orbits tubular; the basal joint of the antennæ is usually much enlarged, the epistoma short, the legs of moderate length, and the fingers acute at the tips.

The following genera are to be added to this subfamily :—*Entomonyx,* Miers, *Picroceroides,* n. gen., and probably *Sisyphus,* Desbonne and Schramm.

Libinia, Leach.

Libénia, Leach, Zool. Miscell., vol. ii. p. 129, 1815.
„   Milne Edwards, Hist. Nat. Crust., vol. i. p. 298, 1834.
„   Miers, Journ. Linn. Soc. Lond. (Zool.), vol. xiv. p. 662, 1879.

The carapace is convex, broadly pyriform, or nearly orbiculate, and its dorsal surface is covered with numerous tubercles, which are sometimes developed as spines ; the præocular tooth or spine is usually distinct. The rostrum is well developed (sometimes rather small), emarginate or bifid at the apex, and so terminating in two teeth or spines. The orbits are small, nearly circular, and well defined, with sometimes an open fissure or hiatus in the superior and inferior margins. The post-abdomen (in the male) is distinctly seven-jointed. The eyes are small and completely retractile. The basal antennal joint is moderately enlarged, and is armed with a tooth or spine at the antero-external angle, behind which, on the exterior margin, is usually another tooth. The flagellum is not concealed by the rostrum. The merus of the exterior maxillipedes is truncated (not toothed) at the distal extremity, and its antero-internal angle is emarginate. The chelipedes (in the adult males) are well developed ; palm elongated, subcylindrical or compressed, but not enlarged ; fingers minutely and evenly denticulated on the inner margins, with a small intermarginal hiatus at base. The ambulatory legs are well developed, sometimes elongated, with the joints subcylindrical, unarmed ; dactyli nearly straight.

This genus is closely allied on the one hand to *Doclea* and on the other to *Libidoclea*, Milne Edwards and Lucas. I have proposed elsewhere [1] to restrict the latter genus to the single species *Libidoclea granaria*, Milne Edwards and Lucas, characterised by having the merus-joint of the exterior maxillipedes armed with a strong tooth on its distal margin, a character never observed in any species either of *Libinia* or *Doclea*.

The following species are referable to the genus *Libinia* as characterised above :—

1. Fissures of the upper and lower orbital margins very narrow or closed ; basal antennal joint with a tubercle or very short spine at its antero-external angle.

  Libinia emarginata, Leach (= *Libinia canaliculata*, Say, and *Libinia affinis*, Randall). East Coast of United States ; West Indies ; West Coast of North America.

  Libinia dubia, Milne Edwards (= *Libinia distincta*, Guér.-Ménév. and B. Capello ; *Libinia inflata*, Streets, var. ?). East Coast of United States ; West Indies ; West Coast of Africa.

  Libinia rhomboidea, Streets ; "East Indies." (The locality is so given by Streets whose description nearly applies to *Libinia dubia*, of which this doubtful species is considered a variety by A. Milne Edwards.)

[1] *Proc. Zool. Soc. Lond.*, p. 28, 1879.

Libinia subspinosa, Streets. Chili.

Libinia rostrata, Bell. Peru.

? *Libinia setosa*, Lockington (= *Libinia semizonale*, Streets, *fide* Lockington). Lower California. (This species may belong to the other section of the genus. I have never seen Street's description of *Libinia semizonale*, which is identified with *Libinia setosa* by Lockington in a MS. note on the margin of my copy of his paper, and I am ignorant of the locality of his types.)

2. Orbits with an open fissure in the superior margin, and a wider hiatus in the inferior margin. Basal antennal joint with a well-developed spine at the antero-external angle :—

Libinia spinosa, Milne Edwards. Brazil, Patagonia, Chili (Gay).

Libinia brasiliensis, Heller (*Libidoclea*). Brazil, Rio de Janeiro.

Libinia gibbosa, A. Milne Edwards. Brazil, Desterro. (The form of the orbits and basal antennal joint is not mentioned by the author.)

Libinia coccinea, Dana (*Libidoclea*). East Coast of Patagonia (30 fathoms).

Libinia gracilipes, n. sp. Coast of Chiloe (45 fathoms).

Libinia smithii, n. sp. Coast of Chiloe (245 fathoms).[1]

Libinia smithii, n. sp. (Pl. IX. fig. 1).

The carapace is subpyriform, rather longer than broad, moderately convex, and is covered with small tubercles and with very long spines, which are disposed as follows :— Four in a longitudinal and median series, of which two are upon the gastric, one upon the cardiac, and one upon the intestinal region, close to the posterior margin of the carapace ; of these four, the cardiac and intestinal spines are longest ; there are also a strong præocular spine, a spine on each hepatic region, and three on each branchial region ; of the branchial spines, the lateral one is extremely long, about half the length of the carapace. The small tubercles of the dorsal surface are situated mostly upon the gastric and branchial regions, and are not very numerous ; there is also a series of small tubercles on the sides of the branchial regions, above the bases of the ambulatory legs, and an oblique series on the pterygostomian regions. The rostrum is about half the length of the carapace, and is composed of two slender spines, which are coalescent for about half their length, and thereafter slightly divergent. The epistoma is short, transverse. The sternum has on each side four obscure transverse ridges ; the post-abdomen has all of its

[1] The *Libinia erpaxa*, A. Milne Edwards, 1878 (= *Doclea orientalis*, Miers, 1879) from the mouth of the Amoor River and Japanese Seas, is doubtfully referred by Milne Edwards to this genus. I have seen no males, but on account of the smallness of the basal antennal joint and other characters, I think it should be retained in *Doclea*. A. Milne Edwards has described a second species, *Libinia bidentata*, from the mouth of the Amoor, which seems to be sufficiently distinguished from *Libinia erpaxa* by the different tuberculation of the carapace.

segments distinct, the last triangulate. The orbits are open above and below, but the postocular lobe is broad and well developed; the eyes are short. The basal antennal joint is rather slender, and is armed with two spines, one on the outer margin and one at the distal and outer angle; the slender flagellum is visible from above at the sides of the rostrum. The ischium of the outer maxillipedes is longitudinally concave on its outer surface; the merus is distally truncated, but strongly notched at its antero-internal angle, with its antero-external angle rounded; the exognath does not quite attain the outer and distal angle of the merus, and is armed with a small tooth on its inner margin near to the distal extremity, which is subacute. The chelipedes (in the male) are rather small and feeble, scarcely exceeding the carapace and rostrum in length, with the joints unarmed; the merus is subcylindrical; the palm very slightly compressed, but not carinated, and shorter than the merus; fingers nearly as long as the palm and meeting along their inner edges, which are obscurely denticulated. The ambulatory legs are slender, with the joints unarmed, and they decrease successively in length; the first pair is more than three times the length of the carapace (without the rostrum), the fifth pair is wanting in the single specimen examined. The body and limbs are everywhere finely and closely granulated and slightly pubescent. Colour (in spirit) light yellowish-brown.

Adult ♂.			Lines.	Millims.
Length of carapace, nearly	.	.	12½	26
Breadth of carapace, .	.	.	12	25
Length of rostrum, .	.	.	6½	13·5
Length of a chelipede,	.	.	21	44·5
Length of first ambulatory leg,	.	.	38½	82

The unique example in the collection (an adult male) was dredged off the coast of Chiloe, in 245 fathoms, in lat. 52° 45′ 30″ S., long. 73° 46′ 0″ W. (Station 311).

This fine species is distinguished from all others of the genus with which I am acquainted by the great development of the spines of the carapace (which, including the præocular spines, are only fourteen in number) and by the greatly elongated and deeply-forked rostrum.

Libinia gracilipes, n. sp. (Pl. IX. fig. 2).

The carapace (in the adult) is moderately convex, but little longer than broad, and is covered with spines and tubercles; in the median line are five spines, of which three (small) are on the gastric region, one (longer) on the cardiac, and one (rather long) on the intestinal region; on either side of the first and second of the median gastric spines there is a tubercle; there is a spine on each hepatic region, and three or four, interspersed with numerous smaller tubercles, on each branchial region; two or three spines or tubercles exist on the sides of the branchial regions, above the bases of the ambulatory legs, and there is an oblique series of small tubercles on each pterygostomian region. The præocular spine is well developed. The rostrum (in the adult) is about one-fourth the length of the carapace, and is cleft through half its length, with the spines slightly divergent. The orbits are widely open above and deeply fissured below; the inferior fissure rather narrower than in the preceding species. The sternum and post-abdomen are nearly as in *Libinia smithii*. The basal antennal joint, as in that species, bears two spines, one on the outer margin, and one at the antero-external angle; the outer maxillipedes also are nearly as in *Libinia smithii*. The chelipedes (in the adult) are moderately elongated, rather less than two and a half times the length of the body, with the joints closely granulated, but not tuberculated; the merus-joint is subcylindrical; the carpus has two obscure ridges on its outer surface; the palm is rather shorter than the merus, slightly compressed, but not carinated; the fingers more than half the length of the palm and denticulated on their inner margins. The ambulatory legs decrease successively in length from the first to the last, the first are more than three times the length of the carapace, the joints in all are subcylindrical and granulated, the dactyli slightly arcuated and shorter than the penultimate joints. Colour (in spirit) yellowish-brown; the carapace is rather thinly pubescent.

Adult ♂.	Lines.	Millims.
Length and breadth of carapace, about	17	36
Length of rostrum, about	5	11
Length of a chelipede,	39½	83·5
Length of first ambulatory leg,	53	112·5

An adult and two smaller males and three small females were dredged off the coast of Chiloe in 45 fathoms, in lat. 46° 53′ 15″ S., long. 75° 12′ 0″ W. (Station 304).

In the younger specimens the tubercles of the carapace are less numerous; the cardiac and intestinal spine, and one (the lateral) branchial spine are always strongly developed.

Perhaps the nearest ally to this species is *Libinia coccinea*, Dana,[1] from the East Patagonian coast, 30 fathoms, from which the Challenger species is distinguished by the much more deeply-cleft rostrum, the existence of well-developed spines on the cardiac region and posterior margin of the carapace, and the longer lateral branchial spines; characters which are constant both in the young and adult. The typical species, *Libinia granaria*, Milne Edwards and Lucas, is distinguished by the peculiar tooth of the distal margin of the merus-joint of the exterior maxillipedes.

Libinia spinosa, Milne Edwards, which has been recorded from Brazil, Patagonia, and the coast of Chili, is apparently distinguished by the much more numerous spines of the carapace and the spinose post-abdomen of the male.

[1] *Libinia coccinea*, Dana, U.S. Explor. Exped., vol. xiii., Crust. i. p. 88, pl. i. fig. 3, 1853.

Pericera, Latreille.

Pericera, Latreille, in Cuvier, Règne Animal, ed. 2, vol. iv. p. 58, 1829.
     „   Milne Edwards (pt.), Hist. Nat. Crust., vol. i. p. 334, 1834.
     „   A. Milne Edwards (pt.), Crust. in Miss. Sci. au Méxique, vol. v. p. 49, 1873.
     „   Miers, Journ. Linn. Soc. Lond. (Zool.), vol. xiv. p. 664, 1879.

The carapace is subpyriform, rather convex, with the dorsal surface uneven, tuberculated or spinose; the lateral margins armed with a series of long spines; the præocular spine is well developed. The rostrum is composed of two well-developed spines, which are not deflexed, and are divergent from the base; the orbits are small and tubular, but do not project as in *Macrocœloma*. The post-abdomen (in the male) is distinctly seven-jointed. The eyes are small, retractile within the orbits. The basal antennal joint is very considerably enlarged, and is armed with one or two small distal spines or tubercles, which are not visible in a dorsal view; the flagellum is sometimes concealed by the rostral spines, sometimes laterally exposed at the sides of the rostrum. The merus of the exterior maxillipedes is distally truncated, with the antero-internal angle emarginate and the antero-external angle rounded or subacute. The chelipedes (in the adult male) are well developed; palm elongated and subcylindrical or somewhat compressed, but not dilated or enlarged; fingers without any or with but a small intermarginal hiatus at base when closed; ambulatory legs moderately elongated, with the joints subcylindrical, without spines; dactyli nearly straight.

The following species are referable to the genus as thus restricted :—

1. Carapace without spines on the dorsal surface :—
    Pericera cornuta (Herbst) (= *Chorinus armatus*, Randall). East Coast of
        United States; West Indies; Brazil; Cape Colony.
    Pericera calata, A. Milne Edwards. Idolos Islands; Havana, to 175 fathoms
        (perhaps a variety of *Pericera cornuta*).

2. Carapace with dorsal spines :—
    Pericera spinosissima, Saussure. West Indies.
    Pericera ovata, Bell. Galapagos Islands.

Pericera cornuta, Milne Edwards.

Pericera cornuta, Milne Edwards, Hist. Nat. Crust., vol. i. p. 335, 1834; Atlas du Règne
        Animal de Cuvier, Crust., ed. 3, pl. xxx. fig. 1.
     „   „   A. Milne Edwards, Crust. in Miss. Sci. au Méxique, pt. 5, vol. i. p. 51, 1873.
Chorinus armatus, Randall, Journ. Acad. Nat. Sci. Philad., vol. viii. p. 108, 1839.

An adult male is in the collection, labelled as from "Simon's Bay, 10–20 faths." This example is of rather small size, as the following measurements show; it differs in nothing from West Indian examples of the species in the collection of the British Museum.

Adult ♂.			Lines.	Millims.
Length of carapace and rostrum,	.	.	35½	75
Breadth of carapace, about	.	.	17½	37·5
Length of a chelipede, nearly	.	.	35	74
Length of first ambulatory leg,	.	.	34	72

Picroceroides, n. gen.

The carapace is narrow and rounded behind, it is constricted behind the orbits, which are tubular and project laterally as in *Macrocœloma*; the width at the orbits about equals the greatest width of the carapace at the branchial regions. The orbits have a long præocular and a short postocular spine, and are emarginate above and below. The rostral spines are slender, divergent, and widely separated at the base. The post-abdomen is seven-jointed in both sexes, and is transversely ridged in the male; the ridges correspond to similar elevations on the sternum, they are rounded and separated by deep depressions. The epistoma is transverse. The antennulary fossettes are small, deep and well defined; the antennæ have the basal joint (as in *Pericera* and *Macrocœloma*) very considerably enlarged, and armed on its exterior surface with a keel or crest placed immediately below the next joint, but the spine of the distal margin of this joint is obsolete and represented merely by a small tubercle; the following joints of the antennæ are slender, and the first mobile joint is partly concealed by the rostral spine. The exterior maxillipedes are small, with the merus-joints distally truncated, and with the antero-external angles rounded and the antero-internal angles emarginate. The chelipedes are moderately elongated and rather slender, with the palms slightly compressed, and more than twice as long as broad; dactyli with an intermarginal hiatus at base. The ambulatory legs are very slender and of moderate length, with the joints subcylindrical, without spines; dactyli nearly straight.

This genus is intermediate in position between *Pericera* and *Macrocœloma*; from the former it is distinguished by the absence of lateral marginal spines of the carapace and the great lateral projection of the orbits, from the latter by the form and development of the rostral and orbital spines and by the absence of the distal spines of the basal antennal joints.

It might with almost equal propriety be regarded as a subgenus of one or the other of these genera.

Picroceroides tubularis, n. sp. (Pl. X. fig. 1).

The carapace is moderately convex, much longer than broad, but little dilated at the branchial regions. The interfrontal space is concave, the gastric regions somewhat elevated and obscurely tuberculated; the cardiac region bears a rounded prominence and the intestinal region (usually) a small spine; the cervical and branchio-cardiac sutures are continuous, and form a longitudinal sinus, separating the branchial from the gastric and cardiac regions. The tubular orbits project laterally to a remarkable degree, and each bears a very long præocular, and a small postocular spine, and has two notches in the inferior and one in the superior margin. The rostral spines (in the adult male) exceed half the length of the carapace; they are very slender, slightly curved, laterally divergent, and widely remote from one another at the base. The sternum (in the male) has three deep transverse fossæ on each side of the post-abdomen, whose segments, also, are narrow and transversely ridged. The antennulary fossettes are very small; the basal joint of the antennæ, which is very greatly dilated, and coalescent with the surrounding parts of the carapace, is without spines, but has a small nearly longitudinal crest near its distal extremity below the following joint; there is also a small tubercle on the distal margin which forms part of the wall of the tubular orbit. The ischium of the outer maxillipedes is longitudinally concave on its outer surface; the merus is distally truncated, very distinctly notched at its antero-internal angle. The chelipedes are slender and somewhat elongated (in the adult male); the merus-joint subcylindrical and unevenly granulated; carpus also granulated; palm somewhat compressed but not carinated, it is granulated on its upper margin, and more distinctly on the inner than on the outer surface, which has a slight longitudinal depression near to the upper margin; the fingers are somewhat curved and meet on their distal halves, the mobile finger has a strong tubercle on its inner margin near to the base. The ambulatory legs are extremely slender, and decrease successively in length; their dactyli are slightly arcuate.

The carapace and limbs (except the chelipedes) are rather thinly pubescent, the margins of the rostrum and of the free peduncular joints of the antennæ have some longer hairs, some of which are clavate. Colour (in spirit) yellowish-brown; tips of the fingers dark brown or black.

Adult ♂.	Lines.	Millims.
Length of carapace, about	7½	16
Breadth at branchial regions,	5	11
Length of rostral spine,	5	11
Length of chelipede,	14	30
Length of first ambulatory leg,	13½	29

Fernando Noronha (shallow water), two males and an adult female; Bahia (shallow water), an adult male and female.

Macrocœloma, Miers.

Macrocœloma, Miers, Journ. Linn. Soc. Lond. (Zool.), vol. xiv. p. 665, 1879.
Pericera (part) auctorum.

The carapace is rather convex, usually broadest at the branchial regions, with the orbits tubular and laterally projecting; the interorbital space broad; the dorsal surface unarmed or tuberculated, or with a few long spines; the margins without a series of elongated lateral spines as in Pericera, but often with a strongly developed lateral epibranchial spine, preceded sometimes by some smaller spines; the præocular spine or tooth is small. The spines of the rostrum are well developed. The post-abdomen in the male is distinctly seven-jointed (in the species I have examined). The eyes are usually completely retractile within the small tubular orbits. The antennæ have the basal joint considerably enlarged and armed with a spine on the distal margin, which is usually visible in a dorsal view, and sometimes with another placed immediately beneath the first (mobile) joint; the mobile part of the antenna is sometimes concealed in a dorsal view by the rostral spines, sometimes visible at their sides; the merus of the exterior maxillipedes is distally truncated, with the antero-internal angle emarginate. The palms of the chelipedes are either somewhat elongated and compressed or shorter and more dilated; fingers with or without an intermarginal hiatus. The ambulatory legs are rather short and moderately robust; dactyli slightly curved.

The following species are referable to this genus:—

1. Lateral epibranchial spines well developed:—
   Macrocœloma trispinosa (Latreille) = Pericera nodipes, Desbonne and Schramm.
     West Indies; Bermuda; Bahia.
   Macrocœloma diacantha (A. Milne Edwards). Majores (to 12 fathoms).
   Macrocœloma camptocera (Stimpson). Florida Straits.
   Macrocœloma heptacantha (Bell). Central America, Puerto Portrero.
   Macrocœloma septemspinosa (Stimpson). West of the Tortugas (to 36 fathoms); Bahia.
   Macrocœloma villosa (Bell) = Pericera fossata, Stimpson. California, Cape St. Lucas; Ecuador, Gulf of Guyaquil.
   Macrocœloma subparallela (Stimps.) = Pericera vilpini, Desbonne and Schramm. St. Thomas; Guadaloupe.
   Macrocœloma diplacantha (Stimpson). St. Thomas.
   Macrocœloma trigona (Dana). Fiji Islands.
   Macrocœloma concava, n. sp. Bahia; Fernando Noronha (7 to 20 fathoms).

2. Lateral epibranchial spines small or obsolete :—

Macrocæloma eutheca (Stimpson) (as figured by A. Milne Edwards). Tortugas ; Santa Cruz, &c. ; Florida Strait, 12 to 115 fathoms.

Macrocæloma lævigata (Stimpson)=*Pericera curvicorna*, Desbonne and Schramm. St. Thomas ; Guadaloupe.

Macrocæloma trispinosa (Latreille).

† *Pisa trispinosa*, Latreille, Encycl. Méth. Hist. Nat., vol. x. p. 142, 1825.
Pericera trispinosa, Milne Edwards, Hist. Nat. Crust., vol. i. p. 336, 1834.
„ „ Guérin, Icon. des. Crust., pl. viii. fig. 3, 1829–44.
„ „ A. Milne Edwards, Crust. in Miss. Sci. au Méxique, pt. 5, p. 52, pl. xv. fig. 2, 1873 ; et synonyma.

An adult and a smaller female is in the collection from Bermuda, and four small specimens (male and three females) from Bahia; collected in shallow water.

I have examined, in the collection of the British Museum, a considerable series of specimens of this reputedly rare species, of different ages and of both sexes. It exhibits considerable variation as regards the form and direction of the rostral spines, and the development of the prominences of the carapace. In a large adult male from the Tortugas, length of carapace and rostrum 25½ lines (54 millims.), the gastric region is very lofty and convex, and the rostral spines straight, but somewhat deflexed, sloping downward in a plane with the anterior dorsal surface of the gastric region ; in smaller individuals the rostral spines are not deflexed, and vary considerably in length, being more than one-half or less than one-third the length of the carapace ; generally they are parallel at the base and very slightly divergent towards the tips, but in one specimen in the Challenger series the divergence is much greater, as in *Macrocæloma diacantha* (A. Milne Edwards), and *Macrocæloma camptocera* (Stimpson), which are perhaps merely varieties of *Macrocæloma trispinosa*.

Macrocæloma septemspinosa (Stimpson).

Pericera septemspinosa, Stimpson, Bull. Mus. Comp. Zoöl., vol. ii. p. 113, 1870.
„ „ A. Milne Edwards, Crust. in Miss. Sci. au Méxique, pp. 59, 200, pl. xv. A. fig. 2, 1873.

Bahia, shallow water, an adult female.

Adult ♀.	Lines.	Millims.
Length of carapace and rostrum,	18½	41
Breadth of carapace, about .	15½	32·5
Length of a chelipede, .	16	34
Length of first ambulatory leg,	14	30

In all of its characters this species is nearly allied to *Macrocæloma heptacantha*[1] from Puerto Portrero, but it may, perhaps, be distinguished, if the figure of *Macrocæloma heptacantha* is to be trusted, by the form of the rostral spines, which in *Macrocæloma heptacantha* are straight and separated by a considerable interspace, whereas in *Macrocæloma septemspinosa* they coalesce at the base and curve laterally away from one another, so that the tips are strongly divergent. It is not a little curious that two species so nearly allied should have been described under names which are practically identical.

Macrocæloma concava, n. sp. (Pl. X. fig. 2).

Carapace longer than broad, moderately convex, and deeply concave upon the hepatic regions; its dorsal surface is armed with about ten spines and tubercles, disposed as follows; three small spines placed in a triangle upon the gastric region, of which the posterior and median one is the largest; behind the median gastric spine are two small spines on the cardiac region and one on the intestinal region close to the posterior margin; there is also a tubercle on the dorsal surface of each branchial region and a rather strong lateral spine. The lateral margins of the carapace are tuberculated, the tubercles being continued in an oblique series over the pterygostomian regions, nearly to the antero-external angle of the buccal cavity. The spines of the rostrum are short, in the adult less than one-fourth the length of the carapace, they are nearly straight, divergent, and separated by a triangulate interspace, but not widely remote at the base as in *Macrocæloma subparallela* (Stimpson). The tubular orbits are laterally much elongated, and bear small spines or tubercles, placed one in front of and one behind the eye. The very largely developed basal antennal joint about equals the base of the rostrum in width, and bears four tubercles or small spines, of which one is placed at the distal extremity, and three on the outer margin. The merus of the outer maxillipedes is somewhat produced at its antero-external angle; the exognath is rather shorter than the endognath. The chelipedes, in the male, are rather slender, but somewhat elongated, exceeding the carapace and rostrum in length; the merus-joint about equals the carpus, and is unevenly granulated and tuberculated on the margins. The carpus has a small tubercle on its inner margin; the palm is slightly compressed, but not dilated nor carinated, and is granulated on its inner surface; the fingers are about half as long as the palm, and bent downwards at an angle with it, they are curved, meeting only toward the tips, and are minutely denticulated on their inner margins; the following legs are short, decreasing successively in length, with the joints subcylindrical, the dactyli terminating in a sharp claw. The body and limbs are covered with a short close

[1] *Pericera heptacantha*, Bell, *Proc. Zool. Soc. Lond.*, p. 173, 1835; *Trans. Zool. Soc. Lond.*, vol. ii. p. 61, pl. xii. fig. 6, 1841.

pubescence and with some longer curled hairs.   Colour (in spirit) yellowish-brown ; tips
of the fingers of the chelipedes, black.

The largest male presents the following dimensions :—

Adult ♂.					Lines.	Millims.
Length of carapace, nearly	.	.	.	.	13	27
Length of rostrum, about	.	.	.	.	2½	5
Breadth of carapace, about	.	.	.	.	11	23
Length of a chelipede, about	.	.	.	.	18½	39
Length of first ambulatory leg,			.		14	30

Bahia, shallow water (an adult female), Fernandho Noronha, 7 to 20 fathoms.   An
adult, smaller and young male, and two small females.

The nearest ally to this species is the *Macrocœloma eutheca* (Stimpson),[1] which
Macrocœloma concava resembles in the great development of the orbits, but in *Macrocœloma concava* the carapace at the branchial regions is much broader, the spines and
tubercles of its dorsal surface more numerous, and the lateral epibranchial spine is
longer.   The spines of the rostrum are described by Stimpson as parallel in *Pericera
eutheca*.   I have observed no specimens in the Challenger series resembling *Pericera
eutheca* in the characters mentioned, but it is possible a larger series would show the
two species to be identical.

Microphrys, Milne Edwards.

Microphrys, Milne Edwards, Ann. d. Sci. Nat. Zool., ser. 3, vol. xvi. p. 251, 1851.
    „   A. Milne Edwards, Crust. in Miss. Sci. au Mexique, p. 59, 1875.
    „   Miers, Journ. Linn. Soc. Lond. (Zool.), vol. xiv. p. 664, 1879, and synonyms.

Carapace broadly subpyriform, and somewhat depressed, with the dorsal surface
uneven and tuberculated, with a small lateral epibranchial spine ; præocular spine
sometimes not developed.   Orbits small, circular, with closed fissures.   Spines of rostrum
slender and more or less divergent.   Post-abdomen, in the male, distinctly seven-
jointed (in the species examined).   Eyes small.   Basal antennal joint considerably
dilated and armed with a long spine at the antero-external angle which is visible in a
dorsal view ; the mobile joints and the flagella are not concealed by the rostral spines.
The merus of the exterior maxillipedes is distally truncated, with the antero-external
angle somewhat produced and rounded, and the antero-internal angle emarginate.   The
chelipedes are moderately developed, with the palm compressed and more or less
enlarged ; fingers with a wide intermarginal hiatus.   The ambulatory legs are rather
short, with the merus and carpus joints sometimes armed with spines ; the dactyli
slightly curved.

[1] *Pericera eutheca*, Stimpson, Bull. Mus. Comp. Zool., vol. ii. p. 112, 1870 ; A. Milne Edwards, Crust. in Miss. Sci.
au Mexique, pt. 5, p. 58, pl. xvs. fig. 1, 1875.

The species are confined to the American coasts and islands adjacent.

The following have been described. I am not, however, convinced of the distinctness of all of them.

Microphrys bicornutus (Latreille) = *Pisa galibica*, Desbonne and Schramm ; *Pisa purpurea*, Desbonne and Schramm ; *Omalacantha hirsuta*, Streets. Bermudas ; coasts and Islands of the Caribbean Sea and Gulf of Mexico (to 37 fathoms) ; Brazil.

Microphrys weddellii, Milne Edwards. Peru ; West Indies.

Microphrys aculeatus (Bell). Galapagos Islands.

Microphrys platisoma (Stimpson) = *Pisoides calatus*, Lockington. California.

Microphrys tennidus (Lockington), *Pisoides tennidus*, Lockington. California.

Microphrys depressa (Lockington) = *Fisheria depressa*, Lockington. California.

Microphrys error, Kingsley, *Microphrys depressa*, Kingsley, nec Lockington. California.

Microphrys bicornutus (Latreille).

Pisa bicornata, Latreille, Encycl. Méth. Hist. Nat., vol. x. p. 141, 1825.

Microphrys bicornatus, A. Milne Edwards, Nouv. Archiv. Mus. Hist. Nat., vol. viii. p. 247, 1872; Crust. in Miss. Sci. au Méxique, p. 61, pl. xiv. figs. 2, 3, 4, 1875, et synonyma.

? *Omalacantha hirsuta*, Streets, Proc. Acad. Nat. Sci. Philad., p. 238, 1871.

„ „ A. Milne Edwards, Crust. in Miss. Sci. au Méxique, *tom. cit.*, p. 65, 1875.

Two adults and a young male of this very variable species were obtained at Bermuda, in shallow water.

In the smaller example the carapace is relatively narrower, the tubercles of its dorsal surface less numerous and the lateral branchial spines are not developed.

The following are the dimensions of the specimens examined :

	Lines.	Millims.
Adult ♂.		
Length of carapace and rostrum,	17	36
Breadth of carapace, about, .	12	25
Length of a chelipede,	21½	45·5
Length of first ambulatory leg,	12⅜	26·5
Young ♂.		
Length of carapace and rostrum, about	6	13
Breadth of carapace, nearly, .	4	8
Length of a chelipede,	5	10·5
Length of first ambulatory leg, about	6	13

Subfamily 2. MITHRACINÆ.[1]

Mithracinæ, Miers, Journ. Linn. Soc. Lond. (Zool.), vol. xiv. p. 666, 1879.

Carapace broadly triangulate or ovate-triangulate, sometimes transverse, with the sides slightly arcuate, interorbital space narrow. Rostrum short or obsolete. Second joint of antennæ not dilated. Chelipedes with the fingers excavated at the tips.

In this subfamily are included those Periceridæ which most nearly approach the Cancridæ in the form of the carapace, the obsolescence of the rostral spines, the small and completely defined orbits, the short epistoma, and the form and development of the chelipedes and ambulatory legs.

Mithrax, Leach.

Mithrax (Leach), Latreille, in Cuvier, Règne Animal, ed. 1, p. 23, 1817.
,, Milne Edwards (part), Hist. Nat. Crust., vol. i. pp. 317, 320, 1834.
,, Miers, Journ. Linn. Soc. Lond. (Zool.), vol. xiv. p. 667, 1879.

The carapace is depressed or moderately convex, sometimes longer than broad, but usually transverse and very broadly rounded at the branchial regions; dorsal surface uneven, tuberculated or spinose. The spines of the rostrum are usually very short, tuberculiform or even obsolete, but rarely well developed and acute. The orbits are small, well defined, armed with tubercles or short spines, or, rarely, entire. The epistoma is transverse, the post-abdomen in the male is distinctly seven-jointed. The eyes are small. The basal antennal joint is armed with a spine or tubercle at its distal extremity, followed usually by one or two on the exterior margin, and is very much enlarged; the following joints are not dilated, and the flagellum is short. The merus-joint of the exterior maxillipedes is usually truncated at the distal extremity, and emarginate at the antero-internal angle, and with the antero-external angle sometimes somewhat produced. The chelipedes, in the adult male, are well developed, and sometimes large and massive, with the palm dilated and compressed, fingers, when closed, with a wide intermarginal hiatus; more rarely they are more slender, with the fingers nearly meeting when closed. The ambulatory legs are of moderate length, sometimes spinose; dactyli nearly straight or slightly curved.

The species are rather numerous, and are, I believe, confined to the coasts of America and islands adjacent; they sometimes inhabit considerable depths.

They have already been enumerated and described by A. Milne Edwards in his fine work above cited. In the list which follows a few are added which have since been described, and a somewhat different arrangement is adopted. The genus *Nemausa*, A. Milne Edwards, is here regarded as a subgenus of *Mithrax*.

[1] The subfamily Othoniinæ, including the two genera *Othonia*, Bell, and *Cyclocœloma*, Miers, is not represented in the Challenger collection.

Subgenus *Nemausa*, A. Milne Edwards.

Nemausa, A. Milne Edwards (genus), Crust. in Miss. Sci. au Méxique, p. 80, 1875.
,,   Miers, *tom. cit.*, p. 666.

Carapace subpyriform, much longer than broad; spines of rostrum well developed, slender, acute. Epistoma scarcely transverse. Basal antennal joint with a long spine at its antero-external angle. Merus of the exterior maxillipedes produced at its antero-external angle. Chelipedes of moderate size; palm rather slender, compressed.

M. A. Milne Edwards refers the following species to this genus, or subgenus as I prefer to regard it :—

Nemausa rostrata, A. Milne Edwards. Gulf of Mexico and Caribbean Sea (to 163 fathoms); Bermuda. (The name *rostrata* has been used by Bell (*vide infra*) for a true *Mithrax*).

Nemausa spinipes (Bell). Galapagos Islands; (Cape?) St. Elena, to 16 fathoms.

Mithrax (Nemausa) rostrata.

Nemausa rostrata, A. Milne Edwards, Crust. in Miss. Sci. au Méxique, pt. 5, vol. i. p. 81, pl. xvii. fig. 4, 1875.

Bermuda, shallow water (a male and female).

In these specimens the spines of the rostrum are somewhat shorter and the tubercles of the carapace are less symmetrically disposed than in A. Milne Edwards' figure; in the smaller specimen there is but one spine upon the sides of the branchial regions, the others being represented by rounded tubercles.

♂.	Lines.	Millims.
Length of carapace and rostrum,	10½	22
Breadth of carapace,	8	17
Chelipedes (deficient).		
Length of first ambulatory leg,	13	27·5

Subgenus *Mithrax*.

Mithrax and *Mithraculus*, Miers, *tom. cit.*, p. 667, 1879, et synonyma.

Carapace very broad and depressed, usually transverse; spines of rostrum very short or obsolete. Basal antennal joint very much enlarged, with short spines at the distal extremity. Chelipedes often large, with the palm compressed and more or less dilated.

1. Carapace with the branchial regions not dorsally sulcated, the lateral margin usually armed with spines.

(1) Spines of rostrum distinct, acute. Palms of chelipedes with or without spines on the superior margin.

    a. Palms of chelipedes in the adult with spines on the upper margin.

        Mithrax spinosissimus (Lamarck). Florida, West Indies (to 100 fathoms).

        Mithrax cornutus (Saussure). West Indies; Florida Straits (to 589 fathoms); Bahia. (Specimens taken by Captain E. Cole, 46 miles south of Key-West, Florida, are in the British Museum Collection).

        Mithrax aculeatus (Herbst). West Indies; Vera Cruz.[1]

    b. Palms of chelipedes without spines.

        Mithrax acuticornis, Stimpson. Near the Quicksands and Tortugas, to 42 fathoms (perhaps, as noted by A. Milne Edwards, the young of *Mithrax cornutus*).

        Mithrax holderi, Stimpson. Tortugas.

        Mithrax armatus, Saussure. West coast of America; Mazatlan.

        Mithrax rostratus, Bell. ?

        Mithrax ursus, Bell (= *Mithrax belli*, Gerstæcker, *nec Cancer ursus*, Herbst). Galapagos Islands; Chili (Brit. Mus.).

(2) Spines of rostrum short, tuberculiform or obsolete. Chelipedes with the palms unarmed.

        Mithrax hispidus (Herbst) = *Mithrax pleuracanthus*, Stimpson, var. South Carolina; Florida Straits; West Indies (to 125 fathoms); Yucatan Channel, near the Jolbos Islands; Brazil (30 to 350 fathoms).

        Mithrax lævimanus, Desbonne and Schramm. Guadeloupe.

        Mithrax tuberculatus, Stimpson. California; Mazatlan.

        Mithrax depressus, A. Milne Edwards. Guadeloupe; Woman Key.

        Mithrax leucomelas, Desbonne and Schramm. Guadeloupe.

        Mithrax verrucosus, Milne Edwards. West Indies.

        Mithrax pygmæus, Bell. Panama (10 fathoms).

        Mithrax triangulatus, Lockington. Gulf of California.

2. Carapace very much depressed, with smooth, shallow interspaces or sulci between the tubercles of the dorsal surface of the branchial regions; the lateral margins tuber-

---

[1] *Mithrax trispinosus*, Kingsley (*Proc. Boston Soc. Nat. Hist.*, vol. xx. p. 148, 1879), from Florida, has three accessory spines on each spine of the rostrum. It is not stated whether the palms of the chelipedes are spinuliferous; they are referred to as "naked." I have observed more or less distinct indications of accessory rostral spinules in *Mithrax spinosissimus* and *Mithrax aculeatus*.

culated (the chelipedes considerably enlarged); *Mithraculus*, White, ? *Teleophrys*, Stimpson.

The following species have been referred to this section or subgenus :—

Mithrax coronatus (Herbst). Central America; West Indies (to 30 fathoms); Brazil.

Mithrax sculptus, Lamarck (= *Mithrax minutus*, Saussure). West Indies; Fernando Noronha (7 to 20 fathoms).

Mithrax nodosus, Bell. Galapagos Islands; Chili (Brit. Mus.).

Mithrax denticulatus, Bell. Galapagos Island; California; Cape St. Lucas; Guayaquil (Brit. Mus.).

Mithrax forceps, A. Milne Edwards. Guiana; West Indies; Bermuda; Bahia; Fernando Noronha ?

Mithrax aculeus, A. Milne Edwards. Guadeloupe.

Mithrax rubei, A. Milne Edwards. Cuba.

Mithrax cinctimanus, Stimpson (= *Mithrax affinis*, Desbonne and Schramm). Gulf of Mexico; Florida Straits; West Indies (to 37 fathoms).

Mithrax areolatus, Lockington. Gulf of California.

Mithrax hirsutipes, Kingsley. Key-West.

? *Mithrax cristulipes* (Stimpson). California; Cape St. Lucas; Bay of Panama.

Mithrax cornutus, de Saussure.

Mithrax cornutus, de Saussure, Rev. et Mag. Zool., ser. 2, vol. ix. p. 501, 1857; Mém. Soc. Phys. d. Genève, vol. xiv. pt. 2, p. 423, 1858.

"   "   A. Milne Edwards, Crust. in Miss. Sci. au Méxique, pt. 5, vol. i. p. 97, pl. xxii., 1875.

Four males of rather small size were obtained in shallow water at Bahia. I subjoin the dimensions of the largest and smallest specimens :—

a. (♂ adult, but not fully grown). This specimen nearly resembles A. Milne Edwards' fine delineation of the large adult male, but the spines of the carapace and limbs are less prominent; there are two, not four teeth on the upper orbital margins, and the tubercles of the upper margin of the palms of the chelipedes do not extend quite so far toward its distal extremity.

♂.			Lines.	Millims.
Length of carapace to base of rostrum, nearly	.	.	13	27
Length of a rostral spine, nearly	.	.	2	4
Greatest breadth of a carapace, nearly	.	.	12	25
Length of a chelipede,	.	.	20½	43·5

b. (♂ young). In this example the palms of the chelipedes are slender, smooth, nearly devoid of tubercles on their upper margins :—

						Lines.	Millims.
Length of carapace to base of rostrum, about	.	.	.	.	.	7½	16
Length of a rostral spine, about	.	.	.	.	.	1½	3
Greatest breadth of carapace, about	.	.	.	.	.	6½	14
Length of a chelipede, .	.	.	.	.	.	9	19

In a specimen of somewhat larger size the chelipedes are slender, and there are a few small tubercles on the upper margin of the palm near to the base.[1]

Mithrax hispidus, var. pleuracanthus, Stimpson.

Cancer hispidus, Herbst, Naturgesch. der Krabben u. Krebse, Heft 8, p. 247, pl. xviii. fig. 100, 1790.

Mithrax hispidus, Milne Edwards, Guérin, Mag. de Zool. (Cl. vii.), 1832–38 ; Hist. Nat. Crust., vol. i. p. 322, 1834.

" " A. Milne Edwards, Crust. in Miss. Sci. au Méxique, pt. 5, p. 93, pl. xxi. fig. 1, 1875.

" pleuracanthus, Stimpson, Bull. Mus. Comp. Zoöl, vol. ii. p. 116, 1870.

" " A. Milne Edwards, Crust. in Miss. Sci. au Méxique, p. 95, pl. xx. fig. 3, 1875.

Brazil, South of Pernambuco, in 30 to 350 fathoms, lat. 9° 5′ 0″ S. to 9° 10′ 0″ S., long. 34° 49′ 0″ W. to 34° 53′ 0″ W. (Stations 122 to 122c). Two small males.

In the larger of these specimens the carapace measures only about 6 lines (12·5 mm.) in length and breadth, but in all its characters resembles large adults, except in having the tubercles of the dorsal surface of the carapace more distinct ; in particular, I may note that the accessory spinule of the second and third antero-lateral marginal teeth is very distinctly developed. In the smallest specimen (length of carapace little over 3 lines or 7 mm.) the accessory spinule is discernible only on the third antero-lateral tooth.

In another small male specimen from the same locality and depth (length nearly 4 lines (8 mm.) the carapace is slightly narrower, more distinctly sulcated, and the teeth of the antero-lateral margins are all of them simple ; this may belong to a distinct species, allied to Mithrax forceps, from fully grown examples of which it is only distinguished by the narrower carapace, or it may even be a variety of that species.

Mithrax forceps (A. Milne Edwards).

Mithraculus forceps, A. Milne Edwards, Crust. in Miss. Sci. au Méxique, pt. 5, vol. i. p. 109, pl. xxiii. fig. 1, 1875.

An adult female was obtained at Bermuda on the shore, and another at Bahia in shallow water.

[1] By A. Milne Edwards Mithrax acuticornis, Stimpson, is regarded as the young of this species. Stimpson's diagnosis, however, scarcely suffices to determine this point ; specimens which have been referred to Mithrax acuticornis in the collection of the British Museum, are distinguished, not merely by the absence of spines from the palms of the chelipedes, but also by the non-spiniliferous wrists. Mithrax cornutus is nearly allied to Mithrax spinosissimus, Saussure, in the spiniferous palms of the chelipedes and in the distinctly developed spines of the rostrum.

These specimens agree with Milne Edwards' description and figure in all particulars, except in having the ridges of the branchial regions somewhat more prominent and defined by deeper intervening depressions; there are obscure indications of a transverse series of five tubercles upon the front of the gastric region in the smaller specimen (that from Bahia) which do not exist in the larger specimen and in the figure of Milne Edwards.

The dimensions of the specimen from Bermuda are as follows :—

Adult ♀.						Lines.	Millims.
Length of carapace and rostrum,						6½	14
Breadth of carapace, nearly						7½	15·5
Length of a chelipede,						9½	20
Length of first ambulatory leg,						9½	20

Several small and not nearly fully grown *Mithraculi* are in the Challenger collection from Fernando Noronha, 7 to 20 fathoms, which may belong to this or to a distinct species; they differ in having the dorsal surface of the carapace much less distinctly tuberculated, and in possessing more slender ambulatory legs. In the largest the carapace is not 3 lines in length (6 mm.).

Mithrax coronatus (Herbst).

Cancer coronatus, Herbst, Naturgesch. der Krabben u. Krebse, vol. i. p. 184, pl. xi. fig. 63, 1790.
Cancer corypho, Herbst, tom. cit., p. 8, pl. iii. 1801.
Mithraculus coronatus, White (part), List Crust. Brit. Mus., p. 7, 1847.
 „ „ A. Milne Edwards, Crust. in Miss. Sci. au Mexique, pt. 5, p. 106, pl. xx. fig. 1, 1875, and references to literature.

A very small male and female collected at Fernando Noronha, in 7 to 20 fathoms, are referred to this species, as distinguished and figured by A. Milne Edwards.

♂ -				Lines.	Millims.
Length of carapace, nearly				3	6
Breadth, nearly				3½	7
Length of a chelipede,				3½	7·5
Length of first ambulatory leg, nearly				4	8

Mithrax sp. (Pl. X. fig. 3).

To render the account of the Challenger specimens of this genus complete, I subjoin a description of a specimen which I cannot refer to any species of the genus, but to which I do not apply any specific name, as it is of extremely small size, and the characters are probably not those of the fully grown individual.

It resembles the *Mithraculus hirsutipes*, Kingsley,[1] from Key West, Florida, in the four-toothed antero-lateral margins of the carapace, the strongly curved dactyle of the ambulatory legs, and in some other characters; but the ambulatory legs are almost naked, not strongly hirsute as in *Mithrax hirsutipes*, the tubercles of the dorsal surface of the carapace are more numerous, and the carpus of the chelipedes tuberculated.

The carapace is about as long as broad, its dorsal surface nearly smooth and covered with scattered tubercles which are disposed as follows; two immediately behind the base of the rostrum, two on the frontal region of the carapace between the orbits, four in a transverse series on the gastric region, one in the centre of the cardiac region, two on the posterior margin of the carapace (placed one on either side of the median line) and several on each branchial region, the largest of which is subspiniform and placed near to the postero-lateral margin; the antero-lateral margins (as already stated) are four-toothed; the first tooth is tuberculiform and placed near to the outer angle of the orbit, the three following spiniform, the last the largest. The rostrum is short, rather deeply bifid; there is a strong tooth or spine above the inner angle of the orbit, but none at its outer angle, and a rather obscure notch in the middle of its upper and lower margins. All the joints of the post-abdomen (in the male) are distinct. The basal antennal joint is considerably dilated and has a strong spine at its antero-external angle, which spine is visible from above, beneath the spine at the inner angle of the orbit, in a dorsal view. The chelipedes are well developed; the merus and also the carpus armed with tubercles above and on the outer surface; the palm enlarged and laterally compressed, but not carinated on its upper and lower margins. The fingers meet only at the denticulated apices and have between them a hiatus when closed, and both dactyl and pollex are armed besides with three tubercles on their inner margins. The ambulatory legs are nearly naked, and are armed (except the dactyli) with strong spines, which are strongest on the upper margins of the merus and carpus-joints, and the dactyli are strongly incurved near the apices. Colour (in spirit) light reddish-brown above, white below; a longitudinal band of white along the middle of the carapace on the dorsal surface. There are a few scattered hairs on the dactyli and penultimate joints of the ambulatory legs.

♂.					Lines.	Millims.
Length and breadth of carapace, about	.	.	.	.	2½	5
Length of a chelipede, nearly	.	.	.	.	3	6

The single specimen (a male) was dredged together with small specimens of *Mithraculus sculptus* at Fernando Noronha, in 7 to 20 fathoms.

In the strongly tuberculated wrists of the chelipedes and in the spinuliferous ambulatory legs, this form resembles *Mithrax ursus*, Bell, near to which species it must, I think, be ranged.

[1] *Proc. Acad. Nat. Sci. Philad.*, p. 147, 1879.

Legion II. PARTHENOPINEA.

Parthenopinea, Dana, U.S. Expl. Exped., vol. xiii., Crust. i., pp. 77, 136, 1852.
„ Miers, Journ. Linn. Soc. Lond. (Zool.), vol. xiv. p. 641, 1879.

Basal antennal joint very small and embedded with the next joint in the narrow hiatus between the front and the inner subocular angle of the orbit; the infraocular space being mainly occupied by the inferior wall of the orbit.

Family IV. PARTHENOPIDÆ.

Parthenopidæ, Miers, Journ. Linn. Soc. Lond. (Zool.), vol. xiv. pp. 641–667, 1879.

Eyes usually retractile within the small circular and well-defined orbits; the inferior wall of the orbit is continued to within a very short distance of the front. The antennæ are very slender; the basal joint does not, as in the Periceridæ, constitute a great part of the inferior orbital margin, but is very small and usually does not reach to the front, and with the next joint occupies the narrow hiatus intervening between the front and inner subocular angle of the orbit. (In the genus *Ceratocarcinus* the antennæ are completely excluded from the orbits.)

This family presents some affinities with the *Oxystomata*, and also as regards the structure of the orbits and the position of the antennæ with certain genera of the Cancridæ.

Subfamily 1. PARTHENOPINÆ.

Parthenopinæ, Miers, Journ. Linn. Soc. Lond. (Zool.), vol. xiv. p. 668, 1879.

Carapace equilaterally or transversely triangulate or elliptical. Rostrum simple. Strongly-marked depressions exist, separating the branchial from the cardiac and gastric regions. Chelipedes greatly developed, with the palm trigonous, fingers acute.

Lambrus, Leach.

Lambrus, Leach, Trans. Linn. Soc. Lond., vol. xi. pp. 308, 310, 1815.
„ Milne Edwards, Hist. Nat. Crust., vol. i. p. 352, 1834.
„ A. Milne Edwards, Crust. in Miss. Sci. au Méxique, pt. 5, p. 146, 1878.
„ Miers, Journ. Linn. Soc. Lond. (Zool.), vol. xiv. p. 668, 1879.

Carapace equilaterally subtriangulate, convex or depressed, with the rostrum usually prominent, triangulate and somewhat deflexed, the lateral margins rounded at the branchial regions and armed with tubercles or spines, which, on the postero-lateral margins are sometimes considerably elongated, orbits small and well defined, with a fissure (which is usually closed) in the superior margin, the interior subocular lobe is sometimes greatly developed. Epistoma usually transverse. The pterygostomian regions (in certain species) are more or less distinctly rigid, as in the genus *Solenolambrus*,

Stimpson. The post-abdomen, in the male, covers the sternum at the base between the fifth pair of legs, and is five or six-jointed, with two or three of the intermediate segments coalescent. Eyes short, robust. Antennules usually obliquely plicated, with the basal joint considerably developed. The basal (or real second) joint of the antennæ is short, not markedly dilated, and does not reach the front; the next joint occupies the interior hiatus of the orbit. The exterior maxillipedes present nothing remarkable; the ischium is not produced at its antero-internal angle; the merus is distally truncated and bears the next joint at its antero-internal angle, and the exognath is slender and straight. Chelipedes (in the male) subequal and either very considerably elongated (typical *Lambrus*), or of moderate length (subgenus *Parthenopoides*, Miers); the merus and palm are elongated and usually spinose or tuberculated; palm trigonous, and armed with a denticulated or spinose crest along the superior and internal margin; fingers short, distally acute, and dentated on the internal margins. Ambulatory legs slender and of moderate length, with the merus-joints sometimes denticulated or tuberculated on the margins; dactyli styliform.

The species of this genus are very numerous and are probably found in all the warmer temperate and tropical seas of the globe, usually in shallow, but sometimes in deeper water.

They have recently been subdivided by Dr. A. Milne Edwards into ten genera.[1]

Of these genera, one, two, or perhaps three (*Rhinolambrus*, A. Milne Edwards, *Platylambrus*, A. Milne Edwards, and *Leiolambrus*, A. Milne Edwards), I think are insufficiently characterized; a fourth, *Parthenolambrus*, A. Milne Edwards, corresponds nearly to my *Parthenopoides*, established in 1879 (when I had not seen the work of A. Milne Edwards). Four of the remaining genera (*i.e.*, *Solenolambrus*, Stimpson, *Mesorhœa*, Stimpson, *Enoplolambrus*, A. Milne Edwards, and *Pisolambrus*, A. Milne Edwards), are not represented in the Challenger collection.

Since Dr. A. Milne Edwards has not indicated with anything like completeness the species which are to be referred to the subgenera established by him, I give below lists, as complete as I can make them, of the species referable to the subgenera represented in the collection of H.M.S. Challenger.

Subgenus *Lambrus*, A. Milne Edwards.

Lambrus, A. Milne Edwards (subgen.) Crust. in Miss. Sci. au Méxique, pt. 5, p. 146, 1878.
    „   Miers, Journ. Linn. Soc. Lond., tom. cit., p. 672, 1879.
Platylambrus, Stimpson (part), Bull. Mus. Comp. Zool., vol. ii. p. 129, 1870.
Rhinolambrus, A. Milne Edwards, (part), Crust. in Miss. Sci. au Méxique, pt. 5, p. 148, 1878.

Carapace moderately convex, and not produced over the bases of the legs, rarely depressed, with the rostrum and epistoma well developed, the lateral epibranchial spine

[1] Crust. in Miss. Sci. au Méxique, pt. 5, p. 146, 1878.

absent or but slightly developed. Chelipedes very considerably elongated, palm and merus usually armed with strong spines.

The pterygostomian regions are sometimes slightly excavated and carinated, but the smooth deep channel characterising the subgenus *Aulacolambrus* is not developed.[1]

The species referable to this subgenus may be divided into two sections as follows :—

1. Merus-joints of ambulatory legs more or less spinose or tuberculated—

Atlantic and Mediterranean Species.

Lambrus macrocheles (Herbst) = *Lambrus mediterraneus*, Roux. Mediterranean; West Africa, to 49 fathoms (Studer).

Lambrus rugosus, Stimpson. Cape Verde Islands (to 20 fathoms).

Lambrus bouvieri, A. Milne Edwards. St. Vincent.

Lambrus setubalensis, B. Capello. Setubal.

Lambrus verrucosus, Studer. Ascension Island (to 60 fathoms).

Lambrus pourtalesii, Stimpson. Florida Straits (to 117 fathoms).

Lambrus fraterculus, Stimpson. Florida Straits (to 68 fathoms).

Indo-Pacific Species.

Lambrus longimanus (Linné ?). Indo-Pacific Region (to 100 fathoms).

Lambrus contrarius (Herbst). "East Indies" (Herbst); Samboangan, Philippines; Mauritius.

Lambrus echinatus (Herbst). India (Coromandel, Pondicherry); Mauritius.

Lambrus laciniatus, De Haan. Japan (5 to 20 fathoms).

Lambrus validus, De Haan. Japan.

Lambrus lamelliger, White (= *Lambrus lamellifrons*, Adams and White, and *Lambrus rumphii*, Bleeker). Indo-Malayan Seas; Philippines.

Lambrus tuberculosus, Stimpson. Hong Kong (to 15 fathoms).

Lambrus intermedius, Miers. Corean Seas; North and West Australia; Shark Bay (?).

Lambrus longispinus, Miers (= *Lambrus spinifer*, Haswell). North and North-east Australia (to 14 fathoms); Shanghai.

Lambrus holdsworthii, Miers. Ceylon.

Lambrus laevicarpus, Miers. Oriental Seas.

Lambrus mossambicanus (Bianconi). Mozambique.

[1] The type of this subgenus is *Lambrus longimanus* (Linné), not, as stated by a *lapsus calami* in Journ. Linn. Soc. Lond., (vol. xiv. p. 672, 1879), *Lambrus crenulatus*, Saussure.

West American Species.

Lambrus hyponcus, Stimpson. Panama, Mexico; Mazatlan. (This species establishes the transition to the next section).

2. Merus-joints of the ambulatory legs without distinct spines or tubercles—

Atlantic and Mediterranean Species.

Lambrus angulifrons (Latreille) = *Parthenope longimana*, Costa. Mediterranean (to 50 fathoms, Heller).

Lambrus serratus, Milne Edwards = *Lambrus crenulatus*, Saussure. West Indies (to 13 fathoms); Bahia.

Lambrus agonus, Stimpson. West Indies and Florida Straits (to 84 fathoms).

Lambrus guérinii, B. Capello. West Indies and Bahia; Mauritius (type). This species, it will be noted, is found in both the Atlantic and Indo-Pacific Regions.

West American Species.

Lambrus depressinoculus, Stimpson. Mexico; Manzanillo.

Indo-Pacific Species.

Lambrus pelagicus, Rüppell. Red Sea.

Lambrus turriger, Adams and White. Philippine Islands; Amboina (15 to 25 fathoms); Borneo; North and North-west Australia (to 36 fathoms).

Lambrus rhombicus, Dana. Vanua Levu, Fiji Islands.

Lambrus gracilis, Dana. Ovalau, Fiji Islands.

Lambrus nodosus, Jacq. and Lucas. North, North-east and West Australia; New Zealand.

Lambrus affinis, A. Milne Edwards. Indo-Pacific Region (Seychelles to New Caledonia).

Lambrus deflexifrons, Miers. Ceylon. (This species perhaps forms a distinct subgenus).

Lambrus contrarius (Herbst).

Cancer contrarius, Herbst, Naturgesch. der Krabben u. Krebse, vol. iv., Heft. 4, p. 8, pl. lx. fig. 3, 1804.

Lambrus spinimanus, Desmarest, Consid. sur la Classe des Crust., p. 36, pl. iii. fig. 1, 1825.

„   contrarius, Milne Edwards, Hist. Nat. Crust., vol. i. p. 354, 1834.

Samboangan, Philippines, 10 fathoms (a female).

The figure of Herbst is a very inexact representation of this species, if the specimens referred to it in the British Museum and Challenger collection are rightly designated. It is much better delineated in Desmarest's figure of *Lambrus spinimanus*, which is cited by Milne Edwards as synonymous with *Lambrus contrarius*.

The Challenger example has the following dimensions :—

Adult ♀.					Lines.	Millims.
Length of carapace and rostrum,	.	.	.	.	18½	41
Breadth of carapace, nearly	.	.	.	.	17½	36·5
Length of a chelipede,	.	.	.	.	44½	94·5
Length of first ambulatory leg, nearly	.	.	.	25	52·5

Lambrus longimanus (Linné).

> ? *Cancer longimanus*, ♀, Linné, Mus. Ludovici Ulricä, p. 441, 1764 ; Syst. Nat., ed. xii., p. 1047, 1766.
> *Lambrus longimanus*, Leach, Trans. Linn. Soc. Lond., vol. xi. p. 310, 1815.
> „ „ Milne Edwards, Hist. Nat. Crust., vol. i. p. 354, 1834 ; Atlas in Latr. Règne Animal de Cuvier, Crust., pl. xxvi. fig. 1.
> „ „ Miers, Ann. and Mag. Nat. Hist., ser. 5, vol. iv. p. 70, 1879.

South of New Guinea, 28 fathoms, in lat. 9° 59′ 0″ S., long. 139° 42′ 0″ E. (Station 188) (two males and two females) ; Amboina, 100 fathoms (a small female)

The specimens thus designated are certainly the *Lambrus longimanus* of Milne Edwards, as figured in the large illustrated edition of the Règne Animal de Cuvier (*loc. cit.*), but in his description of the same species in his Histoire naturelle des Crustacés the lateral margins of the carapace are described as " armés d'épines très-longues et légèrement rameuses," a character inapplicable to any specimens which have come under my observation.

The largest of the Challenger specimens (a female) measures as follows :—

Adult ♀.				Lines.	Millims.
Length of carapace and rostrum, nearly	.	.	.	10½	22
Breadth of carapace, nearly	.	.	.	11	23
Length of a chelipede,	.	.	.	36½	77
Length of first ambulatory leg,	.	.	.	15	31·5

Lambrus affinis, A. Milne Edwards.

> *Lambrus affinis*, A. Milne Edwards, Nouv. Archiv. Mus. Hist. Nat., vol. viii. p. 261, pl. xiv. fig. 4, 1872.

" Torres Strait, August 1874 " (an adult male).

In this species the merus-joints of the ambulatory legs are smooth ; in the following (*Lambrus intermedius*), they are compressed, and the last two pairs are more or less distinctly granulated on the margins.

Adult ♂.	Lines.	Millims.
Length of carapace and rostrum, rather over	. 7	15
Breadth of carapace, nearly . .	. 7	14·5
Length of a chelipede, . . .	. 20	42
Length of first ambulatory leg, nearly .	. 11	23

Lambrus intermedius, Miers (Pl. X. fig. 4).

Lambrus intermedius, Miers, Proc. Zool. Soc. Lond., p. 30, 1879.

Torres Strait, 6 fathoms (Station 187), lat. 10° 36′ 0″ S., long. 141° 55′ 0″ E. (a small male).

This species may be regarded as intermediate in position between *Lambrus affinis* and *Lambrus nodosus*. The figure now given is from the type, from the Corean Seas, in the collection of the British Museum, which, being of much larger size than the Challenger specimen, better represents its characteristic peculiarities. Whether it is to be regarded as a small variety of one or the other above-mentioned species a larger series is needed to determine. I have elsewhere referred to its affinities with *Lambrus nodosus*.

Lambrus turriger, White.

Lambrus turriger, White, Proc. Zool. Soc. Lond., p. 58, 1847.
   „   „   Adams and White, Crust. in Zool. H.M.S. "Samarang," p. 26, pl. v. fig. 2, 1848.
   „   „   Miers, Crust. Rep. in Zool. Col. H.M.S. "Alert," p. 201, 1884.

A fine adult male was dredged at Amboina, in 15 to 25 fathoms.
In this specimen the principal dimensions are as follows :—

Adult ♂.	Lines.	Millims.
Length of the carapace and rostrum,	. 7	15
Greatest breadth of carapace, .	. 6½	13·5
Length of a chelipede, rather over	. 39	83
Length of first ambulatory leg, about	. 14	30

There are several tubercles on the gastric, cardiac, and branchial regions of the carapace, besides the long vertical spines which always characterize this species.

Lambrus guérinii, var. ? Capello.

Lambrus guérinii, F. de B. Capello, Journ. de Sci. Math. Phys. Nat. de Lisboa, vol. iii. p. 264, pl. iiia. fig. 5, 1871.

Bahia, shallow water (an adult female and a smaller male).
These specimens are referred to *Lambrus guérinii* with much uncertainty, since the types of that species were from the Mauritius; they only differ from Capello's figure in

the somewhat greater prominence of the spines of the outer margin of the hand; three of the tubercles of the gastric region, and one on the cardiac and on each branchial region, are more elevated than the others, and from some of the elevated tubercles of the carapace spring long tufts of hairs, which are not shown in the figure of *Lambrus guérinii* (notably from the præocular tubercles and the elevated tubercles of the gastric and branchial regions). These tufts are absent, however, from some Brazilian and West Indian examples in the collection of the British Museum.

The dimensions of the female are :—

Adult ♀.				Lines.	Millims.
Length of carapace and rostrum, about	.	.	.	13	28
Breadth of carapace, .	.	.	.	14½	31
Length of a chelipede, about .	.	.	.	28½	60
Length of first ambulatory leg,	.	.	.	15½	33

Lambrus serratus, Milne Edwards.

Lambrus serratus, Milne Edwards, Hist. Nat. Crust., vol. i. p. 357, 1834.
Platylambrus serratus, A. Milne Edwards, Crust. in Miss. Sci. au Méxique, pt. 5, i. p. 156, pl. xxx. fig. 1, 1878, et synonyma.
Lambrus lupoides, White, List Crust. Brit. Mus., p. 12, 1847.

Bahia, shallow water (an adult male and female).

Adult ♂.				Lines.	Millims.
Length of carapace and rostrum,	.	.	.	9	19
Breadth of carapace, about	.	.	.	10	21
Length of a chelipede, about	.	.	.	28½	61
Length of first ambulatory leg,	.	.	.	12½	26

If the subgenus *Platylambrus*, A. Milne Edwards, be retained, it will probably be best restricted to this species, and, perhaps, *Lambrus granulatus*, Kingsley.

Subgenus *Aulacolambrus*, A. Milne Edwards.

Aulacolambrus, A. Milne Edwards, Crust. in Miss. Sci. au Méxique, pt. 5, p. 147, 1878.

Carapace depressed, with the rostrum and epistoma very short, the lateral epibranchial spine very considerably elongated. The buccal cavity is bordered laterally by a deep, smooth and wide cavity extending from the branchial aperture at the base of the chelipedes nearly to the orbits. The chelipedes are greatly elongated, and the palm and merus are armed with long spines.

(ZOOL. CHALL. EXP.—PART XLIX.—1886.) Ccc 13

The following species are referable to this subgenus :—

> *Lambrus hoplonotus*, Adams and White, and vars. *longioculis*, Miers ; *planifrons*,
>   Miers, and *granulosus*, Miers.   Indo-Pacific Seas (Ceylon to New Caledonia ;
>   to 28 fathoms).
> *Lambrus pisoides*, Adams and White.   Philippines.
> *Lambrus diacanthus*, De Haan.   Japan.
> *Lambrus sculptus*, A. Milne Edwards.   New Caledonia ; Fiji Islands (Coll. Brit.
>   Mus.).
> *Lambrus whitei*, A. Milne Edwards (= *Lambrus carenatus*, White, nec Milne
>   Edwards).   Borneo ; China Sea.
> *Lambrus curvispinus*, Miers. Java Sea.   (This species is perhaps a variety of
>   *Lambrus hoplonotus*).

Lambrus (Aulacolambrus) hoplonotus, Adams and White.

> *Lambrus hoplonotus*, Adams and White, Crust. in Zool. H.M.S. "Samarang," p. 35, pl. vii. fig. 3,
>   1848.
>   „   „   Miers, Ann. and Mag. Nat. Hist., ser. 5, vol. iv. p. 23, 1879.

Off Cape York, 8 fathoms, in lat. 10° 30' 0" S., long. 142° 18' 0" E. ;   Station 186 (an adult male).

This specimen differs from the typical example in having the tubercles of the post-frontal and hepatic regions of the carapace even larger and more flattened ; in the middle line of the carapace, on the gastric and cardiac regions, are two somewhat more elevated and subspiniform tubercles.   The six spines of the outer (or posterior) margin of the chelipedes are very much elongated and of equal length, and are slightly knobbed at the tips.   These characters are perhaps sufficient to permit of the distinction of this specimen as a fourth variety of this protean species, but for the present I will not designate it by a distinct appellation.   Its dimensions are as follows :—

Adult ♂.				Lines.	Millims.
Length of carapace and rostrum,	.	.	.	10½	22·5
Breadth of carapace to base of lateral epibranchial spine,			.	10	21
Length of a chelipede,	.	.	.	25	52·5
Length of first ambulatory leg,	.	.	.	14	30

Lambrus hoplonotus, var. *granulosus*, Miers (Pl. X. fig. 5).

> *Lambrus hoplonotus*, var. *granulosus*, Miers, tom. cit., p. 23, 1879 ; Crust. in Rep. Zool. Coll.
>   H.M.S. "Alert," p. 202, 1884.

Torres Strait, 3 to 11 fathoms (a small but adult female).

Lambrus hoplonotus, var. *longioculis*, Miers.

Lambrus hoplonotus, var. *longioculis*, Miers, *tom. cit.*, p. 23, pl. v. fig. 6, 1879.

South of New Guinea, 28 fathoms, lat. 9° 59′ 0″ S., long. 139° 42′ 0″ E., Station 188 (a female).

This variety is best distinguishable by the subspiniform tubercles of the branchial regions.

Subgenus *Parthenolambrus*, A. Milne Edwards.

Parthenolambrus, A. Milne Edwards, Crust. in Miss. Sci. au Méxique, pt. 5, p. 148, 1878.

Parthenopoides, Miers, Journ. Linn. Soc. Lond., *tom. cit.*, p. 672, 1879.

Carapace often crowded, subtriangulate, with the posterior margin nearly straight, and produced at the postero-lateral angles over the bases of the ambulatory legs. Chelipedes rarely spinose and of moderate length (rarely exceeding two and a half times the length of the carapace).

Lambrus massena, Roux, which is placed by A. Milne-Edwards in his *Rhinolambrus*, must, I think, be included in this subgenus.

The following species are referable to this subgenus :—

Atlantic and Mediterranean Species.

Lambrus massena, Roux = *Parthenope contracta*, Costa, *fide* Heller (?) *Lambrus rugosus*, Stimpson, and var. *gorreensis* (*Spinifer*), Miers, and *atlanticus*, Miers. Mediterranean (70 to 75 metres); Senegambia (9 to 15 fathoms) ; Cape Verde Islands ; Azores, 50 to 90 fathoms.

Lambrus pulchellus, A. Milne Edwards. Cape St. Vincent.

Lambrus bicarinatus, Miers. Senegambia ; Goree Island (9 to 15 fathoms), Canaries.

Lambrus expansus, Miers. Madeira ; Azores, off Fayal (50 to 90 fathoms).

Indo-Pacific Species.

Lambrus calappoides, Adams and White. Philippines ; Tongatabu (18 fathoms); Seychelles, 4 to 12 fathoms.

Lambrus harpax, Adams and White (= *Lambrus*, *Parthenope*, *sandrockii*, Haswell). North and North-east Australia (to 14 fathoms); China, Borneo.

Lambrus turpeius, Adams and White. Oriental Seas.

Lambrus erosus, Miers. Oriental Seas.

Species of Unknown Habitat.

Lambrus trigonus, A. Milne Edwards.

West-American Species.

Lambrus excavatus, Stimpson.   Mexico; Manzanillo.
Lambrus triangulus, Stimpson.   Cape St. Lucas.
Lambrus frons-acutis, Lockington.   California (Simaloa, Santa Catalina).[1]

Lambrus (Parthenolambrus) massena, Roux.

> *Lambrus massena*, Roux, Crust. de la Méditerranée, pl. xxiii., 1830.
> „   „   Milne Edwards, Hist. Nat. Crust., vol. i. p. 356, 1834.
> „   „   Lucas, Animaux Articulés in Explor. Scientif. de l'Algérie, vol. i. (Crust.), p. 10, pl. i. fig. 3 ♀, 1849.
> ? *Lambrus rugosus*, Stimpson, Proc. Acad. Nat. Sci. Philad., p. 220, 1857.
> *Lambrus (Parthenopoides) massena*, Miers, Ann. and Mag. Nat. Hist., ser. 5, vol. viii. p. 207, 1881.

A male and a female of what I have regarded (in the absence of types for comparison) as the typical form of this variable species were collected—the male at St. Vincent, Cape Verde Islands, in July 1873, the female at the Azores, off Fayal, in 50 to 90 fathoms.

♂.				Lines.	Millims.
Length and breadth of carapace and rostrum,	.	.	.	3½	7·5
Length of a chelipede,	.	.	.	5½	11·5
Length of first ambulatory leg, about	.	.	.	4	9

Lambrus massena, Roux, var. atlanticus, Miers.

> *Lambrus massena*, var. *atlanticus*, Miers, *tom. cit.*, p. 208, 1881.

St. Vincent, Cape Verde Islands (with the typical form) two males. The dimensions of the larger, which has lost nearly all the ambulatory legs are as follows :—

♂.			Lines.	Millims.
Length and breadth of carapace and rostrum, .	.	.	4	8·5
Length of a chelipede, nearly .	.	.	6	12·5

Lambrus (Parthenolambrus) expansus, Miers.

> *Lambrus (Parthenopoides) expansus*, Miers, Ann. and Mag. Nat. Hist., ser. 5, vol. iv. p. 25, pl. v. fig. 9, 1879.

Azores, off Fayal, in 50 to 90 fathoms.   Two males and a female were collected.

The colour (even in the spirit-preserved specimens) varies considerably ; the ground-colour in all three specimens is a light pinkish-yellow, with slaty-coloured markings on the chelipedes.   In the smaller male the carapace is covered with similar

---

[1] M. A. Milne Edwards has recently described two species of *Lambrus* from Upolu, one of the Samoan Islands, which I cannot with certainty refer to any one section of this genus; they are *Lambrus gracilipes*, A. Milne Edwards, and *Lambrus pugilator*, A. Milne Edwards ; the latter is perhaps a species of *Aulacolambrus*.

coloured spots, which are wholly absent from the carapace of the larger male and the female; in the latter the frontal and gastric regions and the sides of the branchial regions are of a brick-red colour.

Adult ♂.				Lines.	Millims.
Length of carapace and rostrum, about,	.		.	4½	10
Breadth of carapace, nearly,	.	.	.	6	11
Length of a chelipede,	.	.	.	8	17
Length of first ambulatory leg, nearly		.	.	5	10·5

Lambrus (Parthenolambrus) calappoides (Adams and White).

Parthenope calappoides, Adams and White, Crust. in Zool. H.M.S. "Samarang," p. 34, pl. v. fig. 5, 1848.

,, ,, Miers, Crust. in Rep. Zool. Coll. H.M.S. "Alert," p. 527, 1864.

A male obtained in 18 fathoms, off Tongatabu (Station 172) is referred to this species.

Adult ♂.					Lines.	Millims.
Length of carapace and rostrum,	.	.		.	5	10·5
Breadth of carapace,	.	.	.	.	6½	13·5
Length of a chelipede, nearly	.	.		.	8	16·5
Length of first ambulatory leg,	.	.		.	5	10·5

Cryptopodia, Milne Edwards.

Cryptopodia, Milne Edwards, Hist. Nat. Crust., vol. i, p. 360, 1834.

,, Miers, Journ. Linn. Soc. Lond. (Zool.), vol. xiv. p. 669, 1879.

Carapace very broadly triangulate, with very large lateral clypeiform inferiorly vaulted expansions which completely conceal the ambulatory legs, and are prolonged posteriorly far beyond the base of the post-abdomen; the space between the gastric and the cardiac regions is triangular, concave, and defined by four tubercles or eminences, of which one is placed on the gastric, one on the front of the cardiac, and one on each branchial region. The front is nearly horizontal, subtriangulate, spatuliform and very prominent. The pterygostomian regions are smooth, not ridged. The orbits are very small, nearly circular, with a narrow fissure in the superior margin. The epistoma is well developed; the antennulary fossettes are narrow and somewhat oblique. The post-abdomen, in the male, is five-jointed; the third to fifth segments coalescent. The eyes are very small and retractile. The basal antennal joint is slightly dilated and does not nearly attain the apex of the interior subocular lobe of the orbit; the following joint lies within the interior orbital hiatus. The buccal cavity and exterior maxillipedes are small. The ischium-joint of the endognath of the maxillipedes is not produced at its antero-internal angle; the merus is distally truncated, with the antero-external angle slightly produced, the interior margin notched below the antero-internal angle. The chelipedes are nearly as in *Lambrus;* the merus-joint has a wing-like lobe on the posterior margin near to the distal extremity ; the palms of the chelipedes are elongated, tricarinated, and dentated (as in *Lambrus*) ; fingers short and more or less distinctly dentated on the inner margins. The ambulatory legs are slender and decrease successively but slightly in length, with the fourth, fifth and sixth joints more or less distinctly carinated ; dactyli nearly straight.

The species, which have been enumerated by A. Milne Edwards, occur for the most part in the Oriental region, in water of moderate depth, but a species also occurs in the Gulf of Mexico and another on the coast of California.[1]

To the species mentioned by A. Milne Edwards is to be added : *Cryptopodia spatulifrons,* Miers (with var. *lævimana,* Miers), which occurs on the coasts of North-east and West Australia and Borneo.

Cryptopodia fornicata (Fabricius).

> *Cancer fornicatus,* Fabr., Spec. Insect., vol. ii. Append., p. 502, 1781.
> *Cryptopodia fornicata,* Milne Edwards, Hist. Nat. Crust., vol. i. p. 362, 1834, et synonyma.
> „ „ Miers, Crust. in Rep. Zool. Coll. H.M.S. "Alert," p. 203, 1884.
> *Parthenope (Cryptopodia) fornicata,* De Haan, Crust. in v. Siebold, Fauna Japonica, p. 90, pl. xx. fig. 2 ♂, 1839.

South of New Guinea, in 28 fathoms, lat. 9° 59′ 0″ S., long. 139° 42′ 0″ E. (Station 188), a female of rather small size.

Adult ♂.				Lines.	Millims.
Length of carapace and rostrum,	.	.		9	19
Breadth of carapace, nearly	.	.	.	14	29·5
Length of a chelipede, .	.	.	.	12	25·5
Length of first ambulatory leg,	.	.	.	6½	14

Heterocrypta, Stimpson.

> *Heterocrypta,* Stimpson, Ann. Lyc. Nat. Hist. New York, vol. x. p. 102, 1871.
> „ Miers, Journ. Linn. Soc. Lond. (Zool.) vol. xiv. p. 669, 1871.

This genus is very nearly allied to the preceding (*Cryptopodia*), and it may suffice here to indicate the principal characters by which it is distinguished.

The clypeiform expansions of the carapace are less produced than in the Oriental species of that genus and cover only the bases of the ambulatory legs, and the carapace is scarcely, if at all, posteriorly produced beyond the base of the post-abdomen ; its dorsal lobes and carinæ are more developed. The pterygostomian and subhepatic regions are traversed by a granulated ridge running parallel to the antero-lateral margins of the carapace, which terminates just above the antero-lateral angles of the buccal cavity.

[1] Crust. in Miss. Sci. au Mexique, pt. 5, p. 168, 1878.

I may add that the inferior wall of the orbit is slightly notched; the ischium of the exterior maxillipedes is slightly produced at its antero-internal angle, and the merus of the chelipedes is without the distal and posterior lobe which exists in most of the species of *Cryptopodia*.

The species of this genus are of small size. The following have been described:—

Heterocrypta granulata, Gibbes. East coast of United States, from Virginia to Florida; West Indies; St. Thomas; Bahia.

Heterocrypta macrobrachia, Stimpson. Coast of Mexico and Panama.

Heterocrypta maltzani, Miers = *Heterocrypta marionis*, A. Milne Edwards. Senegambia; Goree Island (9 to 15 fathoms); Bay of Toulon (450 metres); Azores (to 450 fathoms).

Heterocrypta granulata (Gibbes).

Cryptopodia granulata, Gibbes, Proc. Amer. Assoc. Adv. Sci., p. 173, 1850.

Heterocrypta granulata, Stimpson, Ann. Lyc. Nat. Hist. New York, vol. x. p. 103, 1871.

"　　"　　A. Milne Edwards, Crust. in Miss. Sci. au Méxique, pt. 5, p. 166, pl. xxix. fig. 4, 1878.

Bahia, shallow water, September 1873 (a female).

The carapace is more distinctly dentated on the postero-lateral margins, and the chelipedes are more slender than in Milne Edwards' figure of this species, and in these particulars resemble *Heterocrypta macrobrachia*, which is not improbably a mere variety of *Heterocrypta granulata*.

♀.		Lines.	Millims.
Length of carapace and rostrum, about	.	3	6·5
Breadth of carapace, nearly	.	4	8
Length of a chelipede, about	.	6	13
Length of first ambulatory leg,	.	3	6·5

Heterocrypta maltzani, Miers.

Heterocrypta maltzani, Miers, Ann. and Mag. Nat. Hist., ser. 5, vol. viii. p. 209, pl. xiii. fig. 1, September, 1881.

Heterocrypta marionis, A. Milne Edwards, Comptes Rendus, vol. xciii. p. 879, December, 1881; Ann. and Mag. Nat. Hist., ser. 5, vol. ix. p. 38, 1882, "Travailleur" (*planches ined.*).

This species was dredged at the Azores, Fayal, in 50 to 90 fathoms (male and female); and again at the Azores, in 450 fathoms; lat. 38° 38′ 0″ N., long. 28° 28′ 30″ W., Station 75 (an adult male).[1]

[1] Professor A. Milne Edwards having kindly sent to me advanced proofs of figures illustrative of this and other species collected in the "Travailleur" or "Blake" Expedition, I am enabled to satisfy myself of the identity of *Heterocrypta maltzani* and *Heterocrypta marionis*, which were almost contemporaneously described. Professor Milne Edwards' figure is perhaps a truer representation of average-sized examples than mine, which was founded on a specimen of large size.

The latter specimen measures as follows :—

Adult ♂.		Lines.	Millims.
Length of carapace and rostrum,	.	4	8·5
Breadth of carapace,	. .	4	8·5
Length of a chelipede, .	. .	8½	18
Length of first ambulatory leg, .	.	5	10·5

Subfamily 2. EUMEDONIN.E.

Eumedoninæ, Miers, Journ. Linn. Soc. Lond. (Zool.), vol. xiv. p. 670, 1879.

Carapace usually rhomboidal or subpentagonal. Rostrum usually bifid or emarginate. Depressions separating the regions of the carapace obscure or non-existent. Anterior legs of moderate length, chelipedes not trigonous.

To the genera enumerated in the above cited memoir, *Rhabdonotus*, A. Milne Edwards, is perhaps to be added.

Ceratocarcinus, Adams and White.

Ceratocarcinus, Adams and White, Proc. Zool. Soc. Lond., p. 57, 1847 ; Crust. in Zool. H.M.S. "Samarang," p. 33, 1848.
„   Miers, Journ. Linn. Soc. Lond. (Zool.), vol. xiv. p. 670, 1879.

Carapace subhexagonal, about as broad as long, with the dorsal surface moderately convex, spinose or tuberculated. The spines of the rostrum are elongated, acute, and separated by a rather wide interspace, and there is a well-developed lateral epibranchial spine. The orbits are small, circular, excavated below and at the exterior angle, and the subocular lobe joins the front, so as completely to exclude the antennæ from the orbits. The post-abdomen, in the male, is seven-jointed.

The eyes are small, retractile. The basal (or real second) joint of the antennæ is slender and occupies the space between the base of the antennules and the inner subocular lobe. The exterior maxillipedes are small ; the ischium-joint not produced at its antero-internal angle, the merus distally truncated, not produced at the antero-external angle, and scarcely emarginate at the antero-internal angle, where the next joint articulates. The chelipedes are slender and somewhat elongated, with the joints not dilated, the merus and carpus sometimes armed with spines ; the dactyli acute, shorter than the palms and dentated on the inner margins ; the ambulatory legs are slender, with the joints not dilated, the merus sometimes armed with a distal spine ; the dactyli nearly straight.

The species, all of which are Oriental, are small. The following have been described :—

> *Ceratocarcinus longimanus*, Adams and White.   North Borneo ; Balambangen.
> *Ceratocarcinus speciosus*, Dana.   Viti Levu, Fiji Islands.
> *Ceratocarcinus dilatatus*, A. Milne Edwards.   New Caledonia.   (This species is nearly allied to and may be a variety of *Ceratocarcinus longimanus*).
> *Ceratocarcinus spinosus*, Miers.   Oriental Seas.

Ceratocarcinus longimanus, Adams and White.

> *Ceratocarcinus longimanus*, Adams and White, Proc. Zool. Soc. Lond., p. 57, 1847 ; Crust. in Zool. H.M.S. "Samarang," p. 34, pl. vi. fig. 6, 1848.

Arrou (Aröe) Islands, September 18, 1874 (an adult male).

This specimen has the rostral and lateral spines longer, and the prominences of the dorsal surface of the carapace more developed than in White's figure, or in a specimen bearing his name in the collection of the British Museum from Borneo.   In these characters it more nearly approaches *Ceratocarcinus dilatatus*, A. Milne Edwards,[1] from New Caledonia, which indeed is only distinguishable by its broader carapace, and may perhaps be regarded as a variety of *Ceratocarcinus longimanus*.

The dimensions of the Challenger specimen are as follows :—

Adult ♂.	Lines.	Millims.
Length of carapace, . . .	2	4·5
Length of a rostral spine, about .	1	2·5
Breadth of carapace to base of lateral spine,	2	4·5
Length of a chelipede, about .	6	12·5
Length of first ambulatory leg, nearly	6	12·

Ceratocarcinus speciosus, Dana.[2]

This is another nearly allied species, but is apparently well distinguished by its much broader front and distinctly spinulose chelipedes.

---

[1] *Nouv. Archiv. Mus. Hist. Nat.*, vol. viii. p. 256, pl. xiv. fig. a, 1872.
[2] Crust. in U.S. Explor. Exped., vol. xiii. p. 120, pl. vi. fig. 8, 1852.

CYCLOMETOPA or CANCROIDEA.

Cyclométopes, Milne Edwards (pt.), Hist. Nat. Crust., vol. i. pp. 264, 363, 1834.
Cancroidea, Dana (pt.), U.S. Explor. Exped., vol. xiii., Crust. 1, pp. 67, 142, 1852.
Cyclometopa, Miers, Cat. New Zeal. Crust., p. 13, 1876.

Carapace usually transverse, wide in front, with the antero-lateral margins regularly arcuated ; more rarely quadrate or suborbicular, but not rostrated. Epistoma short, transverse. Antennules usually transversely plicated. The exterior maxillipedes, the afferent channels to the branchiæ, the branchiæ, and the verges of the male are as in the Oxyrhyncha.

Legion I. CANCRINEA or CANCROIDEA TYPICA.

Cancrinea or Cancroidea typica, Dana, U.S. Explor. Exped., vol. xiii., Crust. 1, p. 145, 1852.

This section, in Dana's system, includes the whole of the typical Cyclometopa (Cancridæ or Portunidæ), which have almost invariably a littoral or marine habitat, and in which, therefore, the carapace is not antero-laterally convex and largely developed so as to constitute a vaulted respiratory chamber, as in certain Thelphusidæ, which may remain for extended periods out of the water. The buccal cavity is usually well defined, and the flagella of the antennæ are not greatly elongated as in those degraded forms (Corystoidea) which approach the *Anomura*. The dactyli of the ambulatory legs are styliform and unarmed, or in the fifth pair expanded into an ovate natatory organ ; they are rarely, if ever, armed with longitudinally seriate spinules as in those forms (Thelphusidæ) which approach the Catometopa.

Family 1. CANCRIDÆ.

Cancériens, Milne Edwards (tribe), Hist. Nat. Crust., vol. i. p. 368, 1834.
    ,,   A. Milne Edwards (family), Ann. d. Sci. Nat., ser. 4, vol. xviii. p. 41, 1841 ; Nouv. Archiv. Mus. Hist. Nat., vol. i. p. 179, 1865.
Cancridæ and Eriphidæ, Dana, U.S. Explor. Exped., vol. xiii., Crust. 1, pp. 145, 147, 228, 1852.

Carapace, at least in the recent genera, transverse, usually convex, with the antero-lateral margins more or less arcuated, rarely subquadrate. Ambulatory legs all gressorial, with styliform dactyli ; species marine or littoral.

The genus Œthra (the type of the section *Cancériens cryptopodes*, Milne Edwards) is excluded, as being somewhat more nearly related to the Parthenopidæ.[1]

    [1] Miers, Journ. Linn. Soc. Lond. (Zool.), vol. xiv. p. 669, 1879.

The genera comprised in the true Cancroid Cyclometopa are now extremely numerous, and, as is well known, are connected by almost insensible gradations of form and structure. Hence it is difficult, if not impossible, to indicate a system of classification whereby they may be divided into groups at once natural and distinctly defined; and the types represented in the Challenger collection are not sufficiently numerous to permit me to attempt (even did time and experience allow) a new and detailed arrangement of the genera of this group.

The classification of Professor J. D. Dana,[1] although somewhat artificial, is one which is certainly very convenient to systematists, since this author gives a synoptical arrangement, with diagnostic characters, of all the genera known at the time when his work was published, and as it is the one which I have hitherto followed in my own papers, I have thought it best, pending a complete revision of this group, to adhere to this arrangement so far as the sequence of the leading genera is concerned in the present Report, separating merely the somewhat abnormal genus *Trapezia* (which approaches the Catometopa in the form of the carapace) from the remainder of the group; but I refrain from indicating in the present imperfect state of our knowledge any subdivisional characters or subfamilies.

A widely different and in some particulars more natural arrangement has been indicated by A. Milne Edwards,[2] who divides the true Cancridæ into five primary, two transitional, and two accessory groups, characterised mainly by the form of the carapace, and typified respectively by the genera *Cancer*, *Carpilius*, *Xantho*, *Eriphia* and *Trapezia*, and by *Œthra* and *Galene*, and *Pirimela* and *Lioyore*. His monograph of the recent genera of Cancridæ has, however, unfortunately never been completed, and the limits and subdivisional characters of six of the groups have not been defined, but it is probable that the true natural arrangement of the genera lies in the direction indicated by Milne Edwards. The genus *Œthra* (the type of his group Œthridæ) is somewhat more nearly related to the Oxyrhyncha than to the Cyclometopa, and has been arranged, with *Cryptopodia* in the former group, both by Professor S. I. Smith and Dr. W. Stimpson, and by myself.

Section I. **Cancrinæ.**

Cancériens arqués, Milne Edwards, Hist. Nat. Crust., vol. i. pp. 369, 371, 1834.
   ,,   *quadrilatères*, Milne Edwards (pt.), *tom. cit.*, pp. 369, 424, 1834.

Carapace usually convex, with the antero-lateral margins arcuated, and armed with several lobes, teeth or spines. The front is of moderate width and usually does not project over the antennules and bases of the antennæ, which are seldom excluded from the interior hiatus of the orbits.

[1] Crust. in U.S. Explor. Exped., vol. xiii. p. 145, 1852.
[2] *Ann. d. Sci. Nat. (Zool.), sér. 4, vol. xviii. p. 39, 1862 ; Nouv. Archiv. Mus. Hist. Nat., vol. i. pp. 180, 182, 1865.*

The genera included in this section and represented in the Challenger collection are the following :—

A. Endostome or palate without distinct longitudinal ridges defining the openings of the efferent branchial channels (family Cancridæ, Dana).

    a. Fingers of the chelipedes acute or subacute (subfamilies Cancrinæ and Xanthinæ, Dana).

Cancer, Lamarck.	*Actæa*, de Haan.
Carpilius, Desmarest.	*Atergatopsis*, A. Milne Edwards.
Atergatis, de Haan.	*Xantho*, Leach.
Lophactæa, A. Milne Edwards.	*Xanthodes*, Dana.
Lophozozymus, A. Milne Edwards.	*Panopeus*, Milne Edwards.
Medæus, Dana.	*Micropanope*, Stimpson.

    b. Fingers of the chelipedes more or less distinctly excavated at the tips and spoon-shaped (subfamily Chlorodiinæ, Dana).

Etisus, Milne Edwards (pt.).	*Actæodes*, Dana.
Carpilodes, Dana.	*Leptodius*, A. Milne Edwards.
Zozymus, Leach.	*Phymodius*, A. Milne Edwards.[1]

---

[1] A remarkable parallelism in structure (exhibited in the following table, which also indicates a more natural classification) exists between the genera of these two subsections. Were the types in the Challenger collection more numerous, this parallelism could be exhibited in further detail.

1. Carapace widely transverse. Antennules longitudinally or obliquely plicated. The flagella of the antennæ excluded from the orbital hiatus by the enlargement of the basal antennal joint.

    Cancer, Lamarck. | *Etisus*, Milne Edwards.

2. Carapace usually moderately transverse. Antennules obliquely or transversely plicated ; the flagella of the antennæ not excluded from the orbits.

    a. Basal antennal joint produced within the interior hiatus of the orbit ; ambulatory legs not cristated.

        Carpilius, Desmarest. | *Carpilodes*, Dana.

    b. Basal antennal joint not produced within the interior orbital hiatus, but in contact with the front, or with the infero-lateral frontal process.

        * Ambulatory legs carinated or cristated.

Atergatis, de Haan.	*Lophozozymus*, A. Milne Edwards.
Lophactæa, A. Milne Edwards.	*Zozymus*, Leach.

        ** Ambulatory legs not carinated nor cristated.

Medæus, Dana.	*Panopeus*, Milne Edwards.
Actæa, de Haan.	*Micropanope*, Stimpson.
Atergatopsis, A. Milne Edwards.	*Actæodes*, Dana.
Xantho, Leach.	*Leptodius*, A. Milne Edwards.
Xanthodes, Dana.	*Phymodius*, A. Milne Edwards.

B. The endostome or palate with distinct longitudinal ridges defining the apertures of the efferent branchial channels (family Eriphiidæ, Dana ; pt.).

a. The antennæ not excluded from the orbit :—

Eurytium, Stimpson.
Pseudozius, Dana (subgenus *Euryozius*, n.).

Sphærozius, Stimpson.
Pilumnus, Leach.

b. The antennæ excluded from the orbit, whose interior hiatus is closed :—

Eriphia, Latrielle.

A. Endostome not longitudinally ridged (Cancridæ, Dana) :—

Cancer, Lamarck.

Cancer, Lamarck, Syst. Anim. sans Vert., p. 148, 1801.
„ Leach, Trans. Linn. Soc. Lond., vol. xi. p. 320, 1815.
„ A. Milne Edwards, Nouv. Archiv. Mus. Hist. Nat., vol. i. p. 185, 1865.

The carapace is depressed or moderately convex, very widely transverse or elliptical ; its dorsal surface is smooth or uneven, with the regions very obscurely defined ; its antero-lateral margins are regularly arcuated and are divided into ten lobes or teeth, which are sometimes broad, subtruncated, and little prominent, sometimes more prolonged and acute, and which may themselves be armed with accessory denticles ; the postero-lateral margins are shorter than the antero-lateral margins, and defined by a raised line or crest (the postbranchial crest). The front is relatively narrow, and is divided into five lobes or teeth (if the lobes which constitute the superior and interior orbital angle be included) and projects somewhat beyond the orbits, which are small and sometimes dentated ; the teeth defined by two fissures in the superior and two in the inferior margins. The post-abdomen in the male is five-jointed, with the third to the fifth joints coalescent. The eyes are small and are set on very short thick pedicels. The antennules are longitudinally or nearly longitudinally plicated. The basal antennal joint is somewhat enlarged with a distal lobe or tooth, which unites with the front so as to exclude the short flagellum from the orbit. The exterior maxillipedes have the merus-joints usually distally truncated, with the antero-external angle not produced, and they are usually more or less distinctly notched on the inner margin, but the notch is sometimes obsolete. The chelipedes are usually subequal and not very largely developed, and their palms are nearly always longitudinally costated on the exterior surface ; fingers acute and dentated on the inner margins. The ambulatory legs are somewhat elongated, with the dactyli slender and nearly straight (not dilated and flattened as in *Metacarcinus*).

To the details regarding the geographical distribution of this genus given by A. Milne Edwards I may add the following. A species, *Cancer recurvidens* (Spence Bate), occurs at Vancouver Island, which, to judge from the very short description, closely resembles, and may perhaps not be distinguishable from young specimens of *Cancer antennarius*, Stimpson, or may belong to the doubtfully distinct genus *Trichocarcinus*. A variety (*annulipes*, Miers) of *Cancer edwardsii*, Bell, occurs in Trinidad Channel, in the Straits of Magellan.[1]

The details are wanting with regard to the bathymetrical range of most of the species, but the two which inhabit the eastern coasts of North America, *Cancer irroratus*, Say, and *Cancer borealis*, Stimpson, have been shown by Professor S. J. Smith to occur at considerable depths; the former, according to Professor Smith, has been taken by Alexander Agassiz at 65 to 178 fathoms, and the latter is recorded by Smith from 373 fathoms.[2]

Cancer longipes, Bell.

Cancer longipes, Bell, Trans. Zool. Soc. Lond., vol. i. p. 337, pl. xliii., 1835.
,, ,, A. Milne Edwards, Nouv. Archiv. Mus. Hist. Nat., vol. i. p. 199, 1865, and references to literature.

Two females were collected on the shore at Valparaiso. The largest measures as follows :—

Adult ♀.				Lines.	Millims.
Length of carapace,	.	.	.	27½	58·5
Breadth of carapace,	.	.	.	42½	90
Length of left chelipede,	.	.	.	35½	75
Length of first ambulatory leg, about	.	.	.	40	85

The right chelipede in this specimen is small, having evidently been lost and renewed from the coxal joint of the original limb.

Carpilius, Leach.

Carpilius, Leach (MSS. ?).
,, Desmarest, Consid. gén. sur la classe des Crust., footnote, p. 104, 1825.
,, Milne Edwards, (pt.), Hist. Nat. Crust., vol. i. p. 380, 1834.
,, A. Milne Edwards, (pt.), Nouv. Archiv. Mus. Hist. Nat., vol. i. p. 212, 1865.

Carapace very convex, without dorsal tubercles or sulci defining the regions; with the antero-lateral margins regularly arcuated and entire, terminating in a blunt lobe or tooth (the lateral epibranchial tooth) and slightly longer than the postero-lateral margins, which are nearly straight. The front is rather narrow, deflexed, and is divided into

[1] Proc. Zool. Soc. Lond., p. 67, 1881.
[2] Report by U.S. Fish and Fisheries Commission for 1882, p. 343, 1884.

three lobes, the median being prominent, with the anterior margin slightly concave, and the lateral ones rounded and but little developed (or not at all in young examples). The orbits are small, without marginal fissures, and with a rounded lobe or tooth at their exterior angle. The post-abdomen of the male is six-jointed, with the third and fourth joints coalescent. The eyes are set on short thick pedicels. The basal joint of the antennæ is elongated, and enters well within the long and narrow hiatus existing between the front and the orbits, but does not reach the inner and inferior angle of the orbit, and the very small flagellum is also contained within this hiatus. The merus of the exterior maxillipedes is very obliquely truncated at the distal extremity, and not (or scarcely at all) emarginated at the antero-internal angle. The chelipedes (in the adult male) are large and massive, unequal, the palms without crests or tubercles; the fingers of the larger chelipede are armed with one or two large rounded tubercles on the inner margins. The ambulatory legs are moderately elongated, with the joints smooth. Subcylindrical or slightly compressed.

Of the three well-defined species of this genus admitted by Professor A. Milne Edwards (tom. cit., infra, p. 218), two are commonly and widely distributed throughout the Indo-Pacific region and one is common in the West Indies. Little has been recorded concerning their bathymetrical range.[1]

Carpilius maculatus (Linné).

Cancer maculatus, Linn., Mus. Lud. Ulrici, p. 433, 1764; Syst. Nat., ed. 12, p. 1042, 1766.
Carpilius maculatus, Milne Edwards, Hist. Nat. Crust. vol. i. p. 382, 1834; Atlas des Crust.
in Règne Animal de Cuvier, pl. xi. fig. 2.
,, ,, A. Milne Edwards, Nouv. Archiv. Mus. Hist. Nat., vol. i. p. 214, 1865,
where references to literature are given.

Honolulu, on the reefs (an adult male).

Adult ♂.	Lines.	Millims.
Length of carapace, .	31½	67
Breadth of carapace, .	42	89·5

Atergatis, de Haan.

Atergatis, de Haan, (subgenus) Crust. in Siebold, Fauna Japonica, dec. i. p. 17, 1833.
,, A. Milne Edwards, Nouv. Archiv. Mus. Hist. Nat., vol. i. p. 234, 1865, and synonyms.

This genus somewhat resembles Carpilius in the smooth, convex carapace, whose regions are but faintly indicated, (if at all) but it is distinguished by the following characters:—The carapace is usually more transverse and the front less prominent; the antero-lateral margins are defined by a continuous or nearly continuous carina, which, as

[1] Carpilius lividus, Gibbes, is probably, as stated by A. Milne Edwards, founded on a young specimen of Carpilius maculatus (Forskål). Carpilius praetermissus, Gibbes, is probably, as I have noted (Ann. and Mag. Nat. Hist., ser. 5, vol. ii. p. 407, 1878), identical with Liagore rubromaculata, de Haan.

well as the orbital margins, may, however, be interrupted by very minute notches or fissures; the lobe at the exterior orbital angle, and ordinarily the lateral epibranchial lobe, are not developed. The post-abdomen in the male is usually five-jointed, with the third and fourth, and the fifth and sixth segments coalescent. The basal joint of the antennæ is shorter and attains the infero-lateral process of the front, but is not prolonged within the orbital hiatus. The merus of the exterior maxillipedes is much less obliquely truncated at the distal extremity. The chelipedes of the adult males are less massive, and usually subequal. The fourth to the sixth joints of the ambulatory legs are not merely compressed, but somewhat dilated, and carinated.

The species are rather numerous and occur throughout the Indo-Pacific region, in shallow water and on the shore. The sole representative of this genus in the Challenger collection is the *Atergatis floridus* of Linné, which is the longest known, commonest, and most widely distributed member of the genus.

To the species enumerated by A. Milne Edwards, I may add the following :—

> *Atergatis asperimanus*, White = *Atergatis insularis*, Adams and White. Philippines.[1]
>
> *Atergatis montrouzieri*, A. Milne Edwards. New Caledonia. This species is perhaps identical with *Atergatis asperimanus*.
>
> *Atergatis roseus*, var. *alba* (sic.), Kossmann. Red Sea.

This author regards *Atergatis marginatus*, de Haan, *Atergatis scrobiculatus*, Heller, *Atergatis lævigatus*, A. Milne Edwards, and *Atergatis nitidus*, A. Milne Edwards, as mere varieties of *Atergatis roseus*.

Atergatis floridus (Linné).

Cancer floridus, Linn., Syst. Nat., ed. 12, p. 1041, 1766.
Cancer scyröë, Herbst, Naturgesch. der Krabben u. Krebse, vol. iii. p. 20, pl. liv. fig. 2, 1801.
     ,,   ,,   Milne Edwards, Hist. Nat. Crust., vol. i. p. 375, 1834.
Atergatis floridus, de Haan, Crust. in Siebold, Fauna Japonica, pp. 17, 46, 1835.
     ,,   ,,   Dana, Crust. in U.S. Expl. Exped., vol. xiii. p. 159, pl. vii. fig. 4, 1852.
? *Atergatis compressipes*, MacLeay, Annulosa, in Smith, Zool. S. Africa, p. 59, 1849.

Ternate (an adult male and female); Zebu, on the reef (an adult male).

Adult ♂.					Lines.	Millims.
Length of carapace, nearly	.	.	.	.	14	29·5
Breadth of carapace, .	.	.	.	.	20	42

[1] In the lobulated dorsal surface of the carapace this species approaches *Lophozozymus*, but as the antero-lateral margins are entire, I refer it to *Atergatis*. It has been considered by A. Milne Edwards to be synonymous with *Lophozozymus radiatus*, Milne Edwards, but it differs from Milne Edwards' description of *Lophozozymus radiatus* both as regards the antero-lateral margins and the chelipedes. In the *Ann. and Mag. Nat. Hist.* ser. 5, vol. viii. p. 204, 1881, I erroneously referred to *Atergatis lateralis*, White, as being regarded by A. Milne Edwards as synonymous with *Lophozozymus radiatus*.

Lophactæa, A. Milne Edwards.

Lophactæa, A. Milne Edwards, Ann. d. Sci. Nat., ser. 4, Zool., vol. xviii. p. 43, 1862; Nouv. Archiv. Mus. Hist. Nat., vol. i. p. 245, 1865.

Carapace transverse, convex, its dorsal surface very distinctly lobulated in its anterior half, and often granulated. The antero-lateral margins are longer than the concave postero-lateral margins, and not dentated as in Lophozozymus, but defined by a carina or crest, which is more or less distinctly interrupted by narrow fissures. The cervical suture and the depressions between the lobules of the dorsal surface are usually distinct and smooth. The front is deflexed, not very prominent, and is rough, usually about one-third the width of the carapace; its anterior margin is sinuated and nearly straight, with usually a narrow median fissure. The post-abdomen in the male is usually five-jointed, with the third to the fifth segments coalescent. The eyes, antennæ, and exterior maxillipedes differ in no essential particular from the same parts in Atergatis. The chelipedes (in the adult male) are moderately robust and subequal; the merus-joint very short, carpus and palm smooth or granulated on the upper surface, palm sometimes carinated above, dactyli usually carinated on the superior margins. The ambulatory legs are short, laterally compressed; the fourth to the sixth joints dilated and carinated on the superior margins; dactyli slender, nearly straight, and usually longer than the penultimate joints.

This genus can only be distinguished from Atergatis by the lobulated carapace, and from Lophozozymus by the absence of antero-lateral marginal teeth.

The species occur in the Indo-Pacific region, one (Lophactæa lobata) also in the West Indian Seas and Gulf of Mexico, and at Bermuda; one (Lophactæa picta) at the Cape Verde Islands, St. Lucia, and Salamanca, and one (Lophactæa rotundata) at Cape St. Lucas, California (cf. A. Milne Edwards, Crust. in Miss. Sci. au Méxique, pt. 5, pp. 242, 243, 1879, for synonyma).

The following have been described since Milne Edwards' Monograph was published in 1865:—

Lophactæa picta, A. Milne Edwards. St. Lucia; Salamanca.
Lophactæa violacea, A. Milne Edwards. New Caledonia.
Lophactæa helleri, Kossmann. Red Sea.

Lophactæa lobata, A. Milne Edwards.

? Cancer spectabilis, Herbst, Naturgesch. der Krabben u. Krebse, vol. ii. Heft. v. p. 153, pl. xxxvii. fig. 5, 1794.
„ lobatus, Milne Edwards, Hist. Nat. Crust., vol. i. p. 375, 1834.
Lophactæa lobata, A. Milne Edwards, Nouv. Archiv. Mus. Hist. Nat., vol. i. p. 249, pl. xvi. fig. 3, 1865; Crust. in Miss. Sci. au Méxique, pt. 5, p. 242, 1879, et synonyma.

Bermuda, on the shore (an adult male and two females, whereof one bears ova).

In no two of the specimens of this species is the pattern defined by the coloured lines on the dorsal surface of the carapace exactly alike. The colour (in spirit-specimens) is the reverse of that mentioned by A. Milne Edwards in his latest description of this species, the blotches upon the carapace being of a yellowish hue, and the narrow lines defining them of a crimson red, as in the description and figure of Herbst's *Cancer spectabilis*, which very probably represents *Lophactæa lobatus*, but as Herbst was ignorant of the locality of his types, and both description and figure are very insufficient, I follow Professor A. Milne Edwards in retaining for this species the name by which it has been generally known.

The adult male has the following dimensions:—

	Lines.	Millims.
Adult ♂.		
Length of carapace, .	7½	16
Breadth of carapace, about .	10½	22·5

Lophactæa granulosa, (Rüppell).

Xantho granulosa, Rüppell, Beschreib. 24 kurzschwänzigen Krabben des rothen Meeres, p. 24, pl. v. fig. 3, 1830.
Lophactæa granulosa, A. Milne Edwards, Nouv. Archiv. Mus. Hist. Nat., vol. i. p. 247, 1865 : vol. ix. p. 187, 1873, et synonyma.

Honolulu, on the reefs (an adult female).

	Lines.	Millims.
Adult ♀.		
Length of carapace, .	8½	18
Breadth of carapace, .	12½	26

Lophozozymus, A. Milne Edwards.

Lophozozymus, A. Milne Edwards, Ann. d. Sci. Nat., ser. 4, vol. xx. p. 276, 1863 ; Ann. Soc. Entom. France, ser. 4, vol. vii. p. 272, 1867.

The carapace is broader than long, lobulated near the antero-lateral margins, depressed or moderately convex ; its antero-lateral margins are distinctly dentated, but occasionally (subgenus *Lophoxanthus*) the first and second antero-lateral marginal teeth are not developed. The front projects but little, and is divided by a small median notch into two broad and truncated lobes. The orbits are nearly circular, and marked with fissures which are closed, or even obsolete. The post-abdomen in the male is five- to seven-jointed (the third to the fifth segments frequently coalescent). The eyes are set on short, thick peduncles. The basal joint of the antennæ is in contact with the infero-lateral process of the front, but does not enter the interior subocular hiatus. The exterior maxillipedes are of the normal form, nearly quadrate, and distally truncated, with the antero-external angle not produced and the antero-internal angle very slightly (if at all) emarginate ; exognath of moderate width. The chelipedes (in the adult male) are moderately developed, with the merus-joint short, trigonous ; wrist not carinated above, but toothed on the inner margin; palm slightly compressed and often carinated above; fingers toothed on the inner margins. The ambulatory legs are of moderate length, compressed, with the merus, carpus, and penultimate joint carinated on the superior margins.

The species are widely distributed throughout the Oriental region, and occur also on the Californian coasts, and except the type *Lophozozymus epheliticus* (Linn.), are of rather small size. I have described a species, *Lophozozymus (Lophoxanthus) sexdentatus*, from the Senegambian coast.

The single species in the Challenger collection is to be referred to the subgenus *Lophoxanthus*,[1] which is apparently distinguished from the typical *Lophozozymus* merely by the less transverse and generally more depressed carapace, and by the obsolescence of the first and second teeth of the antero-lateral margins, and passes by almost insensible gradations into the typical form, and to which the following species and varieties have been referred :—

> *Lophoxanthus lamellipes* (Stimpson). California and West Coast of Mexico.
> *Lophoxanthus bellus* (Stimpson) = *Xanthodes leucomanus*, Lockington, and
> *Xanthodes hemphilli*, Lockington (vars.). California and Japan (50 fathoms).
> *Lophoxanthus sexdentatus*, Miers. Goree Island, Senegambia (9 to 15 fathoms).

Lophozozymus (Lophoxanthus) bellus, Stimpson, var. *leucomanus*, Lockington (Pl. XI. fig. 1).

> *Xanthodes leucomanus*, Lockington, Proc. Calif. Acad. Nat. Sci., pp. 32, 100, 1876.

The carapace is transverse, nearly flat and plane, except behind the front and the first antero-lateral marginal teeth, where it is rugose and granulated; the rugosities exist also on the front of the gastric and branchial regions; the cervical suture and the sutures of the frontal, gastric and branchial regions are distinct; the frontal margin is nearly straight, with a median notch. Of the antero-lateral teeth, only the three posterior are distinctly developed; these are triangulate and moderately prominent; in front of the first of these is the rudiment of another tooth. The pterygostomian regions are granulated; the sternal surface is nearly smooth. Only five of the post-abdominal segments are distinct (the third and fourth, and the fifth and sixth, are coalescent), the terminal segment is slightly transverse and distally rounded. The basal antennal joint reaches to the infero-lateral frontal process, but not to the apex of the inner subocular angle of the orbit. The outer maxillipedes present nothing remarkable; their narrow exognath reaches to the antero-external angle of the distally truncated merus-joint. The chelipedes are moderately large and robust (either the right or left is the larger). The upper margins of their short and trigonous merus-joints are armed with a series of

---

[1] *Lophoxanthus*, A. Milne Edwards (genus), Crust. in Miss. Sci. au Méxique, pt. 5, p. 256, 1879.

granules, the carpus and palm are rugose and pitted above, the dorsal surface thus appearing reticulated, the pits on the upper surface of the palms disposed in longitudinal series ; the palm of the smaller chelipede is strongly granulated on its outer surface and is longitudinally bisulcated ; these sulci are discernible (but less distinctly) on the larger chelipede ; the fingers are deep brown, denticulated on their inner margins and with the acute apices slightly decussate ; the mobile finger is longitudinally carinated and sulcated above. The ambulatory legs have the merus, carpus, and penultimate joints compressed and more or less distinctly carinated, the carinæ strongest on the upper margins of the merus-joints ; dactyli slender and closely pubescent. Colour (in spirit) yellowish-brown.

	Linea	Millima
Adult ♂.		
Length of carapace, nearly	4	8
Breadth of carapace,	5½	11
Length of larger chelipede, nearly	6	12
Length of second ambulatory leg,	6	12

Several specimens of both sexes were collected in the Japanese Sea, in 50 fathoms, lat. 34° 38′ 0″ N., long. 135° 1′ 0″ E. (Station 233A).[1]

Medæus, Dana.

Medæus, Dana, Amer. Journ. Sci. and Arts, ser. 2, vol. xi. p. 125, 1851 ; U.S. Explor. Exped., vol. xiii., Crust. 1, pp. 149, 181, 1852.

The carapace is moderately transverse and convex, with the surface more or less distinctly lobulated or granulated, and the antero-lateral margins armed with from three to five distinct teeth ; the front projects slightly beyond the orbits, and is usually divided into two lobes by a median notch. The antero-lateral margins are continued, in the typical species, Medæus ornatus, beneath the inferior margin of the orbit as a denticulated crest, which, however, is obsolete in other species which have been referred to this genus. The eyes and orbits are small and nearly as in Xantho. The post-abdomen in the male is five-jointed, with the third to the fifth joints coalescent. The basal antennal joint attains the infero-lateral process of the front, or even, in some species, enters very slightly within the inner orbital hiatus. The exterior maxillipedes present nothing remarkable, being distally truncated and very slightly emarginate at the antero-internal angle. The chelipedes are moderately developed and the palms are usually tuberculated

[1] These specimens are rather doubtfully referred to this variety, originally described from Californian types, on account of the brevity of Lockington's description, but they agree with it in that which seems to be the principal characteristic of this variety, i.e., in the pitted carpus and palm of the chelipedes, which present somewhat the appearance of being covered above with a network of raised lines ; the dactyli, however, are dull, not specially lenceus. Mr. Lockington, in a MS. note in the copy of his paper sent by him to the author, admits the specific identity both of Xanthodes leucomanus and Xanthodes hemphilli with Lophoxanthus bellus. A specimen, sent by Mr. Lockington as "Xanthodes hemphilliana," in the collection of the British Museum, also apparently belongs to the variety leucomanus (cf., his remarks on Xanthodes hemphilliana, tom. cit., p. 100).

on the upper surface; fingers dentated on the inner margins and distally acute. The ambulatory legs are rather slender and the merus-joints are usually, but not invariably, spinuliferous or denticulated above.

This genus is very nearly allied to *Halimede* (de Haan), but is distinguished, as M. A. Milne Edwards has noted, by the coalescence of the third to the fifth post-abdominal segments.

The following species have been referred to this genus :—

  Medæus ornatus, Dana. Hawaiian Islands.
  Medæus elegans, A. Milne Edwards. New Caledonia.
  Medæus nodosus, A. Milne Edwards. New Caledonia and Lifou Islands.
  Medæus spinimanus (Milne Edwards) = *Cancer miniatus*, Desbonne and Schramm, *fide*, A. Milne Edwards. Guadeloupe.
  Medæus haswelli, n. sp. Australia, Twofold Bay, 150 fathoms.
  Medæus simplex, A. Milne Edwards. Madagascar and Samoa Islands (Upolu).

Medæus haswelli, n. sp. (Pl. XI. fig. 2).

The carapace is slightly transverse, rather flat, with the sutures defining the various regions of the dorsal surface very distinct; the body is everywhere very distinctly and evenly granulated; the granules of the dorsal surface somewhat larger and less crowded than those of the inferior surface of the body. The antero-lateral margins of the carapace are somewhat shorter than the postero-lateral margins. The front is rather broad and but little prominent; its anterior margin is obscurely sulcated longitudinally and divided into two lobes by a slight median notch; the antero-lateral margins of the carapace bear only three triangular teeth, the first of which is placed at some distance from the exterior orbital angle, and the second rather nearer to the third than to the first tooth. The infra-orbital prolongations of the antero-lateral margins, characteristic of the type of the genus (*Medæus ornatus*) are here indicated merely by a very obscure line of granules, reaching from the first of the antero-lateral teeth to a point just beneath the exterior angle of the orbit. The orbits are oval, with a moderately prominent inner suborbital lobe and a slightly indicated notch at the outer angle, and another in the middle of the superior margin. The post-abdomen in the male is five-jointed (as in the type), with the third to the fifth joints coalescent, and the terminal joint transverse and semicircularly rounded. The antennæ and maxillipedes present nothing remarkable; the basal antennal joint reaches to the infero-lateral frontal process, and the exognath of the outer maxillipedes attains the antero-external angle of the distally truncated merus-joint. The chelipedes are moderately robust for so small a species, the joints granulated much more distinctly on the outer than on the inner surface; the merus-joint has a series of larger granules on its upper margin, and the carpus a spiniform tooth on its inner surface; the palm of the

larger chelipede is much more distinctly granulated on and near the upper margin and at
the base than elsewhere; the fingers are regularly denticulated on the inner margins and
acute at the tips, and the dactylus sulcated above.   The smaller chelipede is externally more
strongly granulated, with relatively longer and more slender fingers; the ambulatory legs
(of the second pair) are slender, with the merus, carpus, and penultimate joints
denticulated on their upper margins, and the dactyli styliform, straight.   Colour (in spirit)
yellowish-brown; the fingers of the chelipedes are chocolate-brown, the coloration not
extending over the palmar surface.

Adult ♂.						Lines	Millims.
Length of carapace, about	.	.	.	.	.	3	6
Breadth of carapace, about	.	.	.	.	.	4	8·5
Length of larger chelipede, nearly	.	.	.	.	5	10

Two males and a female were trawled off Twofold Bay in 150 fathoms (Station 163a).
All are very small in size, yet the male, from which the description is mainly taken, is
apparently adult.   The ambulatory legs are deficient, except in one specimen, where only
one leg is perfect.

The characteristic disposition of the sulci of the carapace may be better understood
by a reference to the figure than by any detailed description.

The nearest ally of this species is perhaps *Medæus elegans*, A. Milne Edwards, from
New Caledonia,[1] in which, to judge from the figure, the areolation of the carapace is
different and the antero-lateral margins of the carapace longer; in *Medæus granulatus*
the small intermediary spines and tubercles of the antero-lateral margins, characteristic
of *Medæus elegans*, do not exist.   The granulated (not cristiform or nodose) palms of
the chelipedes further distinguish *Medæus granulatus* from all the species known to me.[2]

Actæa, de Haan.

Actæa, de Haan, Crust. in Siebold, Fauna Japonica, dec. i. p. 18, 1833.
  „   Dana, U.S. Explor. Exped., vol. xiii., Crust. 1, p. 148, 1852.
  „   A. Milne Edwards (pt.), Nouv. Archiv. Mus. Hist. Nat., vol. i. p. 259, 1865.
Euxanthodes, Paulson, 1875, *fide* Kossmann.
Psaumis, Kossmann, Crust. in Zool. Ergeb. einer Reise Küstengeb. d. rothen Meeres,
    Malacostraca, p. 26, 1877.[3]

The carapace is transverse and moderately convex and is usually very distinctly
lobulated on the dorsal surface; the lobules usually granulated, with the intervening

---

[1] *Nouv. Archiv. Mus. Hist. Nat.*, vol. ix. p. 211, pl. viii. fig. 1, 1873.

[2] This species also bears some external resemblance to the Australian form I have recently described as *Galene
granulata*, but in *Galene granulata*, as in the typical *Galene biepinosus* (Herbst), the basal antennal joint does not reach
the subfrontal process.   Neither in the species of *Galene* nor in *Medæus haswelli* are ridges developed on the endostome
or palate.

[3] The genus *Banareia*, A. Milne Edwards, may perhaps not be distinct from *Actæa*; I have, however, referred a
species to this genus in the Report on the Crustacea of H.M.S. "Alert," p. 210, pl. xix. fig. c, 1884, *vide* A. Milne
Edwards, *Ann. Soc. Entom. France*, ser. 4, vol. ix. p. 167, 1869.

sutures very distinct. The front is bilobate or indistinctly quadrilobate, the antero-lateral margins are regularly arcuated and divided into lobes which are rounded and often obscurely defined. The orbits are small and are usually marked with two or three closed fissures. The post-abdomen in the male is five- to seven-jointed. The eyes are set on short thick pedicels. The basal antennal joint is short and reaches the infero-lateral process of the front, but not the extremity of the interior subocular lobe of the orbit (*Psaumis*, Kossmann), or more generally, enters slightly within the interior hiatus of the orbit, and then may reach the extremity of the subocular orbital lobe ; the flagellum lies within the interior orbital hiatus. The exterior maxillipedes present nothing remarkable, having the merus-joint quadrate, distally truncated, and slightly emarginate at the antero-internal angle. The chelipedes are subequal and moderately developed, with the wrist and palm usually granulated and the fingers distally acute or subacute, not excavated ; the ambulatory legs are of moderate length and are usually compressed, not distinctly carinated, but the superior margins of the merus-joints are sometimes acute.

The species of this genus (even if *Actæodes* be excluded) are numerous and of small size, and are widely distributed both in the Atlantic and Oriental or Indo-Pacific regions.

The nearest ally to *Actæa* is undoubtedly *Actæodes*, Dana, which is distinguishable merely by the excavated fingers of the chelipedes, and which is united with *Actæa* by Milne Edwards and Kossmann, but in the absence of any complete revision of the numerous genera of the Cyclometopa or Cancroidea since that given by Dana, I have followed his classification, in which they are referred to distinct subfamilies.[1]

The following are species which are referable to this genus, and are either not included in Milne Edwards' monograph of the genus (*tom. cit.*), or have been described since its publication :—

Actæa peronii (Milne Edwards) = *Xantho spinosus*, Hess. Australia.
Actæa parvula (de Haan). Cape of Good Hope ; Natal. This may be a species of *Actumnus*.
Actæa consobrina, A. Milne Edwards. Upolu, Samoan Islands.
Actæa margaritaria, A. Milne Edwards. Cape St. Vincent.
Actæa dovii, Stimpson. San Salvador, Panama.
Actæa hystrix, n. sp. Off Cape York.
Actæa spinifera, Kingsley. Plantation Key.
Actæa glabra (Kossmann). Red Sea.[2]

[1] If (as would probably be done in a more natural system), *Actæodes* be united with *Actæa*, then it will probably be necessary to unite also *Carpilodes* with *Liomera*, *Leptodius* with *Xantho*, *Lophoxyxus* with *Zozymus*, &c.

[2] *Actæodes zantho*, Lockington, according to a MS. note in the author's copy of his paper, is synonymous with *Actæa sulcata*, Stimpson.

Actæa granulata (Audouin).

Cancer granulatus, Audouin, Explic. des Planches in Savigny, "Egypte," Crust., pl. vi. fig. 2.
    „   (*Actæa*) *granulatus*, de Haan, Crust. in Siebold, Fauna Japonica, p. 18, 1835.
Actæa granulata, A. Milne Edwards, Nouv. Archiv. Mus. Hist. Nat., vol. i. p. 275, 1865, et synonyma.

Torres Strait, August, 1874 (an adult male).
This specimen is of a bright orange colour (in spirit).

Adult ♂.				Lines.	Millims.
Length of carapace,	.	.	.	4½	9·5
Breadth of carapace, about	.	.	.	6	13

A small female is in the collection, also obtained in the North Australian Seas, Cape
York (8 fathoms, Station 186), which can hardly be determined with certainty. It is
perhaps referable to *Actæa hirsutissima* (Rüppell), although distinguished from the
typical condition of this species by the less distinctly granulated and areolated carapace.
It has also much resemblance to *Actæa pilosa*, Stimpson, from which it is distinguished
by the less acute fingers of the chelipedes, which are more regularly denticulated on the
inner margins.

♂.		Lines.	Millims.
Length of carapace, rather over	. .	5	11
Breadth of carapace, . .	. .	7½	15·5

Actæa nodulosa, White.

Actæa nodulosa, White, Proc. Zool. Soc. Lond., p. 224, 1847; Crust. Zool. H.M.S.
            "Samarang," p. 39, pl. viii. fig. 4, 1847.
    „   „   A. Milne Edwards, tom. cit., p. 267, 1865.

An adult female was obtained on the reefs at Honolulu.
This specimen (in spirit) is of a bright orange-red colour, and is of larger size than
either of the typical examples, from the Mauritius, in the collection of the British
Museum ; its dimensions are as follows :—

Adult ♀.					Lines.	Millims.
Length of carapace, about	.	.	. .	.	5½	11·5
Breadth of carapace, about	.	.	.	.	8½	18

Actæa nodulosa bears considerable external resemblance to *Actæa granulata*, from
which it is at once distinguishable by the much more transverse carapace and the more
prominent granulated lobes of the antero-lateral margins ; the tubercles of the carapace
are also granulated, and are not confluent as in *Actæa granulata*.

Actæa hystrix, n. sp. (Pl. XI. fig. 3).

The body is naked, except as regards the dactyli of the ambulatory legs, which possess a few hairs. The carapace is but slightly transverse, and its dorsal surface is everywhere covered with close-set, conical, acute granules, which are longer and more spiniform near the antero-lateral margins; the inter-regional sutures and those defining the areolets of the dorsal surface are moderately distinct; the front has its anterior margin deflexed and divided into two lobes by a median notch; the frontal lobes are rounded and granulated, with their external angles little prominent. The three posterior lobes of the antero-lateral margins are distinct, and each is armed with three or four, conical, spiniform granules. The pterygostomian regions are very closely and finely granulated; the sternal surface of the body and the post-abdomen is coarsely punctulated. All the segments of the post-abdomen (in the female) are distinct; the terminal segment is the longest; the short eyes are closely set in the orbits, whose margins are granulated, as are also the eye-peduncles themselves, near to the cornea. The basal antennal joint is somewhat longer than broad, and reaches to the subfrontal process, but not to the apex of the inner suborbital angle; the exognath of the outer maxillipedes reaches to the antero-external angle of the transverse, distally truncated, merus-joint. The chelipedes (in the male) are moderately robust and subequal; the merus, wrist, and palm are covered with numerous close-set, conical, acute granules, which are longer and spinuliform on the upper margins; the inner surface of these joints is granulated; the palm is but little longer than the wrist; the fingers are colourless and regularly denticulated on their inner margins; externally granulated; the dactylus has two or three spinules on its upper surface; the ambulatory legs are of moderate length, closely granulated, the granules lengthening into short spines on and near the upper margins of the merus, carpus, and penultimate joints; the dactyli are armed with a small terminal claw. Colour (in spirit) whitish.

Adult ♂ .		Lines.	Millims.
Length of carapace, nearly	.	3	6
Breadth of carapace, nearly	.	4	8
Length of a chelipede, rather over	.	3	7
Length of second ambulatory leg,	.	3	7

Of this pretty little species but a single female specimen was obtained, off Cape York, in 8 fathoms, lat. 10° 30′ S., long. 142° 18′ E. (Station 186).

In many of its characters it is nearly allied to the West Indian *Actæa acantha* (Milne Edwards),[1] in which species, however, the spinules and granules of the carapace are less numerous and are interspersed with hairs, and the fingers are dark coloured.

[1] *Cf.* A. Milne Edwards, Crust. in Miss. Sci. au Méxique, pt. 5, p. 245, pl. xliii. fig. 1, 1879.

From the Australian form, designated by White *Actæa carcharias*,[1] which is probably to be regarded merely as a marked variety of *Actæa granulata* (Audouin), it is at once distinguished by the non-punctulated carapace, the punctulated post-abdomen, which is smooth in the middle line, and the much longer spinules of the ambulatory legs; the latter character also distinguishes it from the Australian *Actæa calculosa*, Milne Edwards.

Actæa peronii (Milne Edwards).

Xantho peronii, Milne Edwards, Hist. Nat. Crust., vol. i. p. 392, 1834.
    ,,   *spinosus*, Hess, Archiv f. Naturgesch., vol. xxxi. p. 132, pl. vi. fig. 3, 1865.
Actæa peronii, Haswell, Cat. Australian Stalk and Sessile-Eyed Crust., p. 46, 1882.

Off East Moncœur Island, Bass Strait, in 38 to 40 fathoms, Station 162 (several males and females); South Australian coast? "2 to 10 fath., April 1874" (a male).

The colour (in spirit) is pinkish or orange. In the largest examples the tubercles of the carapace and spines of the legs are very prominent, and traces of the former are discernible even on the posterior and postero-lateral parts of the carapace, which, in the smallest examples, is perfectly smooth in its posterior part.

The largest male measures as follows :—

Adult ♂.	Lines.	Millims.
Length of carapace,	4	8·5
Breadth of carapace,	5½	10·5
Length of a chelipede,	5½	10·5
Length of first ambulatory leg,	5½	10·5

Actæa rufopunctata (Milne Edwards), var. *nodosa*.

Actæa nodosa, Stimpson, Ann. Lyc. Nat. Hist., New York, vol. vii. p. 203, 1862.
    ,,   ,, A. Milne Edwards, Nouv. Archiv. Mus. Hist. Nat., vol. i. p. 266, pl. xvii. fig. 6, 1865; Crust. in Miss. Sci. au. Méxique, pt. 5, p. 245, 1879.

Bahia, shallow water, September 1873 (an adult female).

This specimen differs in nothing from Oriental examples of *Actæa rufopunctata*, unless it be in the somewhat more closely and finely granulated prominences of the carapace, and I have observed a specimen from the Mauritius, in the collection of the British Museum, which resembles it in this respect. The carapace, although faded, yet presents distinct traces of the characteristic red colour of *Actæa rufopunctata*.[2]

Adult ♀.	Lines.	Millims.
Length of carapace, rather over	6	13
Breadth of carapace, nearly	9	18·5

[1] Proc. Zool. Soc. Lond., p. 234, 1847.
[2] I have already noted the occurrence of *Actæa rufopunctata* in the Atlantic Region, Proc. Zool. Soc. Lond., p. 68, 1881.

Atergatopsis, A. Milne Edwards.

Atergatopsis, A. Milne Edwards, Nouv. Archiv. Mus. Hist. Nat., vol. i. p. 252, 1865.

Carapace transverse, regularly convex, with the dorsal surface smooth or granulated, but not distinctly lobulated as in *Actæa*. The front (in the adult) is less than one-third the width of the carapace, its anterior margin is sinuated or indistinctly quadrilobated ; the antero-lateral margins of the carapace are longer than the postero-lateral margins, regularly arcuated, neither cristated as in *Atergatis*, nor as distinctly lobated as in *Actæa*. The post-abdomen in the male is usually five-jointed, with the third to the fifth segments coalescent. The orbits, eyes, antennæ, and exterior maxillipedes are nearly as in *Actæa* ; the basal antennal joint reaches the infero-lateral process of the front. The chelipedes, in the adult male, are subequal and well developed, with the carpus bluntly toothed or lobed on the inner margin ; palm rounded, not cristated above ; fingers lobed or toothed on the inner margin, and acute at the distal extremity. The ambulatory legs are of moderate length, with the fourth to the sixth joints usually somewhat compressed, but not greatly dilated or carinated ; dactyli slender and nearly straight.

From *Actæa* this genus is distinguished by the more convex and less distinctly lobulated carapace, with sub-entire, antero-lateral margins, and from *Atergatis* by the non-carinated and non-cristated antero-lateral margins of the carapace, and joints of the ambulatory legs.

The species, which may attain a considerable size, occur throughout the Indo-Pacific region. The following species has been described since the publication of A. Milne Edwards' monograph of the genus :—

> *Atergatopsis amoyensis*, de Haan.   China, Amoy.

Atergatopsis granulatus, A. Milne Edwards.

Atergatopsis granulatus, A. Milne Edwards, Nouv. Archiv. Mus. Hist. Nat., vol. i. p. 255,
                                    pl. xiii. fig. 2, 1865.
          „   „   Miers, Crust. in Rep. Zool. Coll. H.M.S. "Alert," p. 529, 1884.

South of New Guinea, 28 fathoms, lat. 9° 59′ 0″ S., long. 139° 42′ 0″ E. (Station 188), a very small female.

In this specimen the carapace is granulated only near the front and antero-lateral margins ; the chelipedes are granulated as in the adult. The cervical and other sutures of the dorsal surface of the carapace are scarcely distinguishable. The strong compressed tooth of the lower (immobile) finger or pollex is distinctly developed.

♀.	Linea.	Millima.
Length of carapace, about .	3½	7
Breadth of carapace, nearly .	5	10

Xantho, Leach.

Xantho, Leach, Trans. Linn. Soc. Lond., vol. xi. p. 320, 1815.
   „   Milne Edwards (pt.), Hist. Nat. Crust., vol. i. p. 387, 1834.

The carapace is transverse and depressed, often nearly flat above, with the dorsal surface uneven and lobulated, at least near the front and antero-lateral margins. The front projects but slightly beyond the level of the orbits and is usually divided by a very small median notch; the antero-lateral margins are regularly arcuated and are divided in the typical species into four or five dentiform lobes, which are not mere rounded prominences as in *Actæa*, and are not produced and spiniform; in other species the marginal teeth are irregular, the orbits are small, and their margins usually bear indications of two or three closed fissures. The post-abdomen, in the male, is from five-to seven-jointed. The eyes are set on short thick pedicels. The basal antennal joint reaches the infero-lateral frontal process, and the flagellum lies within the interior orbital hiatus. The exterior maxillipedes present nothing remarkable, having the merus-joint distally truncated and usually slightly emarginate at the antero-internal angle. The chelipedes (in the typical species) are often considerably developed in the male, usually unequal, with the palms but slightly compressed, rounded above, and the fingers dentated on the inner margins. The ambulatory legs are of moderate length, and their joints are neither carinated nor spinose.

The species are somewhat numerous and never attain a very large size; they occur commonly on the coasts or islands both of the Atlantic and Indo-Pacific regions, or in shallow water. The Challenger collection contains but a single representative of this genus.

The following species are apparently referable to this genus as defined above, and as restricted by A. Milne Edwards. It is connected by almost insensible gradations on the one hand with *Lophoxanthus* and *Xanthodes*, on the other with *Panopeus* and *Eurypanopeus*.

I. Carapace transverse, the front not very prominent; the antero-lateral margins normally toothed, the teeth not rounded (typical species of the genus).

The genus should perhaps be restricted to include only the species of this section.

   Xantho floridus (Mont.) = *Cancer poressa*, Olivi (?), *Cancer incisus*, Leach.
         Shores of Europe (70 to 75 metres, A. Milne Edwards).
   Xantho rivulosus, Risso = *Cancer hydrophilus*, Herbst (?), Europe, northward to
         Shetlands and Scandinavia, 15 to 40 fathoms (Heller).
   Xantho tuberculatus, Bell.   Mediterranean and South British Coasts.
   Xantho minor, Dana.   Madeira or Cape Verde Islands.
   Xantho pilipes, A. Milne Edwards.   Senegal; Senegambia, Goree Island, 9 to 15
         fathoms.

Xantho impressus (Lamarck). Indo-Pacific Region (Mauritius to New Caledonia).

Xantho hirtipes, Milne Edwards. Red Sea.

Xantho truncatus, de Haan. Japan. In this species the front is somewhat prominent.

Xantho arcuatus, Heller. Tahiti. The basal antennal joint is somewhat elongated and surpasses the infero-lateral frontal process.

Xantho macgillivrayi, Miers. North-East Australia.

Xantho crenatus, Milne Edwards. Peru.

Xantho bidentatus, A. Milne Edwards. Sandwich Islands; Philippines, Samboangan; Samoan Islands (Coll. Brit. Mus.). In this species, only the two posterior teeth of the antero-lateral margins are distinctly developed, but the lobes are dentiform or tuberculiform.

II. Carapace very transverse, with the two or three posterior lobes of the antero-lateral margins developed, the lobes rounded (*Liomera* (pt.), A. Milne Edwards).

This section includes such species of *Liomera* as differ from the typical *Liomera lata*, as figured by Dana (which is regarded by A. Milne Edwards as the young of *Carpilodes cinctimanus*) in having a shorter basal antennal joint, which is not produced along the exterior margin of the infero-lateral frontal process, and in which the fingers of the chelipedes are acute and the ambulatory legs not spinuliferous.

Xantho punctatus, Milne Edwards. Indo-Pacific Region.

Xantho maculatus (Haswell). Endeavour River, North-East Australia.

Xantho granosimanus, A. Milne Edwards. New Caledonia; Seychelles (Coll. Brit. Mus.).

Xantho dispar, Dana (Rio Janeiro (?), Dana), which comes near the species of this section, is probably to be referred to *Heterozius*, A. Milne Edwards.

III. Carapace usually less transverse with the front more prominent, the denticles of the antero-lateral margins of the carapace more numerous and irregular.

Certain of the species of this section approach the genus *Paraxanthus*.

Xantho occidentalis, A. Milne Edwards. Cape St. Vincent.

Xantho denticulatus, White. West Indies; Vera Cruz; Brazil.

Xantho stimpsonii, Smith = *Xantho multidentatus*, Lockington. California, Cape St. Lucas; Mazatlan. Panama, (Coll. Brit. Mus.).

? *Xantho nudipes*, A. Milne Edwards. Indo-Pacific Region, (Seychelles and New Caledonia).

Xantho bidentatus, A. Milne Edwards (Pl. XI. fig. 4).

? *Xantho bidentatus*, A. Milne Edwards, Ann. Soc. Entom. France, ser. 4, vol. vii. p. 266, 1867.

The body and limbs are nearly naked, only the pterygostomian regions and the three last joints of the ambulatory legs being partly clothed with a very short close pubescence ; the joints of the ambulatory legs also are fringed with a few longer hairs. The carapace is transverse, slightly convex, smooth and shining, and is very faintly lobulated on the postfrontal, the hepatic, and the front of the gastric and branchial regions ; the cardiac and the posterior part of the branchial regions are nearly smooth, the cervical and other sutures of the anterior part of the carapace are distinct but not deep. The frontal margin is divided by a small median notch into two truncated and slightly sinuated lobes, which project slightly over the antennulary septa and beyond the inner angles of the orbits, which are not at all prominent ; the orbital margins are entire, the antero-lateral margins of the carapace are rather shorter than the postero-lateral margins, and are armed only with two distinct teeth, the fourth and fifth of the normal series ; in front of these, two very obscure prominences indicate the position of the second and third antero-lateral teeth. The post-abdomen, in the male, is five-jointed, the third to the fifth joints coalescent. The eyes are small; the basal joint of the antennæ is but little longer than broad, and its antero-internal angle is in contact, as is usual in the genus, with the infero-lateral process of the frontal margin ; the ischium-joint of the outer maxillipedes is marked on its outer surface with a longitudinal, impressed, and slightly sinuated line ; the merus-joint is somewhat transverse, slightly emarginate at its antero-internal angle; the exognath does not reach to the distal extremity of the merus-joint. The chelipedes, in the male, are smooth and polished ; the merus or arm very short, trigonous ; carpus with an angulated prominence on its inner surface, palm smooth, both on its outer and inner surfaces, with the upper margin rounded, its outer surface rather convex, with scattered punctulations ; fingers denticulated on their inner margins and with acute apices ; the ambulatory legs are relatively rather slender and compressed.   Colour (in spirit) yellowish-brown ; the fingers dark-brown or nearly black ; the colour of the lower finger extends for a very short distance inwards over the inner and outer surfaces of the palm.

	Lines.	Millims.
Adult ♂.		
Length of carapace, rather over	4	9
Greatest breadth,	6	12·5
Length of larger chelipede, about	8	17
Length of second ambulatory leg, nearly	6½	14

The unique specimen (a male) was obtained on the beach at Samboangan, in the Philippines.

This species is characterised by the smooth and shining surface of the carapace, which, viewed with the microscope, is seen to be closely and minutely punctulated, by the obsolescence of the three first teeth of the antero-lateral margins, the very obscure lobulation of the dorsal surface, and the form of the chelipedes.

In the single specimen in the Challenger collection the right chelipede is considerably the larger and more robust, but does not differ markedly in form from the smaller one.

It has much the aspect of a species of *Liomera*, but the basal antennal joint does not enter so far within the inner orbital hiatus as in the typical species of that genus. It much resembles *Xantho punctatus*, Milne Edwards, as figured by A. Milne Edwards,[1] which I formerly referred to the genus *Liomera*, but is distinguished by the more distinctly developed although not prominent teeth of the antero-lateral margins.

A. Milne Edwards' description of this species is very brief, and its identification is therefore somewhat uncertain. If the Challenger species be distinct it may be designated by the name I originally applied to it, *Xantho lævidorsalis*.

Xanthodes, Dana.

Xanthodes, Dana, Proc. Acad. Nat. Sci. Philad., p. 75, 1852; U.S. Explor. Exped., vol. xiii., Crust. 1, p. 148, 1852.

In this very doubtful genus the carapace is more convex and deflexed near the front and antero-lateral margins than in the preceding (somewhat approaching certain species of *Pilumnus*), but posteriorly depressed; the antero-lateral margins are not thin-edged or cristiform, and their teeth are tuberculiform, or even spinuliform. The basal antennal joint is short and barely reaches the infero-lateral process of the front. The post-abdomen, in the male, is five-jointed.

It includes the following species, as at present restricted:—

A. *Oriental Species.*

Xanthodes lamarckii (Milne Edwards) = *Xantho cultrimanus*, White, and *Xanthodes granosomanus*, Dana. Indo-Pacific Region.

Xanthodes nitidulus, Dana. Paumotu Archipelago; New South Wales, Port Jackson (Coll. Brit. Mus.).

Xanthodes notatus, Dana. Paumotu Archipelago; Sandwich Islands; New Caledonia.

Xanthodes depressus, White. Philippines.

Xanthodes elegans, Stimpson. Japan, Simoda.

Xanthodes pachydactylus, A. Milne Edwards. New Caledonia. This species, in the well-developed antero-lateral teeth, approaches *Xantho*.

Xanthodes scabra (Fabricius). Souda Islands; East Australia; "India-orientalis" (Fabricius).

Xanthodes atromanus, Haswell. ?

[1] *Nouv. Archiv. Mus. Hist. Nat.*, vol. ix. p. 199, pl. vii. fig. 6, 1873.

B. West American Species.

Xanthodes taylori, Stimpson = *Xanthodes spinituberculatus* (Lockington). California.
Xanthodes xantusii, Stimpson. California.
Xanthodes insculpta, Stimpson. California.
Xanthodes (?) *angustus*, Lockington. California.

C. Atlantic Species.

Xanthodes eriphioides, A. Milne Edwards. Cape St. Vincent.
Xanthodes melanodactylus, A. Milne Edwards. Cape St. Vincent; Cape Verde Islands and Azores (50 to 90 fathoms); Madeira, Ascension Island; Goree Island, Senegambia (Coll. Brit. Mus.).
Xanthodes rufopunctatus, A. Milne Edwards. Cape St. Vincent; Maio. (Very near the preceding.)
Xanthodes bidentatus, A. Milne Edwards. Grenada (92 fathoms). This species should, perhaps, be placed in the genus *Xantho*, and if so, a new name must be adopted for it.

Xanthodes melanodactylus, A. Milne Edwards.

Xanthodes melanodactylus, A. Milne Edwards, Nouv. Archiv. Mus. Hist. Nat., vol. iv. p. 60, pl. xvii. figs. 1–3, 1868.

Numerous specimens were obtained at St. Vincent, Cape Verde Islands, and also off Fayal, in 50 to 90 fathoms, and a very small example off Gomera, Canary Islands, in 75 fathoms.

I have elsewhere remarked on the variability of the colour of this species.[1]

In nearly all of the specimens in the Challenger collection the figures are dark-coloured, and many have the carapace varied with dusky spots. In one example from St. Vincent (a male of large size) the chelipedes are rose-coloured, and the fingers a slaty-pink.

An adult male from St. Vincent has the following dimensions:—

Adult ♂.	Lines.	Millims.
Length of carapace,	3	6·5
Breadth of carapace, nearly	5	10

Panopeus, Milne Edwards.

Panopeus, Milne Edwards (pt.), Hist. Nat. Crust., vol. i. p. 403, 1834.
„ A. Milne Edwards, Crust. in Miss. Sci. au Mexique, pt. 5, p. 306, 1880.

Carapace transverse, depressed, shaped nearly as in *Xantho*, with the dorsal surface often marked with transverse prominences which are minutely granulated. The cervical

[1] Ann. and Mag. Nat. Hist., ser. 5, vol. viii. p. 212, 1881.

and mesogastric sutures distinct. The front projects but slightly beyond the level of the orbits, and is divided by a median notch into two truncated or sinuated lobes, which are separated by another notch from the antero-internal angle of the orbit, whose margins are interrupted by fissures or notches, and there is a distinct emargination immediately below the tooth at the exterior orbital angle. The antero-lateral margins are shorter than the postero-lateral margins, and are regularly divided into five teeth, of which the first or exterior orbital tooth is united more or less completely with the next; there is often a small tubercle on the subhepatic region below the interspace, between the first and second antero-lateral teeth. The endostome or palate is not longitudinally cristated. The post-abdomen in the male is usually five-jointed, with the third to the fifth of the normal segments coalescent, and it completely covers the sternum at base. The eyes are small. The basal antennal joint reaches the infero-lateral process of the front, but not the extremity of the subocular lobe of the orbit, which is often prominent and spiniform, and the flagellum occupies the inner orbital hiatus. The merus-joint of the exterior maxillipedes is often somewhat produced at its antero-external angle. The chelipedes in the adult male are moderately developed, and are nearly as in *Xantho*. The ambulatory legs are of moderate length, with the joints neither carinated nor spinose, and the dactyli styliform and straight.

The species, which never attain a very large size, have been described by A. Milne Edwards in his work above referred to, and, so far as known, with two exceptions, inhabit the American coasts. One species, *Panopeus xanthiformis*, A. Milne Edwards, has been recorded from rather deep water (73 to 118 fathoms). Two, *Panopeus africanus*, A. Milne Edwards, and *Panopeus blanchardi*, A. Milne Edwards, occur on the West African coasts.

To the species enumerated by A. Milne Edwards I may add the following :—

? *Panopeus lævis*, Dana. Brazil (Coll. Brit. Mus.).

Panopeus packardi, Kingsley. Florida ; Key-West.

Panopeus subverrucosus (*Oxius subverrucosus*, White). ? This species
is represented in the British Museum Collection by a carapace only.[1]

Panopeus herbstii, var. *serratus*.

Panopeus serratus, Saussure, Rev. et Mag. Zool., ser. 2, vol. ix. p. 502, 1857 ; Mém. Soc. Phys. d. Genève, vol. xiv. (3), p. 433, pl. i. fig. 7, 1858.

Here is somewhat doubtfully referred a small female from Bermuda, obtained in shallow water.

In this specimen the carapace is convex, with several transverse striæ upon the

[1] The Australian *Panopeus acutidens*, Haswell, is to be referred to the genus or subgenus *Heteropanope*, Stimpson, and is perhaps identical with *Heteropanope dentatus* (Adams and White).

gastric and branchial regions; and the first and second teeth of the antero-lateral margins are less distinctly separated than in larger specimens, but in other particulars it agrees well with the description of Saussure.

♀.					Lines.	Millims.
Length of carapace, nearly	.	.	.	.	3	6
Breadth of carapace,	.	.	.	.	4	8·5

Micropanope, Stimpson.

Micropanope, Stimpson, Bull. Mus. Comp. Zoöl., vol. ii. p. 139, 1870.
     ,,   A. Milne Edwards, Crust. in Miss. Sci. au Méxique, pt. 5, p. 324, 1880.

I have examined no specimens of this genus except the one in the Challenger collection, referred below to *Micropanope spinipes*, A. Milne Edwards; I must, therefore, express myself with some hesitation upon its limitation, but probably it may be conveniently restricted to those (usually deep-water) species which are enumerated and figured by A. Milne Edwards, and which are distinguished from *Panopeus* and *Eurypanopeus* by the more or less distinctly spiniform teeth of the antero-lateral margins, and by the spiniferous ambulatory legs. In the former character, and in the granulated or spinuliferous palms of the chelipedes, and in their small size, these species resemble *Pilumnus*. In the single specimen examined, there exists no ridge upon the endostome or palate.

Of these species, four, *Micropanope sculptipes*, Stimpson; *Micropanope pugilator*, A. Milne Edwards; *Micropanope pusilla*, A. Milne Edwards, and *Micropanope lobifrons*, A. Milne Edwards, have been dredged at various West-Indian localities, and in the Florida Straits, in depths varying from 15 to 170 fathoms; one species, *Micropanope spinipes*, A. Milne Edwards, occurs at the Abrolhos, Brazil, in 30 fathoms, and at Bahia, in shallow water.

Micropanope spinipes (?), A. Milne Edwards.

? *Micropanope spinipes*, A. Milne Edwards, Crust. in Miss. Sci. au Méxique, pt. 5, p. 326, pl. liv. fig. 2, 1880.

I thus designate, though with some hesitation, a male obtained at Bahia in shallow water, associated with *Pilumnus floridanus* and *Pilumnus fragosus*.

This specimen agrees in the proportions of the carapace and limbs, and particularly in the absence of granulations or spinules on the outer surface of the palms of the chelipedes, with Milne Edwards' description and figure of this species (the only one of the genus recorded from the Brazilian coast), but it is distinguished by having the first of the antero-lateral marginal teeth well developed and perfectly distinguishable from the tooth or spine at the outer angle of the orbit, although smaller than the second antero-lateral tooth.

Adult ♂.						Lines.	Millims.
Length of carapace, nearly	.	.	.	.	.	3	6
Breadth of carapace, about	.	.	.	.	.	4	8·5
Length of a chelipede, nearly	.	.	.	.	.	5	10
Length of second ambulatory leg,	.	.	.	.	.	5½	12

Etisus, Milne Edwards.

Etisus, Milne Edwards (pt.), Hist. Nat. Crust., vol. i. p. 410, 1834.
„   Dana, U.S. Explor. Exped., vol. xiii., Crust. 1, pp. 149, 183, 1852.

In this genus, which, as Dana has shown, represents in the subfamily or section Chlorodiinæ the genus *Cancer* in the Cancrinæ, the carapace is usually moderately convex, widely transverse, more or less distinctly lobulated or uneven on the dorsal surface, with the antero-lateral margins divided into from five to eight lobes, teeth or spines, between which are sometimes some smaller teeth or spines. The frontal lobes are truncated, with a median notch or closed fissure, and are separated by a wider hiatus from the interior angle of the orbit, which is usually prominent or rounded. The orbital margins have two or three marginal notches or fissures, and their inner subocular angles are dentiform or acute. The post-abdomen in the male is usually five or six-jointed, with two or three of the intermediate segments coalescent and forming a single joint. The antennules are obliquely or transversely plicated. The basal joint of the antennæ is enlarged, and is considerably produced at its extero-distal angle, so as to completely exclude the flagellum from the orbit. The ischium of the exterior maxillipedes is not produced at its antero-internal angle, the merus is scarcely, if at all, emarginate at its antero-internal angle. The chelipedes in the adult males are usually unequal and considerably developed, with the merus somewhat elongated, carpus with a strong spine or tooth on its inner margin, palm not carinated, fingers toothed or tuberculated on the inner margins and meeting only at the apices, which are strongly excavated. The ambulatory legs are of moderate length, with the joints sometimes compressed but not carinated, and often armed with spines or spinules.

The species, which often attain a considerable size, occur on the shores and islands of the Indo-Pacific region.[1]

The following are species referable to the genus *Etisus* as at present restricted :—

Etisus lævimanus, Randall, common throughout the Indo-Pacific Region, from the Red Sea and Mozambique to the Polynesian Islands : *Etisus utilis*, Jacquinot and Lucas, which occurs from Singapore eastward to New Caledonia, and at Cochin-China and the

[1] The West-Indian *Etisus occidentalis* of White (List Crust. Brit. Mus., p. 20, 1847), is not properly referable to this genus, but is synonymous with *Leptodius floridanus*, Gibbes.

Philippines; *Etisus punctatus*, Jacquinot and Lucas, from Mangareva Island; *Etisus deflexus*, Dana, from the Fijis; *Etisus dentatus*, Milne Edwards, from Natal and the Mauritius to New Caledonia and the Fijis; also perhaps *Etisus anaglyptus*, Milne Edwards, from Australia and the Philippines, and *Etisus rhynchophorus* (A. Milne Edwards), from Japan, adult examples of which are distinguished from the typical species of *Etisodes*, to which genus they are referred by A. Milne Edwards, by the broader transverse carapace and the greater dilatation of the basal antennal joint, which is not so greatly reduced at its antero-external angle as in *Etisus*.

Etisus lævimanus, Randall.

Etisus lævimanus, Randall, Journ. Acad. Nat. Sci. Philad., vol. viii. p. 115, 1839.
       „   „   Dana, Crust. in U.S. Explor. Exped., vol. xiii. p. 185, pl. x. fig. 1, 1832.
       „   „   A. Milne Edwards, Nouv. Archiv. Mus. Hist. Nat., vol. ix. p. 23, 1873.
       „   „   Miers, Crust. in Rep. Zool. Coll. H.M.S. " Alert," p. 217, 1884.
       „   *macrodactylus*, Jacquinot and Lucas, Crust. in Voy. au Pole Sud Zool., vol. iii. p. 30, pl. ix. fig. 2, 1853.
       „   *convexus*, Stimpson, Proc. Acad. Nat. Sci. Philad., p. 31, 1858.
       „   *maculatus*, Heller, Sitzungsb. math.-nat. Cl. k. Akad. Wiss. Wien, vol. xliii. (1) p. 332, 1861.
       „   „   *fide* Hilgendorf, Monatsber. d. k. preuss. Akad. d. Wiss. Berlin, p. 791, 1878.

Tonga Islands, Tongatabu, on the reefs (an adult male and smaller female); Sandwich Islands, Honolulu, reefs (a male).

The colour of this species is subject to very considerable variation. The ground colour of the three specimens in the Challenger collection is yellowish-brown (in spirit), the adult male from Tongatabu has the carapace punctulated with spots of a brownish-red.

The male from Tongatabu measures as follows :—

Adult ♂.				Lines.	Millims.
Length of carapace,	.	.	.	15	32
Breadth of carapace,	.	.	.	24	51

The female, which, although of small size, bears ova, as follows :—

Adult ♀.				Lines.	Millims.
Length of carapace,	.	.	.	9½	20
Breadth of carapace,	.	.	.	14	30

Carpilodes, Dana.

Carpilodes, Dana, Amer. Journ. Sci. and Arts, ser. 2, vol. xii. p. 126, 1851; U.S. Explor. Exped., vol. xiii., Crust. 1, p. 149, 1852.
     „   A. Milne Edwards, Nouv. Archiv. Mus. Hist. Nat. vol. i. p. 224, 1865, et synonyms.

In this genus, which represents in the Chlorodiinæ in Dana's classification the genus *Liomera* in the Xanthinæ, the carapace is very widely transverse, longitudinally convex, and usually (but not invariably) distinctly lobulated in its anterior half. The front is somewhat deflexed, with its anterior margin sinuated and usually with a small median notch. The antero-lateral margins are not cristated, and are divided into rounded lobes (not teeth), of which only the three posterior are usually distinct. The orbits are small, and the fissures of their upper margins very slightly marked, the interior subocular angle is not produced. The post-abdomen of the male is usually five-jointed, with the three intermediate segments coalescent and forming a single joint. The eyes are set on short thick pedicels. The basal antennal joint is somewhat elongated, and produced along the exterior margin of the infero-lateral frontal process so as to enter partly within the interior orbital hiatus. The merus of the exterior maxillipedes is usually transverse and distally truncated. The chelipedes, in the male, are subequal and moderately developed, with the palms rounded, not carinated above, and the fingers excavated at the distal extremity. The ambulatory legs are of moderate length, with the joints (except the dactyli) slightly compressed, but not carinated; the dactyli styliform and straight.

The species of *Carpilodes* are small, and rather numerous, and, with one doubtful exception, occur in the Oriental region; but the range of two or three species extends eastwards to the West American coasts. None, I believe, have been ascertained to occur in deep water.

The following species are probably referable to this genus, as I have characterised it above, besides the species enumerated by A. Milne Edwards in 1865. They were either referred by him to *Actæa* and to *Liomera*, or have been described since that date:—

> *Carpilodes cinctimanus* (Adams and White). Indo-Pacific Region, eastwards to the west coast of North America.
> ? *Carpilodes longimanus* (A. Milne Edwards) = *Cancer nigerrimus*, Desbonne and Schramm. West Indies.
> *Carpilodes monticulosus*, A. Milne Edwards. New Caledonia.
> *Carpilodes margaritatus*, A. Milne Edwards. New Caledonia.
> *Carpilodes edwardsii*, Kossmann. Red Sea.
> *Carpilodes granulosus*, Haswell. Torres Strait.
> *Carpilodes bellus* (Dana). Samoa Islands; Paumotu; Philippines.
> ? *Carpilodes dia* (White) = *Actæodes cavipes*, Dana (?). Philippines; Polynesian Islands. In this species the basal antennal joint enters the interior orbital hiatus as in the typical *Carpilodes*, but the species is distinguished by the peculiar lunate crests of the ambulatory legs. It should, perhaps, be separated as a distinct genus.

Carpilodes bellus (Dana).

> *Actæodes bellus*, Dana, Proc. Acad. Nat. Sci. Philad., p. 78, 1852; U.S. Explor. Exped., vol. xiii., Crust. 1, p. 196, pl. xi. fig. 2, 1852.
> *Actæa bella*, A. Milne Edwards, Nouv. Archiv. Mus. Hist. Nat., vol. i. p. 261, 1865.

Philippines, Samboangan, beach.   A small male.

In the specimen referred to this species (which agrees very well with the description and figure of Dana) the basal joint of the antennæ is produced a little within the hiatus, between the inner subocular lobe of the orbit and the lateral subfrontal process, it is therefore referred to *Carpilodes* rather than to *Actæodes*, with the species of which genus it altogether corresponds in external appearance.

In the structure of the antennæ this specimen agrees with examples referred to *Carpilodes bellus* from the Samoa and Keeling or Cocos Islands, in the collection of the British (Natural History) Museum.

				Lines.	Millims.
Adult ♂.					
Length of carapace,	.	.	.	3	6·5
Breadth of carapace, about		.	.	4½	9·5

Zozymus, Leach.

> *Zozimus*, Leach, Dict. d. Sci. Nat., vol. xii. p. 75, 1818.
> *Zozymus*, Milne Edwards (pt.) Hist. Nat. Crust., vol. i. p. 383, 1834.

This genus represents in the section or subfamily Chlorodiinæ, Dana, the genus *Lophozozymus* in the Xanthinæ; it can in fact only be distinguished from *Lophozozymus* by the excavated finger tips of the chelipedes, and by the more numerous and uniform bosses or lobules of the carapace and exterior surface of the carpus and wrist of the chelipedes.

The species, as at present restricted, are few in number, and inhabit the Indo-Pacific region; they are *Zozymus æneus* (Linné), common throughout the Indo-Pacific region, from the Red Sea and Mauritius to the Polynesian Islands; *Zozymus gemmula*, Dana, from Sooloo, with which *Zozymus pilosus*, A. Milne Edwards, from New Caledonia, is perhaps identical, and which is distinguished by the somewhat granulated tubercles of the carapace, and the non-carinated palms of the chelipedes; *Zozymus pumilus*, Jacquinot and Lucas, from the Marianna Islands.[1]

Zozymus æneus (Linné).

> *Cancer æneus*, Linné, Mus. Lud. Ulrici, p. 451, 1764; Syst. Nat., ed. 12, p. 1048, 1766.
> *Zozymus æneus*, Milne Edwards, Hist. Nat. Crust., vol. i. p. 385, 1834, et synonyma.
> ,,   ,,   Dana, U.S. Explor. Exped., vol. xiii., Crust. 1, p. 192, pl. x. fig. 3, 1852.

Kandavu, Fiji Islands, on the reefs (two adult males).

							Lines.	Millims.
Adult ♂.								
Length of carapace, .	.	.	.	.	.	.	24½	52
Breadth of carapace to base, about	.	.	.	.	.	.	34	72·5

[1] *Zozymus lævis*, Dana, from Balabac Passage, probably does not belong to this genus.

Actæodes, Dana.

Actæodes, Dana, Amer. Journ. Sci. and Arts., ser. 2, vol. xii. p. 126, 1851; U.S. Explor. Exped., vol. xiii., Crust. 1, p. 149, 1852.

The genus *Actæodes*, which is not sustained by A. Milne Edwards and by some other authors, in Dana's system represents in the Chlorodiinæ the genus *Actæa* in the Xanthinæ, which it resembles in the transverse and distinctly lobulated carapace and in the rounded lobes of the antero-lateral margins, also in the structure of the orbits, antennæ, maxillipedes, and in other essential characters; it is, in fact, only distinguished from *Actæa* by the structure of the chelipedes, whose fingers are excavated in the adult at the apices, and is connected with that genus by almost insensible structural gradations.

The species are small, and inhabit the coasts and islands of the Indo-Pacific region; one species, *Actæodes faba*, Dana, occurs in the Atlantic region. Most of them are apparently somewhat local in distribution, but the only species (*Actæodes tomentosus*) collected by the naturalists of the Challenger Expedition, ranged from the Red Sea and the Mauritius eastwards to the Sandwich Islands.

Besides the species described by Dana which are placed by Milne Edwards in *Actæa*, the following are, I think, referable to *Actæodes*:—

Actæodes nodipes, Heller, perhaps = *Actæodes speciosus*, Dana. Red Sea.
Actæodes polyacanthus, Heller. Red Sea.
Actæodes frugifer (White). Philippine Islands.
Actæodes pubescens (Milne Edwards). Mauritius.
Actæodes variolosus (A. Milne Edwards). Upolu, Samoa Islands.

Actæodes tomentosus (Milne Edwards).

Zozymus tomentosus, Milne Edwards, Hist. Nat. Crust., vol. i. p. 385, 1834; Crust. in Latreille, Règne Animal de Cuvier, ed. 3, Atlas, pl. xi. *bis*, fig. 2.
Actæodes tomentosus, Dana, U.S. Explor. Exped., vol. xiii., Crust. 1, p. 197, 1852.
Actæodes affinis, Dana, Proc. Acad. Nat. Sci. Philad., p. 78, 1852; U.S. Explor. Exped., tom. cit., p. 197, pl. xi. fig. 3, 1852.
Actæa tomentosa and *affinis*, A. Milne Edwards, Nouv. Archiv. Mus. Hist. Nat., vol. i. pp. 262, 263, 1865.

A series of specimens was collected at Samboangan, Philippines, on the beach.

	Lines.	Millims.
Adult ♂.		
Length of carapace,	6	12·5
Breadth of carapace,	9½	20

The basal antennal joint in this species is united in the adult somewhat broadly with the infero-lateral process of the front at its antero-internal angle, but it does not enter the interior orbital hiatus.

Leptodius, A. Milne Edwards.

Leptodius, A. Milne Edwards, Ann. d. Sci. Nat., ser. 4, vol. xx. p. 283, 1863 ; Nouv. Archiv. Mus. Hist. Nat., vol. ix. p. 221, 1873.
Chlorodius (sect. 2), Dana, U.S. Explor. Exped., vol. xiii., Crust. 1, p. 207, 1852.
Xanthodius (subgenus), Stimpson, Ann. Lyc. Nat. Hist. New York, vol. vii. p. 52, 1859.

In this genus the carapace is widely transverse, somewhat depressed, and very distinctly lobulated in its anterior half, posteriorly nearly plain ; the frontal margin is truncated and slightly sinuated, with a median notch or fissure ; the antero-lateral margins are longer than the postero-lateral margins, and are normally divided into four teeth or lobes, besides the exterior orbital tooth, which is usually obsolete ; behind the last antero-lateral tooth are occasionally one or two smaller teeth on the postero-lateral margin ; the marginal teeth are usually dentiform, not rounded as in *Actæodes*. The post-abdomen in the male is five-jointed ; the three intermediate segments coalescent and forming a single joint. The superior margins of the orbits are marked with two slight notches and fissures. The basal antennal joint reaches the infero-lateral process of the front or even enters slightly within the interior orbital hiatus. The merus of the exterior maxillipedes is truncated or slightly notched at the distal extremity, and its antero-internal angle is slightly emarginate or sinuated; its antero-external angle is often slightly produced.

The chelipedes in the adult male are moderately developed ; the merus or arm is short and almost invariably covered in a dorsal view by the antero-lateral margins, carpus dentated on the inner margin, palm not cristated above, fingers excavated at the distal extremity. The ambulatory legs are moderately developed, with the fourth to the sixth joints slightly compressed, but not carinated or spinose ; dactyli styliform.

This genus may be considered the representative in the Chlorodiinæ of *Xantho* in the Cancrinæ ; it is also nearly allied to *Carpilodes*, *Actæodes*, and to the genus which follows (*Phymodius*).

From *Carpilodes* and *Actæodes* it is usually to be distinguished by the dentiform or tuberculiform antero-lateral marginal teeth, not to speak of other characters.

The species are somewhat numerous ; several inhabit the Indo-Pacific region. Two species, *Leptodius macandreæ* and *Leptodius punctatus*, occur in the West Atlantic or at the Canaries and Cape Verde Islands. The others inhabit the American coasts from Florida to Brazil, and from California to Chili. One species (*Leptodius cooksoni*) occurs at the Galapagos. None, I believe, have been recorded from very deep water. They may be distributed under two subgeneric divisions or sections as follows :—

I. Endostome or palate without any trace of longitudinal carina. *Leptodius*.

A. *Indo-Pacific Species (excluding the West Coast of America and Islands adjacent)*.

Leptodius exaratus (Milne Edwards) = *Cancer inæqualis*, Audouin; *Xantho affinis*, de Haan ; *? Xantho quinque-dentatus*, Krauss, var.; *Chlorodius edwardsii*, Heller ; *Lagostoma nodosa*, Randall (?); *Chlorodius sanguineus*, Milne Edwards, var.; *Xantho distinguendus*, de Haan, var.   Indo-Pacific Region. Very common. (The West Indian and Brazilian *Chlorodius floridanus*, Gibbes, is perhaps not specifically distinct from this species).

Leptodius gracilis (Dana).   Indo-Pacific Region.

Leptodius eudorus (Herbst).   New Zealand.

Leptodius lividus (de Haan).   Japanese and Australian Seas.

(The three foregoing species are perhaps mere varieties of *Leptodius exaratus*, for which species, therefore, Herbst's designation *eudorus* may have to be used).

Leptodius nudipes (Dana).   Polynesian Islands ; New Zealand.

Leptodius hombronii (Jacquinot and Lucas).   ?

Leptodius crassimanus, A. Milne Edwards.   New Caledonia.

B. *Atlantic Species*.

Leptodius convexus, A Milne Edwards.   Cape Verde Islands.

Leptodius punctatus, Miers.   Senegambia (9 to 15 fathoms), West Africa ; Cape Verde Islands.

Leptodius macandreæ, Miers.   Canary Islands.

Leptodius floridanus (Gibbes) = *Chlorodius limosus*, Desbonne and Schramm, and *Etisus occidentalis*, White.   West Indies ; Florida ; Brazil.

Leptodius dispar (Stimpson).   Cuba.

Leptodius agassizii, A. Milne Edwards.   Florida, 12 to 18 fathoms.   Differs from others referred to this genus in the spinuliferous legs.

Leptodius (?) caribæus (Desbonne and Schramm).   Guadeloupe.   This species differs from others, according to A. Milne Edwards, in the longer merus of the chelipedes and the seven-jointed post-abdomen.

C. *West American Species*.

Leptodius occidentalis (Stimpson) = *Chlorodius fischeri*, Lockington.   Panama ; Mexico ; California.

Leptodius cooksoni, Miers.   Galapagos Islands.

Leptodius lobatus, A. Milne Edwards.   Chili, " S. America " (Coll. Brit. Mus.).

II. Endostome with the longitudinal carinæ partially developed. *Xanthodius*, Stimpson.

Leptodius americanus (Saussure).   West Indies ; Florida.

Leptodius sternberghii, Stimpson = *Actæodes mexicanus*, Lockington.   Panama ; California.

Leptodius hebes, Stimpson.   Cape St. Lucas, California.

Leptodius exaratus (Milne Edwards), var. *sanguineus*.

> *Chlorodius sanguineus*, Milne Edwards, Hist. Nat. Crust., vol. i. p. 402, 1834. (?)
>   „   „   Dana, U.S. Explor. Exped., vol. xiii., Crust. I, p. 207, pl. xi. fig. 11, 1852.
> *Leptodius exaratus*, var. *sanguineus*, Miers, Proc. Zool. Soc. Lond., p. 134, 1877, et synonyma.

Several specimens were collected at Samboangan, mostly obtained on the beach ; one (the largest male) on the reefs, in 10 fathoms ; the dimensions of this specimen are as follows :—

		Lines.	Millims.
Adult ♂.			
Length of carapace,	.	10	21
Breadth of carapace,	.	16	34
Length of a chelipede,	.	18	38

The variety thus designated is generally, but perhaps not invariably, characterised by the presence of one or even two additional marginal teeth behind the fifth and last of the normal antero-lateral teeth ; these additional teeth are situated on the postero-lateral margin of the carapace. It is certainly the form figured by Dana as *Chlorodius sanguineus*, but I am not positively convinced of its identity with the *Chlorodius sanguineus* of Milne Edwards. The colour varies from reddish to greenish-yellow ; the fingers are dark or light brown.

Leptodius punctatus, Miers.

> *Leptodius punctatus*, Miers, Ann. and Mag. Nat. Hist., ser. 5, vol. viii. p. 214, pl. xiii. fig. 3, 1881.

St. Vincent, Cape Verde Islands, July 1873 (a good series of specimens).

The Challenger examples are all of much smaller size than the type from Senegambia, now in the collection of the British Museum, and are mostly of a greenish-yellow colour (in spirit). The antero-lateral teeth of the carapace are connected one with another by a thin crest or keel which borders the antero-lateral margins, and is much less distinctly developed in the larger Senegambian specimen.

The largest example (a female) in the Challenger series measures as follows :—

Adult ♀.	Lines.	Millims.
Length of carapace, nearly	5½	11
Breadth of carapace, nearly	8	16·5

Phymodius, A. Milne Edwards.

Phymodius, A. Milne Edwards, Ann. d. Sci. Nat., sér. 4, Zool. xx. p. 283, 1863; Nouv. Archiv.
Mus. Hist. Nat., vol. ix. p. 217, 1873.
Chlorodius (section 1), Dana, U.S. Explor. Exped., vol. xiii., Crust., p. 205, 1852.

This genus is very nearly allied to the foregoing (*Leptodius*) from which it is distinguished merely by the narrower, nearly hexagonal carapace, which is lobulated not only in front, but also (less distinctly) in its posterior part, by the more elongated chelipedes, whose merus-joints in the typical species are prolonged beyond and are not covered by the antero-lateral margins of the carapace, and by the more or less distinctly granulated or spinuliferous merus, carpus, and penultimate joints of the ambulatory legs. The basal antennal joint is united largely with the infero-lateral process of the front in the typical species. The chelipedes are more or less distinctly tuberculated or spinuliferous.

Phymodius is distinguished from *Chlorodius*, as restricted by recent authors, only by the more distinctly lobulated carapace and chelipedes.

The following species are referred to this genus :—

Phymodius angulatus (Milne Edwards), with vars. *gracilis*, Dana, and *curtimanus*,
Dana. Indian Ocean to Polynesian Islands.

Phymodius monticulosus (Dana) = *Chlorodius areolatus*, Adams and White (nec
Milne Edwards); *Xantho peuce*, White; *Chlorodius obscurus*, Jacq. and
Lucas. Indo-Pacific Region.

Phymodius rugipes (Heller). Red Sea.

Phymodius maculatus (Stimpson). Tortugas ; West Indies.

Phymodius monticulosus (Dana).

Xantho peuce, White, List Crust. Brit. Mus., p. 125, 1847.
Chlorodius areolatus, Adams and White, Crust. in Zool. H.M.S. "Samarang," p. 41, pl. xi.
fig. 3, 1848; nec Milne Edwards.
Chlorodius monticulosus, Dana, Proc. Acad. Nat. Sci. Philad., p. 79, 1852, U.S. Explor.
Exped., vol. xiii., Crust. 1, p. 206, pl. xi. fig. 9, 1852.
? *Chlorodius obscurus*, Jacquinot and Lucas, Crust. in Voy. au Pole Sud. Zool., t. iii. p. 26,
pl. iii. fig. 4, 1853, bad.
? *Phymodius obscurus*, A. Milne Edwards, Nouv. Archiv. Mus. Hist. Nat., vol. ix. p. 220, 1873.

Samboangan (beach). Several specimens of different sizes were collected.

The colour of the carapace (in dried or spirit specimens) varies from reddish or chocolate-brown to greyish-yellow.

Adult ♀.	Lines.	Millims.
Length of carapace, nearly	6	12
Breadth of carapace, about	7½	16

In this species or well-marked variety the tubercles of the dorsal surface are stronger and more distinct than in the almost equally common *Phymodius ungulatus*, and they are more or less distinctly granulated or rugose, not (as often in *Phymodius ungulatus*) simply punctulated ; the chelipedes have the palm and wrist very distinctly tuberculated, the tubercles becoming spinous in the younger animal, and the finger in the young is spinous at the base ; whereas in *Phymodius ungulatus*, the chelipedes are smooth or but obscurely tuberculated, and the fingers are without spines at the base.[1]

B. Endostome longitudinally ridged (Eriphiidæ, Dana) :—

Eurytium, Stimpson.

Eurytium, Stimpson, Ann. Lyc. Nat. Hist. New York, vol. vii. p. 56, 1859.
„ A. Milne Edwards, Crust. in Miss. Sci. au Méxique, pt. 5, p. 332, 1880.

This genus can only be distinguished from *Panopeus* and *Eurypanopeus* by the broader transverse carapace, which is nearly smooth, not lobulated, on the dorsal surface, and in the typical species is convexly arcuated in a longitudinal, not a transverse direction. The ridges on the endostome or palate are distinct and well developed, and the ducts leading to the male copulatory appendages pass to them from the fifth joints of the ambulatory legs beneath the margin of the sternum ; a character in which, as Dr. Stimpson notes, this genus is related to the Ocypodoidea.

Professor A. Milne Edwards, however, has shown that short sternal channels for the copulatory organs exist in certain species of *Eurypanopeus*, and I may add, that both in *Panopeus* and in *Eurypanopeus*, the ridges of the endostome or palate are occasionally partially developed.

Only two species are referred by A. Milne Edwards to this genus, *Eurytium limosum* (Say), which occurs on the eastern coasts of the American continent from New York to Rio de Janeiro, and was obtained by the Challenger Expedition at Bermuda, and *Eurytium affine* (*Panopeus affinis*, Streets and Kingsley), which is found on the Californian coast and is intermediate between this genus and *Panopeus*, having the ridges of the endostome well developed, but the carapace shaped nearly as in *Panopeus* and *Eurypanopeus*.

[1] In a large series of specimens of *Phymodius ungulatus*, in the collection of the British Museum, I have observed only one or two specimens which approach *Phymodius monticulosus* in these characters.

Eurytium limosum (Say).

Cancer limosus, Say, Journ. Acad. Nat. Sci. Philad., vol. i. p. 446, 1817.

Eurytium limosus, Stimpson, Ann. Lyc. Nat. Hist. New York, vol. vii. p. 56, 1859.

  „   „   A. Milne-Edwards, Crust. in Miss. Sci. au Méxique, pt. 5, p. 332, pl. lx.
          fig. 2, 1880, and references to literature.

Hungry Bay, Bermudas. A male and female were obtained in the Mangrove Swamps. In these specimens the notch in the middle of the frontal margin is obsolete ; in other particulars they agree with the excellent description and figure of M. A. Milne-Edwards of this species.

The dimensions of these specimens are as follows :—

♂.	Lines.	Millims.
Length of carapace, .	6	12·5
Breadth of carapace, .	8½	17·5

♀.	Lines.	Millims.
Length of carapace, nearly	8	16·5
Breadth of carapace,	11½	24·5

Pseudozius, Dana.

Pseudozius, Dana (pt.), Amer. Journ. Sci. and Arts, ser. 2, vol. xii. p. 127, 1851 ; U.S. Explor. Exped., vol. xiii., Crust. 1, p. 229, 1852.

Carapace markedly transverse, depressed, with the dorsal surface nearly smooth, not lobulated, and with the teeth of the antero-lateral margins sometimes developed, sometimes obsolete ; the sutures of the dorsal surface (except sometimes the postfrontal or mesogastric suture), are not distinguishable. The front is moderately deflexed, usually less than one-third the width of the carapace, and its anterior margin is sinuated or obscurely bilobate or quadrilobate ; the orbits are small, with the margins entire. The ridges of the endostome or palate are wholly or partly developed. The post-abdomen in the male is distinctly seven-jointed (in the species I have examined). The eye-peduncles are set on very short pedicels. The antennules are transversely plicated. The basal peduncular joint of the antennæ is short, and does not attain the infero-lateral angles of the front. The ischium of the outer maxillipedes is not produced at its antero-internal angle ; the merus is distally truncated, with the antero-internal angle scarcely, or not at all, emarginate, and the antero-external angle usually not prominent. The chelipedes in the adult male are moderately robust ; carpus bluntly toothed on its inner margin, palm nearly smooth, rounded above, fingers toothed on the inner margins and distally acute. The ambulatory legs are of moderate length, with the joints smooth, slightly compressed, but not carinated.

This genus is very nearly allied to *Menippe*, de Haan,[1] from which it is scarcely

[1] Corresponds nearly to *Pseudocarcinus*, A. Milne Edwards ; I propose to restrict this latter designation to the *Pseudocarcinus gigas* (Lamarck).

distinguished except by the narrower, smoother, depressed carapace, which is not lobulated, has not the cervical suture developed, and has the antero-lateral margins less distinctly toothed.

Eurycarcinus, A. Milne Edwards,[1] is also very nearly allied to this genus, and should perhaps not be separated from it; the typical species, *Eurycarcinus grandidieri*, from Zanzibar, which I think is identical with *Galene natolensis*, Krauss, is distinguished merely by the more convex carapace, with sinuated but not quadrilobate front and more distinct antero-lateral marginal teeth; the basal antennal joint (in specimens in the museum collection), reaches the inner subocular tooth, but not the infero-lateral angle of the front.

The species (with one exception, *Pseudozius bouvieri*, which I have placed below in a separate subgenus), inhabit the Oriental region. Nothing has been recorded of their bathymetrical distribution.

Subgenus *Pseudozius*.

Carapace moderately transverse, the frontal lobes not prominent (in the species I have examined), the orbital margins not notched, the ridges of the endostome reach the anterior margin of the buccal cavity. The merus of the exterior maxillipedes is not produced at its antero-external angle.

> *Pseudozius caystrus* (Adams and White) = *Pseudozius planus*, Dana. Philippine Islands; Paumotu Archipelago.
>
> *Pseudozius inornatus*, Dana. Sandwich Islands. Perhaps a variety of *Pseudozius caystrus*.
>
> *Pseudozius microphthalmus*, Stimpson. Bonin Islands. The diagnosis is in few words, but the species is apparently scarcely distinct from *Pseudozius caystrus*.
>
> *Pseudozius sinensis*, A. Milne-Edwards. China.

Subgenus *Euryozius*, nov.

Carapace very transverse, the median lobes of the front prominent, rounded, orbital margins notched at the antero-external angle. The ridges of the endostome or palate but partially developed. Merus of the exterior maxillipedes slightly produced at its antero-external angle.

> *Pseudozius bouvieri* (A. Milne Edwards) = *Pseudozius mellissii*, Miers, var. (?). Cape Verde Islands; St. Helena; Ascension Island.

[1] *Ann. Soc. Entom. France*, ser. 4, vol. vii. p. 276, 1867.

Pseudozius bouvieri, var. *mellissii* (Pl. XII. fig. 3).

1 *Xantho bouvieri*, A. Milne Edwards, Rev. and Mag. Zool., vol. xxi. p. 377, 1869.
Pseudozius mellissii, Miers, Ann. and Mag. Nat. Hist., ser. 5, vol. viii. p. 432, 1881.

Ascension Island (an adult female).

This specimen was received among the fishes of the Challenger collection.

Carapace naked, transverse, much broader than long, its surface punctulated, the punctulations numerous and crowded in front, sparser posteriorly, and nearly obsolete near the postero-lateral and posterior margins. Some larger pits occur here and there near the antero-lateral margins, and the upper margins of the carpus and hand of the chelipedes are also punctulated. The front is four-lobed; the two median lobes are prominent and rounded, and separated by a median notch; the outer lobes (or inner orbital angles) are very little prominent, and separated from the median lobes only by a rather shallow sinus. The antero-lateral margins are longer than the postero-lateral margins, and are defined along the greater part of their length by a low, obliquely-striated, entire line or crest; after which follow, at the broadest part of the carapace, two small but distinctly defined teeth. All the segments of the post-abdomen are distinct in both sexes; the two basal segments cover the whole width of the sternal surface between the bases of the fifth ambulatory legs. The chelipedes are large and robust; the arm or merus-joint short, trigonous, with the margins unarmed; carpus large, with its inner margin produced into a broad, squarely truncated lobe; the palm has its upper margin rounded, the lower margin straight, thin-edged and entire; inner surface with a rounded prominence; fingers black, dentated on their inner margins, with acute apices, the coloration not extending over the inner or outer surface of the palm. Ambulatory legs slender, smooth, and nearly naked, with the penultimate and terminal joints canaliculated, the longitudinal channels on the penultimate joints not always extending along the whole length of the joints; the legs of the fourth pair are longer than those of the third pair. The colour in spirit is reddish or bluish-pink.

Adult ♂.		Lines.	Millims.
Length of carapace, about	.	12	25
Breadth of carapace,		18	38

In a larger male from St. Helena, in the collection of the British Museum, the carapace is even more transverse.

It is with much hesitation that this form is identified with *Xantho bouvieri*, A. Milne Edwards, which is very briefly described, and has not, I believe, been figured, nor am I certain from the short description whether *Xantho bouvieri* be properly a *Pseudozius*, since it is not stated whether the ridges of the endostome are at all developed.

The specimens from St. Helena and Ascension Island differ in the distinctly four-lobed front from A. Milne Edwards' description.

Sphærozius, Stimpson.

Sphærozius, Stimpson, Proc. Acad. Nat. Sci. Philad., p. 35, 1858.

Body subglobose ; carapace convex, smooth, shining, and neither lobulated nor sulcated on the dorsal surface, except on the postfrontal or epigastric region, where there is the usual longitudinal median suture, which reaches the front and is posteriorly bifurcated. The antero-lateral margins are shorter than the postero-lateral margins, and are divided behind the postero-lateral angles into four serratures or teeth, which are not spiniform, as in *Pilumnus*. The front is slightly deflexed, about one-third the width of the carapace, and is divided by a median notch ; its anterior margin is not separated by a notch from the interior angle of the orbit. The orbital margins are small, entire. The longitudinal ridges of the endostome or palate are not very prominent. The post-abdomen in the male, and in the single specimen I have seen of a female, is distinctly seven-jointed. The eyes are set on short thick pedicels. The basal joint of the antennæ is very short, and does not nearly reach the front ; the following joint is slender and just attains the infero-lateral frontal angle. The ischium of the endognath of the exterior maxillipedes is not produced at its antero-internal angle ; the merus is distally truncated, with the antero-external angle rounded and not produced, and the antero-internal angle scarcely emarginate. The straight exognath reaches the antero-external angle of the merus of the endognath. The chelipedes are robust, unequal ; carpus bluntly toothed on its inner margin, palm rounded above ; fingers distally acute and dentated on the inner margins. Ambulatory legs slender and of moderate length, with the joints smooth, hairy ; dactyli straight, with a well-developed terminal claw.

The nearest ally to this genus is, I think, *Pilumnopeus*, A. Milne Edwards,[1] from which it is mainly distinguished by the smooth and narrower carapace, which is not lobulated, the absence of the exterior orbital hiatus, and the shorter basal antennal joint. *Pilumnopeus* should not, perhaps, be regarded as generically distinct.

The following species have been referred to *Sphærozius* by its author :—

    Sphærozius dispar (Dana). Sooloo Sea.
    Sphærozius nitidus, Stimpson. Hong-Kong ; Kobé, Japan, 50 fathoms.

Sphærozius nitidus (Pl. XII. fig. 4).

Sphærozius nitidus, Stimpson, Proc. Acad. Nat. Sci. Philad., p. 35, 1858.

Kobé, Japan, lat. 34° 38′ 0″ N., long. 135° 1′ 0″ E. (Station 233A), in 50 fathoms (a female).

    [1] *Ann. Sc. Entom. France*, ser. 4, vol. vii. p. 277, 1867.

To the description of Stimpson I may add the following:—

This specimen (preserved in spirit) is yellowish; the carapace and chelipedes are thickly punctulated with minute purple spots, which are very indistinct on the upper surface of the palm, and are obliterated on its outer surface; the fingers purplish-brown, the coloration not extending over the inner and outer surface of the palm. Stimpson says nothing with regard to the coloration of his types.

The carapace and chelipedes are smooth and naked except for a few inconspicuous hairs on the interior surface of the carpus; in this particular this species differs altogether from *Pilumnus*, but the ambulatory legs are slightly hairy.

♂.		Lines.	Millims.
Length of carapace,		2½	5
Breadth of carapace,		3	6·5

Pilumnus, Leach.

Pilumnus, Leach, Trans. Linn. Soc. Lond., vol. xi. p. 321, 1815.
    „ Milne-Edwards (pt.), Hist. Nat. Crust., vol. i. p. 415, 1834.
Acanthus, Lockington, Proc. Calif. Acad. Nat. Sci., vol. vii. pt. i. p. 32, 1876.
Parapilumnus, Eupilumnus (subgenera), Kossmann, Malacostraca in zool. Ergebn. einer Reise Küstengeb. d. rothen Meeres (erste Hälfte), p. 38, 1877 ; *not Eupilumnus*, Kingsley, Proc. Acad. Nat. Sci. Philad., p. 397, 1879.

Carapace longitudinally convex, little broader than long, with the regions indistinctly defined; it is thickly covered with hair on the dorsal surface, which extends also over the chelipedes and ambulatory legs, but not lobulated as in *Lobopilumnus*; the antero-lateral margins are regularly arcuated, shorter than the postero-lateral margins, and are normally armed with short spines in place of the usual antero-lateral teeth; the orbital margins are sometimes entire, sometimes spinuliferous; the front is rather narrow and its anterior margin is usually spinuliferous or granulated and divided into two rounded lobes by a median emargination, exterior to each of which is usually a smaller lobe or tooth. The endostome is usually very distinctly longitudinally carinated. The post-abdomen of the male is distinctly seven-jointed, and its base usually occupies the whole width of the sternum, between the fifth ambulatory legs. The eyes are of moderate length. The basal antennal joint is short and slender, and barely reaches the infero-lateral frontal process, or sometimes falls short of it; the next joint lies within the interior orbital hiatus. The merus of the endognath of the exterior maxillipedes is distally truncated, and in the typical species is not narrower than the ischium-joint. The chelipedes are moderately robust and usually unequal, with the merus-joint short and trigonous, the carpus and palm more or less granulated and spinuliferous on the superior margin and exterior surface; fingers distally acute or subacute, and dentated on the inner margins; ambulatory legs of moderate length, slightly compressed, but not carinated, with the fourth to the sixth joints sometimes armed with spinules; dactyli slender, styliform and nearly straight.

The species of this genus are always of small size, and occur very commonly both in the Atlantic and Indo-Pacific regions in shallow, or more rarely in deeper water. The described forms are very numerous; but it is probable that were a revision of the types and of sufficient series possible, it would necessitate the reunion of many of these recently described with the types of the earlier authors. As Professor A. Milne-Edwards, the highest authority on the Brachyura, justly remarks, "It has not yet been proved that the number of spines which cover the carapace and limbs may not vary in different examples of the same species."[1]

In the arrangement of the species which follows, I have found it necessary, in the present state of our knowledge, to arrange them roughly in geographical and chronological order, without regard to their affinities. I must regard the subgeneric distinctions recently proposed by Dr. Kossmann (*tom. cit. supra*, p. 38) and which are cited, but apparently not adopted, by Milne-Edwards, as of doubtful value, and there are not a few species, which, from the descriptions at present published, could not be classed with certainty in any one of the three subgenera he has established.

European and East Atlantic Species.

Pilumnus hirtellus, (Linné) = *Cancer ferrugineus*, Herbst (?). European Seas (to 35 fathoms, Heller).

Pilumnus spinifer, Milne Edwards. Mediterranean; Black Sea; Azores, 50 to 90 fathoms (?).

Pilumnus villosus, Risso = *Pilumnus spinulosus*, Kepler (*fide* Heller). Mediterranean; Black Sea.

Pilumnus affinis, Capello. Setubal, Portugal. (Very near *Pilumnus hirtellus*.)

Pilumnus teixeirianus, Capello. Setubal.

Pilumnus tridentatus, Maitland. Holland; Black Sea. Perhaps not a true *Pilumnus*; it resembles *Pilumnoplax* and *Heteroplax* in the form of the marginal teeth (Coll. Brit. Mus.).

Pilumnus calculosus, Dana. Madeira (?)

Pilumnus africanus, A. Milne Edwards. Gorea and Angola, St. Vincent, Cape Verde Islands.

South African Species.

Pilumnus verrucosipes, Stimpson. South Africa; Simon's Bay (11 fathoms); Senegambia, 9 to 15 fathoms.

[1] Crust. in Miss. Sci. au Méxique, pt. 5, p. 260, 1880.

Pilumnus granulatus, Krauss. Natal. (The *Pilumnus xanthoides*, Krauss, from Natal, which is included by Dr. Kossmann in this genus, differs from the typical *Pilumni* in the form of the antero-lateral marginal teeth.)

West Atlantic Species.

Pilumnus aculeatus (Say). East coast of United States; West Indies (36 fathoms, A. Milne Edwards).

Pilumnus gemmatus, Stimpson. West Indies; Florida (to 37 fathoms). In some particulars this species approaches the genus *Heteractæa*.

Pilumnus reticulatus, Stimpson. St. Thomas.

Pilumnus floridanus, Stimpson. Tortugas; Brazil, off Barra Grande, 30 to 350 fathoms; Bahia (?).

Pilumnus lacteus, Stimpson. West Indies (to 37 fathoms, A. Milne Edwards).

Pilumnus nudifrons, Stimpson. West Indies; Sombrero Key (to 125 fathoms); West Barbados (to 134 fathoms, A. Milne Edwards).

Pilumnus granulimanus, Stimpson. Cuba.

Pilumnus melanecanthus, Kingsley. Florida.

Pilumnus dasypodus, Kingsley. Florida.

Pilumnus caribæus, Desbonne and Schramm. Guadeloupe.

Pilumnus vinaceus, A. Milne Edwards. Florida (to 37 fathoms, A. Milne Edwards). Martinique.

Pilumnus gracilipes, A. Milne Edwards. Barbados, 100 fathoms (A. Milne Edwards).

Pilumnus quoyi, Milne Edwards. Rio de Janeiro; Guiana.

Pilumnus urinator, A. Milne Edwards. Santa Cruz, 245 fathoms.

Pilumnus miersii, A. Milne Edwards. West Indies.

Pilumnus tessellatus, A. Milne Edwards. Brazil, Desterro.

Pilumnus fragosus, A. Milne Edwards. St. Thomas; Bahia, 7 to 20 fathoms.

Pilumnus brasiliensis, n. sp. Bahia, 7 to 20 fathoms.

West American Species.

Pilumnus xantusii, Stimpson. California, Cape St. Lucas.

Pilumnus depressus, Stimpson. California, Cape St. Lucas.

Pilumnus stimpsonii = Pilumnus marginatus, Stimpson.[1] California, Cape St. Lucas.

Pilumnus spinohirsutus (Lockington). California.

Pilumnus limosus, Smith. Peru; Panama.

[1] The name *marginatus* is preoccupied by Stimpson himself, for an Oriental species.

Indo-Pacific Species.

Pilumnus vespertilio (Fabricius) = *Pilumnus ursulus*, Adams and White ; and *Pilumnus mus*, Dana. Indo-Pacific Region (to 17 fathoms).

Pilumnus tomentosus, Latreille. Off Port Philip, Australia, 33 fathoms (Station 161) ; Bass Strait, 38 fathoms (Station 162).

Pilumnus peronii, Milne Edwards. Seas of Asia.

Pilumnus forskålii, Milne Edwards = *Cancer incanus*, Forskål. Coasts of Egypt. This is possibly a variety of *Pilumnus vespertilio*, since Dr. Kossmann unites *Pilumnus ursulus* with this species, but in specimens from the Red Sea, referred to *Pilumnus forskålii* in the collection of the British Museum, not only are the tubercles of the palms of the chelipedes more distinctly longitudinally seriate, but also the interior margin of the front is armed with longer spinules.

Pilumnus lanatus, Latreille. East Indies ; North and East Australia, to 11 fathoms (H.M.S. "Alert") ; Tasmania.

Pilumnus asper (Rüppell). Red Sea.

Pilumnus minutus, de Haan = *Pilumnus hirsutus*, Stimpson. Japan and Corean Straits (to 50 fathoms) ; North China Sea ; New Caledonia ; North-East Australia ; Holborn Island (20 fathoms).

Pilumnus scabriusculus, Adams and White. Eastern Seas ; Samboangan, Philippines.[1]

Pilumnus globosus, Dana. Tahiti and Paumotu Islands ; Japan (50 fathoms).

Pilumnus lævimanus, Dana. Balabac Passage ; North of Borneo ; New Caledonia.

Pilumnus lævis, Dana. Mangsi Islands, Straits of Balabac.

Pilumnus tenellus, Dana. Sooloo Sea, or Straits of Balabac.

Pilumnus rufopunctatus, Stimpson. East and South Australia.

Pilumnus fissifrons, Stimpson. East Australia.

Pilumnus forficigerus, Stimpson. Oosima, Japan (30 fathoms).

Pilumnus lapillimanus, Stimpson. North China Sea (25 fathoms).

Pilumnus marginatus, Stimpson. Loo-Choo.

Pilumnus dorsipes, Stimpson. Hong Kong (10 fathoms).

Pilumnus (?) *pugilator* (A. Milne Edwards). Lifu Island ; North-East to East Australia (to 17 fathoms).

Pilumnus fragifer, A. Milne Edwards. Indian Ocean (perhaps a *Pilumnopeus*).

Pilumnus elatus, A. Milne Edwards. Upolu, Samoan Islands.

Pilumnus ovalis, A. Milne Edwards. Sandwich Islands.

[1] *Pilumnus dilatipes*, Adams and White, cannot be referred to this genus. Not only are the palatal ridges nearly obsolete, but the ambulatory legs are strongly cristated. It should probably be constituted the type of a distinct genus, *Lophopilumnus*.

Pilumnus deflexus, A. Milne Edwards. Australia.

Pilumnus cærulescens, A. Milne Edwards. New Caledonia.

Pilumnus barbatus, A. Milne Edwards. New Caledonia.

Pilumnus cursor, A. Milne Edwards. New Caledonia; North-East Australia; Samoan Islands.

Pilumnus longipes, A. Milne Edwards. New Caledonia. (Perhaps a *Prionoplax*).

Pilumnus purpureus, A. Milne Edwards. New Caledonia.

Pilumnus actumnoides, A. Milne Edwards. New Caledonia.

Pilumnus vermiculatus, A. Milne Edwards. New Caledonia.

Pilumnus nitidus, A. Milne Edwards. New Caledonia.

Pilumnus cristimanus, A. Milne Edwards. New Caledonia.

Pilumnus savignyi, Heller. Red Sea.

Pilumnus vauquelinii (Audouin), Heller. Red Sea. This species resembles *Pilumnus tridentatus* in the form of the antero-lateral marginal teeth.

Pilumnus brachytrichus, Kossmann. Red Sea. (Perhaps, according to Dr. Kossmann, synonymous with *Pilumnus tomentosus*).

Pilumnus longicornis, Hilgendorf. East Africa; Inhambane; Tongatabu (18 fathoms).

Pilumnus dehaani, Miers. Japan (in shell of *Balanus*); Philippines (18 fathoms).

Pilumnus bleekeri, Miers. New Guinea.

Pilumnus labyrinthicus, Miers. North and North-East Australia (to 14 fathoms, H.M.S. "Alert").

Pilumnus seminudus, Miers. North and North-East Australia.

Pilumnus semilanatus, Miers. North and East Australia (to 15 fathoms, H.M.S. "Alert").

Pilumnus pulcher, Miers. North Australia.

Pilumnus normani, n. sp. Ki Islands (140 fathoms), Station 192.

Pilumnus monilifera, Haswell. Tasmania.

Pilumnus terræ-reginæ, Haswell. North-East Australia; Port Molle.

Pilumnus vestitus, Haswell. Port Jackson; Port Stephens.

Pilumnus (?) *glaberrimus*, Haswell. Port Jackson.

Pilumnus (?) *inermis*, Haswell. Port Jackson. (I consider this and the preceding species as doubtfully belonging to *Pilumnus*, on account of the obscurely defined teeth of the antero-lateral margins. *Pilumnus integer*, Haswell, from Port Jackson, which has the antero-lateral margins entire, cannot, I think, be retained in this genus nor can *Pilumnus fimbriatus*, Milne Edwards, for which I have established the genus *Cryptocœloma*.)

Pilumnus africanus, A. Milne Edwards (Pl. XIII. fig. 1).

Pilumnus africanus, A. Milne Edwards, Ann. Soc. Entom. France, ser. 4, vol. vii. p. 280, 1867.

St. Vincent, Cape Verde Islands, July 1873 (a male and a female of large size, with ova).

The specimens in the Challenger collection agree in nearly all particulars with the description of A. Milne Edwards. This species is distinguished from *Pilumnus hirtellus*, of which it may be a variety, by the much more developed spinules of the outer and upper surface of the chelipedes, which spinules are arranged in longitudinal series, and from *Pilumnus villosus* by the unarmed superior margin of the orbit.

The colour (of specimens dry, and in spirits) is reddish-brown, the pubescence yellow; the spinules of the antero-lateral margins and chelipedes and the fingers are black.

The antero-lateral margins are normally six-spined, but the small spine which immediately follows that at the exterior orbital angle (the second of the series) is obsolete on one side and very small on the other in the larger (female) specimen.

The dimensions of both are given below :—

		Lines.	Millims.
♂.			
Length of carapace,	.	5	10·5
Breadth of carapace,	.	6	12·5
Adult ♀.			
Length of carapace,	.	6½	13·5
Breadth of carapace, nearly		8	17

Pilumnus spinifer (?), Milne Edwards.

? *Pilumnus spinifer*, Milne Edwards, Hist. Nat. Crust., vol. i. p. 420, 1834.
    ,,   ,,   *cf.* Savigny, Descr. de l'Egypte, Crust. Atlas, pl. v. fig. 4.
?   ,,   ,,   Heller, Crust. des südlichen Europa, p. 73, 1863.

Off Fayal, 50 to 90 fathoms. Several specimens of both sexes.

The specimens very doubtfully referred to this species are of rather small size, and differ from Milne Edwards' description, but not from the figure cited, in having the superior margins of the orbits entire, without spines. The tubercles of the chelipedes only tend to become spinuliferous near the superior margins. There are but three spines on the antero-lateral margins, exclusive of the small spinule at the exterior angle of the orbit, as in the description and figure.[1]

								Lines.	Millims.
Adult ♂.									
Length of carapace,	.	.	.	.	.	.	.	4½	9·5
Breadth of carapace,	.	.	.	.	.	.	.	5½	11·5

[1] Dr. C. Heller (Sitzungsb. math.-nat. Cl. k. Akad. Wiss. Wien, vol. xliii. (i.) p. 345, 1861) at first referred Savigny's *Pilumnus* to a new species, designated by Heller *Pilumnus savignyi*, on account of the non-spinuliferous superior margins of the orbits, but in 1863, in the work quoted above, it is referred to *Pilumnus spinifer*, Milne Edwards.

The Challenger specimens are at once distinguished from *Pilumnus tridentatus*, Maitland, as figured by Hoek,[1] by the presence of the small spine at the exterior angle of the orbit, by the distinctly spiniform antero-lateral marginal teeth, by the absence of transverse lines upon the dorsal surface of the carapace, and by other characters.

Pilumnus brasiliensis, n. sp. (Pl. XIII. fig. 2).

The body and limbs are rather thinly clothed with longish hairs; the carapace is convex, broader than long, shaped nearly as in *Pilumnus aculeatus*, and the front is moderately prominent, with the median lobes truncated and armed with short spinules on the anterior margins; the lateral lobes obsolete. The antero-lateral margins of the carapace are armed with three strong spines besides the spine at the outer angle of the orbit; between the outer orbital spine and the first antero-lateral spine is a smaller spinule, and the first antero-lateral spine bears a small accessory spinule on its posterior margin near the base. The dorsal surface of the carapace is without granulations or spinules; the subhepatic region is finely granulated; there is no spinule on the pterygostomian region. The superior and inferior margins of the orbit are armed with strong spinules, which are of unequal size; two on the upper margin and three on the inferior margin (near the inner hiatus of the orbit) being longer than the others. The antennæ and maxillipedes present nothing remarkable. The chelipedes (in the female) are nearly equal, but the right is a little larger than the left; the trigonous merus-joint is granulated on its upper and lower margins and bears a strong acute tooth near the distal angle of the upper margin, which terminates in a spine; the wrist and palm are armed with spinules over the whole of the upper and outer surface, the spinules longer on the upper surface and especially on its inner margin; those of the palm are arranged in eight longitudinal series; the fingers are dentated on their inner margins; the upper is sulcated and granuli-spinulous above at the base. The ambulatory legs are of moderate length and rather slender; the carpus or antepenultimate joint is armed with spinules; the straight and slender dactyli terminate in a small and not strongly curved claw. Colour (in spirit) light yellowish-brown; the hairs yellowish; the fingers of the chelipedes are of a slaty hue, the colour not extending over the inner and outer surface of the palm.

The dimensions are as follows :—

♀.	Lines.	Millims.
Length of carapace, nearly	4	8·5
Breadth of carapace, nearly	5	10·5
Length of a chelipede, nearly	5	10·5
Length of second ambulatory leg,	6	15

[1] *Tijd. Ned. Dierk. Vereenig.*, vol. ii. p. 243, pl. xiv. figs. 12, 16, 1876.

The single specimen (which is a female, perhaps not fully adult), was obtained near Bahia, in 7 to 20 fathoms, with *Pilumnus floridanus*.

This species evidently belongs to the same section of the genus as *Pilumnus aculeatus* (Say), *Pilumnus vinaceus*, A. Milne Edwards, and *Pilumnus caribæus*, Desbonne. It is more nearly allied to the last-mentioned species than to any other of the genus with which I am acquainted, agreeing with it and differing from other allied American species (as described by authors) in having the outer surface of the palms of both chelipedes wholly covered with strong, longitudinally seriate, spiniform tubercles, but it differs in the smooth (not granulated or spinuliferous) carapace, and in the absence of a pterygostomian spine.

From *Pilumnus gracilipes*, A. Milne Edwards, it is apparently distinguished by the well-developed spinules of the upper orbital margin and by the absence of tubercles or spines from the subhepatic and pterygostomian regions.[1]

Pilumnus floridanus, Stimpson (Pl. XIII. fig. 3).

Pilumnus floridanus, Stimpson, Bull. Mus. Comp. Zoöl., vol. ii. p. 141, 1870.
    „   „   A. Milne Edwards, Crust. in Miss. Sci. au Méxique, pt. 5, p. 287, 1880.

Here is referred an adult ova-bearing female, dredged off the coast of Brazil, in from 30 to 350 fathoms, lat. 9° 5′ 0″ S. to 9° 10′ 0″ S., long. 34° 49′ 0″ W. to 34° 53′ 0″ W. (Stations 122–122c).

This specimen scarcely differs from Stimpson's description in any particular, except in the absence of a transverse series of longer hairs on the frontal region. Several such hairs exist, however, near the front and antero-lateral margins.

From *Pilumnus quoyi*, Milne Edwards, found at Rio de Janeiro,[2] this species would seem to be distinguished by the spinose ambulatory legs.

Three males collected near Bahia, in shallow water, are also doubtfully referred to this species. They are distinguished from the typical *Pilumnus floridanus* only by the absence of the tubercles from that part of the outer surface of the palm of the larger chelipede which lies nearest to the base of the lower (immobile) finger. In all other particulars, as (e.g.) the non-spinuliferous lobes of the front and upper margins of the orbits, the form of the carapace, and the moderately robust and elongated spinuliferous and hairy ambulatory legs, they agree with the typical form of the species. In one (the smallest) of these specimens, the tubercles cover a larger part of the outer surface of the palm than in the others. It is worthy of note that the type of Stimpson's description was a female; I think it therefore possible, that the partial absence of the tubercles of the palm of the larger chelipede may be characteristic of the male sex.

[1] The spinules of the upper orbital margins are foreshortened in an anterior view, and are not shown, therefore, in the magnified view of the front (pl. xiii. fig. 2a).
[2] A. Milne Edwards, tom. cit., p. 289, pl. l. fig. 5, 1880.

The dimensions of the largest male are as follows :—

Adult ♂.	Lines.	Millims.
Length of carapace, nearly	3½	7·5
Breadth of carapace, about	4½	9·5
Length of a chelipede, nearly	6	12·5
Length of second ambulatory leg,	6½	13·5

A small male *Pilumnus* is in the collection from St. Thomas (shallow water), which I cannot certainly refer to any species of this very difficult genus, yet do not venture to constitute the type of a distinct species. It is allied to *Pilumnus quoyi*, Milne Edwards, as described and figured by A. Milne Edwards, in many characters, and particularly in the rounded tubercles of the exterior surface of the palms of the chelipedes, which, however, become spinuliform on the superior margins ; it differs from that form in having a few very small spinules interspersed among the hairs of the ambulatory legs, &c. *Pilumnus miersii*, A. Milne Edwards, to which it is also nearly allied, has an additional antero-lateral spine. *Pilumnus lacteus*, Stimpson, which it resembles in the whitish pubescence of the body (interspersed with longer hairs), is distinguished by the smooth exterior surface of the palms of the chelipedes.

Adult ♂.	Lines.	Millims.
Length of carapace, about	2	4·5
Breadth of carapace, about	2½	5·5

Pilumnus fragosus, A. Milne Edwards, var.

> *Pilumnus fragosus*, A. Milne Edwards, Crust. in Miss. Sci. au Mexique, pt. 5, p. 296, pl. lii. fig. 1, 1880.

Bahia (7 to 20 fathoms). Two females.

In these specimens the tubercles of the carapace and of the chelipedes are less flattened and less distinctly pedunculated than in the figure of Milne Edwards, whose types were from the island of St. Thomas. There exists a second row of less elevated tubercles behind the tubercles of the protogastric and epigastric lobes which are not shown in the figure cited. I think it therefore possible that the Challenger specimens may prove to be specifically distinct.

The largest example measures as follows :—

Adult ♀.	Lines.	Millims.
Length of carapace,	3½	7·5
Breadth of carapace, about	4½	9·5

This specimen bears ova, and is, therefore, fully adult ; some longer hairs are interspersed among the short greyish-brown pubescence of the carapace and limbs ; these hairs are slightly clavate.

Pilumnus minutus (?), de Haan, var. *hirsutus.*

1 *Pilumnus minutus*, de Haan, Crust. in v. Siebold, Fauna Japonica, pp. 19, 50, pl. iii. fig. 2, 1835.
Pilumnus hirsutus, Stimpson, Proc. Acad. Nat. Sci. Philad., p. 37, 1858.
,,   ,,   Miers, Proc. Zool. Soc. Lond., p. 31, 1879.

Specimens are referred to this species from Kobé, Japan, lat. 34° 38′ 0″ N., long. 135° 1′ 0″ E., 50 fathoms (Station 233A).

In the specimens I have examined there is commonly a small accessory spinule on the antero-lateral margins, situated behind the spine at the outer angle of the orbit.

	Lines.	Millims.
Adult ♂.		
Length of carapace, nearly	4	8
Breadth of carapace,	5	10

As the original description of this species by de Haan is very brief, I append the following, based on the largest male in the Challenger series.

The carapace is moderately convex and transverse, and, as well as the ambulatory legs, is covered with a short greyish-brown pubescence, interspersed with longer hairs; the regions are not defined, and the dorsal surface is not distinctly granulated. The front is divided by a median notch into two somewhat obliquely truncated lobes, whose anterior margins are not spinuliferous. The orbits are moderately large; their superior margins have very obscure indications of a notch or sinus, their inferior margins are denticulated, the denticles becoming more prominent and spinuliform at the interior subocular lobe, and are armed with four small spiniform teeth, the second of which in this specimen (but not usually), bears a very small accessory spinule; there is no spinule upon the subhepatic region. The chelipedes are very unequal, with the merus short, trigonous, and denticulated on the superior and anterior margins, the carpus armed with granules, which tend to become spinuliform, on the superior and exterior surface; palm (in the larger chelipede) granulated above and at the base, but smooth over nearly the whole of its outer surface; in the smaller chelipede it is covered, both above and externally, with conical, acute granules or small spines; dactyli granulated above at the base (but very obscurely in the larger chelipede); fingers toothed (the dactylus rather obscurely), on the inner margins. The ambulatory legs are clothed, especially on the last three joints, with greyish pubescence, interspersed with longer hairs, but are not armed with spines or spinules.

In 1879 I regarded Stimpson's species, *Pilumnus hirsutus*, as distinct from *Pilumnus minutus*, de Haan, on account of the acute antero-lateral marginal teeth, and the denticulated inferior orbital margins, but I am now inclined to regard *Pilumnus hirsutus* as a variety only of *Pilumnus minutus*; the diagnosis of de Haan's species is, however, very short and the figure insufficient.

Pilumnus globosus, Dana.

Pilumnus globosus, Dana, Proc. Acad. Nat. Sci. Philad., p. 81, 1852; U.S. Explor. Exped., vol. xiii., Crust. 1, p. 236, pl. xiii. fig. 10, 1852.

Kobé, Japan, 50 fathoms (Station 233A), lat. 34° 38′ 0″ N., long. 135° 1′ 0″ E. (two females).

The specimens referred to this species have the body more thickly covered with yellowish-brown hair than the types described by Dana and a specimen in the British Museum collection, and the under as well as the upper surface of the palm is fringed with long hairs.

♀.	Lines.	Millims.
Length of carapace,	3	6·5
Breadth of carapace,	4	8·5

Pilumnus dehaani, Miers (Pl. XIV. fig. 1).

Pilumnus dehaanii, Miers, Proc. Zool. Soc. Lond., p. 32, 1879.

Philippines, lat. 11° 37′ 0″ N., long. 123° 31′ 0″ E., 18 fathoms (Station 208). A small female.

♀.	Lines.	Millims.
Length of carapace, nearly	2	5
Breadth of carapace,	2½	5·5

The figure, which I think it useful to give of this species (which may prove to be a variety of *Pilumnus* (?) *squamosus*, de Haan, although differing in the points I have indicated in the original description) is from the large type specimen in the collection of the British Museum. In the very small Challenger specimen the tubercles of the chelipedes, although conical, are less prominent.

Pilumnus forficigerus, Stimpson,[1] which is a nearly allied form, dredged at Oosima in 30 fathoms, is apparently distinguished by the less numerous tubercles of the palm of the smaller chelipede, and also by the smooth carpi of the chelipedes.

Pilumnus scabriusculus (?), Adams and White.

? *Pilumnus scabriusculus*, Adams and White, Crust. Zool. H.M.S. "Samarang," p. 44, pl. ix. fig. 5, 1848.

Philippines, Samboangan, an adult male, found on the beach.

In this specimen the interregional sutures of the dorsal surface of the carapace, which are very indistinctly indicated in the figure given by Adams and White (and in a specimen which I suppose to be typical, in the collection of the British Museum) are obliterated.

[1] *Proc. Acad. Nat. Sci. Philad.*, p. 36, 1858.

In neither of the specimens I have seen is the front subtriangular as in the original description, which I think to be erroneous in this particular.

The species is apparently well characterised by the uniformly granulated carapace, which is covered by a fuscous pubescence which arises in tufts of two or three setæ, from each of the granules of the dorsal surface, and by the wide, denticulated or granulated teeth of the antero-lateral margins, as shown in the figure referred to.

In the specimens I have examined the front is four-lobed; the median lobes subtruncated and somewhat more prominent than the lateral lobes, which are small and dentiform, and situated just inside of the inner canthus of the orbit.

Another specimen in the British Museum collection which was designated (though not by White), *Pilumnus scabriusculus*, certainly does not belong to this species, but to the variety of *Pilumnus longicornis*, described below, having a deeply incised four-lobed front, and prominent spiniform antero-lateral marginal teeth.

The Challenger specimen, which is rather the larger, has the following dimensions :—

	Lines.	Millims.
Adult ♂.		
Length of carapace, . .	10½	22
Breadth of carapace, nearly .	13	27

The pubescence of the carapace and limbs is yellow; the merus-joints of the ambulatory legs are thin-edged and acute, but are not distinctly carinated as in the species designated *Pilumnus dilatipes*, Adams and White.

Pilumnus normani, n. sp. (Pl. XIV. fig. 2).

The carapace is broader than long, longitudinally convex, and covered with a short, thick, close pubescence, from amongst which spring numerous longer hairs. The front is but slightly deflexed, and is divided by a deep median sinus; the median lobes are broad, subtruncated, and obscurely granulated on the margins, the lateral lobes very small and dentiform. The surface of the carapace, beneath the pubescence, is smooth, not granulated. The orbits are obscurely granulated on their upper margins, and more distinctly on the lower margins, where the granules tend to become spinuliform as they approach the inner subocular spine. The antero-lateral margins are armed with three strong, simple spines, behind the spine at the exterior angle of the orbit, which is smaller than the antero-lateral spines; there are no spines on the subhepatic and pterygostomian regions, which are slightly pubescent, but one or two small granules on the subhepatic region. The basal antennal joint does not nearly reach the infero-lateral process of the front. The left chelipede is wanting, the right (in the male) is moderately robust; merus-joint granulated on its upper and antero-inferior margins, and armed on the upper margin with a spine immediately behind the small distal spine;

carpus clothed externally with rather long hairs, and covered on the whole of its outer surface with small spinules, and with a stronger spine on the inner margin; palm clothed with longish hairs on its upper surface, and on the outer surface near the base, spinulose on its upper margin, and granulated on the whole of its outer surface; the granules are longitudinally seriate, and become spinuliform near the articulation with the wrist; the fingers are denticulated on their inner margins, the dactyl obscurely sulcated and granulated above at the base; the ambulatory legs are somewhat elongated, and clothed with a short pubescence and with longer hairs. Colour of the short pubescence, greyish-yellow (in spirit), the longer hairs yellowish; fingers of chelipedes chocolate-brown, the coloration not extending over the inner or outer surface of the palm.

Adult ♂.	Lines.	Millims.
Length of carapace, .	5½	11·5
Breadth of carapace, about .	6¾	14
Length of a chelipede, about	7½	16
Length of second ambulatory leg,	10	21

Near the Ki Islands, Station 192, lat. 5° 49′ 15″ S., long. 132° 14′ 15″ E., 140 fathoms (an adult male).

The nearest ally to this species with which I am acquainted is *Pilumnus purpureus*, A. Milne Edwards,[1] from New Caledonia, from which *Pilumnus normani* is distinguished by the smooth, not granulated, dorsal surface of the carapace. A strong spinule exists on the inner margin of the carpus in *Pilumnus normani*, which is not shown in Milne Edwards' figure of *Pilumnus purpureus*.

From most of its congeners it may be distinguished by the distant and greatly elongated antero-lateral marginal spines.

Pilumnus longicornis, Hilgendorf, var.

[1] *Pilumnus longicornis*, Hilgendorf, Monatsber. d. k. preuss. Akad. d. Wiss. Berlin, p. 794, pl. i. figs. 8, 9, 1878.

In the Challenger specimen referred to this species the carapace is convex, rather broader than long, its surface roughened with small granules, from which spring rather long fulvous hair. Similar hairs cover the dorsal and posterior (or outer) surface of the chelipedes and ambulatory legs; this pubescence, however, is not so thick as to conceal completely the granulations of the carapace and limbs. The cervical and cardiaco-branchial sutures are distinct. The front equals in width about one-third the greatest width of the carapace, and is divided into four lobes; the median lobes broad, subtruncated, and separated by a rather deep median suture; the lateral lobes small and

[1] *Nouv. Archiv. Mus. Hist. Nat.*, vol. ix. p. 246, pl. x. fig. 5, 1873.

dentiform. The orbit has two very obscurely indicated sutures in its upper margins; its lower margin is subentire, without distinct spinules. The antero-lateral margins are shorter than the postero-lateral, and are armed with three spiniform teeth; there is no distinct tooth or spine at the outer angle of the orbit, and the small tooth which in this genus usually exists on the subhepatic region is also deficient, but the subhepatic region is granulated. The terminal segment of the male post-abdomen narrows somewhat to the distal extremity, which is bluntly rounded. The eye-peduncles are of the form usual in the genus, moderately short and thick. The antennæ terminate in rather long flagella, their basal joint barely reaches to the subfrontal process, and does not nearly attain the inner suborbital angle; the merus-joint of the outer maxillipedes is rather broader than long, and is distinctly excavated at its antero-internal angle. The chelipedes are nearly of equal size, but the right is a little larger than the left; the trigonous arm or merus-joint has a strong curved spine near the distal end of its upper margin, the wrist is covered on its upper and outer surface with large but not very numerous granules, which in some places are spinuliform, and there is a spine on the inner margin of the wrist; the palm, also, is covered on its upper and outer surface with conical acute granules, which on the upper margin are more distinctly spinuliform, and are disposed in longitudinal series over the whole of the outer surface, but on the lower margin are more numerous and crowded; the fingers are short, armed on their inner margins with rather indistinct serrated teeth; the dactylus longitudinally sulcated and granulated at the base. The ambulatory legs are rather robust and of moderate length; the merus-joints in the first three pairs are armed with a blunt tooth or lobe behind the upper and distal angle, which is also produced, dentiform and acute; the dactyli are styliform, more slender than the preceding joints, and armed with a small claw. The ground colour of the carapace and limbs (in spirit) is light reddish-brown, but the pubescence with which the body and legs are clothed is of a whitish-yellow hue; the fingers are chocolate-brown; the coloration not extending over any part of the inner and outer surface of the palm.

Adult ♂.								Lines.	Millims.
Length of carapace,	.	.	.	.	.	.	.	7	15
Breadth of carapace, about	.	.	.	.	.	.	.	9	19
Length of larger chelipede, when extended as far as its conformation will allow,								9	19
Length of second ambulatory leg, about .	.	.	.	.	.	.		1 inch	25

The single specimen in the Challenger collection (an adult male), was collected off Nukalofa, Tongatabu, in 18 fathoms (Station 172).

I have examined in the British Museum collection a specimen apparently of this species, without definite locality, referred to above as having been erroneously designated *Pilumnus scabriusculus*, White. The pubescence is everywhere much thinner than in the Challenger specimen, and is, in fact, nearly absent from the dorsal surface of the carapace.

From the typical form of this species, as figured by Hilgendorf, the Challenger type can only be distinguished by the less distinctly toothed orbital margins, the strongly and distinctly granulated palm of the larger chelipede (the granules being arranged, as in the smaller chelipede, in longitudinal series over the whole of the outer surface), and the stronger tooth or spine at the distal extremity of the superior margins of the meral joints of the ambulatory legs.[1]

Pilumnus vestitus, Haswell (Pl. XIV. fig. 3).

Pilumnus vestitus, Haswell, Proc. Linn. Soc. N.S.W., vol. vi. p. 753, 1882; Cat. Aust. Crust., p. 68, 1882.

Port Jackson (shore) an adult female :—

Adult ♀.	Lines.	Millims.
Length of carapace, nearly	6	12
Breadth of carapace,	7	15

Mr. Haswell's description in the Australian Catalogue is very brief, and I therefore append the following description of the specimen in the Challenger collection referred to this species.

Carapace moderately convex and transverse, covered, as well as the limbs, with yellow hairs. The front is divided by a median notch into two rather prominent rounded lobes, which are separated from the rounded interior angle of the orbit by a sinus. The superior margin of the orbit is subentire, the inferior margin is granulated, the interior subocular angle of the orbit is but slightly prominent and dentiform. The subhepatic and pterygostomian regions are granulated. The antero-lateral margins are armed with three acute spines behind the spine at the exterior angle of the orbit. The chelipedes (in the female) are nearly equal, the left somewhat the larger; the short, trigonous merus-joint is armed with a short spine at the distal extremity of its superior margin, and with another spine just posterior to this ; the wrist and palm are armed with conical acute granules or short spines, which are not very numerous, and are smaller near the inferior margin of the palm and the base of the immobile finger, which is regularly and distinctly dentated ; the dactylus is more obscurely dentated on the interior margin and is granulated above at the base ; the ambulatory legs are slightly compressed, and the fifth and sometimes the fourth and sixth joints are armed with spinules on their superior margins.

---

[1] *Pilumnus dromipes*, Stimpson (Proc. Acad. Nat. Sci. Philad., p. 37, 1858), from Hong-Kong, which, to judge from the brief description, is apparently a nearly allied form, is apparently distinguished by the sulcated latero-inferior margin of the carapace.

Pilumnus tomentosus (?), Latreille (Pl. XIV. fig. 4).

1 *Pilumnus tomentosus*, Latreille, Encyd. Méth., vol. x. p. 125, 1825.
   „   „   Milne Edwards, Hist. Nat. Crust., vol. i. p. 418, 1834.

Specimens from Bass Strait, are referred, but very doubtfully, to this species. They were collected off East Moncœur Island in 38 fathoms (Station 162). Also a female from the entrance to Port Philip, Australia, 38 fathoms (Station 161).

Pilumnus tomentosus is but very briefly characterised, and has, I believe, never been figured, therefore its identification must remain uncertain. The species I thus designate is, it would appear, one of the commonest on the South and South-Eastern Australian coasts. The colour is variable; the specimens from East Moncœur Island are light reddish-brown with yellowish pubescence, that from Port Philip is darker brown, and others in the collection of the British Museum blackish-brown, as in the description of Milne Edwards; the pubescence is always short and rather dense. The frontal lobes are rather prominent and rounded, the orbital margins are not distinctly denticulated above; there exists, besides the spinules mentioned by Milne Edwards, a small spinule on the subhepatic region, which is visible between the exterior orbital and first antero-lateral spine. The surface of the carapace is sometimes smooth, but usually bears several spinuliform granules near the antero-lateral marginal teeth, these granules and the spinules of the upper surface of the wrist and palm in the Challenger specimens from Bass Strait (Station 162) are red coloured.

		Lines.	Millims.
Adult ♂.			
Length of carapace, about	.	8	17
Breadth of carapace, about	.	10½	22

Pilumnus rufopunctatus, Stimpson (Pl. XIV. fig. 5).

Pilumnus rufopunctatus, Stimpson, Proc. Acad. Nat. Sci. Philad., p. 36, 1858.

South Australian coast, 2 to 10 fathoms, April 1874. (A small ova-bearing female).

This species is apparently distinguishable from the one I have referred to as *Pilumnus tomentosus*, by the broader carapace, less prominent front, the existence of several denticules on the upper margin of the orbit, and by the closely approximated granules on either side of the median line of the carapace, on the front of the gastric region.

The pubescence of the carapace is short, close, and rather dense, especially near the front and antero-lateral margins; the dorsal granules are less numerous than in Stimpson's description. The granulations of the larger chelipede, which extend over the outer surface, are absent from the inferior margin. If, as is possible, this species be the true *Pilumnus tomentosus*, Latreille, the Challenger specimens designated *Pilumnus tomentosus* must receive a new specific designation.

Adult ♀.					Lines.	Millims.
Length of carapace, nearly	.	.	.	.	4	8
Breadth of carapace, about	.	.	.	.	5½	11·5

In this specimen the pubescence of the carapace is not so dense as in adult males in the British Museum (H.M.S. "Alert").

Pilumnus pulcher, Miers.

> *Pilumnus pulcher*, Miers, Crust. in Rep. Zool. Coll. H.M.S. "Alert," p. 219, pl. xxii. fig. A, 1884.

Torres Strait, August 1874 ; a fine adult male. Off Cape York, 8 fathoms, lat. 10° 30′ 0″ S., long. 142° 18′ 0″ E. (Station 186) a very small male.

Adult ♂.			Lines.	Millims.
Length of carapace, nearly	.	.	13½	28·5
Breadth of carapace,	.	.	15½	33

The pubescence in the small male is whitish, not yellow as in the adult.

Pilumnus seminudus, Miers.

> *Pilumnus seminudus*, Miers, Crust. in Rep. Zool. Coll. H.M.S. "Alert," p. 222, pl. xxi. fig. C, 1884.

An adult male was obtained in Torres Strait with the preceding species.

Adult ♂.		Lines.	Millims.
Length of carapace, about		4½	9
Breadth of carapace.	.	6	13

Pilumnus labyrinthicus, Miers.

> *Pilumnus labyrinthicus*, Miers, Crust. in Rep. Zool. Coll. H.M.S. "Alert," p. 224, pl. xxii. fig. C, 1884.

Near Cape York, lat. 10° 30′ 0″ S., long. 142° 18′ 0″ E., in 8 fathoms (Station 186). A small male.

The vermicular ridges of the carapace in their number and disposition closely resemble those of the larger example, which is the type of my figure and description. The coloration is similar except as regards the fingers, which are white.

♂.					Lines.	Millims.
Length of carapace,	.	.	.	.	3	6·5
Breadth of carapace, nearly	.	.	.	.	4	8

Eriphia, Latreille.

Eriphia, Latreille, Crust. in Règne Animal de Cuvier, ed. 1, vol. iii. p. 18, 1817.
,,   Milne Edwards, Hist. Nat. Crust., vol. i. p. 425, 1834.

Carapace moderately convex, but little broader than long, with the antero-lateral margins much shorter than the postero-lateral margins; the frontal region broad, but the front not prominent; the antero-lateral margins (and usually the front) are spinuliferous, and the parts adjacent are usually armed with tubercles or short spines. The post-abdomen in the male is usually distinctly seven-jointed. The eye-peduncles are short and deeply set in the orbits, whose prominent margins are often granulated; it is broadly united with the front at its interior angle, so that the antenna is completely excluded from the orbit. The epistoma is transverse; the ridges of the endostome or palate are strongly defined. The basal antennal joint barely reaches the inferior margin of the front, but the flagellum is more developed than in many genera of Cancroidea. The ischium of the endognath of the exterior maxillipedes is not produced at its antero-internal angle; the merus is truncated but not emarginate at the distal extremity, and scarcely, if at all, emarginate at its antero-internal angle. The chelipedes (in the adult male) are considerably developed and usually unequal; fingers distally acute or subacute; ambulatory legs of moderate length, with the joints neither carinated nor spiniferous; dactyli styliform, nearly straight.

The following are species of this genus :—

Eriphia spinifrons (Herbst). Mediterranean and Black Seas; Canary Islands; Madeira; Japan (?); Singapore (?) (Coll. Brit. Mus.).
Eriphia lævimana, Milne Edwards = *Eriphia rugosa*, Milne Edwards (*nom. ined.* ?); *Eriphia fordii*, MacLeay, and *Eriphia smithii*, MacLeay, var.; *Eriphia trapeziformis*, Hess. Indo-Pacific Region.
Eriphia scabricula, Dana. Indo-Pacific Region.
Eriphia gonagra (Fabricius). Florida to Brazil (Rio de Janeiro).
Eriphia armata, Dana. East Patagonia.
Eriphia squamata, Stimpson. West American Coasts (California to Chili); (perhaps a variety of *Eriphia gonagra*).
Eriphia granulosa, A. Milne Edwards. Chili.

Eriphia gonagra (Fabricius).

    Cancer gonagra, Fabricius, Entom. Syst., vol. ii. p. 466, 1793; Suppl., p. 337, 1798.
    Eriphia gonagra, Milne Edwards, Hist. Nat. Crust., vol. i. p. 426, pl. xvi. figs. 16, 17, 1834.
       ,,   ,,   A. Milne Edwards, Crust. in Miss. Sci. au Méxique, pt. 5, p. 358, pl. xvi.
            fig. 4, 1880.

Bermuda (a small adult male) :—

Adult ♂.	Lines.	Millims.
Length of carapace,	5	10·5
Breadth of carapace,	7	14·5

Section II. Trapeziinæ.

    Cancériens quadrilatères, Milne Edwards (pt.), Hist. Nat. Crust., vol. i. pp. 369, 424, 1834.
    Trapézides, A. Milne Edwards, Ann. d. Sci. Nat., ser. 4, Zool., vol. xviii. p. 41, 1862; Nouv.
      Archiv. Mus. Hist. Nat., vol. i. p. 183, 1865.

Carapace depressed and nearly quadrilateral, smooth, with the postero-lateral angles truncated, the dorsal regions not defined; the antero-lateral margins are straight, form a right angle with the front, and are entire or have but one tooth (the lateral epibranchial tooth) developed. The front is horizontal, broad, lamellate, and projects over the antennules and bases of the antennæ, which are widely excluded from the orbits.

The genera comprised in this section are the following :—

Trapezia, Latreille (= *Grapsillus*,	*Tetralia*, Dana.
MacLeay).	*Quadrella*, Dana.[1]

Trapezia, Latreille.

    Trapezia, Latreille, Fam. Nat. du Règne Animal, p. 269, 1825; Crust. in Règne Animal de
      Cuvier, ed. 2, vol. iv. p. 44, 1829.
      ,,   Milne Edwards, Hist. Nat. Crust., vol. i. p. 427, 1834.
    Grapsillus, MacLeay, Annulosa in Smith, Zool. South Africa, p. 67, 1838.

The carapace is nearly quadrate, depressed, smooth, with the antero-lateral margins parallel, the postero-lateral margins convergent backwards from the lateral epibranchial spine, which is not always developed, the regions on the dorsal surface are not indicated; the front projects beyond the orbits, and is divided into several lobes or teeth. The post-abdomen in the male is usually five-jointed. The short eye-peduncles are not concealed by the shallow orbits, whose margins are without fissures ; there is often a tooth or

---

[1] The Crustacean from Providence Island (19 fathoms), referred to without specific or generic designation in my Report on the Crustacea of H.M.S. "Alert" (Zool. Coll., p. 536, footnote, 1884), belongs to this rare genus, and is even perhaps identical with Dana's species, *Quadrella coronata.*

spine at the inferior and inner lobe of the orbit, which lobe attains the front, so as to exclude the antennæ from the orbital cavity. The peduncular joints of the antennæ are small and slender; the basal joint does not nearly reach the front. The exterior maxillipedes are small, the ischium-joint not produced at the antero-internal angle; the merus with the antero-internal angle obliquely truncated, scarcely, if at all, emarginate, the antero-external angle not produced. The chelipedes in the adult male are subequal, and considerably developed; merus with the antero-external angle dentated or spinose toward the distal extremity; carpus without spines, palms somewhat compressed, smooth; fingers distally acute, with small teeth or granules near the base; the dactylus sometimes with a larger, granulated, sub-basal lobe on the inner margin; the ambulatory legs are small, with the joints slightly compressed, the dactyli hairy and rather shorter than the penultimate joints.

The species are small, and commonly inhabit the coral reefs of the Indo-Pacific region. They are connected one with another by almost insensible gradations of form and colour, hence their discrimination is extremely difficult, as shown by Dr. R. Kossmann,[1] who, indeed, would apparently regard all, except perhaps *Trapezia digitalis*, Latreille, as varieties of a single species.

In the following enumeration of the described species I have grouped them mainly according to the presence of the lateral epibranchial tooth, by the form of the chelipedes, and according to their colour-variations; but I would not be understood as regarding all as distinct and well-characterised types.

1. Carapace without lateral spines or teeth :—

    (1) Carapace neither spotted nor areolated.

        Trapezia digitalis, Latreille = *Trapezia leucodactyla*, Rüppell. Red Sea.

        Trapezia formosa, Smith. Pearl Islands, Bay of Panama. (Perhaps not specifically distinct from *Trapezia digitalis*).

    (2) Carapace areolated.

        Trapezia speciosa, Dana. Paumotu Archipelago, Carlshoff Island.

    (3) Carapace spotted.

        Trapezia bella, Dana. Paumotu Archipelago, Carlshoff Island.

2. Carapace with a spine or tooth in the middle of the lateral margins (the lateral epibranchial tooth) :—

    (1) Carapace and limbs of a uniform colour, without spots or areolæ.

[1] Zool. Ergebn. einer Reise Küstengeb. d. rothen Meeres, Malacostraca, pp. 11–15, 1877.

a. Palms of chelipedes subcristate above, externally hairy. Lateral marginal teeth of carapace usually acute in the adult.

> *Trapezia cymodoce* (Herbst), Miers, 1878 = *Trapezia ferruginea*, Latreille; *Trapezia dentifrons*, Latreille; *Trapezia hirtipes*, Jacquinot and Lucas. Oriental Region. The citations of *Trapezia cymodoce* and *dentifrons* for this species are a little doubtful; it is certainly the form figured by MM. Jacquinot and Lucas as *Trapezia hirtipes.*

b. Palms of chelipedes rounded above, externally glabrous. Lateral marginal teeth of carapace blunt or almost obsolete in the adult.

> *Trapezia cærulea*, Rüppell = *Trapezia miniata*, Jacquinot and Lucas (?); *Trapezia subdentata*, Gerstæcker; *Grapsillus subinteger*, MacLeay. Oriental Region. I was probably wrong in designating this species, in 1878, by Latreille's name, *Trapezia ferruginea*, as that author describes the superior margins of the palms of the chelipedes as acute. The citations of *Trapezia miniata*, *Trapezia subdentata*, and *Grapsillus subinteger* for it, are, on account of the insufficiency of the authors' descriptions, somewhat uncertain.

> The following species have a uniformly coloured carapace, and may be identical with one or other of the two preceding groups :—

> *Trapezia dentata*, MacLeay. Cape of Good Hope. (Perhaps identical with *Trapezia cymodoce.*)

> *Trapezia fusca*, Jacquinot and Lucas. Noukahiva, Marquesas Islands. (Perhaps identical with *Trapezia cærulea.*)

> *Trapezia corallina*, Gerstæcker. Veragua.

> *Trapezia nigrofusca*, Stimpson. Cape St. Lucas, California. (Perhaps identical with *Trapezia corallina*, but is dark brown, not red, and the front is entire, not denticulated.)

(2) Carapace of a uniform colour, the limbs spotted.

> *Trapezia guttata*, Rüppell. Red Sea.

(3) Carapace marked with pink spots.

> *Trapezia rufopunctata* (Herbst) = *Grapsillus maculatus*, MacLeay; *Trapezia tigrina*, Eydoux and Souleyet; *Trapezia acutifrons*, A. Milne Edwards, var. Oriental Region.

(4) Carapace areolate, with pink reticulating lines.

> Trapezia areolata, Dana = Trapezia reticulata, Stimpson (?);
> Trapezia areolata, var. inermis, A. Milne Edwards; Trapezia
> areolata, var. lævis.   Polynesian Islands ; Loo-Choo.   This
> species varies much in the acuteness of the lateral marginal
> teeth and in the size of the areoles of the carapace, which
> may extend over the chelipedes, and perhaps over the
> ambulatory legs (cf. Stimpson).

> Trapezia flavo-punctata, Eydoux and Souleyet = Trapezia
> latifrons, A. Milne Edwards (?).   Sandwich Islands ; New
> Caledonia.   Distinguished from Trapezia areolata by the
> areolated ambulatory legs, and by having the inferior
> margins of the palms of the chelipedes armed with granules
> or small tubercles.

Trapezia cymodoce (?), (Herbst).

† Cancer cymodoce, Herbst, Naturgesch. der Krabben u. Krebse, vol. iii. (2), p. 22, pl. li.
   fig. 5, 1801.
Trapezia ferruginea, Latreille, Encycl. Méth. Hist. Nat., vol. x. p. 695, 1825.
   „   hirtipes, Jacquinot and Lucas, Crust. in Voy. au Pole Sud. Zool., vol. iii. p. 44. pl. iv.
           fig. 14, 1853.
   „   cymodoce, Miers, Ann. and Mag. Nat. Hist., ser. 5, vol. ii. p. 409, 1878, et synonyma.

Samboangan, Philippines, 10 fathoms, on the reefs, an adult male and female ; off
Nukalofa, Tongatabu, 18 fathoms, in lat. 20° 58' 0" S., long. 175° 9' 0" E. (Station
172), an adult and a fully grown and a smaller female.

As I have noted above, the carapace and limbs in this species are of a uniform
reddish or bluish-brown, without reticulating lines or spots ; the lateral marginal teeth of
the carapace are acute and well developed in the adult ; the palms of the chelipedes are
subcristated on the superior margins, and hairy on the exterior surface.

Adult ♂ .		Lines.	Millims.
Length of carapace, nearly .		7	14·5
Breadth of carapace, about .		7½	16·5
Length of a chelipede, nearly		15	31·5
Length of first ambulatory leg,	.	9	19

Trapezia guttata, Rüppell, var. (Pl. XII. fig. 1).

cf. Trapezia guttata, Rüppell, Beschreib. u. Abbild. 24 kurzschwänzigen Krabben als Beitrag
   zur Naturgesch. des rothen Meeres, p. 25, 1830.

Kandavu, Fiji Islands (on the reefs), a small female.

In this variety, to which I think it unnecessary to apply a distinctive appellation, the carapace and chelipedes are of a uniform reddish or yellowish-brown. The ambulatory legs, but not the chelipedes, are covered with small, distinct, red or brownish spots.

In Rüppell's specimens from the Red Sea, it would appear from the diagnosis, which is, however, very short, that all the legs are covered with small dark brown spots.

♀.		Lines.	Millims.
Length of carapace, nearly.	.	4	8
Breadth of carapace, about	.	4½	10

Trapezia areolata, var. inermis, A. Milne Edwards.

Trapezia areolata, var. inermis, A. Milne Edwards, Nouv. Archiv. Mus. Hist. Nat., vol. ix.
p. 259, pl. x. fig. 6, 1873.

Kandavu, Fiji Islands (a female with ova).

In this form, distinct areolæ, formed of pink reticulating lines, exist on the dorsal surface of the carapace and on the chelipedes, but not on the ambulatory legs; the denticules of the front and the lateral marginal teeth of the carapace are blunt in the adult.

♀.		Lines.	Millims.
Length of carapace, about		4	9
Breadth of carapace, nearly		5	10
Length of a chelipede,		8	16·5
Length of first ambulatory leg,		5	10·5

Trapezia rufopunctata (Herbst).

Cancer rufopunctatus, Herbst, Naturgesch. der Krabben u. Krebse, vol. iii. (1) p. 54, pl. xlvii.
fig. 6, 1799.
Trapezia rufopunctata, Latreille, Encycl. Méth. Hist. Nat., vol. x. p. 695, 1825.
    "   "   A. Milne Edwards, Nouv. Archiv. Mus. Hist. Nat., vol. ix. p. 258, 1873,
                et synonyma, except Trapezia flavopunctata, Eydoux and Souleyet.
    "   "   Miers, Crust. in Rep. Zool. Coll. H.M.S. "Alert," p. 536, 1884.

Samboangan, reefs (10 fathoms) an adult male and female.

These specimens have the frontal lobes deeply cut, as in the figure of Herbst, but the reddish spots which cover the body and legs are somewhat larger. In the female the spots of the smaller (right) chelipede are much larger and fewer in number than those of the left chelipede, which is of interest as showing how little reliance is to be placed upon the number and size of these spots for distinguishing the species or varieties.[1]

[1] In this species, the inferior margins of the palms of the chelipedes are sometimes granulated as in Trapezia flavopunctata, Eydoux and Souleyet, but Trapezia flavopunctata is distinguished (if the figure is to be trusted) by the coloration of the carapace, which is reticulated (not spotted), with red lines which circumscribe yellow spots.

The dimensions of this specimen, which is of larger size than the male, are :—

Adult ♀.					Lines.	Millims.
Length of carapace,	.	.	.	.	8½	18
Breadth of carapace,	.	.	.	.	9½	20
Length of left chelipede,	.	.	.	.	17½	36·5
Length of first ambulatory leg,	.	.	.	9½	20

An adult ova-bearing female from Honolulu, 18 fathoms, has the spots of the carapace and legs much less numerous, the front less deeply incised.

Adult ♀.					Lines.	Millims.
Length of carapace, .	.	.	.	.	6	12·5
Breadth of carapace, about	.	.	.	.	7½	15·5
Length of a chelipede, nearly	.	.	.	.	13	27
Length of first ambulatory leg,	.	.	.	.	7½	16

It belongs to the variety which I erroneously referred, in the Report on the Crustacea of H.M.S. "Alert" (tom. cit., p. 536), to *Trapezia guttata*, Rüppell.

Trapezia rufopunctata (Herbst), var. *intermedia*, nov. (Pl. XII. fig. 2).

Thus are designated a male and two female examples in the collection from Honolulu (reefs), which differ from any specimens of *Trapezia rufopunctata* I have seen in the coloration of the palms of the chelipedes, which are transversely marked on the upper surface with irregular, reddish lines, which tend to unite and form reticulations ; the inferior half of the inner and outer surfaces of the palms are uniformly reddish-brown.

The body and limbs, except the palms and fingers of the chelipedes, are marked with red spots; the anterior margin of the front is sinuated; the teeth but very little developed ; the median tooth of the lateral margins of the carapace (the lateral epibranchial tooth) blunt or rudimentary ; the anterior margin of the merus of the chelipedes is armed with from five to eight serratures, whose margins are crenulated, the palms of the chelipedes are compressed, with the superior margins rounded or subacute, but not carinated. The ground colour of the carapace and limbs is yellowish, fingers light chocolate-brown.

Adult ♀.					Lines.	Millims.
Length of carapace, nearly .	.	.	.	.	6½	13·5
Breadth of carapace, .	.	.	.	.	8	17
Length of a chelipede, nearly	.	.	.	.	14½	30·5
Length of first ambulatory leg,	.	.	.	.	9½	20

Family II. PORTUNIDÆ.

Portuniens, Milne Edwards (tribe), Hist. Nat. Crust., vol. i. p. 432, 1834.
,, A. Milne Edwards (family), Archiv. Mus. Hist. Nat., vol. x. p. 311, 1861.
Portunidæ and *Platyonychidæ*, Dana, U.S. Explor. Exped., vol. xiii. Crust. I, pp. 267, 290, 1852.

Carapace (in the recent forms at least), depressed, usually more or less hexagonal, never very convex. Fifth pair of ambulatory legs nearly always natatorial, with the dactylus ovate, expanded, and lamellate.

Species marine or littoral.

The genera of this family have been grouped under the following primary sections :—

I. Portuninæ.   II. Podophthalminæ.

Section I. **Portuninæ**.

Portuniens normaux, A. Milne Edwards, Ann. d. Sci. Nat., ser. 4, vol. xiv. p. 225, 1860; Archiv. Mus. Hist. Nat., vol. x. pp. 311, 312, 1861.

Carapace moderately transverse, and usually widest at the last antero-lateral marginal spine. Front horizontal, and not spatuliform. Orbits and eye-peduncles of moderate length ; the spine or tooth at the exterior angle of the orbit does not project laterally beyond the antero-lateral marginal teeth ; of these more than one (usually from five to nine), are developed.

In the following classification, which is a modification of that proposed by A. Milne Edwards, the recent genera of this section are subdivided into four groups or subfamilies.

Subfamily 1. LUPINÆ.

Lupiens, Carupiens, Lupocyclicus, A. Milne Edwards, Archiv. Mus. Hist. Nat., vol. x. p. 311, 1861.

Carapace transverse, with the antero-lateral margins oblique, and usually armed with more than five (from five to nine) marginal teeth. Basal antennal joint moderately developed. Chelipedes usually elongated. Fifth legs natatorial, with the terminal joint expanded and ovate.

1. Flagellum occupying the interior orbital hiatus.

Genera :—*Neptunus*, de Haan (= *Euctenota*, Gerstæcker ; *Posidon*, Herklots, and ? *Arenæus*, Dana), with the subgenera *Amphitrite*, de Haan (= *Pontus*, de Haan), *Achelous*, de Haan, and *Hellenus*, A. Milne Edwards ; *Callinectes*, Stimpson ; *Xiphonectes*, A. Milne Edwards ; *Scylla*, de Haan ; ? *Lupocyclus*, Adams and White (with subgenus *Parathranites*, nov.) ; ? *Carupa*, Dana.

2. Antennal flagellum excluded from the orbital hiatus by a process of the basal antennal joint.

Genera :—*Cronius*, Stimpson ; *Goniosoma*, A. Milne Edwards (= *Oceanus*, de Haan ; *Charybdis*, de Haan) ; *Thalamonyx*, A. Milne Edwards.

Subfamily 2. THALAMITINÆ.

Thalamitiens, A. Milne Edwards (pt.), Archiv. Mus. Hist. Nat., vol. x. p. 311, 1861.

Carapace widely transverse, antero-lateral margins usually straight, forming a right angle with the front, with few (four or five) teeth. Basal antennal joint greatly developed, widely transverse, flagellum widely excluded from the orbit. Chelipedes moderately elongated. Fifth legs natatorial, as in the Lupinæ.

Genera :—*Thalamita*, Latreille ; *Thalamitoides*, A. Milne Edwards ; ? *Hedrophthalmus*, Nauck.

Subfamily 3. CARCININÆ.

Carciniens, Polytiens, Lissocarciniens, A. Milne Edwards, Archiv. Mus. Hist. Nat., vol. x. p. 311, 1861.

Carapace but slightly (if at all) transverse, with the antero-lateral margins arcuated and armed with five or fewer marginal teeth. Basal antennal joint moderately developed. Chelipedes rather small. Fifth legs natatorial (except in *Carcinus*).

1. Antennæ not wholly excluded from the orbits.

Genera :—*Portunus*, Fabricius (*Liocarcinus*, Stimpson, subgenus ?) ; *Bathynectes*, Stimpson (= *Thranites*, Bovallius) ; *Nectocarcinus*, A. Milne Edwards ; *Carcinus*, Leach ; *Platyonychus*, Latreille (= *Anisopus*, de Haan) ; *Portumnus*, Leach (= *Xaiva*, MacLeay) ; *Polybius*, Leach ; *Lissocarcinus*, Adams and White.

2. Antennæ wholly excluded from the orbits by the union of the front with the inferior orbital margin.

Genus :—*Cænophthalmus*, A. Milne Edwards.

Subfamily 4. CAPHYRINÆ.

Carapace but slightly (if at all) transverse, with the antero-lateral margins arcuated, and armed with only four or five teeth. Basal antennal joint moderately developed.

Fifth legs raised somewhat upon the dorsal surface of the body and subprehensile, with the dactylus styliform and slightly curved.

Genus :—*Caphyra*, Guérin (= *Camptonyx*, Heller). This genus is not represented in the Challenger collection.

Subfamily 1. LUPINÆ.

Neptunus, de Haan.

Neptunus, de Haan, Crust. in v. Siebold, Fauna Japonica, decas i. p. 7, 1833.
   ,,   A. Milne Edwards (pt.), Archiv. Mus. Hist. Nat., vol. x. p. 314, 1861.
Achelous, Amphitrite, Pontus, de Haan, *tom. cit.*, pp. 8, 9, 1833.
? *Psilon*, Herklots, Addit. Faun. Carcin. African occidentalis, p. 3, 1851.
? *Arenæus*, Dana, Amer. Journ. Sci. and Arts, ser. 2, vol. xii. p. 130, 1851 ; U.S. Explor. Exped., vol. xiii., Crust. 1, p. 290, 1852.
Euctenota, Gerstæcker, Archiv f. Naturgesch., vol. xxii. p. 131, 1856.
Hellænus, A. Milne Edwards (subgenus), Crust. in Miss. Sci. au Méxique, pt. 5, pp. 210, 221, 1879.

Carapace usually transverse, depressed, or even nearly flat, with the dorsal surface granulated, rarely tuberculated, and usually marked with several transverse lines, of which one is prolonged inwards for a short distance from the lateral epibranchial spine. The front projects but little, and is divided into from five to eight lobes or teeth, of which the two exterior constitute the interior lobes of the orbits. The antero-lateral margins are arcuated or nearly straight, and are divided into eight nearly equal and well-developed teeth, including the tooth or lobe at the exterior angle of the orbit, and excluding the ninth or lateral epibranchial spine, which is usually much longer than the preceding antero-lateral teeth ; the postero-lateral angles of the carapace are either rounded or angulated or spiniferous. The orbits are widely open above, their superior margins usually marked with fissures, and sometimes dentated, and the infraocular lobe is produced and dentiform or spiniform. The post-abdomen of the male is usually five-jointed (several of the segments coalescent). The eye-peduncles are of moderate size. The basal (or real second) joint of the antennal peduncle is moderately dilated and barely reaches the frontal margin ; the flagellum occupies the interior orbital hiatus. The ischium of the outer maxillipedes is usually not at all produced at its antero-internal angle. The endostome is longitudinally ridged (except in the species constituting Dana's genus *Arenæus*). The chelipedes are well developed, sometimes considerably elongated ; the merus is armed with spines on its anterior and sometimes on its posterior margin. The carpus and palm are usually longitudinally ridged, and the ridges may terminate in spines : the palm is elongated and somewhat prismatic ; the dactyli straight, incurved at the apices, and armed with numerous unequal lobes or tubercles on the inner margins. The ambulatory legs are considerably elongated, with the penultimate and terminal joints compressed and longitudinally sulcated ; in the fifth legs these joints are usually smooth, and as in other Portunidæ, considerably dilated ; the terminal joint is oval and lamelliform.[2]

The species of the genus *Neptunus* thus defined are very numerous, and occur in all the warmer temperate and tropical seas of the globe. *Neptunus sayi* is a pelagic and widely-distributed species, occurring commonly on the floating gulf-weed. Most of the species are, I believe, sub-littoral or shallow-water forms ; several, however, have been recorded from depths varying between 20 and 50 fathoms, and the remarkable *Neptunus* (*Hellenus*) *spinicarpus* occurs usually at much greater depths, and has been taken by the Challenger off the coast of Brazil in (probably) 350 fathoms.

A. *Neptuni* with the spine on the interior margin of the carpus of the chelipedes normally developed :—

  a. Carapace broadly transverse ; antero-lateral margins forming with the frontal margin a regular curve with long radius. Lateral epibranchial spine much longer than the preceding tooth.

Subgenus, *Neptunus.*

Neptunus arquis, A. Milne Edwards (pt.), Archiv. Mus. Hist. Nat., vol. x. p. 316, 1861.[3]

The following are species belonging to this subgenus, which are not referred to by A. Milne Edwards, or have been described since the publication of his monograph in 1861 :—

  Neptunus mexicanus (Gerstæcker, as *Euctenota*) = *Arenæus bidens*, T. J. Smith. Mexico ; Nicaragua.

  Neptunus (?) *pudica* (Gerstæcker). Coast of Brazil. (Perhaps a species of the subgenus *Amphitrite.*)

  Neptunus trituberculatus, Miers. China and Japan.[3]

---

[1] In the species of this genus, the merus of the exterior maxillipedes is of very variable form ; it may be (as in the typical species of *Neptunus*, *Neptunus pelagicus*) obliquely truncated at the distal extremity, with the antero-external angle rounded and not at all produced, the antero-internal angle slightly produced and rounded, or, as in the subgenus *Achelous*, de Haan (type *Achelous spinimanus*), more elongated, truncated and produced at neither angle, or, as in the species of the subgenus *Amphitrite*, de Haan (type *Amphitrite gladiator*), the antero-external angle of the merus may be considerably produced and acute. Sometimes, as in *Pontus*, de Haan (type *Pontus convexus = Neptunus sieboldi*, A. Milne Edwards), the antero-external angle is a right angle and the merus-joint is quadrate, sometimes, as in *Euctenota*, Gerstæcker (type *Euctenota mexicana*), it is produced and somewhat rounded at the distal extremity.

[2] From this section is excluded the genus *Callinectes*, distinguished by the ⊥-shaped post-abdomen of the male which is regarded as distinct by A. Milne Edwards in his later work, Crust. in Miss. Sci. au Méxique.

[3] *Neptunus madagascariensis*, Hoffmann, is very probably a species of *Callinectes*.

Neptunus (Neptunus) pelagicus (Linné).

Cancer pelagicus, Linné, Mus. Lud. Ulrici, p. 434, 1764 ; Syst. Nat., ed. 12, p. 1042, 1766.
Lupa pelagica, Audouin, in Savigny, Descr. de l'Egypte, Crust. Atlas, pl. iii. fig. 1.
Neptunus pelagicus (partim), A. Milne Edwards, Archiv. Mus. Hist. Nat., vol. x. p. 320, 1861.
„   „   Miers, Ann. and Mag. Nat. Hist., ser. 4, vol. xvii. p. 221, 1876.

Philippine Islands near Masbate (Station 203), lat. 11° 6′ 0″ N., long. 123° 9′ 0″ E., in 20 fathoms (an adult male.)

Adult ♂.	Lines.	Millims.
Length of carapace, . . .	28	59·5
Breadth to base of lateral epibranchial spine,	49½	105

Neptunus (Neptunus) sayi, A. Milne Edwards.

Lupa pelagica, Say, Journ. Acad. Nat. Sci. Philad., vol i. p. 97, 1817, see Linné.
Neptunus sayi, A. Milne Edwards, Archiv. Mus. Hist. Nat., vol. x. p. 317, pl. xxix. fig. 2, 1861, and references to literature ; Crust. in Miss. Sci. au Méxique, pt. 5, p. 210, 1878.

A large series of specimens of this common pelagic species was taken from the gulf-weed in the Western North Atlantic, in April 1873 ; an adult female in the North Atlantic, May 1876 ; also an adult male south of Nova Scotia, in lat. 43° 3′ 0″ N., long. 63° 39′ 0″ W., at Station 49, where the depth was 85 fathoms.

The convex, marbled carapace, short lateral epibranchial spine, and the absence of a spine on the posterior margin of the merus of the chelipedes are characteristic of this species.

In the smallest examples in the Challenger series (length of carapace 4½ lines, 9·5 mm., breadth to base of lateral epibranchial spines about 7 lines, 14·5 mm.), the full number of lateral marginal spines is developed, the posterior or epibranchial spine being relatively no larger than in the adult ; the frontal teeth are somewhat less prominent and distinct.

The smallest ova-bearing female has the following dimensions :—

♀.	Lines.	Millims.
Length of carapace, about	7	15
Breadth to base of epibranchial spine, nearly .	11	23

The largest male has the following dimensions :—

Adult ♂.	Lines.	Millims.
Length of carapace, about . . .	15½	33
Breadth to base of lateral epibranchial spine, about .	25½	54·5

Neptunus (Neptunus) sanguinolentus.

> *Cancer sanguinolentus,* Herbst, Naturgesch. der Krabben u. Krebse, vol. i. p. 161, pl. viii. figs. 56, 57, 1782.
>
> *Lupa sanguinolenta,* Milne Edwards, Hist. Nat. Crust., vol. i. p. 451, 1834 ; Atlas in Règne Animal de Cuvier, ed. 3, Crust., pl. x. figs. 1, 8.
>
> *Portunus (Neptunus) sanguinolentus,* de Haan, Crust. in v. Siebold, Fauna Japonica, p. 8, 1833.
>
> *Neptunus sanguinolentus,* A. Milne Edwards, Archiv. Mus. Hist. Nat., vol. x. p. 319, 1861 ; and references to literature.

Specimens were obtained at the following localities :—On the South Australian Coast, 2 to 10 fathoms, April 1874 (a young male) ; Ternate, October 15, 1874 (a female) ; Sandwich Islands, off Honolulu, 16 to 20 fathoms (a fine adult male), also several specimens from the market at Honolulu ; Hawaii (a male).

	Lines.	Millims.
Adult ♂.		
Length of carapace, .	30	63·5
Breadth to base of lateral epibranchial spine, nearly .	53	112

The areolated colour-markings of the carapace may vary in size and distinctness, but the three large, equidistant, irregularly oval and usually ocellated spots which ornament the cardiac and branchial regions are very constant and characteristic of this species.

> b. Carapace narrower ; antero-lateral margins forming with the frontal margin an angle or a curve with short radius. Lateral epibranchial spine well developed, as in the typical *Neptuni.*

Subgenus *Amphitrite.*[1]

> *Neptunus angulatus,* A. Milne Edwards, Archiv. Mus. Hist. Nat., vol. x. p. 325, 1861.

The following are species which have been described since the publication of the monograph of A. Milne Edwards in 1861, or are not included in that memoir :—

> *Neptunus exasperatus* (Gerstæcker). Puerto Cabello, Venezuela. (Perhaps a species of the subgenus *Achelous.*)
>
> *Neptunus transversus* (Stimpson). Mexico ; Manzanillo.
>
> *Neptunus tumidulus* (Stimpson). West of Tortugas and Conch Reef, 37 to 40 fathoms.
>
> *Neptunus acuminatus* (Stimpson). Panama.
>
> *Neptunus panamensis* (Stimpson) = *Achelous paucispinis,* Lockington. Panama ; California.
>
> *Neptunus vocans,* A. Milne Edwards. Cape Verde Islands.
>
> *Neptunus ventralis,* A. Milne Edwards. Guadeloupe.

[1] This name had been previously used by Müller, for a genus of Polychætous Annelida.

Neptunus sulcatus, A. Milne Edwards. Guadeloupe, 17 fathoms.

Neptunus inæqualis, Miers. Senegambia, Goree Island.

Neptunus amnicola, Rochebrune. Senegambia, rivers and marshes. (Perhaps a species of the subgenus *Neptunus*.)

Neptunus edwardsii, Rochebrune. Senegambia, rivers.

Neptunus pallidus, Rochebrune. Senegambia, lakes and marshes.

Neptunus tomentosus, Haswell. Port Jackson.

Neptunus spinipes, n. sp. Philippines, 18 fathoms.[1]

Neptunus (Amphitrite) hastatus (Linné).

> *Cancer hastatus*, Linné, Syst. Nat., ed. 12, p. 1046, 1766.
> *Lupa dufourii* (Desmarest), Roux, Crust. de la Méditerranée, pl. xliv. figs. 1, 6, 1830.
> *Neptunus hastatus*, A. Milne Edwards, Archiv. Mus. Hist. Nat., vol. x. p. 327, 1861, et synonyma.

Four young males obtained off Teneriffe or Gomera, in 75 or 78 fathoms, are referred to this species.[2]

These specimens are very prettily mottled with purple on a yellowish ground ; the frontal teeth are less prominent and acute than in adult examples of *Neptunus hastatus*.

In the smallest example, length only about $2\frac{1}{2}$ lines (5 mm.), the full number of antero-lateral marginal teeth are developed.

The largest specimen measures as follows :—

δ .		Lines.	Millims.
Length of carapace, about	. . . .	$4\frac{1}{2}$	9·5
Breadth of carapace to base of lateral epibranchial spine,		$7\frac{1}{2}$	15·5

Neptunus (Amphitrite) hastatoides (Fabricius).

> *Portunus hastatoides*, Fabricius, Entom. Syst. Suppl., p. 368, 1798.
> *Portunus (Amphitrite) hastatoides*, de Haan, Crust. in v. Siebold, Fauna Japonica, pp. 9, 39, pl. i. fig. 3 ♂, 1833.
> *Neptunus hastatoides*, A. Milne Edwards, Archiv. Mus. Hist. Nat., vol. x. p. 332, 1861.

South of New Guinea, lat. 9° 59′ 0″ S., long. 139° 42′ 0″ E., in 28 fathoms (Station 188), four males and one female ; Hong Kong, 10 fathoms (a good series) ; Kobé, Japan, 8 to 50 fathoms (an adult male) ; Japanese Seas, lat. 34° 18′ 6″ N., long. 133° 35′ 0″ E. (Station 233B), 15 fathoms (three adult males).

[1] The *Portunus pontica* of Fabricius (Entom. Syst. Suppl., p. 369, 1798) as described and figured by Herbst, is apparently a species of this subgenus. It is referred to by H. Milne Edwards (Hist. Nat. Crust., vol. i. p. 457, 1837).

[2] There is some discrepancy between the label on the outside of the bottle and the written parchment note of the locality placed within it with the specimens. In the same bottle was an example of the common *Nautilograpsus minutus*, which generally occurs on the floating gulf-weed, perhaps, therefore, the depth is inaccurately noted.

This species is easily recognised by its flattened pubescent carapace, whose postero-lateral angles are spinuliform; the merus or arm of the chelipedes bears two spines on its posterior margin. There is commonly, but not invariably, a dark-coloured spot at the distal extremity of the terminal joint of the fifth leg.

Adult ♂.		Lines.	Millims.
Length of carapace,		7½	16
Breadth of carapace to base of lateral epibranchial spine,		11½	24·5

Adult males and females in the series have, attached to the ventral surface of the abdomen, a Rhizocephalous parasite allied to or identical with *Sacculina*.

Neptunus (Amphitrite) tuberculosus, A. Milne Edwards.

Neptunus tuberculosus, A. Milne Edwards, Archiv. Mus. Hist. Nat., vol. x. p. 333, pl. xxxi. fig. 5, 1861.

Arrou Island, September 18, 1874 (a small male).

In this specimen the teeth of the antero-lateral margins are somewhat unequally developed, usually alternately larger and smaller; the spines of the chelipedes are very small, that near the wrist being reduced to a tubercle, as in the figure of Milne Edwards.

♂.		Lines.	Millims.
Length of carapace,		4½	10
Breadth of carapace to base of lateral epibranchial spine, nearly		6½	13

Neptunus (Amphitrite) rugosus, A. Milne Edwards.

Neptunus rugosus, A. Milne Edwards, Archiv. Mus. Hist. Nat., vol. x. p. 335, pl. xxxiii. fig. 3, 1861.

A small male, obtained in Torres Strait (August 1874), and an adult male dredged at the Philippines, lat. 11° 37′ 0″ N., long. 123° 31′ 0″ E., in 18 fathoms (Station 208), are referred to this species; also a female dredged in the Celebes Sea, in 10 to 20 fathoms, lat. 6° 54′ 0″ N., long. 122° 18′ 0″ E. (Station 212).

The two latter are distinguished from the types of Milne Edwards' description and figure, from Shark Bay, West Australia (preserved in the collection of the British Museum), by the much less strongly and distinctly tuberculated carapace, the much shorter lateral epibranchial spine, and the longer chelipedes, distinctions which I suppose to be due to the greater age of the Challenger specimens.

This species is distinguished (as pointed out by Professor A. Milne Edwards) from all of its congeners, except *Neptunus tenuipes*, de Haan, by having only five frontal teeth, and from *Neptunus tenuipes* by the dentiform, not rounded, postero-lateral angles of the carapace. Except in this latter important character, the adult Challenger specimens more nearly resemble *Neptunus tenuipes* than *Neptunus rugosus*, and a sufficient series of specimens would perhaps establish the identity of the two species.

The dimensions of the specimen from the Philippines are as follows :—

Adult ♂.	Lines.	Millims.
Length of carapace,	8	17
Breadth to base of lateral epibranchial spine,	11	23·5

Neptunus (Amphitrite) gladiator (Fabricius).

> *Portunus gladiator*, Fabricius, Entom. Syst. Suppl., p. 368, 1798.
> ,, (*Amphitrite*) *gladiator*, de Haan, Crust. in v. Siebold, Fauna Japonica, pp. 9, 39, pl. i. fig. 3♂, 1833.
> *Neptunus gladiator*, A. Milne Edwards, Archiv. Mus. Hist. Nat., vol. x. p. 330, 1861, et synonyma.

South of New Guinea, lat. 9° 59′ 0″ S., long. 139° 42′ 0″ E., in 28 fathoms (Station 188). Two adult males.

Adult ♂.	Lines.	Millims.
Length of carapace,	18	38·5
Breadth to base of lateral epibranchial spine, nearly	27½	58

In this species the postero-lateral angles of the carapace are rounded, not dentiform, the median are shorter than the submedian frontal teeth, and there are two spines on the posterior margin of the merus of the chelipedes.

Neptunus (Amphitrite) gladiator, var. *argentatus* (White).

> *Amphitrite argentata*, White, List Crust. Brit. Mus., p. 126, 1847, descript. nulla.
> *Neptunus argentatus*, A. Milne Edwards, tom. cit., p. 334, pl. xxxi. fig. 4, 1861.

Of this variety, which is scarcely distinguishable from the typical *Neptunus gladiator*, except by the more rudimentary frontal teeth and somewhat narrower carapace, and which I think to be founded on younger examples of Fabricius' species, a female was dredged in the Celebes Sea in 10 to 20 fathoms, lat. 6° 54′ 0″ N., long. 122° 18′ 0″ E. (Station 212).

♀.	Lines.	Millims.
Length of carapace,	8	17
Breadth of carapace,	11½	23·5

I should add, that in all the specimens I have seen of this variety, the terminal joint of the fifth legs is marked with a very distinct purple or reddish spot, which is very obscurely indicated in, or absent from the specimens of the typical *Neptunus gladiator* I have examined, which have, however, a similar spot on the preceding (penultimate) joint of the same legs, but as these colour-markings apparently vary in other species (as *e.g., Neptunus (Amphitrite) hastatoides*) they can scarcely be regarded as evidencing the distinctness of *Neptunus gladiator* and *argentatus*.

Neptunus (Amphitrite) spinipes, n. sp. (Pl. XV. fig. 1).

The carapace is rather narrow and convex for a species of this genus, its surface is everywhere rather obscurely granulated, and is covered with inequalities or rounded prominences, which are most distinctly developed on the cardiac and branchial regions. The front is armed with six teeth, the median and submedian projecting much more than the lateral teeth, and separated from one another by a concave interspace; the median and lateral (or inner orbital) teeth are small and dentiform, the submedian teeth larger and triangulate. The antero-lateral marginal teeth are small (the two following the exterior orbital tooth the smallest), and are dentiform rather than spiniform; the lateral epibranchial spine (in the largest example), scarcely exceeds one-fifth the greatest width of the carapace; the postero-lateral margins are concave, and form nearly a right angle at their junction with the straight posterior margin, but the postero-lateral angles are not spinuliferous, as in *Neptunus hastatoides*; the orbits are deep, in a dorsal view almost semicircularly concave; the sternal surface of the body is polished and very distinctly granulated; the post-abdomen (in the male) is very narrow in its distal half, and composed of only five distinct joints, the second and third, and the fourth and fifth segments having coalesced. The eyes are large and short; the basal antennal joint barely reaches the subfrontal process, and does not reach to the apex of the inner subocular lobe of the orbit; the outer maxillipedes are rather thinly clothed with longish hairs, which also extend over the subhepatic and pterygostomian regions of the body; the merus of the outer maxillipedes is narrowed towards its distal extremity, which is rounded. The chelipedes are subequal, very obscurely granulated, and covered with a close whitish pubescence; the merus is armed with three spines on its anterior margin, and two on its posterior margin (whereof one is at the distal extremity), the carpus with a spine on its inner margin, and one on its outer surface, the palm with two spines on its supero-internal margin (whereof the distal one is small and placed just above the base of the dactyl), and one at the base near the articulation with the wrist; both dactyl and pollex are straight, closely denticulated on their inner margins, and incurved at the tips. The second to the fourth legs are slender, compressed, and have the inferior margins of the penultimate joints and dactyli fringed with close-set hairs; the fifth (swimming) legs are moderately robust, the merus-joints are armed with a spine at the distal end of its posterior margin, and the terminal joints rather narrow-ovate and ciliated on the margins.

Colour (in spirit) brownish-pink, the close short pubescence with which the carapace and limbs are covered, whitish.

Adult ♂.						Lines.	Millims.
Length of carapace, nearly	.	.	.	.	.	7	14·5
Greatest breadth of carapace, nearly	.	.	.	.	9	18·5
Length of a chelipede,	.	.	.	.	.	15	32
Length of first ambulatory leg,	.	.	.	.	14	30

Philippines, 18 fathoms, lat. 11° 37' 0" N., long. 123° 31' 0" E. (Station 208). Two males and a young female were collected.

This species is distinguished from nearly all its Oriental congeners known to me by the presence of a spine at the distal end of the inferior margin of the merus-joint of the fifth ambulatory legs. It is allied in this particular to *Neptunus sebæ* (Milne Edwards),[1] and several other American species, from which it is distinguished by the existence of two spines on the posterior margin of the merus of the chelipedes, the form of the frontal teeth, &c.

It is also nearly allied to the Oriental *Achelous whitei*, A. Milne Edwards,[2] from which it is distinguished not merely by the longer epibranchial spine, but also by the different arrangement of the spines on the palm of the chelipedes, &c.

  c. Carapace as in the subgenus *Amphitrite* or even narrower; lateral epibranchial spine no longer or very little longer than the preceding tooth of the antero-lateral margin.

Subgenus *Achelous*, de Haan.

Achelous, de Haan, A. Milne Edwards, Archiv. Mus. Hist. Nat., vol. x. p. 340, 1861.

To the species enumerated by A. Milne Edwards, the following is to be added :—

  Neptunus unispinosus, Miers. North Australia (Prince of Wales Channel, Torres Strait).

Neptunus (Achelous) whitei (A. Milne Edwards).

  Achelous whitei, A. Milne Edwards, Archiv. Mus. Hist. Nat., vol. x. p. 343, pl. xxxi. fig. 6, 1861.

South of New Guinea, 28 fathoms, lat. 9° 59' 0" S., long. 139° 42' 0" E. (Station 188). A large series was collected.

[1] *Vide* A. Milne Edwards, Archiv. Mus. Hist. Nat., vol. x. p. 329, pl. xxviii. fig. 2, 1861.
[2] Tom. cit., p. 343, pl. xxxi. fig. 6, 1861.

An adult male measures as follows :—

Adult ♂.	Lines.	Millims.
Length of carapace,	11	23·5
Breadth to base of lateral epibranchial spine,	15½	32·5

This species, together with *Neptunus (Achelous) spinimanus* and *Neptunus uni-spinosus*, might be placed with almost equal justice in the subgenus *Amphitrite*, since in these forms the lateral epibranchial spine is distinctly longer than the preceding.[1]

Neptunus (Achelous) granulatus (Milne Edwards).

Lupa granulata, Milne Edwards, Hist. Nat. Crust., vol. i. p. 454, 1834.
Portunus (Amphitrite) gladiator, de Haan (*not* Fabricius), Crust. in v. Siebold, Fauna Japonica, p. 65, pl. xviii. fig. 1, 1837; *not* p. 39, pl. i. fig. 5.
Achelous granulatus, A. Milne Edwards, Archiv. Mus. Hist. Nat., vol. x. p. 344, 1861, et synonyma.

Samboangan, Philippines, 10 fathoms (a male); Admiralty Islands, 16 to 25 fathoms (a male).

♂.	Lines.	Millims.
Length of carapace,	7½	15·5
Breadth to base of lateral epibranchial spine,	9½	20·5

Neptunus (Achelous) unispinosus (Miers).

Achelous granulatus, var. *unispinosus*, Miers, Crust. in Rep. Zool. Coll. H.M.S. "Alert," p. 230, pl. xxiii. fig. B, 1884.

The carapace is rather narrow and moderately convex, with obscure rounded pro-minences which are granulated, and is everywhere covered with a short close pubescence; the postero-lateral angles are rounded.   The front is six-toothed; the median and sub-median somewhat more prominent than the lateral teeth, the median teeth smaller, but not less prominent than the submedian teeth; all are distally rounded, not acute at the apices.   The antero-lateral marginal teeth are small and acute; the lateral epibranchial spine small, but very distinctly longer than the preceding antero-lateral teeth.   The post-abdomen of the male is composed of five to seven distinct segments, and is narrowed from the base of the penultimate segment; the eyes are large, and are borne on very short peduncles.   The basal antennal joint is short and rather broad, but reaches the lateral subfrontal process; the ischium-joint of the outer maxillipedes is longitudinally sulcated; the merus subtruncated at its distal extremity.   The chelipedes (in the male) are subequal and rather short; the merus-joint (arm) has three or four

[1] In a small specimen of *Achelous whitei* in the collection of the British (Natural History) Museum, the lateral spine is as long as in many species of the subgenus *Amphitrite*. I prefer, however, to retain these species in the position assigned to them by A. Milne Edwards for this reason, if for no other, that were *Amphitrite spinimanus* removed from *Achelous* it would be necessary to rename this subgenus, of which it is the original type.

spinules on its anterior margin, and one at the distal extremity of its posterior margin ; the wrist has a small spinule on its inner margin and another on its outer surface ; the palm is longitudinally costated on its upper and outer surface, and is armed with two small spinules, one, basal, near the articulation with the wrist, and one close to the distal end of the upper margin, near the ˙articulation with the dactylus ; the dactyl and pollex are barely as long as the palm and are denticulated on their inner margins. The ambulatory legs are very slender and compressed ; the dactyli longer than the penultimate joint ; the posterior or swimming legs are of moderate length, their merus-joints smooth, without spinules, the terminal joints rather narrow, ovate. Colour (in spirit) light yellowish-brown, the pubescence whitish. The dimensions of a Challenger specimen are :—

♂.	Lines.	Millims.
Length of carapace, nearly .	6	12
Breadth to base of lateral epibranchial spine, nearly .	8	16·5
Length of a chelipede,	11	23
Length of first ambulatory leg, about	9½	20·5

Two small males were collected in the Torres Strait, one of these at Station 186, in 8 fathoms, lat. 10° 30′ 0″ S., long. 142° 18′ 0″ E.

This species is not easily confounded with any of its Oriental congeners. From *Neptunus (Achelous) granulatus*, as I have shown, it is distinguished by having but a single spine on the posterior margin of the arm, by the less spiniform antero-lateral marginal teeth, &c. Its nearest ally in the subgenus *Achelous* is apparently the American *Neptunus (Achelous) depressifrons* ; in which species, however, the lateral epibranchial spine is shorter, and the spine on the upper margin of the palm of the chelipedes is placed further back from the distal extremity, &c. From such species of the subgenus *Amphitrite* (also American), as are allied to it in the rounded postero-lateral angles of the carapace, unarmed merus-joint of the swimming legs, &c. (*e.g.* *Neptunus (Amphitrite) gibbesii*), it is distinguished by the form of the frontal teeth, and the shorter lateral epibranchial spine.

Neptunus (Achelous) depressifrons (Stimpson).

Amphitrite depressifrons, Stimpson, Ann. Lyc. Nat. Hist. New York, vol. vii. p. 58, 1859.
Achelous depressifrons, Stimpson, tom. cit., p. 223, 1860.
    ,,   ,,   A. Milne Edwards, Archiv. Mus. Hist. Nat., vol. x. p. 342, 1861 ; Crust. in Miss. Sci. au Méxique, pt. 5, p. 230, pl. xl. fig. 4, 1879.

Bermuda (shore). An adult male.

Adult ♂.	Lines.	Millims.
Length of carapace, nearly .	7	14·5
Breadth of carapace to base of lateral epibranchial spine, nearly .	9½	19·5

In this specimen the palms and merus-joints of the chelipedes are more elongated than in the figure of A. Milne Edwards, and the spine at the distal extremity of the posterior margin of the merus-joint is obsolete. The lateral epibranchial spine does not at all exceed the preceding spine in length, and in this character, as well as in those of the chelipedes, this specimen differs from authentically-named examples of *Neptunus depressifrons* in the collection of the British (Natural History) Museum. It may be that these characters will prove, upon examination of a sufficient series, sufficient to separate the Bermuda swimming-crab as a distinct species from *Neptunus depressifrons*.

B. *Neptuni* with the spine on the inner margin of the carpus abnormally developed (in the adult male exceeding the palm in length):—

Subgenus *Hellenus*, A. Milne Edwards.

Hellenus, A. Milne Edwards (pt.), Crust. in Miss. Sci. au Méxique, pt. 5, pp. 210, 221, 1879.

In the single species which follows, to which I propose to restrict this subgenus, there is scarcely any trace of a spine or tooth at the postero-lateral angles of the carapace, and I am inclined to think that the form of the postero-lateral angles cannot be regarded as affording a character of importance sufficient to distinguish the subgenus.

Neptunus (Hellenus) spinicarpus (Stimpson).

Achelous spinicarpus, Stimpson, Bull. Mus. Comp. Zoöl., vol. ii. p. 148, 1870.
Neptunus (Hellenus) spinicarpus, A. Milne Edwards, Crust. in Miss. Sci. au Méxique, pt. 5, p. 221, pl. xl. fig. 1, 1879.

Brazil, off Barra Grande, lat 9° 5′ 0″ S. to 9° 10′ 0″ S., long 34° 49′ 0″ W. to 34° 53′ 0″ W., in from 32 to 400 fathoms (Stations 122 to 122c). An adult male and two adult and one young females are in the collection.

In none of the females does the spine on the inner margin of the wrist, although greatly developed, reach the distal extremity of the upper margin of the palm; in the adult male it is far larger, reaching nearly to the middle of the dactylus, or mobile finger, and is considerably dilated and strongly compressed, and bordered on its outer margin with a fringe of long hairs; the carapace also is provided with stronger rounded prominences on the cardiac and branchial regions in the male than in the females.

The dimensions of this specimen are as follows :—

Adult ♂.						Lines.	Millims.
Length of carapace,	.	.	.	.	.	8½	18
Breadth to base of lateral epibranchial spine,		.	.	.		12	25
Length of a chelipede,	.	.	.	.	.	22½	45·5
Length of spine on the inner margin of the wrist,		.	.	.		9	19
Length of first ambulatory leg,	.	.	.	.	.	17	36

Xiphonectes, A. Milne Edwards.

Xiphonectes, A. Milne Edwards, Nouv. Archiv. Mus. Hist. Nat., vol. ix. p. 157, 1873.

This genus, established by A. Milne Edwards for certain small species which have much the appearance of *Neptuni* which have not attained their full development, is distinguished from *Neptunus* (subgenus *Amphitrite*) by the following characters:—

The carapace is narrower than is usual in *Amphitrite*, the lateral epibranchial spine even more developed, and the lateral (or submedian) frontal lobes more prominent. The antero-lateral marginal teeth are spiniform, more remote from one another, and very rarely attain the number eight (exclusive of the lateral epibranchial spine) characteristic of *Amphitrite*. The orbits and eyes are very large.

An examination of the series in the British (Natural History) Museum and Challenger collections, shows, I think, that the various described species of this genus must be regarded as varieties of one widely distributed type, which occurs throughout the Indo-Pacific region in water of only moderate depth (to 18 fathoms).

Xiphonectes longispinosus (Dana).

Amphitrite longispinosa, Dana, Proc. Acad. Nat. Sci. Philad., p. 84, 1852; U.S. Explor. Exped., vol. xiii. Crust. 1, p. 277, pl. xvii. fig. 2, 1852.

Amphitrite vigilans, Dana, tom. cit., p. 278, pl. xvii. fig. 3, var.

Neptunus vigilans and longispinosus, A. Milne Edwards, Archiv. Mus. Hist. Nat., vol. x. pp. 336, 337, 1861.

? Xiphonectes leptocheles, A. Milne Edwards, Nouv. Archiv. Mus. Hist. Nat., vol. ix. p. 159, pl. iv. fig. 1, 1873, var.

Neptunus vigilans, Richters, Decapoda in Möbius Beitr. zur. Meeresfauna der Mauritius und der Seychellen, p. 152, 1880.

Xiphonectes vigilans, var. obtusidentatus, Miers, Crust. in Rep. Zool. Coll. H.M.S. "Alert," p. 558, pl. xlviii. fig. A, 1884.

Tongatabu, 18 fathoms (Station 172), lat. 20° 58′ 0″ S., long. 175° 9′ 0″ E. Two small males.

These specimens offer a curious intermixture of the characters cited as characteristic of each of the supposed species, *Xiphonectes longispinosus*, *Xiphonectes vigilans*, and *Xiphonectes leptocheles*. In the larger specimen, the chelipedes are comparatively robust as in the typical *Xiphonectes longispinosus*, in the smaller they are slender and elongated as in *Xiphonectes leptocheles*. In the larger specimen the arm or merus has four spines on its anterior (or inner) margin, as in the typical *Xiphonectes vigilans*, but the six spines of the antero-lateral margins are equidistant, as in *Xiphonectes longispinosus*. In the smaller specimen there are five spines on the anterior margin of the arm. In both specimens the post-abdomen is composed of only five distinct segments; in the larger specimen, with robust chelipedes, the sixth segment is more nearly of the form figured by

Milne Edwards as characteristic of his *Xiphonectes leptocheles* than in the smaller specimen with slender chelipedes.[1]

The dimensions of the specimens are as follows:—

♂.			Lines.	Millims.
Length of carapace, about	.		4	9
Breadth of carapace, nearly	.	.	4½	9·5
Length of lateral spine, nearly	.	.	3	6
Length of a chelipede, about	.	.	9	19
Young ♂.				
Length of carapace, nearly			3	6
Breadth of carapace, nearly	.		3½	7·5
Length of lateral spine, about	.		2½	5
Length of a chelipede, about	.		8	17·5

Scylla, de Haan.

Scylla, de Haan, Crust. in v. Siebold, Fauna Japonica, decas i. p. 11, 1835.
  ,,   A. Milne Edwards, Archiv. Mus. Hist. Nat., vol. x. p. 347, 1861.

This genus is very nearly allied to *Neptunus*, but is distinguished by its smoother, more convex carapace, which is not marked by the transverse raised lines which are characteristic of that genus, except by one on each side which are prolonged inwards for a short distance from the base of the lateral epibranchial tooth. The front is armed with six lobes or teeth, which are much more prominent in the young than in the adult animal. The antero-lateral margins are oblique, as in the typical *Neptuni*, and are armed with nine lobes or teeth, but the last (lateral epibranchial) tooth is not at all longer than those which precede it. The epistoma is distinctly developed. The post-abdomen in the male is subtriangulate, with the lateral margins nearly straight and convergent from the anterior margin of the third segment, which, with the fourth and fifth, is consolidated into a single joint. The eye-peduncles are of moderate thickness. The basal antennal joint is somewhat dilated, and barely reaches the front, it is prolonged at its antero-external angle into a lobe, which does not, as in *Gonisoma*, exclude the flagellum from the orbital hiatus. The merus of the exterior maxillipedes is obliquely truncated at the antero-internal angle, slightly rounded at the distal extremity, with the antero-external angle but little prominent and subacute. The chelipedes in the adult male are large and massive, and sometimes unequally developed; merus, carpus, and palm, are armed with spines, but the exterior surface of the carpus and palm are not longitudinally cristated as in *Neptunus*, and the palm is shorter and more turgid, the fingers strongly toothed on the inner margins, and the dactylus distinctly arcuated. The ambulatory

[1] In the Report on the Crustacea of H.M.S. "Alert," before seeing the Challenger specimens, I regarded the three described species of *Xiphonectes* as distinct.

legs present nothing remarkable ; the merus-joint of the fifth (natatory) legs is not armed with a spine on its inferior margin.[1]

The only recent species of this genus is the very common and widely distributed Indo-Pacific species which follows :—

Scylla serrata (Forskål).

Cancer serratus, Forskål, Descript. animalium quæ in itinere orientali observavit P. Forskål, p. 90, 1775.
Portunus (*Scylla*) *serratus*, de Haan, Crust. in v. Siebold, Fauna Japonica, p. 44, 1835.
Scylla serrata, A. Milne Edwards, Ann. d. Sci. Nat., ser. 4, Zool., xiv. p. 252, pls. i., ii., 1860; Archiv. Mus. Hist. Nat., vol. x. p. 349, 1861, et synonyma.

Tahiti, Papiete.   An adult female and a smaller male.

♂.					Lines.	Millims.
Length of carapace, about	.	.	.	.	27	57·5
Breadth of carapace, nearly	.	.	.	.	37½	79

Lupocyclus, Adams and White.

Lupocyclus, Adams and White, Crust. in Zool. H.M.S. "Samarang," p. 47, pl. xii. fig. 4, 1848.
„   A. Milne Edwards, Archiv. Mus. Hist. Nat., vol. x. p. 387, 1861.

Carapace transverse, but little broader than long ; the antero-lateral margins armed with five regular and even teeth, between which may alternate several smaller teeth ; the last antero-lateral tooth may be (subgenus *Parathranites*) elongated and laterally porrected ; the front is narrow, moderately prominent, and four to six-lobed. The orbital fissures are very distinct. The longitudinal ridges of the endostome or palate are distinctly defined ; the post-abdomen may have, as in so many Portunidæ, two or three of the intermediate segments coalescent. The eyes are set on very short pedicels. The basal antennal joint is free, neither united with the front nor with the interior wall of the orbit, it is narrow or moderately dilated, not produced at the antero-external angle, and lies within the interior orbital hiatus. The ischium of the exterior maxillipedes is not produced at its antero-internal angle ; the merus is somewhat elongated, and rounded or subtruncated at the distal extremity, but little, if at all, emarginated at the antero-internal angle. The chelipedes are subequal and moderately slender, the joints are armed with spines, and the palms are externally costated. The ambulatory legs of the second, third, and fourth pairs are slender, and the dactyli are styliform ; in the fifth legs the penultimate and terminal joints are dilated, and the terminal joints ovate, as in other Portunidæ.

[1] In this genus the longitudinal ridges of the epistoma are sometimes partially or obscurely developed. The epistoma is transversely sulcated, but this sulcus, upon which much stress is laid by A. Milne Edwards as a generic character, is not always very distinct.

From the European and East American genus *Bathynectes*, Stimpson,[1] with which *Thranites*, Bovallius, is I think, synonymous,[2] this genus is distinguished by the more elongated and less squarely truncated apex of the merus of the exterior maxillipedes, and by the more dilated and ovate dactylus of the fifth ambulatory legs.

Besides the Australian and Malayasian *Lupocyclus rotundatus*, Adams and White, the only species referable, as far as I know, to this genus is the *Lupocyclus philippinensis*, Nauck, and the one described below, which differs from the type and approaches *Bathynectes* in the more elongated and lateral epibranchial tooth of the carapace, and the short chelipedes, on which account I refer it to a new subgenus.

Lupocyclus (Parathranites) orientalis, n. sp. (Pl. XVII. fig. 1).

The carapace is hexagonal, broader than long; its upper surface is rather coarsely granulated and tuberculated; the tubercles disposed as follows:—one on the gastric and two on the cardiac region, and on either side of the gastric and cardiac tubercles usually three smaller prominences, in a longitudinal series; the front is rather narrow (as in *Lupocyclus rotundatus*), and has four small spinuliform teeth, the two median of which, in the adult, are usually more prominent than the lateral teeth, which are at the inner angle of the orbit. The antero-lateral margins are arcuated and divided into five teeth, the first four of which are broad and equidistant (shaped much as in *Platyonychus bipustulatus*), the last is spiniform and laterally porrected; the concave postero-lateral margins terminate in a small spinule at the postero-lateral angles of the carapace; the upper margins of the orbits are divided by two narrow fissures (as in *Lupocyclus rotundatus*), the inferior margin is emarginate below the tooth at the exterior orbital angle. The pterygostomian regions are granulated. The narrow transverse epistoma is shaped nearly as in *Lupocyclus rotundatus*. The terminal segment of the post-abdomen is longer than broad, and narrower at its base than the penultimate segment. The eyes are rather small. The basal joints of the antennules are broadly dilated, as in *Lupocyclus rotundatus*, and are distally truncated or slightly concave. The slender and narrow basal joint of the antennæ is not in contact with the basal antennulary joint or with the margin of the wide orbital hiatus, and does not attain the apex of the inner subocular lobe of the orbit; the two following joints are slender and rather short, the flagellum moderately elongated. The outer maxillipedes are granulated on their exterior surface; the ischium is elongated and longitudinally sulcated, the merus narrowed to its distal extremity, which is rounded; the narrow and straight exognath reaches the distal extremity of the merus-joint. The chelipedes are of moderate size; the trigonous merus-joint is armed commonly with a single spine on its anterior margin, and two on

---

[1] *Bull. Mus. Comp. Zoöl.*, vol. ii. p. 145, 1870.

[2] *Öfversigt k. Vetensk.-Akad. Förhandl.*, vol. xxxiii. No. 9, p. 59, pls. xiv., xv., 1876; op. cit., vol. xxxviii. No. 2, p. 9, pl. ii., 1881.

its posterior margin, of which one is near, and the other at, the distal extremity; the carpus has a long spine on its inner margin, and its outer surface is externally granulated and bears a small spine and two or three spinules; the palm is about as long as the merus and is externally granulated and bears three spines, one of which is placed near the base, one at, and the third near, the distal extremity; the fingers about equal the palm and are denticulated nearly as in *Lupocyclus rotundatus*. The ambulatory legs are slender and smooth, with the penultimate joints compressed, the dactyli compressed and not greatly exceeding the penultimate joints in length. The swimming legs have the penultimate and terminal joints dilated and compressed, the terminal joint, as already stated, regularly ovate, and distally rounded. Colour (in spirit) yellowish-brown.

The dimensions of an adult male are as follows:—

Adult ♂.					Lines.	Millims.
Length of carapace, .	.	.	.	.	7	15
Breadth to base of lateral spines,	.	.	.	8½	18
Length of chelipede,	.	.	.	.	11½	24·5
Length of first ambulatory leg,	.	.	.	14	29·5
Length of swimming leg, .	.	.	.	9½	20

Several males and females were collected near the Ki Islands, in 140 fathoms (Station 192), in lat. 5° 49′ 15″ S., long. 132° 14′ 15″ E., and an adult male, north of the Admiralty Islands, in 150 fathoms (Station 219), in lat. 1° 54′ 0″ S., long. 146° 39′ 40″ E.

In the male from the Admiralty Islands the two median frontal lobes are somewhat more prominent than in the specimens dredged at Station 192. The tubercles of the dorsal surface of the carapace are most distinct in the smallest specimens.

This species is at once distinguishable from the *Lupocyclus philippinensis*, Semper (ined.), Nauck, by the five-toothed antero-lateral margins of the carapace and the shorter chelipedes.

Cronius, Stimpson.

Cronius, Stimpson, Ann. Lyc. Nat. Hist. New York, vol. vii. p. 225, 1860.
    „   A. Milne Edwards, Crust. in Miss. Sci. au Méxique, pt. 5, p. 231, 1879.

This genus includes certain species formerly referred to *Amphitrite*, *Achelous*, and *Goniosoma*, in which the antero-lateral margins of the carapace are armed with eight alternately larger and smaller teeth, besides the lateral epibranchial tooth, which scarcely exceeds in length the larger teeth of the antero-lateral margins. As in many species of *Goniosoma*, the frontal margin is six to eight-lobed. The endostome or palate is obscurely longitudinally ridged, and the post-abdomen in the male is five-jointed (in the specimens I have examined). The flagellum of the antennæ is excluded, as in *Goniosoma*, from the interior orbital hiatus by a process of its extero-distal angle, but this process is not in contact with the inferior margin of the front. The chelipedes and ambulatory legs present nothing remarkable; the merus-joints of the fifth legs are armed with a spine placed near the distal extremity of the inferior margin.

Cronius is connected with *Goniosoma* through such species as *Goniosoma erythrodactylum*, which has fewer antero-lateral marginal teeth, but one or two rudimentary as in *Cronius*.

The following species are included in *Cronius* :—

> *Cronius ruber* (Lamarck) = *Amphitrite edwardsii*, Lockington.   Gulf of Mexico ; West Indies ; Vera Cruz ; Brazil ; Panama (East Coast) ; Mazatlan.
>
> *Cronius milleri* (A. Milne Edwards).   Cape Verde Islands (St. Vincent) ; Senegambia, Goree Island, 9 to 15 fathoms.
>
> *Cronius bispinosus*, n. sp.   Brazil, Bahia.

Cronius bispinosus, n. sp. (Pl. XV. fig. 2).

I propose to designate thus a small species which is distinguished from all its congeners by having but two spines on the palm of the chelipedes, and from all, except perhaps *Cronius edwardsii* (Lockington), by having the third frontal tooth in the adult wholly confluent with and indistinguishable from the external frontal (or inner orbital tooth). In nearly all other characters this species closely resembles the well-known *Cronius ruber*,[1] but it is further distinguished by the unarmed merus-joint of the fifth or swimming legs from that species.   The carapace is shaped nearly as in *Cronius ruber;* it is moderately convex, pubescent, and has straight, transverse, smooth lines or low ridges on the gastric and cardiac regions, and a similar but curved ridge passing inwards along the front of the branchial regions from the lateral epibranchial spine.   The frontal lobes are somewhat broader, less prominent and more obtuse than in *Cronius ruber*, the median ones are separated from one another by a much shallower interspace, the next on each side are as broad as long, and subtruncated or but slightly rounded at the distal extremity, the third, as already stated, are wholly confluent with the outer teeth, and they form together a broad truncated lobe.   The antero-lateral margins are armed with eight alternately larger and smaller spines, as in *Cronius ruber*, but the first two teeth are somewhat blunter and less spiniform than in that species ; the ninth (or lateral epibranchial spine) is distinctly longer than any of the preceding.   As in *Cronius ruber*, the basal antennal joint terminates in a short spine or tooth.   The outer maxillipedes closely resemble those of that species.   The chelipedes are subequal ; the merus or arm has but four spines on its anterior margin ; the small spines which usually exist at the distal extremity of both the anterior and posterior margins of the arm in

---

[1] *Vide* A. Milne Edwards, Crust. in Miss. Sci. au Méxique, pt. 5, p. 232, 1879.

Cronius ruber, are wanting in *Cronius bispinosus;* the wrist is armed with two spines as in *Cronius ruber*, the palm also has but two spines, one basal, close to the articulation with the wrist, and one on the upper margin, placed some distance behind the distal extremity.

The ambulatory and swimming legs are nearly as in *Cronius ruber*, but the spine which exists in *Cronius ruber*, at the distal end of the posterior margin of the merus-joint of the swimming legs, is absent in *Cronius bispinosus.* Colour light reddish or yellowish-brown.

	Lines.	Millims.
♀.		
Length of carapace,	8	17
Breadth of carapace to base of lateral epibranchial spine,	10½	22
Length of a chelipede,	14½	31
Length of second ambulatory leg,	12½	26·5

Two adult females were collected at Bahia in shallow water.[1]

Goniosoma, A. Milne Edwards.

Goniosoma, A. Milne Edwards, Ann. d. Sci. Nat., ser. 4, Zool., xiv. p. 263, 1860 ; Archiv. Mus. Hist. Nat., vol. x. p. 367, 1861.

Oceanus, Charybdis, de Haan, Crust. in v. Siebold, Fauna Japonica, decas i. pp. 9, 10, 1833, names previously used.[2]

Carapace depressed, more or less hexagonal, usually marked, as in other Portunidæ, with raised lines, one of which is prolonged inwards from the base of the lateral epibranchial tooth towards the gastric region ; the antero-lateral margins are oblique, slightly arcuated or nearly straight, they form an obtuse angle with the front and orbits, and are usually armed with five or six teeth (inclusive of the exterior orbital but exclusive of the lateral epibranchial tooth), one or two of which may be rudimentary. The front is nearly always divided into eight lobes or teeth (including the tooth at the interior angle of the orbit). The orbits are large, open above and have usually two fissures in the upper margin and another in the lower margin, the interior subocular lobe of the orbit is sometimes, but not always, produced and spiniform. The longitudinal ridges of the endostome are usually distinctly developed. The post-abdomen is composed of five or (rarely) of six joints (two or three of the intermediate segments being consolidated). The eyes are set on short, thick pedicels. The antennæ have the basal joints enlarged

---

[1] It may be of interest to note here that in a very small male *Cronius milleri* from Senegambia (length only 3½ lines, or little over 7 mm.), in the collection of the British Museum, the submedian and external frontal teeth are confluent, as in *Cronius bispinosus*, but in this specimen the four spines of the palm of the chelipedes are all perfectly distinguishable. I have examined a small male example of *Goniosoma bispinosum* from Brazil, in the British Museum, from the collection of my late grandfather, J. Miers, F.R.S.

[2] It should be noted that the name *Charybdis* used in 1809 by Péron and Lesueur, is spelt somewhat differently from de Haan's designation *Charybdis ;* I think it better, however, to retain Milne Edwards' designation for the species of this genus, which has been generally adopted.

(but not so greatly as in *Thalamita*), and their extero-distal angles are considerably produced, the process thus constituted fills the interior hiatus of the orbit and thus excludes the flagellum from the orbit, and is usually in contact with the front. The ischium of the endognath of the exterior maxillipedes is not produced at its antero-internal angle ; the merus has its antero-external angle more or less produced, and rounded or subacute, it is distally truncated, and its antero-internal angle is obliquely truncated or slightly emarginate. The chelipedes (in the adult male) are subequal and often considerably developed, as in other Portunidæ ; the joints are armed with spines ; the palms are often longitudinally costated on their exterior surface, and the fingers are straight and armed with unequal lobes or teeth on their inner margins. The ambulatory legs present nothing remarkable ; the merus-joints of the fifth (or natatory) legs are usually armed with a spine placed near the distal extremity of the inferior margin.

The described forms are rather numerous and occur commonly in the littoral or shallow waters of the Indo-Pacific region, but I am not aware that any have been recorded from water of a greater depth than 50 fathoms.

The following are species of this genus which have been described since the publication of A. Milne Edwards' Monograph in 1861 :—

> *Goniosoma kellerii*, A. Milne Edwards. Indian Archipelago (Nicobars) ; New
>   Caledonia.
> *Goniosoma acutum*, A. Milne Edwards. Japan.
> *Goniosoma longifrons*, A. Milne Edwards Samoan Islands, Upolu.
> *Goniosoma acutifrons*, de Man. Timor.
> *Goniosoma spiniferum*, Miers. North-East Australia, Port Molle ; South
>   Formosa (Coll. Brit. Mus.). This species is perhaps a variety of *Goniosoma*
>   *variegatum.*

The *Goniosoma dubium*, of Hoffman, from Réunion and Timor, is, according to de Man (Notes Leyden Mus., v. p. 151, 1883), identical with *Goniosoma orientale*, Dana.

Goniosoma variegatum (Fabricius).

> *Portunus variegatus*, Fabricius, Entom. Syst. Suppl., p. 364, 1798.
>   „   „   Milne Edwards, Hist. Nat. Crust., vol. i. p. 465, 1834.
> *Charybdis variegatus*, de Haan, Crust. in v. Siebold, Fauna Japonica, pp. 10–42, pl. i.
>   fig. 2, 1835.
> *Goniosoma variegatum*, Miers, Proc. Zool. Soc. Lond., p. 33, 1879 ; Crust. in Rep. Zool. Coll.
>   H.M.S. "Alert," p. 232, 1884, et synonyma.

I have already, in the Report cited above, given a detailed description of the species I suppose to be *Goniosoma variegatum* (Fabricius).

Certain specimens in the Challenger series apparently belong to a new variety of this form which I propose to designate :—

Goniosoma variegatum, var. *bimaculatum*, nov. (Pl. XV. fig. 3).

The specimens thus designated come in nearly all of their characters very near to the form I regard as the typical *Goniosoma variegatum* (Fabricius)[1] (*Goniosoma callianassa*, A. Milne Edwards), which they resemble in the prominent and laterally-projecting lateral epibranchial tooth, in the turgid three-spined palms of the chelipedes, &c., but are distinguished by having the median and submedian frontal teeth nearly wholly coalescent, and forming together a broad lobe with slightly concave anterior margin, and in having a very distinct black spot on each branchial region.

Japan, Kobé, 50 fathoms. Four adult males were collected.

Adult ♂.	Lines.	Millims.
Length of carapace, about	9½	20
Breadth of carapace, nearly	13	27

Goniosoma cruciferum[2] (Fabricius), A. Milne Edwards.

Cancer cruciatus, Herbst, Naturgesch. der Krabben u. Krebse, ii. Heft. v. p. 155, pl. viii. fig. 53, pl. xxxviii. fig. 1, 1794.

Portunus crucifer, Fabricius, Entom. Syst. Suppl., p. 364, 1798.

Goniosoma cruciferum, A. Milne Edwards, Archiv. Mus. Hist. Nat., vol. x. p. 371, 1861, and references to literature.

Hong Kong, 7 fathoms, from the trawl (a fine adult male).

In this specimen the frontal teeth are all prominent and well developed ; the first of the antero-lateral marginal teeth (or exterior orbital tooth) has a distinctly excavated apex.

Adult ♂.	Lines.	Millims.
Length of carapace, .	30	63·5
Breadth of carapace, about	40½	86·5

Goniosoma ornatum, A. Milne Edwards.

Portunus (*Thalassita*) *truncatus*, de Haan, Crust. in v. Siebold, Fauna Japonica, pp. 10, 43, pl. ii. fig. 3 (*junior*), pl. xii. fig. 3, adult ♂, ♀, 1835, *nec Portunus truncatus*, Fabricius.

Goniosoma ornatum, A. Milne Edwards, Archiv. Mus. Hist. Nat., vol. x. p. 376, 1861.

Arafura Sea, lat. 8° 56′ 0″ S., long. 136° 5′ 0″ E., in 49 fathoms (Station 190). An adult ova-bearing and an immature female.

[1] Fabricius' description of his *Portunus variegatus* is very brief, and the identification of the species, therefore, difficult. H. Milne Edwards, however (Hist. Nat. Crust., vol. i. p. 465, 1834), supposes it to be "une espèce très-voisine" of *Goniosoma callianassa* (Herbst), and as Fabricius' name has priority, I adopt it in preference to Herbst's designation. By de Haan (Crust. in v. Siebold, Fauna Japonica, p. 42, pl. i. fig. 2, 1835), *Portunus variegatus*, Fabricius, and *Cancer callianassa*, Herbst, are regarded as distinct species, but A. Milne Edwards (Archiv. Mus. Hist. Nat., vol. x. p. 382, 1861), unites the form figured by de Haan as *variegatus* with *Goniosoma callianassa*, although he uses the last-mentioned name for the species.

[2] I retain Fabricius' designation *cruciferum* for this species, although Herbst's name, *cruciatus*, has distinct priority, because the former is not only the one that has been used by all systematists, but also is clearly indicative of the remarkable coloration, in reference to which the term *cruciatus* is inapplicable.

These specimens differ from what I will regard as the typical condition of the species (represented in de Haan's figures of the adult) in the more slender chelipedes, which are less distinctly granulated and bear longer spines. I may add, that the posterior tooth of the antero-lateral margin is more distinctly spiniform ; the orbital margins are granulated, as is the basal joint of the antennæ. The larger of the two females in the Challenger collection measures as follows :—

· Adult ♀.							Lines.	Millims.
Length of carapace, nearly	.	.	.	.	.	.	13	27·5
Breadth of carapace, .	.	.	.	.	.	.	17½	37

Thalamonyx, A. Milne Edwards.

Thalamonyx, A. Milne Edwards, Nouv. Archiv. Mus. Hist. Nat., vol. ix. p. 168, 1873.

This genus was established by A. Milne Edwards for two small forms, which I think are not specifically distinct, and which are perhaps best placed between *Goniosoma* and *Thalamita*, but also present evident affinities with *Portunus*, *Bathynectes*, and *Lissocarcinus*. The description given above of *Goniosoma* will apply, in nearly all particulars, to *Thalamonyx*, but the carapace is narrower, and the antero-lateral margins less oblique than in many species of that genus. The front is only four-lobed, as in many species of *Thalamita*; the median lobes prominent, much wider than the lateral lobes, and divided by a small median notch or fissure. As in *Goniosoma* the basal antennal joint is produced at its antero-external angle into a process which occupies the interior hiatus of the orbit, but does not widely exclude the flagellum from it, as in *Thalamita*. The chelipedes are subequal and rather small. The merus-joint of the fifth (natatory) legs has a spine on its inferior margin, near the distal extremity; the penultimate and terminal are dilated, as in *Goniosoma* and *Thalamita*.

Thalamonyx danæ, var. *gracilipes*, A. Milne Edwards.

Charybdis (?), Dana, U.S. Explor. Exped., vol. xiii., Crust. 1, p. 287, pl. xvii. fig. 13, 1852.
Goniosoma danæ, A. Milne Edwards, Nouv. Archiv. Mus. Hist. Nat., vol. v. p. 153, pl. vii. figs. 6, 7, 1869.
Thalamonyx danæ, A. Milne Edwards, op. cit., vol. ix. p. 170, 1873.
Thalamita gracilipes, A. Milne Edwards, tom. cit., vol. ix. p. 169, pl. iv. fig. 3, 1873, var.

Tongatabu, 18 fathoms (Station 172) ; a small male and two females were collected.

♀.						Lines.	Millims.
Length of carapace, about	.	. .	.	.	.	3½	7·5
Breadth of carapace, about	.	.	.	.	.	4½	9·5

The male measures as follows :—

♂ .	Lines.	Millims.
Length of carapace, about	2½	5·5
Breadth of carapace, about	3	6·5

Neither of the females bears ova. These specimens should perhaps be designated as a distinct variety from either *Thalamita dauæ* or *Thalamita gracilipes*, A. Milne Edwards, since the carapace is not only very distinctly granulated but also pubescent. The chelipedes are slender, and the palms are granulated as in the latter form.

Subfamily 2. THALAMITINÆ.

Thalamita, Latreille.

Thalamita, Latreille, Crust. in Règne Animal de Cuvier, ed. 2, vol. iv. p. 33, 1829, footnote.
,,   A. Milne Edwards, Archiv. Mus. Hist. Nat., vol. x, p. 354, 1861.
Thalamites quadrilobata, Milne Edwards, Hist. Nat. Crust., vol. i. p. 459, 1834.

Carapace widely transverse and depressed, with the raised lines of the dorsal surface usually very distinctly defined, those which originate from the base of the lateral epibranchial spine often extending in a nearly continuous line from side to side ; anteriorly the carapace is truncated ; the front is divided into four, six, or eight lobes or teeth ; the antero-lateral margins are short, and set at nearly a right angle with the front, and are nearly always divided into four or five teeth, one of which may be rudimentary. The orbits are not large, and are marked with two fissures in the upper margin, and with a notch or fissure in the lower margin ; the interior subocular angle is usually rounded and not prominent. The post-abdomen (as usual in the Portunidæ) is five-jointed ; the third to the fifth segments consolidated into a single piece. The eyes are set on very short, thick pedicels. The basal joint of the antennæ is very greatly developed and widely transverse, is united with the front along nearly the whole of its interior (or anterior) margin, and entirely fills the interior orbital hiatus, from which the flagellum is usually separated by the whole width of the lateral frontal tooth. The ischium of the endognath of the exterior maxillipedes is not produced at its antero-internal angle ; the merus is distally truncated, obliquely truncated or very slightly emarginated at its antero-internal angle, and with the antero-external angle rounded or subacute and but little produced. The chelipedes in the adult male are subequal and moderately developed, and are armed with spines ; and (as in so many other Portunidæ) the carpus and palm are usually longitudinally costated on their exterior surface ; the dactyli (as in *Neptunus*) are nearly straight, with the tips incurved, and are armed with unequal rounded teeth on their interior margins. The ambulatory legs present nothing remarkable ; the fifth (natatory) legs have, as usual, the penultimate and terminal joints considerably dilated.

The species never attain a very large size, and occur very commonly throughout the Indo-Pacific region, and also less frequently on the eastern shores of the Atlantic, and only, I believe, in shallow water (to 20 fathoms). Many are separated by characters of slight importance, and Dr. R. Kossmann,[1] goes so far as to unite, under the single designation *Thalamita prymna*, all the species with eight-lobed front (*Thalamita prymna, Thalamita stimpsoni, Thalamita picta, Thalamita spinimana, Thalamita cœruleipes, Thalamita crenata, Thalamita danæ*), which are distinguished by A. Milne Edwards in his Monograph, several of which, in the present Report, are retained, provisionally at least, as distinct.

The following are species which have been described since the publication of A. Milne Edwards' Report on the Portunidæ :—

Thalamita speciosa, Miers.   Samoan Islands.

Thalamita hellerii, Hoffmann.   Madagascar ; Nossy Faly Island.   (Perhaps a variety of *Thalamita crenata*.)

Thalamita integra, var. *africana*, Miers.   Canary Islands ; Goree Island, Senegambia, 9 to 15 fathoms.

Thalamita quadrilobata, Miers.   Seychelles, 4 to 12 fathoms.

Thalamita intermedia, n. sp.   Torres Strait.

Thalamita sexlobata, n. sp.   Tongatabu, off Nukalofa, 18 fathoms.

§. *Thalamitæ with Four-lobed front.*

Thalamita admete (Herbst).

> *Cancer admete*, Herbst, Naturgesch. der Krabben u. Krebse, iii. Heft. iii., p. 40, pl. lvii. fig. 1, 1803.
> *Thalamita admete*, Latreille, Crust. in Règne Animal de Cuvier, ed. 2, vol. iv. p. 33, 1829.
>   ,,   ,, Milne Edwards, Hist. Nat. Crust., vol. i. p. 439, 1834 ; Règne Animal de Cuvier, ed. 3, Crust. Atlas, pl. ix. fig. 2 (?).
>   ,,   ,, A. Milne Edwards, Archiv. Mus. Hist. Nat., vol. x. p. 350, 1861.

Levuka, Fiji Islands (on the reefs), a small female.

In this specimen the chelipedes are slender ; the fourth antero-lateral tooth, although smaller than the rest, is yet well developed.   The ambulatory legs are prettily banded with fuscous brown.

Two males obtained at Tongatabu, on the reefs, approach the variety designated *Thalamita savignyi* by A. Milne Edwards, in having the palms of the chelipedes granulated between the spines of the upper surface.

Adult ♂.	Lines.	Millims.
Length of carapace (of the largest specimen), ...	8½	18
Breadth of carapace, ...	14½	31

[1] Malacostraca in Zool. Ergebn. einer Reise Küstengeb. d. rothen Meeres, p. 47, 1877.

Thalamita integra, Dana.

Thalamita integra, Dana, Proc. Acad. Nat. Sci. Philad., p. 85, 1852 ; U.S. Explor. Exped., vol. xiii., Crust. 1, p. 281, pl. xvii. fig. 6, 1852.

„ „ A. Milne Edwards, Archiv. Mus. Hist. Nat., vol. x. p. 358, 1861.

Honolulu. Several specimens, males, were taken on the reefs, and one purchased in the market.

Adult ♂.	Lines.	Millims.
Length of carapace, ...	14½	30·5
Breadth of carapace, about	22	47

The smoothness of the basal antennal joint and of the exterior surface of the palms of the chelipedes are sufficient, apparently, to distinguish adult examples of this form from its near ally, *Thalamita admete*.[1]

Thalamita sima, Milne Edwards.

Thalamita sima, Milne Edwards, Hist. Nat. Crust., vol. i. p. 460, 1834.

„ „ A. Milne Edwards, Archiv. Mus. Hist. Nat., vol. x. p. 359, 1861.

„ „ Miers, Crust. in Zool. Coll. H.M.S. "Alert," p. 231, 1884.

Portunus (*Thalamita*) *arcuatus*, de Haan, Crust. in v. Siebold, Fauna Japonica, p. 10, 1833, *sine descriptione*, and p. 43, pl. ii. fig. 2, and pl. xiii. fig. 1, 1835.

South Australian Coast, 2 to 10 fathoms, April 1874 (an ova-bearing female) ; Torres Strait, lat. 10° 30′ 0″ S., long. 142° 18′ 0″ E., in 8 fathoms, Station 186 (another female bearing ova). This latter, which is the largest example, measures as follows :—

Adult ♀.	Lines.	Millims.
Length of carapace, ...	7½	15·5
Breadth of carapace, nearly	11	23

§§ *Thalamitæ with Six-lobed front.*

I place in this new section of the genus two species which are of much interest as being intermediate between the typical *Thalamitæ* and the genus or subgenus *Thalamitoides*, A. Milne Edwards.[2]

They agree with the latter in their six-lobed front, but are distinguished by the much lesser width of the carapace and less numerously spined palms of the chelipedes, points wherein they altogether resemble the other species of *Thalamita*. Should intermediate forms occur, it may be found more convenient to unite the species of this section to *Thalamitoides*.

[1] In some small specimens, however, and particularly in small examples of the variety designated *africana*, by myself, from Goree Island, Senegambia, traces of ridges and granulations appear on the exterior surface of the palms of the chelipedes.

[2] *Nouv. Archiv. Mus. Hist. Nat.*, vol. v. p. 146, 1869.

Thalamita intermedia, n. sp. (Pl. XVI. fig. 1).

The carapace is broadly transverse, and is covered with a close, whitish pubescence; the transverse ridges which cross its dorsal surface are not more distinct than in *Thalamita admete,* to which species and to *Thalamita savignyi, Thalamita intermedia* is nearly allied. Of the six lobes of the front the median are smallest, and separated by a narrow and rather deep incision; the submedian and lateral are subequal, the latter slightly overlapping the former; the lateral lobes project somewhat less than the others. The orbits have two distinct fissures in the upper and one in the lower margin. The five spines of the antero-lateral margin are all well developed, but the three anterior are very slightly larger than the fourth and fifth. The basal antennal joint is very distinctly granulated; the maxillipedes present nothing remarkable. The chelipedes in the male are subequal, the merus or arm with three spines on its anterior margin, of which the two nearest to the distal extremity are largest; wrist with a strong spine on its inner margin and three small spinules on its outer surface; palm with three or four spines, disposed alternately in two series, on its upper surface, and with three granulated ridges on its outer surface, between which are other granules, as in *Thalamita savignyi;* the fingers are somewhat shorter than the palm, and irregularly denticulated on their inner margins; the ambulatory legs slender and slightly compressed; the fifth legs, shaped much as in *Thalamita admete* and *Thalamita savignyi,* with a spine near the distal end of the inferior margin of the merus-joint, and with the inferior margin of the penultimate joint armed with a very distinct series of small spinules. Colour (in spirit) pinkish-brown; pubescence whitish.

Adult ♂.							Lines.	Millims.
Length of carapace, nearly	.	.	.	.	.	.	6	14
Breadth of carapace,	.	.	.	.	.	.	10	21
Length of a chelipede,	.	.	.	.	.	.	13	27·5
Length of first ambulatory leg,	.	.	.	.	.	.	12	25

The unique specimen (a male) was dredged in the Torres Strait, in 8 fathoms, lat. 10° 30′ 0″ S., long. 142° 18′ 0″ E. (Station 186).

Thalamita sexlobata, n. sp. (Pl. XVI. fig. 2).

The carapace is transverse and closely pubescent; the transverse ridges of its dorsal surface distinct but not prominent. The median lobes of the front are slightly rounded, smaller than the submedian lobes, and are separated by a small and shallow notch, the submedian are broader than either the median or lateral lobes, and slightly overlap the former, from which they are separated by a scarcely appreciable notch; the lateral lobes are separated from the submedian lobes by a rather wide interspace, and are not very prominent, their inner margin is continued as an oblique carina for a short distance behind the frontal margin. The upper margin of the orbit is marked with two fissures, and the lower margin with a small notch. The antero-lateral marginal teeth (the fourth excepted) are subequal, the first a very little larger than the others; the fourth is represented by a mere rudiment; the postero-lateral angles of the carapace are rounded. The basal antennal joint is armed with a minutely granulated ridge. The chelipedes in the female are pubescent and rather slender; the merus-joint is armed with two spines (and a smaller denticle) on its anterior margin, and a strong triangular lobe or tooth at the distal end of its inferior (not posterior) margin; its posterior margin is rounded and entire; the wrist has a strong spine on its inner margin, and three small spinules on its outer surface; the palm has four spines, one basal near the articulation with the wrist, one distal at the base of the dactylus, one (longer) behind this on the inner and upper margin, and one on the upper surface; the fingers are somewhat longer than the palm, and are armed, as usual in the genus, with irregularly alternating teeth; the ambulatory legs are of moderate length (the third the longest), compressed and slightly pubescent; the merus of the fifth or swimming legs is armed on its inferior margin with a strong spine placed at some distance behind the distal extremity; the penultimate joint, as already stated, is unarmed; the terminal joint is ciliated, ovate, and mucronate at its distal extremity.

	Lines.	Millims.
Adult ♀.		
Length of carapace, about	4½	9·5
Breadth of carapace,	6½	13·5
Length of a chelipede,	8½	18
Length of third ambulatory leg, about	8	17·5

Off Nukalofa, Tongatabu, 18 fathoms (Station 172). An adult and a smaller female.[1]

This species is very nearly allied to the foregoing, which it resembles in its six-lobed front, and in the form of the carapace, but it is distinguished by the following characters: the carapace is narrower; the frontal lobes, as a reference to the figure will show, are of different shape; the fourth antero-lateral spine is rudimentary; the basal antennal joint is nearly smooth, and the inferior margin of the penultimate joint of the fifth ambulatory legs is not denticulated or spinuliferous as in *Thalamita intermedia*.

§§§ *Thalamitæ with Eight-lobed front.*

Thalamita prymna.

‡ *Cancer prymna*, Herbst, Naturgesch. der Krabben u. Krebse, iii. Heft. iii. p. 41, pl. lvii. fig. 2, 1803.

§ *Portunus (Thalamita) prymna* (var. β.), de Haan, Crust. in v. Siebold, Fauna Japonica, pp. 10, 43. pl. xii. fig. 2, 1833.

Thalamita prymna, Milne-Edwards, Hist. Nat. Crust., vol. i. p. 461, 1834.

„   „   A. Milne Edwards, Archiv. Mus. Hist. Nat., vol. x. p. 360, 1861, et synonyma.

Tongatabu (on the reefs), an adult female; Ternate, 15th October, 1874, an adult male.

[1] The figure is restored from the two specimens, neither being perfect.

In the specimen from Ternate, the third of the frontal teeth is less distinctly truncated, narrower and more rounded at its extremity than in other adult examples I have examined.

Adult ♂.						Lines.	Millims.
Length of carapace, nearly	.	.	.	.	.	18	37·5
Breadth of carapace, .	.	.	.	.	.	24½	52

A female from the Philippine Islands (Samboangan) is in the collection, which is in some degree intermediate between this species and the newly-allied and perhaps not distinct *Thalamita stimpsoni*, A. Milne Edwards. In this specimen the frontal lobes altogether resemble those of *Thalamita prymna*, but the basal antennal joint bears a line of strong granules in place of the well-developed spines of that species. The fourth antero-lateral marginal tooth is reduced to a mere insignificant rudiment, not discernible except by close examination with a lens.

♀.				Lines.	Millims.
Length of carapace,	.	.	.	12	25
Breadth of carapace,	.	.	.	18	38

Thalamita stimpsoni, A. Milne Edwards.

> *Thalamita stimpsoni*, A. Milne Edwards, Archiv. Mus. Hist. Nat., vol. x. p. 362, pl. xxxv. fig. 4, 1861.

Samboangan, 10 fathoms (from the reefs). A small female, bearing ova.

The fourth antero-lateral marginal tooth in this specimen is on one side wholly absent, and on the other represented by a mere insignificant rudiment. In all other characters, and particularly in the characteristic form and disposition of the frontal lobes, this specimen agrees with the description and figure cited.

Adult ♀.						Lines.	Millims.
Length of carapace, .	.	.	.	.	.	5½	11·5
Breadth of carapace, .	.	.	.	.	.	8½	18

Two small and immature females are in the collection from Nukalofa, Tongatabu, 18 fathoms (Station 172), which are perhaps young of this species. They differ, however, in the form of the frontal lobes; the lobes are smaller and less prominent, and the next on each side narrower, rounded, and very distinctly separated from the submedian lobes by a deep fissure. The larger has the following dimensions:—

♀.				Lines.	Millims.
Length of carapace, .	.	.	.	3	6·5
Breadth of carapace, .	.	.	.	4½	9·5

Thalamita crenata, Rüppell.

Thalamita crenata, Rüppell, Beschreib. u. Abbild. 24 kurzschwänzigen Krabben d. rothen Meeres, p. 6, pl. i. fig. 2, 1830.

„ „ Milne Edwards, Hist. Nat. Crust., vol. i. p. 461, 1834.

„ „ A. Milne Edwards, Archiv. Mus. Hist. Nat., vol. x. p. 365, 1861, and references to literature.

Albany Island, Cape York (an adult ova-bearing female).

Adult ♀.					Lines.	Millims.
Length of campace, about	.	.	.	.	18½	39
Breadth of carapace, about	.	.	.	.	25½	53·5

This form, when fully grown, is one of the largest of the group. *Thalamita danæ*, Stimpson, and *Thalamita cæruleipes*, Jacquinot and Lucas, are perhaps mere varieties of this species.

Subfamily 3. CARCININÆ.

Portunus, Fabricius.

Portunus, Fabricius (pt.), Entom. Syst. Suppl. (?), p. 363, 1798.

„ Milne Edwards (pt.), Hist. Nat. Crust., vol. i. p. 439, 1834.

„ A. Milne Edwards (pt.), Archiv. Mus. Hist. Nat., vol. x. p. 392, 1861.

Liocarcinus (subgenus), Stimpson, Bull. Mus. Comp. Zoöl., vol. ii. p. 146, 1870, footnote.

Carapace moderately transverse, depressed, with the lateral margins arcuated, and armed with five regular and even teeth; the front of moderate width, and divided into several lobes or teeth, which are not very prominent; and the last tooth scarcely, if at all, exceeds the preceding one in length; the orbits are not large, and possess two fissures in the superior, and one or two in the inferior margin. The ridges of the endostome or palate are not always developed. The post-abdomen is five-jointed (the third to the fifth segments consolidated). The eyes are set on very short, thick pedicels. The basal joint of the antenna is short and but little dilated, it is generally united with the frontal margin, but does not always completely fill the interior hiatus of the orbit. The ischium of the exterior maxillipedes is not produced at its antero-internal angle, the merus is distally truncated or very slightly rounded, with the antero-internal angle scarcely, if at all, emarginate, and the antero-external angle not produced. The chelipedes are subequal and moderately developed; the arm or merus is smooth, without spines, carpus or wrist with a strong spine on its inner margin, palm with a spine near to the distal extremity of the superior margin, and usually somewhat indistinctly costated on the exterior surface. Ambulatory legs of moderate length, with the dactyli styliform; fifth or natatory legs with the penultimate and terminal joints dilated and compressed, the terminal joint ovate, as in other Portunidæ.

Of the species enumerated by A. Milne Edwards, one, *Portunus longipes*, Risso (= *Portunus dalyelli*, Spence Bate), has since been referred to *Bathynectes*. The species designated by Kinahan, *Portunus carcinoides*, is perhaps identical with *Portunus corrugatus*.

The species of this genus occur commonly in the European and North Atlantic Seas, in water of moderate depth, to 40 fathoms, or rarely at much greater depth (*e.g.*, *Portunus pusillus*, whose occurrence in the Mediterranean dredgings of the "Travailleur" at a depth of 450 mètres is recorded by A. Milne Edwards); one species, *Portunus guadulpensis*, occurs in the West Indian Seas.

Portunus corrugatus, the only species collected by the naturalists of H.M.S. Challenger, is also the form with the widest geographical range ; it occurs, as noted below, not only in the European Seas and at the Azores and Cape Verde Islands, but also in the South Australian Seas and at Japan.

Portunus corrugatus (Pennant).

Cancer corrugatus, Pennant, Brit. Zool., vol. iv. p. 5, pl. v. fig. 9, 1777.

Portunus corrugatus, Leach, Edin. Encycl. Lond., vol. vii. p. 390, 1814 ; Trans. Linn. Soc. Lond., vol. xi. p. 315, 1815 ; Malacostraca Podophthalmata Britanniæ, pl. vii. figs. 1, 2, 1815.

„ „ Milne Edwards, Hist. Nat. Crust., vol. i. p. 443, 1834.

„ „ de Haan, Crust. in v. Siebold, Fauna Japonica, p. 40, 1835.

„ „ Bell, Brit. Stalk-eyed Crust., p. 94, 1853.

„ „ A. Milne Edwards, Archiv. Mus. Hist. Nat., vol. x. p. 401, pl. xxxvi. fig. 3, 1861.

„ „ Heller, Crust. des südlichen Europa, p. 86, 1863.

„ „ Miers, Proc. Zool. Soc. Lond., p. 33, 1879.

? „ *carcinoides*, Kinahan, Dubl. N. H. Rev. Proc. of Soc., vol. iv. p. 66, pl. ix. fig. 3, 1857.

„ *strigilis*, Stimpson, Proc. Acad. Nat. Sci. Philad., p. 38, 1858.

„ „ A. Milne Edwards, *tom cit.*, p. 402, 1861.

„ *subcorrugatus*, A. Milne Edwards, Archiv. Mus. Hist. Nat., *tom. cit.*, p. 402, pl. xxxvi. fig. 2, 1861, var.

Off Fayal, Azores, 50 to 90 fathoms (Station 75 ?), a small female ; Cape Verde Islands, St. Vincent, July 1873 (an immature female) ; Australia, Victoria, Port Philip, 33 fathoms, Station 161 (an adult male and three small specimens) ; Bass Strait, off East Moncœur Island, 38 fathoms, Station 162 (an adult male and three small specimens), April 1874.

The largest Australian male measures as follows :—

Adult ♂ .		Lines.	Millims.
Length of carapace,	.	6½	13·5
Breadth of carapace, nearly	.	8	16·5

Portunus corrugatus is now for the first time, I believe, recorded from the southern hemisphere; its occurrence in the Japanese Seas was noted by de Haan so long ago as 1835, and again by Dr. Stimpson in 1858 (as *Portunus strigilis*), and by myself in 1876. The form designated by A. Milne Edwards *Portunus subcorrugatus*, from the Red Sea, is probably, as I have already noted, a variety of *Portunus corrugatus.*

Platyonychus, Latreille.

Platyonychus, Latreille (pt.), Nouv. Dict. d. Hist. Nat., vol. xxvii. p. 4, 1818.
,, Milne Edwards, Hist. Nat. Crust., vol. i. p. 435, 1834.
,, A. Milne Edwards (pt.), Archiv. Mus. Hist. Nat., vol. x. p. 410, 1861.
Anisopus, de Haan (subgen.), Crust. in v. Siebold, Fauna Japonica, p. 12, 1833.

Carapace depressed and transverse; the dorsal surface without tubercles or transverse ridges; the front is rather narrow, and armed with three or four lobes or teeth; the antero-lateral margins arcuated, and armed with five teeth or spines, including the lobe at the exterior orbital angle and the lateral epibranchial tooth, which is no longer than the preceding tooth; the orbits are rather widely open above, and have one or two fissures in the superior, and one in the inferior margin. The ridges of the endostome are obsolete. The post-abdomen is usually distinctly seven-jointed, but in *Platyonychus ocellatus* it is five-jointed, with the third to the fifth segments consolidated. The eyes are of moderate length. The basal joint of the antennæ is short and not dilated, and occupies, but does not wholly fill, the interior hiatus of the orbit, and is not united at its distal extremity with the front. The exterior maxillipedes are rather large; their ischium joint is not produced at the antero-internal angle; the merus is obliquely truncated at the distal extremity, with the antero-external angle rounded and not at all produced. The chelipedes (in the adult males) are subequal and not greatly elongated, with the palms externally more or less distinctly longitudinally costated; the fingers elongated, and armed on the interior margins with large, triangular lobes, alternating with smaller teeth. The ambulatory legs are of moderate length, with the dactyli styliform and compressed; the fifth or natatory legs have, as usual, the penultimate and terminal joints compressed and dilated; the terminal joint not lanceolate as in *Portumnus*, but broadly ovate and rounded at the distal extremity.

I have elsewhere [1] proposed to restrict this genus to the forms with broader carapace and more broadly dilated and ovate dactyli of the fifth or natatory legs, and to separate, under Leach's designation *Portumnus*, the other species included by A. Milne Edwards in *Platyonychus*. Thus restricted, the genus *Platyonychus* will include the following species, the first two of which are, I believe, confined to shallow water:—

[1] Crust. in Zool. H.M.SS. "Erebus" and "Terror," p. 2, 1874.

Platyonychus ocellatus (Herbst).   East Coast of United States ; Gulf of Mexico,
New Zealand (Dr. J. Hector).[1]

Platyonychus bipustulatus, Milne Edwards (= *Anisopus punctatus*, de Haan ;
Portunus catharus, White ; *Platyonychus purpureus*, Dana).   Indo-
Malaysian, and Australian Seas ; New Zealand ; Chili ; West Patagonia,
Trinidad Channel.

Platyonychus iridescens, n. sp.   Ki Islands, 140 fathoms (Station 192).

Platyonychus bipustulatus, Milne Edwards.

> *Platyonychus bipustulatus*, Milne Edwards, Hist. Nat. Crust., vol. i. p. 437, pl. xvii. figs. 7–10,
> 1834.
> ,,   ,,   A. Milne Edwards, Archiv. Mus. Hist. Nat., vol. x. p. 413, 1861,
> and references to synonyms.
> *Corystes (Anisopus) punctatus*, de Haan, Crust. in v. Siebold, Fauna Japonica, pp. 13, 44, pl. ii.
> fig. 1, 1835.

Off the South Australian Coast, 2 to 10 fathoms, April 1874.   Three small specimens
(two males and a female).

	Lines.	Millims.
Adult ♂.		
Length of carapace,	9½	20
Breadth of carapace to base of lateral epibranchial tooth,	11	23·5

Platyonychus iridescens, n. sp. (Pl. XVII. fig. 2).

The carapace in this fine species is shaped nearly as in *Platyonychus ocellatus* and
Platyonychus bipustulatus; it is moderately convex, broader than long, with the antero-
lateral margins arcuated and shorter than the concave postero-lateral margins ; its surface
is rather coarsely granulated, except near the front, and in the posterior parts, near the
posterior margin, the shallow depressions indicating the cervical suture are as distinct as
in *Platyonychus ocellatus.*   The front (as in that species) is armed with three spines
(the lateral spines being those of the inner angle of the orbit), the median spine, although
somewhat more prominent than the lateral spines, is not nearly so large as in *Platyonychus
ocellatus.*   The orbits have a notch in their upper margins, and a spine at the inner
subocular angle.   The antero-lateral margins are armed with five rather distant spiniform
teeth ; the first (or exterior orbital tooth) is rather prominent, the second and third smaller
and rather more closely approximated than are the other antero-lateral teeth.   The
pterygostomian regions are granulated, and are marked (as in *Platyonychus ocellatus*)
with an oblique granulated line extending to the antero-lateral angles of the buccal
cavity.   The segments of the post-abdomen are distinct, in the male the first and second
segments are small, the third transversely carinated ; the sixth segment is longest, with

[1] *Trans. New Zeal. Inst.*, vol. ix. p. 473, pl. xxviii. fig. 1, 1877.

the sides slightly arcuated, and it is strongly excavated at its distal extremity, where it receives the very small transverse terminal segment, which is distally rounded. The eyes are large and thick, and their corneæ occupy a large part of the inferior surface of the peduncle. The antennules are long, with the basal joints considerably enlarged, and they terminate in two short flagella. The short oblong basal joints of the antennæ occupy the whole of the space between the inner subocular spine and the basal joints of the antennules; the second and third joints are slender and short, the flagella (in the male) rather long, composed of upwards of thirty joints. The outer maxillipedes have the ischium (as in *Platyonychus ocellatus*) longitudinally canaliculated; the merus is longer than broad, obliquely truncated at its distal extremity, with the antero-external angle rounded, the antero-internal angle not (as in *Platyonychus ocellatus*) greatly produced, but terminating in a small rounded lobe as in *Platyonychus bipustulatus*. The chelipedes (in the adult) are robust and elongated; the merus-joint is armed with three to five spinules on the distal half of its anterior margin; the carpus is granulated above, and is armed with three spines (one on its inner surface, one on its outer, and one, distal, above the articulation with the wrist); the palm is about as long as the merus, longitudinally concave, and very coarsely granulated, on its upper surface more finely granulated, with two longitudinal ridges on the outer surface, and with four or five spinules on the distal half of its upper and inner margin; the fingers are longer than the palm; the dactyl has five or six strong spines on its upper margin, both fingers are incurved at the tips, and strongly dentated on their inner margins, the teeth disposed nearly as in *Platyonychus bipustulatus*. The second to fourth legs are compressed, nearly smooth, with the dactyli slenderer and more elongated than in *Platyonychus ocellatus* and *Platyonychus bipustulatus*; the fifth legs are nearly as in *Platyonychus ocellatus*, with the antepenultimate joint carinated above, the penultimate and terminal joints greatly compressed, the terminal joint large and ovate. Colour (in spirit) yellowish-brown; the legs, and especially the chelipedes, are strongly and beautifully iridescent (hence the specific name).

The largest male and female have the following dimensions:—

	Lines.	Millims.
Adult ♀ .		
Length of carapace and rostrum,	23	48·5
Breadth of carapace,	28	59·5
Length of a chelipede, nearly	43	91
Length of first ambulatory leg,	36	76·5
Length of fifth ambulatory leg,	29¼	61
Adult ♂ .		
Length of carapace and rostrum, about	18	38·5
Breadth of carapace,	21½	45
Length of a chelipede, about	35	74
Length of first ambulatory leg,	30	63·5
Length of fifth ambulatory leg,	27	56·5

An adult and two immature males, and two adult females, were dredged near the Ki Islands, in 140 fathoms, lat. 5° 49′ 15″ S., long. 132° 14′ 15″ E. (Station 192).

In the adult female the post-abdomen is rather narrow, broadest at the sixth joint. In the small males the carapace is more convex and uneven, the eyes very large, the marginal spines of the carapace are relatively longer, and the penultimate and terminal joints of the fifth legs are fringed with longer hairs. One of these has the following dimensions :—

Young ♂.	Lines.	Millims.
Length of carapace and rostrum,	6½	13·5
Breadth of carapace,	7½	15·5
Length of a chelipede,	11	23·5
Length of fifth leg, .	8½	18

This very remarkable *Platyonychus* is allied to the East American *Platyonychus ocellatus* in the form of the frontal spines, and to the Oriental *Platyonychus bipustulatus* in the structure of the merus of the outer maxillipedes, but differs from both in the strongly-granulated and spinuliferous palm and dactyl of the chelipedes, the more slender dactyli of the ambulatory legs, the iridescent reflections of the limbs, &c.

Lissocarcinus, Adams and White.

Lissocarcinus, Adams and White, Crust. in Zool. H.M.S. "Samarang," p. 45, 1848.
   „   A. Milne Edwards, Archiv. Mus. Hist. Nat., vol. x. p. 417, 1861.

Carapace depressed, smooth, transverse, little, if at all, broader than long, with the frontal margins and the antero-lateral margins of the carapace thin, acute, the lateral margins either subentire or divided into five lobes or teeth. The front projects somewhat beyond the level of the orbits, and is either subentire or divided by a median notch into two lobes, which may be separated on each side by a slight incision from the interior angle of the orbit. The ridges of the endostome are usually developed, but indistinct. The post-abdomen of the male in *Lissocarcinus orbicularis*[1] is five-jointed, with the third to the fifth segments coalescent. The eyes are very short. The basal antennal joint is dilated, but not transverse, occupies the interior hiatus of the orbit, and is slightly produced at its antero-external angle. The ischium of the endognath of the exterior maxillipedes is not produced at its antero-internal angle, the merus is longer than broad, obliquely truncated along the interior margin, and is not greatly produced at its antero-external angle. The chelipedes are short, subequal ; merus trigonous, carpus with a spine on its interior margin, palm short, either smooth or costated externally, fingers strongly dentated on the inner margins, the dactylus sometimes carinated above. Ambulatory legs very short, with the dactyli styliform ; fifth or

---

[1] This is the only species of which I have examined males.

natatory legs with the penultimate and terminal joints moderately dilated, the terminal joints either ovate or ovate lanceolate.

This genus is evidently very nearly allied to *Carupa*, Dana, and transitional forms may hereafter occur, which may render it desirable to unite *Carupa* with *Lissocarcinus*. *Carupa* has, however, a less prominent front, and a narrower basal antennal joint.

Of the species already described, one, *Lissocarcinus boholensis*, Nauck, is found at Bohol, one, *Lissocarcinus polybioides*, occurs in the Indo-Malaysian and Australian Seas, the other, *Lissocarcinus orbicularis*, Dana, throughout the Indo-Pacific region.

In the collection of H.M.S. Challenger is an apparently undescribed form from the Celebes Sea (10 to 20 fathoms) which I will designate *Lissocarcinus lævis*.

Lissocarcinus polybioides, Adams and White.

<blockquote>

Lissocarcinus polybioides, Adams and White, Crust., in Zool. H.M.S. "Samarang," p. 46, pl. xi. fig. 5, 1848.

„   „   A. Milne Edwards, Archiv. Mus. Hist. Nat., vol. x. p. 417, 1861.
</blockquote>

South Australian Coast, 2 to 10 fathoms, April 1874.   An adult ova-bearing female.

Adult ♀.	Lines.	Millims.
Length of carapace and front, nearly	6	12
Breadth of carapace,	6	12

Lissocarcinus lævis, n. sp. (Pl. XVII. fig. 3).

The carapace is broader than long, in its median portion moderately convex, and to the naked eye it appears smooth and shining, but under the microscope punctulated and minutely granulated. The front is four-lobed; the median lobes scarcely more prominent than the lateral ones, broad and truncated, separated by a small median notch, the lateral lobes (or inner orbital angles) small and dentiform. Orbits entire. Antero-lateral margins shorter than the postero-lateral margins and armed with five teeth, whereof the first is very small and little prominent, the second, third, and fourth subequal and not acute, the fifth small and dentiform; the postero-lateral margins are strongly concave; the eye-peduncles are very short. The antennulary fossettes are transverse. The basal joint of the antennæ is much enlarged, longer than broad; its inner and distal angle fills the inner orbital hiatus, and is produced into a strong tooth or lobe, which projects considerably beyond the inner subocular lobe; the two following joints are small and subequal; the ischium of the outer maxillipedes is longitudinally canaliculated on its outer surface, the merus is narrower at its distal extremity (which is truncated) than at the base. The chelipedes (in the female) are subequal; merus without teeth or spines, but with a small tubercle on its anterior margin near the distal extremity; carpus smooth, but with a strong spine on its inner margin ; palm smooth and rounded on its outer and upper surfaces, without either teeth, spines, or carinæ; fingers as long as the palm, acute, and regularly toothed on their inner margins. The three following legs are smooth and very slender, with straight styliform dactyli ; the posterior or swimming legs moderately robust; merus without spines, but obscurely dentated at the distal inferior angle; penultimate joint thin and compressed, terminal joint ovate, the margins almost destitute of cilia.   Colour (in spirit) yellowish-brown, carapace rather obscurely marked with irregular purplish spots and blotches, as in *Lissocarcinus orbicularis.*

Adult ♀.		Linea.	Millims.
Length of carapace,	.	5½	11·5
Breadth of carapace,	.	6½	13·5
Length of a chelipede, about	.	7½	16
Length of first ambulatory leg,	.	7	15

The single specimen (an adult ova-bearing female) was dredged in the Celebes Sea, south of Mindanao, in lat. 6° 54′ 0″ N., long. 122° 18′ 0″ E., in 10 fathoms (Station 212). The nearest ally to this species in the genus is *Lissocarcinus polybioides,* Adams and White, from which *Lissocarcinus lævis* is distinguished not only by the less prominent, more distinctly truncated front, but also by the different form of the anterolateral marginal teeth.

Lissocarcinus lævis in some particulars is intermediate between the genera *Lissocarcinus* and *Thalamonyx.*   The Challenger species has a broader, less prominent front than the other species of *Lissocarcinus,* and herein resembles *Thalamonyx,* from which it is distinguished by the absence of distinct spines from the palms of the chelipedes and from the inferior margin of the merus-joint of the fifth (natatory) legs.

Lissocarcinus boholensis of Semper (*ined.*) and Nauck, from Bohol, in the Philippines, is (although very briefly characterised), distinguishable from *Lissocarcinus lævis* by the more rounded frontal teeth and the rougher chelipedes.

Section II. **Podophthalminæ.**

Portuniens anormaux, A. Milne Edwards, Ann. d. Sci. Nat., ser. 4, vol. xiv. p. 283, 1860;
Archiv. Mus. Hist. Nat., vol. x. pp. 311, 419, 1861.

Carapace transverse and widest anteriorly, with the front very narrow, spatuliform and inflexed.   Orbits extremely large.   Eye-peduncles very greatly elongated, occupying nearly the whole width of the carapace in front.

This section includes only the genera *Podophthalmus,* Lamarck, and *Euphylax,* Stimpson.[1]

[1] According to Dr. E. Nauck (Das Kaugerüst der Brachyuren, Zeitschr. f. wiss. Zool., vol. xxxiv.), the genus *Hedrophthalmus,* Nauck, is allied to *Podophthalmus,* but it will not enter into the Podophthalminæ as above characterised.

Podophthalmus, Lamarck.

Podophthalmus, Lamarck, Syst. Anim. sans Vert., vol. v. p. 152, 1801.

„ Milne Edwards, Hist. Nat. Crust., vol. i. p. 465, 1834.

„ A. Milne Edwards, Archiv. Mus. Hist. Nat., vol. x. p. 419, 1861.

In this long known but remarkable genus the carapace is depressed and very broadly transverse, anteriorly truncated, and widest in front; the exterior angles of the orbits terminate in a strong spine, and posterior to this, upon the lateral margins (in *Podophthalmus vigil*), is another spine; the lateral margins converge very rapidly to the posterior margin. The front or space included between the bases of the eye-peduncles is linear (as in some species of *Gelasimus*), but below the eyes it is transversely dilated and prolonged for a short distance beneath the bases of the eye-peduncles. The orbits are enormously developed, and extend along the whole of the anterior margin of the carapace. But very obscure traces (if any) exist of the longitudinal ridges of the endostome. The post-abdomen (in the male) is five-jointed, with the third to the fifth segments consolidated. The eye-peduncles are very slender and extremely elongated. The antennules are placed beneath the bases of the eye-peduncles, in large quadrate fossettes, into which the considerably elongated flagella cannot be retracted. The basal joint of the peduncle of the antennæ is short and moderately dilated, it occupies the interior hiatus of the orbit, and reaches the lateral process of the front. The ischium of the exterior maxillipedes is not produced at its antero-internal angle; the merus-joint is obliquely truncated at the distal extremity, and its antero-external angle is somewhat produced and broadly rounded. The chelipedes (in the adult male) are subequal and very large; their merus and carpus-joints are armed with spines, the palms are considerably elongated and longitudinally carinated on the exterior surface; fingers nearly straight and unequally dentated, as in other genera of Portunidæ. The ambulatory legs present nothing abnormal; the merus-joint of the fifth natatory legs is somewhat compressed, and has a spine on its inferior margin near the distal extremity; the penultimate and terminal joints are dilated, as in other Portunidæ. The single recent species of this genus (*Podophthalmus vigil*) is distributed throughout the whole Indo-Pacific region.

Podophthalmus vigil (Fabricius).

Portunus vigil, Fabricius, Entom. Syst. Suppl., p. 363, 1798.

Podophthalmus vigil, Leach, Zool. Miscell., vol. ii. p. 149, pl. cxviii., 1815.

„ „ Milne Edwards, Hist. Nat. Crust., vol. i. p. 467, 1834, et synonyma; Crust. in Règne Animal de Cuvier, atlas, pl. ix. fig. 1.

„ „ A. Milne Edwards, Archiv. Mus. Hist. Nat., vol. x. p. 420, 1861.

Honolulu, on the reefs (a fine adult male).

The principal dimensions of this specimen are as follows:—

Adult ♂.	Lines.	Millims.
Length of carapace,	22½	47
Breadth of carapace to base of antero-lateral marginal spines,	43	91

Legion II. CYCLINEA.

Cyclinea, Dana, U.S. Explor. Exped., vol. xiii., Crust. 1, p. 294, 1852.

This section was established by Dana for the single genus *Acanthocyclus*, which is intermediate in structure and position between the typical Cancroidea, the Plagusiinæ, and, as pointed out by Dr. Strahl,[1] the genera *Bellia* and *Corystoides*, which in Dana's system constitute a distinct subtribe, Bellidea, of the Crustacea Anomura. The nearest ally to *Acanthocyclus* is, I think, *Bellia*, which resembles *Acanthocyclus* in the more or less orbiculate carapace, in the form of the front, chelipedes, and ambulatory legs, but is distinguished by the narrower, more elongated merus of the exterior maxillipedes, by the broader post-abdomen of the male, and the less distinctly defined buccal cavity. Perhaps, nevertheless, as in Dr. Strahl's arrangement, this genus should be placed in the same section of the Brachyura as *Acanthocyclus*.

The Cancroid genus *Cymo*, and the genus *Crossotonotus*, which is placed by A. Milne Edwards in the Catometopa (which have a more or less orbiculate carapace) are distinguished from *Acanthocyclus* by the form of the front, the well-developed flagellum of the antennæ, &c.

Acanthocyclus, Milne Edwards and Lucas.

Acanthocyclus, Milne Edwards and Lucas, Crust. in d'Orbigny, Voy. dans l'Amérique Méridionale, Zool., vol. vi. p. 29, 1843.
Plagusetis, Heller, Verhandl. d. k. k. zool.-bot. Gesellsch. Wien, vol. xii. p. 522, 1862.

In this genus the carapace is subcircular, moderately convex, with the cervical and cardiac-branchial sutures distinct, the lateral margins are arcuated, or, in adult examples of large size, the carapace is somewhat quadrate, with the angles rounded. The antero-lateral margins are dentated. The front is rather narrow, with a prominent median lobe, which projects considerably beyond the interior angles of the orbits, which are small and without fissures. The post-abdomen (in the male) is narrow and five-jointed, with the third to the fifth segments consolidated. The epistoma is very small. No longitudinal ridges are developed upon the endostome. The eye-peduncles are short. The antennulary fossæ are very small, and receive no more than the bases of the antennules. The basal antennal joint is short, moderately dilated, and fills the interior orbital hiatus;

[1] *Monatsber. d. k. preuss. Akad. d. Wiss. Berlin*, p. 714, fig., 1861.

the flagellum is obsolete. The ischium-joint of the endognath of the exterior maxillipedes is much longer than the merus, and very slightly produced at its antero-internal angle, the merus is small, distally truncated, very slightly emarginated or obliquely truncated at the antero-internal angle, and the antero-external angle is not produced. The chelipedes (in the adult males at least) are unequal and well developed, or rather large, the joints are smooth, without spines; palm of moderate length, rounded above; fingers rather shorter than the palm, with the teeth of the inner margins small or sometimes indistinctly developed; in the larger chelipede the dactyl is arcuated, and the fingers have between them a wide intermarginal hiatus. Ambulatory legs of moderate length, with the joints smooth without spines; dactyli terminating in a strong arcuate claw.

The type species of this genus, *Acanthocyclus gayi* (= *Acanthocyclus villosus*, Strahl; *Plagusetes elatus*, Heller) occurs commonly on the shores of Chili and Patagonia, and in the Magellan Straits.

Acanthocyclus gayi, Milne Edwards and Lucas.

Acanthocyclus gayi, Milne Edwards and Lucas, Crust. in d'Orbigny, Voy. dans l'Amérique
         Méridionale, vol. vi. p. 30, pl. xv. fig. 1, 1843.
    ,,   ,,   Miers, Proc. Zool. Soc. Lond., p. 69, 1881, and references to literature.

A male and also an adult female of extraordinary size were taken at Valparaiso on the beach, and a small male in Messier Channel, Patagonia, in January 1876 (in fresh water).

The dimensions of the adult female are as follows:—

Adult ♀.	Lines.	Millims.
Length of carapace, nearly .	14½	30·5
Breadth of carapace,	15½	33
Length of larger chelipede when extended as far as its conformation will allow,	21½	45·5

Legion (?) III. CORYSTOIDEA.

Corystiens, Milne Edwards, Hist. Nat. Crust., vol. ii. p. 139, 1837.
Corystoidea, Dana (subtribe), U.S. Explor. Exped., vol. xiii., Crust., p. 297, 1852.

The Corystoidea, which are placed by Milne Edwards between the Leucosiidae and Dorippidae at the end of the Brachyura, in the classification proposed by Dana constitute a distinct subtribe between the Cyclometopa and Catometopa, and are regarded by Dr. Claus[1] as a distinct family of the Cyclometopa, and placed by him between the Portunidae and Thelphusidae. This is perhaps their most natural position. As limited

[1] Grundzüge der Zoologie, p. 636, 1880.

by Dana, they constitute a somewhat heterogeneous and not very well-defined group, generally distinguished by the extension of the exterior maxillipedes over the anterior margin of the buccal cavity, and the elongated flagella of the exterior antennæ, and include forms related to widely separated families, as, for instance, *Trichocarcinus*, Miers = (*Trichocera*, de Haan), which is scarcely distinguishable from the genus *Cancer*.

The Challenger collection includes but two genera of this group ; the first, *Hypopeltarium* (= *Peltarion*, Auctorum), on account of its orbiculate carapace, may be regarded as establishing the transition from the Corystoidea to the Cyclinea, the other, *Gomeza*, in the elongated carapace and the greatly developed flagella of the antennæ, is altogether a typical representative of the Corystoidea.

Hypopeltarium, n. gen.

Peltarion, Jacquinot and Lucas, Crust. in Voy. au Pole Sud. Zool., vol. iii. p. 80, 1853 ; name preoccupied.

Carapace nearly orbiculate, about as broad as long, moderately convex, with the dorsal surface uneven ; the lateral margins and the margins of the orbits armed with spinules. The front is narrow, about one-sixth the greatest width of the carapace, and three-spined, the median spine sometimes notched at the distal extremity. There are two slight incisions in the superior margins of the orbits, and a much deeper and wider hiatus in the inferior margin, near the exterior orbital angle. The epistoma is but slightly transverse. The anterior margin of the buccal cavity is not distinctly defined ; the ridges of the endostome are partially developed. The post-abdomen in the male is narrow and five-jointed, with the third to the fifth segments consolidated. The eye-peduncles are rather slender and of moderate length. The antennules are longitudinally plicated. The antennæ occupy the interior hiatus of the orbits ; their basal joints are short, but slightly enlarged, scarcely any wider than the following joint, which is longer than the basal joint and slightly compressed ; the antennal flagella, as in other Corystidæ, are elongated, but not half as long as the carapace. The maxillipedes are normal ; the ischium is not produced at its antero-internal angle, the merus is subtruncated and spinuliferous at the distal extremity, and rounded at the antero-external angle ; the antero-internal angle is not emarginated. The chelipedes (in the adult male) are well developed, but not very large ; merus trigonous without spines ; carpus with a strong spine at its antero-internal angle ; palm compressed, short, spinuliferous above ; fingers robust, scarcely as long as the palm, and rather obscurely dentated on the inner margins, distally acute ; the dactylus spinuliferous on the superior margin. The ambulatory legs are moderately elongated, with the joints, except the dactyli, more or less granulated or spinuliferous ; dactyli styliform, slender, and much longer than the penultimate joints.

The single species of this genus (*Hypopeltarium spinosulum*) is common in the

Straits of Magellan, at the Falklands, and on the coasts of Chili and Patagonia, in shallow water (to 45 fathoms).

It is nearly allied to *Atelecyclus*, which is found both on the shores of Europe and Chili, but is distinguished by its narrower, three-spined front, the spinuliferous, not dentated, antero-lateral margins of the carapace, and the shorter, more truncated merus-joints of the exterior maxillipedes.

Hypopeltarium spinosulum (White).

Atelecyclus spinosulus, White, Ann. and Mag. Nat. Hist., ser. 1, vol. xii. p. 345, 1843.
Peltarion spinulosum, White, List Crust. Brit. Mus., p. 52, 1847.
       „   „   Dana, U.S. Explor. Exped., vol. xiii., Crust. 1, p. 304, pl. xviii. fig. 6, 1852.
       „   „   Cunningham, Trans. Linn. Soc. Lond., vol. xxvii. p. 494, 1871.
       „   „   Miers, Proc. Zool. Soc. Lond., p. 68, 1881.
Peltarion magellanicus, Jacquinot and Lucas, Crust. in Voy. au Pole Sud. Zool., vol. iii. p. 83, pl. viii. fig. 1, 1853.

Off the coast of Chiloe, 45 fathoms, Station 304 (a small male); Port Stanley, Falklands (a fine male taken in the trawl); near the same locality, 5 to 12 fathoms, Station 315 (five males and a female); Station 316, 4 to 5 fathoms (a small male).

	Lines.	Millims.
Adult ♂.		
Length of carapace, and breadth, about	22½	47

Gomeza, Gray.

Gomeza, Gray, Zool. Miscell., p. 39, 1831.
Œidea, de Haan (subgenus, in part), Crust. in v. Siebold, Fauna Japonica, p. 15, 1833.

Carapace elongated, dorsally convex, and granulated, with the lateral margins rounded and armed with teeth or spines along their whole length ; the front is narrow, compressed, laminated; it projects over the bases of the antennules, and is emarginate or biemarginate at the distal extremity. The orbits are small, and are armed with a prominent tooth or spine at their interior angle, and with a smaller spinuliferous lobe at the interior sub-ocular angle. The epistoma is distinct and transverse. The longitudinal ridges of the endostome are developed, but do not reach the anterior margin of the buccal cavity. The post-abdomen of the male is five to seven-jointed. The eyes are short and completely retractile within the orbits. The antennules are longitudinally plicated ; the antennæ terminate in long flagella, sometimes exceeding the body in length, their basal joints are moderately dilated and reach the margin of the front, they occupy the space between the antennulary fossettes and the orbits from which the antennal flagella are excluded. The merus-joints of the endognathi of the exterior maxillipedes are either distally truncated or narrowed and rounded, and bear the next joint at their antero-internal angles. The chelipedes (in the adult) are of moderate length, usually somewhat unequal, with the merus trigonous ; carpus armed with a strong spine on its inner margin ; palm short, compressed ; dactyli distally acute and dentated on the inner margins. The ambulatory legs are rather slender and of moderate length, with the joints smooth ; dactyli in all slender, styliform.

One species of this genus, *Gomeza bicornis*, is rather common in the Indo-Malaysian, Japanese, and Australian Seas, another, *Gomeza serrata*, Dana, occurs on the coasts of Chiloe, Patagonia, and in the Straits of Magellan, in shallow water (to 30 fathoms).

Gomeza bicornis, Gray.

> *Gomeza bicornis*, Gray, Zool. Miscell., p. 39, 1831 ; Crust. in Griffith, Animal Kingdom of Cuvier, vol. xiii. p. 296, pl. xxiv. fig. 1, 1833 ; List Crust. Brit. Mus., p. 52, 1847.
> *Corystes* (*Gdidea*) *viginti-spinosa*, de Haan, Crust. in v. Siebold, Fauna Japonica, p. 44, pl. ii. fig. 5, 1835.
> *Gomeza viginti spinosa*, A. Milne Edwards, Nouv. Archiv. Mus. Hist. Nat., vol. x. p. 52, pl. iii. fig. 5, 1874.

Celebes Sea, 10 fathoms (Station 212), lat. 6° 54′ 0″ N., long. 122° 18′ 0″ E. A small female.

	Lines.	Millims.
♀.		
Length of carapace, about	7½	15
Breadth of carapace,	5	10·5

Legion IV. THELPHUSINEA.

> *Thelphousiens*, Milne Edwards, Hist. Nat. Crust., vol. ii. p. 7, 1837.
> *Thelphusiens*, Milne Edwards, Ann. d. Sci. Nat., Zool. xx. p. 207, 1853.
> *Thelphusinea* or *Cancroidea Grapsidica*, Dana, U.S. Explor. Exped., vol. xiii., Crust. 1, pp. 145, 292, 1852.

This section of the Cyclometopa includes in Dana's system those terrestrial or fluviatile crabs which are intermediate in structure between the typical Cancroidea (Cancridæ) and certain Catometopa (Gecarcinidæ). They resemble these latter in their terrestrial habits, and approach them in many particulars of structure, *e.g.*, in the form of the carapace, which is more or less dilated at the branchial regions, and in the usually spinuliferous dactyli of the ambulatory legs, but as in other Cancroidea the male verges pass directly through the basal joint of the fifth ambulatory legs and not through sternal ducts, and the carpal joint of the endognath of the exterior maxillipedes is articulated with the merus at or near its antero-internal angle or at the apex, not at the antero-external angle.

This group includes the single family Thelphusidæ.

Thelphusa, Latreille.

Thelphusa, Latreille, Nouv. Dict. d. Hist. Nat., ed. 2, vol. xxxiii. p. 500, 1819.
,, Milne Edwards, Hist. Nat. Crust., vol. ii. p. 10, 1837; Ann. d. Sci. Nat., ser. 3, Zool. xx. p. 209, 1853.
,, A. Milne Edwards, Nouv. Archiv. Mus. Hist. Nat., vol. v. p. 163, 1869, et synonyma.
,, Kingsley, Proc. Acad. Nat. Sci. Philad., p. 35, 1880.

Carapace transverse, with the antero-lateral margins arcuated and usually armed with a single tooth, placed at a short distance behind the tooth or lobe at the exterior angle of the orbit, sometimes moderately convex, with a more or less distinct post-frontal crest, sometimes very convex, and the post-frontal crest obsolete. Front deflexed, of moderate width. Orbits usually without marginal fissures, and with the interior sub-ocular lobe but little developed. Epistoma narrow, transverse. Endostome not longitudinally ridged. Post-abdomen (in the male) usually distinctly seven-jointed. Eyes of moderate size. Antennules transversely plicated and usually almost concealed beneath the front. The basal joint of the antennæ is very short, and does not always reach the frontal margin; the short flagellum lies within the anterior hiatus of the orbit. The ischium-joint of the exterior maxillipedes is not produced at its antero-internal angle; the merus is small, usually distally truncated, with the antero-external angle rounded, the antero-internal angle (at or near which the next joint articulates with the merus) slightly emarginate. The chelipedes (in the males) are well developed and usually unequal, with the merus trigonous; carpus with a strong spine on its inner margin, palm rounded above; fingers usually not so long as the palm, distally acute, and dentated on the inner margins. The ambulatory legs are rather long, with styliform dactyli, which are armed with spinules disposed in longitudinal series.

The described species of *Thelphusa* are very numerous; one, the common *Thelphusa fluviatilis*, occurs not only on the shores of the Mediterranean, but also in Asia Minor, Syria and Persia; the others are found in all the warmer temperate and tropical regions of the Old World, extending southward to the Cape of Good Hope, Madagascar and Australia, but not to New Zealand; a species (*Thelphusa chilensis*) occurs in Chili.

The genus may be divided into three sections or subgenera, characterised mainly by the form (when developed) of the postfrontal crest; the Challenger collection includes a species belonging to each of these sections, which are connected with one another by insensible gradations.

I. Carapace moderately convex, but little dilated in front of the branchial regions, with the epibranchial tooth usually distinctly developed. Postfrontal crest distinct, but not continuous. Typical *Thelphusæ*.

The species are mostly Mediterranean and Asiatic.

II. Carapace moderately convex, usually antero-laterally dilated, with sometimes one or more lateral teeth developed in front of the epibranchial tooth. Postfrontal crest extending in an unbroken line from the median incision to the antero-lateral margins. *Potamonautes,* MacLeay.

The species are mostly African.

III. Carapace dorsally very convex and dilated in front of the branchial regions, with the post-frontal crest and epibranchial tooth obsolete or nearly obsolete. *Geothelphusa,* Stimpson.

Species mostly Indo-Malaysian, Japanese and Australian.

Thelphusa nilotica, Milne Edwards, which has a continuous post-frontal crest, and the carapace armed with lateral marginal spines behind the epibranchial tooth, should perhaps be regarded as the type of a fourth section of the genus.

The following are species which were inadvertently omitted from the most recent list of species of the genus (that published by Mr. Kingsley in 1880) or have been since described :—

> *Thelphusa borneensis,* v. Martens. Borneo.
> *Thelphusa dubia,* B. Capello. West Africa.
> *Thelphusa kuhlii,* de Man. Java.
> *Thelphusa limula,* Hilgendorf. Senegambia.
> *Thelphusa longipes,* A. Milne Edwards. Pulo Condore, China Sea.
> *Thelphusa madagascariensis,* A. Milne Edwards. Madagascar.
> *Thelphusa pealiana,* Wood-Mason. Assam.
> *Thelphusa sumatrensis,* Miers. Sumatra.[1]

Thelphusa sinuatifrons (?), Milne Edwards, var. (Pl. XVIII. fig. 1).

> *Thelphusa sinuatifrons,* Milne Edwards, Ann. d. Sci. Nat., ser. 3, Zool. xx. p. 211, 1853 (?).
>   „   „   A. Milne Edwards, Nouv. Archiv. Mus. Hist. Nat., vol. v. p. 177, pl. x. fig. 2, 1869 (?).
>   „   „   Miers, Ann. and Mag. Nat. Hist., ser. 5, vol. v. p. 305, 1880.

Philippines, Mindanao, Pasananca (in the river). An adult male and female are referred to this species.

The female agrees nearly with the description of A. Milne Edwards in the obscurely trisinuated frontal margin, and in other characters.

The male presents a very curious and remarkable variation in the form of the front, which I am inclined, however, to regard merely as an individual peculiarity, since it differs in no other particulars from the other example from the same locality. The front,

---

[1] I have quite recently proposed the names *Thelphusa cuningii* and *Thelphusa leichardti,* as alternative designations for specimens which are allied to but may be distinct from *Thelphusa crassa,* A. Milne Edwards, and have described a new variety, *johnstoni,* from Kilimandjaro, Eastern Africa, of *Thelphusa depressa,* Krauss.

instead of presenting an obscure median and two lateral sinuses is divided into two lobes by a deep, triangular median notch (see the figure). I have never observed such a variation to anything like the same degree in any other species of the genus.[1]

These specimens have the following dimensions :—

		Lines.	Millims.
Adult ♂ .			
Length of carapace, nearly	.	13	27
Breadth of carapace, nearly .	.	16½	34·5
Adult ♀ .			
Length of carapace, nearly	.	14	29
Breadth of carapace,	.	17½	38

Thelphusa (Potamonautes) perlata, Milne Edwards.

> *Thelphusa perlata*, Milne Edwards, Hist. Nat. Crust., vol. ii. p. 13, 1837 ; Ann. d. Sci. Nat., ser. 3, Zool. xx. p. 209, 1853.
> „   „   A. Milne Edwards, Nouv. Archiv. Mus. Hist. Nat., vol. v. p. 179, pl. ix. fig. 3, 1869.
> „   „   MacLeay, Annulosa in Smith, Zool. of South Africa, p. 64, 1849, *Potamonautes* in text.

South Africa, Wellington and Cape Town (in the rivers). A good series is in the collection.

The largest male has the following dimensions :—

		Lines.	Millims.
Adult ♂ .			
Length of carapace,	.	22	46·5
Breadth of carapace,	.	30½	64·5
Length of large chelipede, .	.	47½	100·5

The postfrontal crest is distinct and continuous in the smallest specimen in the series, a male, whose carapace is not 4 lines (8·5 mm.) in length, and 5 lines (10 mm.) in width.

Thelphusa (Geothelphusa) dehaanii, White.

> *Cancer (Thelphusa) berardii*, de Haan, Crust. in v. Siebold, Fauna Japonica, p. 52, pl. vi. fig. 2, 1835.
> *Thelphusa dehaanii*, White, List Crust. Brit. Mus., p. 30, 1847.
> „   „   Milne Edwards, Ann. d. Sci. Nat., ser. 3, Zool. xx. p. 212, 1853.
> „   „   A. Milne Edwards, Nouv. Archiv. Mus. Hist. Nat., vol. v. p. 174, 1869, and references to synonyma.

Japan, in paddy fields near Lake Biwa, May 1875 (a male) ; at Hakona, at an elevation of 2,500 feet above the sea-level, May 1875 (two small males) ; near Kobé, June 1875 (three males).

[1] An obscure median excavation of the front is, however, occasionally seen, as for instance in an adult male *Thelphusa africana*, A. Milne Edwards, in the collection of the British (Natural History) Museum.

Scarcely any indications of a postfrontal crest exist in this species, but the proto-gastric prominences are obscurely developed.   Ordinarily there is no lateral epibranchial tooth, the antero-lateral margins of the carapace being simply granulated, but occasionally, as in the specimen collected near Lake Biwa, it is distinctly defined.   This, the largest specimen, has the following dimensions :—

	Lines.	Millims.
Adult ♂.		
Length of carapace, about .	9½	20·5
Breadth of carapace, .	12	25·5

CATOMETOPA or OCYPODIIDEA.

Quadrilatera, Latreille (pt.), Fam. Nat. du Règne Anim., p. 269, 1825.
Catométopes, Milne Edwards (pt.), Hist. Nat. Crust. vol. i. p. 264, vol. ii. p. 1, 1834–37.
Ocypodidæ, Milne Edwards (pt.) Ann. d. Sci. Nat., ser. 3, Zool. xviii. p. 140, 1852.
Grapsoidea, Dana, U.S. Explor. Exped., vol. xiii., Crust., pp. 67, 306, 1852.
Catometopa, Miers, Cat. New Zeal. Crust., p. 32, 1876.

Carapace broad in front, often subquadrate, sometimes subglobose, truncated or arcuated anteriorly, but not rostrated.   Epistoma short, often almost linear.   Afferent branchial channel as in the Oxyrhyncha.   The carpal joint of the exterior maxillipedes inserted at the summit, or more usually at the antero-external angle of the merus, very rarely at its antero-internal angle.   Branchiæ usually fewer than nine in number.   The male verges are inserted either in the sternum itself or in the basal joints of the last pair of legs, thence passing through channels in the sternum beneath the post-abdomen.

Family I. GEOCARCINIDÆ.

Gécarciniens, Milne Edwards, Hist. Nat. Crust., vol. ii. p. 16, 1834.
Geocarcinacea, Milne Edwards (pt.), Ann. d. Sci. Nat., vol. xviii. p. 200, 1852.
Geocarcinidæ, Dana, U.S. Explor. Exped., vol. xiii., Crust. 1, p. 374, 1852.

Carapace dorsally very convex, and especially dilated over and in front of the branchial regions, with the antero-lateral margins usually entire and very strongly arcuated ; the front of moderate width and strongly deflexed.   Orbits and eye-peduncles of moderate size.   The post-abdomen of the male usually covers at the base the whole width of the sternum, between the bases of the ambulatory legs.   The carpal joint of the exterior maxillipedes is sometimes inserted at the antero-external angle of the merus, sometimes at the apex or near its antero-internal angle, and may be completely concealed beneath the merus. The chelipedes in the adult male are very robust and usually unequally developed. The dactyli of the ambulatory legs are nearly always granulated and armed with spines disposed in longitudinal series.[1]

Geocarcinus, Leach.

Geocarcinus, Leach, Trans. Linn. Soc. Lond., vol. xi. p. 322, 1815.
        „   Milne Edwards, Hist. Nat. Crust., vol. ii. p. 26, 1837; Ann. d. Sci. Nat., ser. 3, Zool. xx. p. 201, 1853.

Carapace transverse and anteriorly convex, with the mesogastric and cervical sutures strongly defined, and the branchial regions antero-laterally convex and greatly developed; the antero-lateral margins not distinctly dentated. The front is deflexed and narrow or of moderate width, anteriorly truncated, nearly reaches the anterior margin of the buccal cavity, and almost completely covers the antennules. Orbits of moderate size, with the superior margins subentire, no distinct tooth at the exterior angle, and with a well-marked hiatus in the inferior margin, near the interior subocular lobe, which is in contact with the front, and excludes the short antennæ from the orbit. No distinct ridges are developed upon the endostome or palate. The post-abdomen in the male is usually distinctly seven-jointed. The eyes are set on rather short, thick pedicels. The antennæ are very short and occupy the angle formed by the contact of the inner subocular lobe of the orbit with the front, their basal joint is slightly dilated and does not reach the frontal margin. The exterior maxillipedes bulge out externally, and enclose a lozenge-shaped interspace; ischium and merus are broad, the ischium not produced at the antero-internal angle, merus distally rounded and prolonged so as almost entirely to conceal the following joints, it has sometimes a fissure in the antero-lateral margin; the following joint is articulated with the merus on its inner surface. The chelipedes (in the adult male) are considerably developed and usually unequal; merus trigonous and somewhat elongated; carpus without a spine on the interior margin; palm rounded above; fingers distally acute and dentated on the inner margins. The ambulatory legs are robust and somewhat elongated, with the penultimate and terminal joints, and sometimes the antepenultimate joint, compressed, and armed with spinules disposed in longitudinal series.

The species included in this genus are not numerous and inhabit, for the most part, the West Indian Islands, Brazil, Central America and California. One species, *Geocarcinus lagostoma*, has apparently a very extensive range, i.e., from Australasia (?) to the Cape

[1] In *Uca* they are compressed and unarmed.

of Good Hope and West African coast, and thence to Ascension Island and Bermuda (see below).

The following are species which have been described since Milne Edwards' account of the genus in 1853 :—

>Geocarcinus planatus, Stimpson. California, near Cape St. Lucas.
>
>Geocarcinus depressus, Saussure. West Indies, Hayti. This species is perhaps not distinct from Geocarcinus lagostoma.[1]

Geocarcinus lagostoma (?) (Pl. XVIII. fig. 2).

[1] Geocarcinus lagostoma, Milne Edwards, Hist. Nat. Crust., vol. ii. p. 27, 1837; Ann. d. Sci. Nat., ser. 3, Zool. xx. p. 203, 1853.

Two adult females, from Ascension Island, are referred to this species.

The largest specimen has the following dimensions :—

Adult ♀.	Lines.	Millims.
Length of carapace,	33½	71
Breadth of carapace,	43½	92

A small male from Bermuda differs so slightly from small examples of *Geocarcinus lagostoma* that I do not venture to separate it as a distinct species. It is distinguished principally by the form of the merus of the outer maxillipedes, which, although notched as in *Geocarcinus lagostoma*, has the antero-external lobe less produced and rounded, so that the notch is more nearly terminal than in the specimens of *Geocarcinus lagostoma* I have examined. The spinules of the inferior margin of the penultimate joints of the ambulatory legs, which are well developed in small specimens of *Geocarcinus lagostoma*, are, in the specimen from Bermuda, nearly obsolete. In the coloration it nearly resembles specimens of *Geocarcinus lateralis*.

This specimen may perhaps be referable to *Geocarcinus depressus*, Saussure,[2] in which, however, the maxillipedes (as figured) are of rather different form.

The specimen from Bermuda is of the following dimensions :—

♂.	Lines.	Millims.
Length of carapace,	15½	32·5
Breadth of carapace,	18½	39

Geocarcinus lagostoma has been but briefly described, and never, I believe, figured. I think it useful, therefore, to figure the adult female from Ascension Island in the Challenger collection, to facilitate its identification by future authors. The adult males

[1] Geocarcinus barbatus, Pœppig, from Chili, is thought by Milne Edwards (tom. cit., p. 205, 1853), to be a species of Cardisoma; Geocarcinus ruvus, Pœppig, also from Chili, is perhaps not a Geocarcinus.

[2] Vide Mém. Soc. Phys. d. Genève, vol. xiv., pt. ii. p. 439, pl. ii. fig. 14, 1858.

differ little from the females in the series I have examined. They sometimes, but not invariably, have larger and more robust chelipedes.[1]

Cardiosoma, Latreille.

Cardiosoma, Latreille (Cardisoma), Encycl. Méth. Hist. Nat., vol. x. p. 685, 1825.
,, Milne Edwards, Hist. Nat. Crust., vol. ii. p. 22, 1837; Ann. d. Sci. Nat., ser. 3, Zool. xx. p. 203, 1853.
,, S. J. Smith (Cardiosoma), Trans. Connect. Acad., vol. ii. p. 142, 1870.

Carapace transverse, elevated, and sometimes very convex anteriorly; with the branchial regions antero-laterally convex, and very greatly developed, as in Geocarcinus. The antero-lateral margins are sometimes armed with a small tooth placed at a short distance from that at the exterior angle of the orbit. Front deflexed, and usually broader than in Geocarcinus; it nearly reaches the anterior margin of the buccal cavity, and conceals, in part, the antennules. Orbits large and widely open, with the margins entire; the interior subocular lobe is separated by a wide hiatus from the frontal margin, and this hiatus is occupied by the antennæ. Endostome without distinct longitudinal ridges. Post-abdomen (in the male) distinctly seven-jointed. Eye-peduncles of moderate size and thickness; they do not nearly fill the orbital cavities. The basal joint of the antennæ is short and somewhat dilated, and does not usually quite reach the frontal margin; the flagellum is very short. The exterior maxillipedes do not meet along their inner margins, but enclose a lozenge-shaped interspace; the ischium and merus-joints of the endognathi are rather broad and truncated, the merus even rather concave at the distal extremity; the carpal-joint is articulated with the merus at its antero-external angle. The chelipedes are usually unequal, and the larger one sometimes (Cardiosoma guanhumi) enormously developed; merus more or less trigonous; carpus usually with a spine on its interior margin; palm often shorter than the fingers, which, as usual, are more or less distinctly dentated on the inner margins. Ambulatory legs robust, and more or less elongated; the merus-joints with the superior margins acute, and armed with a subterminal spine; dactyli as in Geocarcinus, armed with spinules ranged in longitudinal series.

The species of this genus which, like Geocarcinus, are terrestrial or subterrestrial, are not numerous, and their discrimination is often difficult.[2]

One, Cardiosoma carnifex (Herbst) = ? Cardiosoma obesum, Dana, Cardiosoma

---

[1] The types of these species were from the Australasian seas, whence also there is a good series of specimens in the British Museum collection, obtained during the voyage of H.M.SS. "Erebus" and "Terror"; the Museum also possesses a specimen designated as from the Cape of Good Hope, and another from the West African Coast (Fraser). With these specimens the Challenger examples apparently agree in all particulars. In adult specimens of large size the branchial regions are very considerably dilated, and the granulated line which defines their antero-lateral margins is partially or even entirely obsolete. In the smallest examples this line is very distinct, but the lateral series of spinules on the dactyli of the ambulatory legs is not developed, and they are therefore armed with only four (marginal) series of spinules.

[2] Several of the species here placed provisionally as synonymous with Cardiosoma carnifex, are regarded by M. de Man (Notes Leyden Mus., ii. pp. 31–36, 1879) as distinct species, but A. Milne Edwards is inclined upon the whole to doubt their specific distinctness (Nouv. Archiv. Mus. Hist. Nat., vol. ix. p. 264, 1873).

hirtipes, Dana, *Cardiosoma urvillei*, Milne Edwards, is widely distributed throughout the Oriental region ; another, *Cardiosoma armatum*, Herklots, occurs on the West Coast of Africa and at the Cape Verde Islands ; a third, *Cardiosoma guankumi*, Latreille = *Cardiosoma diurnum*, Gill (*fide* Smith), and *Cardiosoma quadratum*, Saussure (*fide* von Martens), is common at the West Indies and Florida Keys, and occurs also in Brazil and at the Cape Verde Islands (Stimpson) ; a fourth, *Cardiosoma cressum*, Smith (= *Cardiosoma latimanus*, Lockington ?), inhabits the West Coast of America and Lower California ; and a fifth, *Cardiosoma* (?) *barbiger* (Pœppig) referred to by Milne Edwards as *Cardiosoma barbatus*, occurs in Chili.

Cardiosoma guanhumi, Latreille.

> *Cardiosoma guanhumi* (Latreille), Milne Edwards, Hist. Nat. Crust., vol. ii. p. 24, 1837 ; Atlas in Règne Animal de Cuvier, ed. 3, pl. xx. fig. 1 ; Ann. d. Sci. Nat., ser. 3, Zool. xx. p. 204, 1853.
> *Cardiosoma guanhumi*, S. J. Smith, Trans. Connect. Acad., vol. ii. p. 143, pl. v. fig. 3, 1870, where references to synonyma.

Bermuda, April 1873 (two adult males).

The largest example has the following dimensions :—

Adult ♂.	Lines.	Millims.
Length of carapace, about	46	98
Breadth of carapace,	56½	120

Cardiosoma armatum, Herklots.

> *Cardiosoma armatum*, Herklots, Addit. Faun. Carcin. Africæ occidentalis, p. 7, pl. i. figs. 4, 5, 1851.
> „ „ Milne Edwards, Ann. d. Sci. Nat., ser. 3, Zool. xx. p. 205, 1853.
> „ „ de Man, Notes Leyden Mus., ii. No. 5, p. 32, 1879.

Cape Verde Islands, Porto Praya, St. Jago (an adult female).

Adult ♀.	Lines.	Millims.
Length of carapace,	32½	69
Breadth of carapace,	40½	86

This species is always distinguishable from the Oriental *Cardiosoma carnifex* by the more strongly spinuliferous merus and granulated palm of the chelipedes.[1]

Cardiosoma carnifex (Herbst).

> *Cancer carnifex*, Herbst, Naturgesch. der Krabben u. Krebse, vol. ii. p. 163, pl. xli. fig. 1 ♂, 1794.
> *Cardiosoma carnifex* (Latreille), Milne Edwards, Hist. Nat. Crust., vol. ii. p. 23, 1837 ; Ann. d. Sci. Nat., ser. 3, Zool. xx. p. 204, 1853.
> „ „ A. Milne Edwards, Nouv. Archiv. Mus. Hist. Nat., vol. ix. p. 264, 1873.
> „ „ de Man, Notes Leyden Mus., ii. No. 5, p. 31, 1880.

Specimens were in the Challenger collection from the following localities :—
Tracey Island, Nares Harbour, Admiralty Islands (three adult males and an adult

[1] *Cf.* Hilgendorf, *Monatsber. d. k. preuss. Akad. d. Wiss. Berlin*, p. 801, 1878.

female); Kandavu, Fiji Islands (an adult female); Tahiti, September 30, 1875 (two adult males and two females); Papiete (Tahiti), September and October, 1875 (three adult males and an adult female).

An adult male from Tracey Island measures as follows:—

Adult ♂.		Lines.	Millims.
Length of carapace, about	.	31½	67
Breadth of carapace, about	.	40	85

The Challenger examples fall into two very distinct series which may thus be characterised :—In the first, to which belong the specimens from Tahiti, the carapace is moderately tumid at the branchial regions; the postfrontal and postorbital prominences of its dorsal surface are not very prominent; the exterior orbital tooth is prominent, although small, and is followed rather closely by the epibranchial tooth; the antero-lateral margins are defined by a very distinct raised line, which extends halfway along the postero-lateral margins, and the exterior subocular angle of the carapace is about a right angle. These are apparently the form distinguished by M. de Man as the typical *Cardiosoma carnifex*. In the second form, to which belong the specimens from the Admiralty Islands and Kandavu, the carapace is much more swollen and arched at the branchial regions, the postfrontal and postorbital prominences are much more prominent, the exterior orbital tooth less prominent, the lateral epibranchial tooth more remote from that at the outer angle of the orbit; the raised line defining the antero-lateral margins of the carapace is shorter, and the exterior subocular angle more acute. I may add that the merus of the exterior maxillipedes generally narrows more decidedly to its base in the typical *Cardiosoma carnifex*.

This form may, I think, be identified with *Cardiosoma hirtipes* of de Man (*tom. cit.,* p. 34), though perhaps not of Dana, but the basal antennal joint is usually somewhat excavated, and the chelipedes in the male are often unequally developed.

Family II. OCYPODIDÆ.

Ocypodien, Milne Edwards, Hist. Nat. Crust., vol. ii. p. 39, 1837.
Macrophthalmidæ (pt.), Dana, U.S. Explor. Exped., vol. xiii., Crust. 1, pp. 308, 312, 1852.
Ocypodinæ, Milne Edwards (pt.), Ann. d. Sci. Nat., ser. 3, Zool. xviii. p. 140, 1852.

Carapace usually moderately convex, cancroid or trapezoidal, with the antero-lateral margins straight or arcuated, but the branchial regions not greatly dilated, as in the Geocarcinidæ; the front of moderate width, or very narrow. Orbits and eye-peduncles sometimes of moderate size, sometimes very greatly developed. The post-abdomen does not always cover the sternum at the base between the bases of the fifth ambulatory legs. The carpal joint of the endognath of the exterior maxillipedes is inserted at the antero-internal, or, rarely, at the antero-external angle of the merus. The chelipedes (in the adult males) are usually of moderate size, sometimes rather slender and very considerably elongated; the dactyli of the ambulatory legs are styliform and are not armed with strong spines as in the Geocarcinidæ.

The species are for the most part small and littoral or shallow-water forms, but occasionally occur in deep water.

Subfamily 1. CARCINOPLACINÆ.

Carapace transverse, usually convex, and more or less cancroid in form, with the antero-lateral margins arcuated, spinose, or dentated, rarely entire. Frontal region usually of moderate width, orbits rather small. The post-abdomen at the base usually covers the whole width of the sternum between the bases of the fifth ambulatory legs. The fifth joint of the exterior maxillipedes articulates at the antero-internal angle of the merus. Chelipedes in the adult male usually subequal, and sometimes considerably elongated. Ambulatory legs moderately elongated, slender, with the dactyli styliform, sometimes compressed.

The genus *Pseudorhombila* must be taken as the type of this family, since in it the characters are most evident, and it is the only genus included in the Histoire Naturelle des Crustacés; moreover the designation has priority over *Carcinoplax*, which is the name proposed by Milne Edwards in 1852, for the species of *Curtonotus*, de Haan, *Curtonotus* having been previously used in the Coleoptera.

The genera have been divided into the following sections, which are apparently connected by insensible gradations :—

1. Euryplacinæ.

Euryplacinæ, Stimpson, Bull. Mus. Comp. Zoöl., vol. ii. p. 150, 1870.
Eucratopsinæ, Stimpson, tom. cit., vol. ii. p. 151, 1870.

In this section the antero-lateral margins of the carapace are dentated or spinose, and the post-abdomen in the male does not entirely cover the sternum at the base. In the typical genera, which are nearly related to the Cancroidea through *Panopeus* and *Galene*, the last segment of the sternum is exposed at the anterior corners only; these genera are, *Euryplax*, Stimpson, *Panoplax*, Stimpson, *Eucratopsis*, S. J. Smith, and perhaps *Glyptoplax*, Smith (this genus is placed by Milne Edwards in the Cancroidea near *Panopeus*).

In other genera, e.g., *Speocarcinus*, Stimpson, *Eucratoplax*, A. Milne Edwards, and *Prionoplax*, Milne Edwards (if this genus truly belongs here), the post-abdomen of the male is much narrower at the base, and a large part of the posterior segment of the sternum is exposed.

The genera of this section, except perhaps the typical species of *Prionoplax*, are all American, and are not represented in the Challenger collection.

2. Carcinoplacinæ.

Carcinoplacinæ, Milne Edwards, Ann. d. Sci. Nat., ser. 3, Zool. xviii. p. 164, 1852.

In this section the antero-lateral margins of the carapace are usually dentated or spinose, and the post-abdomen completely conceals the sternum at the base (except rarely in *Pilumnoplax*).

The genera referred to the section Carcinoplacinæ are :—*Carcinoplax*, Milne Edwards (= *Curtonotus*, de Haan); *Pseudorhombila*, Milne Edwards; *Geryon*, Kröyer (= *Chalæpus*, Gerstaecker); *Eucrate*, de Haan; *Litocheira*, Kinahan (= *Brachygrapsus*, Kingsley); *Pilumnoplax*, Stimpson; *Heteroplax*, Stimpson; *Bathyplax*, A. Milne Edwards; *Frevillea*, A. Milne Edwards; and *Camptoplax*, Miers; also perhaps *Cotoptrus*, A. Milne Edwards, and *Libystes*, A. Milne Edwards, where the form of the post-abdomen in the male is not known, and *Camptandrium*, Stimpson, which is regarded by Dr. Stimpson as the type of a distinct family (Camptandriidæ).

3. Rhizopinæ.

Rhizopinæ, Stimpson, Proc. Acad. Nat. Sci. Philad., p. 95, 1858.

In this section the antero-lateral margins of the carapace are entire or subentire, and the post-abdomen in the male very rarely covers the whole width of the sternum at the base. The genera included in it are *Scalopidia*, Stimpson (= *Hypophthalmus*, Richters); *Rhizopa*, Stimpson ; *Typhlocarcinus*, Stimpson; *Ceratoplax*, Stimpson; *Notonyx*, A. Milne Edwards ; *Xenophthalmodes*, Richters ; and perhaps *Cryptocœloma*, Miers.

The species are all of small size and are almost exclusively Oriental or Indo-Pacific forms.

Carcinoplacinæ.

Carcinoplacinæ, Milne Edwards, Ann. d. Sci. Nat., ser. 3, Zool. xviii. p. 164, 1852.

Geryon, Kröyer.

Geryon, Kröyer, Nat. Hist. Tidskr., ser. I, vol. I. p. 20, 1837.
? *Chalæpus*, Gerstaecker, Archiv f. Naturgesch., vol. xxii. p. 118, 1856.

Carapace cancroid in form, moderately convex, slightly broader than long ; the antero-lateral margins shorter than the postero-lateral margins, and armed normally with three spines, including the spine or tooth at the exterior angle of the orbit, which is sometimes less developed ; the intermediate spines (the second and fourth of the normal series) are sometimes present but rudimentary. Front slightly deflexed, of moderate width, and divided into two or four lobes or teeth. Epistoma short, transverse. The ridges of the endostome or palate are faintly indicated or obsolete (in the species I have examined). The post-abdomen in the male is distinctly seven-jointed, and covers the whole width of the sternum at the base between the bases of the fifth ambulatory legs. The eye-peduncles are short and robust. The antennules are transversely plicated. The basal antennal joint is slender, and occupies the interior hiatus of the orbit, and its distal extremity is free, not united to the front. The merus of the exterior maxillipedes is shorter than the ischium, truncated or slightly rounded at the distal extremity, and the next joint is articulated at its antero-internal angle. The chelipedes (in the male) are subequal and moderately robust ; merus trigonous ; carpus with a spine or tooth on its interior surface ; palm short, rounded above and below ; fingers distally acute and denticulated on the inner margins. The ambulatory legs are somewhat elongated, with the joints subcylindrical ; dactyli slender, styliform, but slightly compressed and not ciliated.

The genus *Geryon* is very nearly allied both to *Pseudorhombila* and *Pilumnoplax*, and to the Cancroid genus *Galene ;* it is distinguished from them by the considerable development of the lateral marginal spines of the carapace, and from *Pseudorhombila*, as figured by Milne Edwards, by the more slender basal antennal joint, which does not reach the front.

The species, which occur at considerable depths in the North Atlantic, may attain a large size. The following have been described :—

　　Geryon tridens, Kröyer.　Danish and Scandinavian Coasts; off Valentia, Ireland
　　　　(80 to 808 fathoms).
　　Geryon quinquedens, Smith.　Nova Scotia and East Coast of the United States
　　　　(to 740 fathoms).
　　Geryon longipes, A. Milne Edwards.　Mediterranean and North Spanish Coasts
　　　　(to 700 mètres).
　　Geryon (?) incertus, n. sp.　Off the Bermudas (435 fathoms, Station 33).[1]

Geryon (?) incertus, n. sp. (Pl. XVI. fig. 3).

This species is represented in the collection by the carapace of a single specimen (probably immature), since it is of very small size, and as the fifth ambulatory legs are deficient, I am somewhat uncertain as to its generic identification. It may belong to a genus of the Portunidæ, near to *Bathynectes*, where I originally placed it.

The carapace is little broader than long; its surface is very uneven and marked with transverse granulated ridges, of which there are two on the gastric region, placed one on either side of the median line, and one (a continuous ridge) crossing the

[1] On account of the form of the basal antennal joint, which is said to resemble that of *Galene*, de Haan, the genus *Chalæpus*, Gerstæcker, is identified with *Geryon*, rather than with *Pseudorhombila*, but the identification is somewhat uncertain; the typical species *Chalæpus tripinosus* (Herbst), is said to occur in the East Indies.

middle of the carapace and originating on each side from the base of the posterior antero-lateral marginal tooth, from behind which, also, a granulated ridge extends along the postero-lateral margins of the carapace nearly to the postero-lateral angles, which are rounded, not spiniform, and not prominent. The orbit has two small incisions in its superior margin. The antero-lateral margins are shorter than the postero-lateral margins and are armed with four teeth ; of which the first, second, and fourth are prominent and spiniform, the third is almost obsolete on one side and exists on the other merely as a small tubercle. The front is rather prominent, about one-third the width of the carapace, and divided by a small median notch into two broad and slightly sinuated lobes, which are rounded at the lateral angles. The post-abdomen (in the female) is sub-triangulate, with the segments short, the fourth to the sixth apparently partly coalescent, the last segment rounded at the distal extremity. The antennules are transversely plicated ; the basal joint very large. The basal antennal joint is slender, and longer than the two following peduncular joints, but it is free and not conjoined with the infero-lateral process of the front, as in that genus ; the flagellum is somewhat elongated. The ischium of the exterior maxillipedes is longer than broad, with the inner margin convexly arcuated ; the merus is almost as long as broad (and much shorter than the ischium), it is subtruncated at its distal extremity, with the antero-external angle rounded and not at all prominent, and is scarcely at all emarginate at the antero-internal angle, where the next joint is articulated with it. No limbs remain attached to this specimen, but two ambulatory legs, which occurred in the same tube, and probably belong to it, are moderately elongated and very slender, with the joints smooth ; the dactylus styliform, nearly straight, elongated, slightly longer than the penultimate joint (see figs. 3b, 3c). Colour (in spirit) whitish.

Young ? .		Lines.	Millims.
Length and breadth of carapace,	.	2½	5
Length of ambulatory leg, about	.	5½	11·5

Off the Bermudas, in 435 fathoms, in lat. 32° 21′ 30″ N., long. 64° 35′ 55″ W., Station 33 (a young female).

From all the described species of this genus, *Geryon incertus* is distinguished by the form of the front, and from the typical species of *Bathynectes*, Stimpson, not only by this character, but also by the structure of the basal antennal joint, which is free and not united with the lateral subfrontal process.

Pilumnoplax, Stimpson.

Pilumnoplax, Stimpson, Proc. Acad. Nat. Sci. Philad., p. 93, 1858.

Carapace moderately transverse, and longitudinally slightly convex, or rather depressed ; the antero-lateral margins are very short, much shorter than the postero-lateral margins, and are armed with three or four teeth, which are rarely as spinuliform as in *Pilumnus*, to which this genus is somewhat nearly allied. The front is of moderate width, slightly deflexed, and distally truncated, and often has a small median notch in the anterior margin. The orbits are of moderate size, and their margins are sometimes notched, but not spinuliferous. The epistoma is very narrow and transverse. The ridges of the endostome or palate are very obscurely defined. The post-abdomen in the male is distinctly seven-jointed, and is broadest at the base, where it covers the whole, or nearly the whole, width of the sternum, between the bases of the fifth ambulatory legs. The eyes are set upon rather short, stout pedicels. The antennules are transversely or somewhat obliquely plicated. The basal antennal joint is short and rather slender, and usually does not reach the infero-lateral process of the front: the flagellum arises from within the interior orbital hiatus. The exterior maxillipedes are closely applied to the buccal cavity; their ischium and merus-joints are distally truncated, the merus shorter than the ischium, and bearing the next joint at its antero-internal angle, which is very slightly, if at all, emarginated. The chelipedes are either subequal or unequal in the male, and if unequal, the palms may be dissimilarly tuberculated; the merus-joint of the larger chelipede is trigonous, and prolonged little, if at all, beyond the antero-lateral margins of the carapace; the carpus has a spine on its antero-internal margin ; the palm is compressed, rounded above, and tuberculated or smooth on the exterior surface; the fingers are distally acute and denticulated on the inner margins. Ambulatory legs moderately elongated, with the joints slender and unarmed; the dactyli usually styliform, but slightly compressed, and ciliated on the margins.

The species are all of small size, and inhabit the Chinese, Japanese, and Australian Seas, in water of moderate depth. *Pilumnoplax heterochir*, Studer, is a deep-water species occurring in the Challenger collection, near the Cape of Good Hope, on the Agulhas Bank and at Nightingale Island (Tristan da Cunha group), in 100 to 150 fathoms; and I have described a variety (*atlantica*) of the Oriental *Pilumnoplax sulcatifrons*, Stimpson, from Senegambia (Goree Island). In the Challenger collection is another deep-water species, *Pilumnoplax abyssicola*, n. sp., from the Fiji Islands (315 fathoms).

The following is, I believe, a complete list of the species which have been assigned to this genus, but not improbably some others which have been referred to allied genera may hereafter be included in it :—

Pilumnoplax sulcatifrons, Stimpson. Hong Kong ; Port Molle, North-East Australia (var. *australiensis*, Miers), and Goree Island, Senegambia (var. *atlantica*, Miers).

Pilumnoplax longipes, Stimpson. Oosima, Japan.

Pilumnoplax sculpta, Stimpson. Oosima, Japan.

Pilumnoplax ciliata, Stimpson. Limoda, Japan.

Pilumnoplax vestita (de Haan). Japan; North and North-East Australia (var. *sexdentata*, Haswell).

Pilumnoplax heterochir (Studer). Off Cape of Good Hope and Agulhas Bank (to 150 fathoms); Nightingale Island, Tristan da Cunha (100 fathoms).

Pilumnoplax abyssicola, n. sp. Fiji Islands (315 fathoms).

The nearest ally to this genus is perhaps *Pseudorhombila*, Milne Edwards, of which I formerly regarded *Pilumnoplax* a subgenus,[1] but the species of *Pilumnoplax* may be distinguished by their much smaller size, and by the narrower basal antennal joint and compressed and ciliated dactyli of the ambulatory legs. From the species of *Pilumnus*, and, I think, *Eucrate*, de Haan, *Pilumnoplax* is distinguished by the less convex carapace with shorter antero-lateral margins, and more slender, longer ambulatory legs, not to mention other more important but generally less constant characters.[2]

Pilumnoplax heterochir (Studer) (Pl. XIX. fig. 1).

Pilumnoplax heterochir, Studer, Abhandl. d. k. Akad. d. Wiss. Berlin, Abh. ii. p. 11, pl. i. fig. 3, 1882.

Pseudorhombila (*Pilumnoplax*) *zornianae*, Miers, Narr. Chall. Exp., vol. i. pt. ii. p. 587, 1880.

The carapace is little broader than long, somewhat depressed, and granulated near the front and antero-lateral margins, and with short, obscurely defined, transverse ridges on the gastric region and on the front of the branchial regions. The antero-lateral margins are much shorter than the postero-lateral margins, which are straight and converge to the postero-lateral angles, and the former are divided into three teeth, of which the first is broad and obtuse, and the second and third dentiform and acute; behind these, on the postero-lateral margins, there is usually a small tuberculiform rudiment of a fourth tooth. The interorbital frontal carina is entire or slightly notched in the middle line and granulated; the frontal margin is divided into two lobes by a shallow, triangulate median notch. The orbit has two small notches in its superior margin, but none at its exterior angle or on its inferior margin. The third segment of the post-abdomen in the male is the broadest, and laterally angulated; the sixth segment is slightly longer than the fifth, the seventh segment slightly transverse and distally rounded. The eyes are of moderate size. The basal antennal joint is short and slender, and does not attain the lateral subfrontal process. The merus of the outer maxillipedes is shorter than the ischium-joint, distally truncated, with a rather prominent antero-external angle. The chelipedes (in the male) are unequal, the right usually the larger; the

[1] Crust. in Rep. Zool. Coll. H.M.S. "Alert," p. 241, 1884.

[2] As has been noted under *Pilumnus*, the European species described by Maitland as *Pilumnus tridentatus*, may belong to *Pilumnoplax* or to *Heteroplax*, which latter genus is, according to Dr. Stimpson's diagnosis, separated from *Pilumnoplax* by characters of scarcely more than subgeneric value. Besides the species mentioned by Dr. Stimpson, I have described one, *Heteroplax* (?) *nitidus*, from the Corean Seas.

merus and carpus-joints are granulated, and the carpus has usually two spiniform teeth on its inner margin; the palm (in the larger chelipede) is robust and somewhat turgid, granulated near the base, and elsewhere smooth; the fingers dentated on the inner margins and rather shorter than the palm; in the smaller chelipede both wrist and palm are externally covered with numerous, crowded, acute granules, and the fingers are relatively longer; the ambulatory legs are slender, with the superior margins of the fourth to the sixth joints granulated; dactyli styliform, slightly hairy, and terminating in a short claw. The male verges are inserted very near the bases of the fifth legs. Colour (in spirit) yellowish-brown; fingers black or brownish.

Adult ♂.				Lines.	Millims.
Length of carapace, about	.	.	.	3	6·5
Breadth of carapace, nearly	.	.	.	4	8

Off Nightingale Island, Tristan da Cunha Group, October 17, 1873, in 100 fathoms (a good series of specimens); Agulhas Bank, off Cape Agulhas, South Africa (Station 142), in lat. 35° 4' 0" S., long. 18° 37' 0" E., in 150 fathoms (several specimens).

This species is distinguished from all its congeners by the dissimilar development and tuberculation of the right and left chelipedes.[1]

Pilumnoplax abyssicola, n. sp. (Pl. XIX. fig. 2).

The carapace is everywhere closely granulated, nearly glabrous, with a scanty pubescence near the margins. The front is about one-third the greatest width of the carapace, its anterior margin is straight, entire, without a median notch, and is not transversely sulcated; the antero-lateral margins of the carapace are shorter than the postero-lateral, and armed with three teeth behind the exterior angle of the orbit, which is not at all prominent; the first tooth is very small (on one side scarcely discernible), the second and third spiniform and acute; the postero-lateral margins are straight, and converge to the postero-lateral angles of the carapace; the orbital margins are entire; the inner subocular lobe is small and not prominent. The epistoma is very short and transverse; the pterygostomian regions are finely granulated and somewhat pubescent; the post-abdomen of the male is triangulate, broad at the base, where it covers the whole of the sternal surface, its segments are distinct and short, the terminal segment subtriangulate. The eyes are borne on short, thick pedicels. The basal (or real second) joint of the peduncles of the antennæ is rather slender, and does not reach the subfrontal process. The merus of the outer maxillipedes is nearly quadrate, distally truncated, and not produced at the antero-external angle; the narrow straight exognath does not quite reach

[1] The foregoing description was drawn up, and the figure was outlined, before I had identified the Challenger specimens with Dr. Studer's species, and both may still be useful for the identification of this remarkable deep-water form, which in the preliminary account of the Brachyura published in the Narrative of the Challenger Expedition I had designated *Pseudorhombila (Pilumnoplax) normanni*.

the antero-external angle of the merus-joint. The chelipedes are moderately robust, with
the joints very finely granulated; the merus short, trigonous, with an obscure tooth on
its upper margin, near the distal extremity; carpus with a spine on its inner margin;
palm without spines or tubercles, finely granulated, rounded above the fingers, distally
acute and denticulated on the inner margins. The ambulatory legs are slender and
somewhat elongated, with the joints slightly compressed, but not dilated, pubescent, and
clothed on the margins with some longer hairs; the dactyli are slender, styliform, and
about as long as the penultimate joints. Colour (in spirit) yellowish-white; fingers of
the chelipedes chocolate-brown.

♂.							Lines.	Millims.
Length of carapace, about	.	.	.	.	.	.	4	9
Breadth of carapace, about	.	.	.	.	.	.	4½	10
Length of a chelipede,	.	.	.	.	.	.	6½	13·5
Length of second ambulatory leg,	.	.	.	.	.	.	9	19

Off Matuku, Fiji Islands, in 315 fathoms (Station 173), in lat. 19° 9′ 35″ S., long.
179° 41′ 50″ E. (a male, perhaps not fully grown).

The bases of the male verges (external genital appendages) originate very near the
bases of the fifth ambulatory legs.

Pilumnoplax vestita, var. *sexdentata* (Haswell).

> ? *Cancer* (*Curtonotus*) *vestitus*, de Haan, Crust. in v. Siebold, Fauna Japonica, p. 51, pl. v.
>   fig. 3 ♀, 1835.
> ? *Carcinoplax vestitus*, Milne Edwards, Ann. d. Sci. Nat., ser. 3, Zool., vol. xviii. p. 164, 1852.
> ? *Eucrate sexdentata*, Haswell, Cat. Australian Crust., p. 86, 1882.
>   *Pseudorhombila vestita*, var. *sexdentata*, Miers, Crust. in Rep. Zool. Coll. H.M.S. "Alert,"
>   p. 240, 1884.

Japan, off Yokoska, 10 fathoms (a female, not fully adult?); Japanese Seas, 15 fathoms,
in lat. 34° 18′ 0″ N., long. 133° 35′ 0″ E., Station 233b (a young female).

The specimens referred to this species resemble de Haan's description of the male in
that the pubescence of the chelæ covers only the upper part of the palm and the base of
the dactylus; the spines of the antero-lateral margins are more strongly defined than in
his figure of the female; the small spine of the outer surface of the wrist is obsolete, and
the ambulatory legs (of the first three pairs especially) less hairy. It may be, therefore,
that the Challenger specimens belong to a distinct but nearly-allied species.[1] The largest
specimen has the following dimensions :—

♀.							Lines.	Millims.
Length of carapace,	.	.	.	.	.	.	4½	9·5
Breadth of carapace,	.	.	.	.	.	.	5½	11·5

[1] Whether or not this species be identical with the very briefly-described *Eucrate sexdentata*, Haswell, must remain
uncertain. Haswell's types were from the North-Eastern Coast of Australia (Port Denison, 20 fathoms). The Challenger
specimens certainly belong to the same species as those from the Arafura Sea, referred to in my Report on the Crustacea
of H.M.S. "Alert," as *Pseudorhombila vestita*, var. *sexdentata* (Haswell).

Bathyplax, A. Milne Edwards.

Bathyplax, A. Milne Edwards, Bull. Mus. Comp. Zoöl., vol. viii., No. 1, p. 16, 1880.

In this genus the carapace is slightly transverse, longitudinally convex and finely granulated on the dorsal surface; the branchio-cardiac sutures are distinct. The antero-lateral margins are arcuated and armed with two spines (the hepatic and lateral epibranchial spines). The front is deflexed, and is more than one-third the width of the carapace; its straight anterior margin projects over the bases of the antennules. The orbits are very small and rudimentary. The epistoma is very narrow and transverse. The longitudinal ridges of the endostome or palate are rather indistinctly defined in the specimens I have examined. The post-abdomen of the male is short, distinctly seven-jointed, and occupies at the base the whole width of the sternum between the bases of the fifth ambulatory legs. The eye-peduncles are very short and nearly immobile, and the corneæ in the typical form are not developed. The flagella of the antennules are elongated and transversely plicated. The basal antennal joint is much larger than the next joint, and reaches the infero-lateral process of the front; the elongated flagella arise from within the interior hiatus of the orbits. The exterior maxillipedes are short, and their exognath is rather broad; the ischium of the endognath is not produced at its antero-internal angle; the merus is distally truncated; its antero-external angle is not produced, and its antero-internal angle, where the next joint articulates, is emarginated. The chelipedes are dissimilar and of moderate length; the merus is short, trigonous, with a spine on its superior margin, and it has on the inner surface a transverse stridu-lating ridge near the distal extremity; the carpus has a spine or tubercle on its inner surface; the palms are short and compressed, and the left palm (but not the right) has lobe or tubercle on the inner surface, near the superior margin; the dactyli are com-pressed, dentated on the inner margins, and distally acute. The ambulatory legs are slender and somewhat elongated; their dactyli styliform and straight.

Of the single species described there is in the Challenger collection an interesting variety which I have designated—

Bathyplax typhlus, var. oculiferus, nov. (Pl. XX. fig. 3).

cf. Bathyplax typhlus, A. Milne Edwards, Bull. Mus. Comp. Zoöl., vol. viii., No. 1, p. 16, 1880.

South of Pernambuco, off the coast of Brazil, in 30 to 400 fathoms (Stations 122 to 122c), lat. 9° 5′ 0″ to 9° 10′ 0″ S., long. 34° 49′ 0″ to 34° 53′ 0″ W. (An adult female bearing ova.)

The dimensions of this specimen are as follows :—

Adult ♀.							Lines.	Millims.
Length of carapace,	.	.	.	.	.	.	8	17
Greatest breadth of carapace,	.	.	.	.	.	.	9	19
Length of right chelipede,	.	.	.	.	.	.	15½	33
Length of second ambulatory leg,	.	.	.	.	.	17½	37

This variety differs from the type of the species, as described by Milne Edwards, in one character only, and that which constituted the most distinctive peculiarity of the type, i.e., in having the small ocular peduncles provided with distinct, small, terminal cornea. I may add, that the ambulatory legs are not only hispid with short hairs, but also fringed with longer hairs. In all other characters, as, e.g., in its being furnished with a stridulating ridge at the distal extremity of the merus-joint of the chelipedes, and in the curious dissimilarity of the right and left chelæ, this specimen agrees with the typical form of the species.[1]

Litocheira, Kinahan.

Litocheira, Kinahan, Journ. Roy. Dublin Soc., vol. i. p. 121, 1858.
? Brachygrapsus, Kingsley, Proc. Acad. Nat. Sci. Philad., p. 203, 1880.

Carapace broader than long, somewhat quadrilateral, with the sides nearly straight ; the antero-lateral margins armed with a tooth or spine behind the exterior angle of the orbit. The front is straight or slightly arcuated, and (in the species I have examined) it is rather broad, usually exceeding half the width of the carapace, and the orbital margins are entire. The epistoma is transverse. The ridges of the endostome are distinctly developed (in the species I have examined). The post-abdomen in the male and the basal segments cover the whole width of the sternum, between the bases of the fifth ambulatory legs. The eye-peduncles are robust and of moderate length, the cornea large. The basal joint of the antennæ is slender and rather longer than the following joint, and usually does not reach the infero-lateral process of the front ; the antennal flagellum is moderately elongated. The exterior maxillipedes meet, or nearly meet, along their inner margins; their ischium-joints are not produced at the antero-internal angles; the merus-joints are distally truncated, and the antero-lateral angles (where the next joint articulates) are slightly emarginate ; the antero-external angles not greatly produced. The chelipedes in the adult male are subequal and of moderate length, with the merus-joints

[1] It is worthy of note that the specimens described by Milne Edwards from the collections obtained in the Expedition of the U.S.S. "Blake," under the superintendence of Professor A. Agassiz, in 1877 to 1879, were dredged at Frederickstadt and Santa Lucia at a much greater depth (483 to 451 fathoms). The Rev. A. M. Norman (in Wyville-Thomson, Depths of the Sea, p. 176), mentions a somewhat analogous modification of the ocular peduncles in Ethusa granulata, where the eyes are smooth and rounded in specimens dredged in 110 to 370 fathoms, but are firmly fixed in their sockets, and assume the functions of a rostrum, in the specimens (of more northerly habitat) dredged in 542 and 705 fathoms.

trigonous, palms somewhat inflated and rounded above, and rounded or subcarinated below, fingers distally acute and denticulated on the inner margins. The ambulatory legs are rather slender and of moderate length ; dactyli styliform or lanceolate.

The following species are to be referred to this genus :—

> *Litocheira bispinosa*, Kinahan.   Port Philip (15 fathoms); Bass Strait; Port Curtis; King George's Sound (Coll. Brit. Mus.).
> *Litocheira* (?) *lævis* (= *Brachygrapsus lævis*, Kingsley).   New Zealand.
> *Litocheira kingsleyi*, n. sp.   Agulhas Bank, 150 fathoms (Station 142).

Litocheira kingsleyi, Miers (Pl. XXI. fig. 1).

Brachygrapsus kingsleyi, Miers, Narr. Chall. Exp., vol. i. pt. ii. p. 587, 1880.

The body and limbs are rather thinly pubescent. Carapace quadrate, very slightly broader than long, with the sides nearly straight, longitudinally it is slightly convex and arcuate, it has a prominent spiniform tooth at the outer angle of the orbit, and a second at a short distance behind it on the lateral margin. The front is nearly half the width of the carapace, curves slightly downwards, and is divided by a median notch into two rounded lobes, external to which, at the inner angle of the orbit, is an inconspicuous tooth. The margins of the rather large orbits are granulated, without fissures. The epistoma is transversely linear. The post-abdomen of the male is rather narrow, with the sides slightly convergent, as in *Nautilograpsus*, the segments are thinly pubescent, the third to the sixth segments coalescent, the terminal segment small and subtriangulate ; the ocular peduncles are moderately robust and hairy, with the corneæ terminal. The antennules are transversely plicated. The basal antennal joint is short, and does not quite reach the infero-lateral process of the front, the next joint more slender and slightly longer, the terminal peduncular joint short, the flagellum about fourteen-jointed. The exterior maxillipedes are formed nearly as in *Trapezia*; ischium longer than broad, with an obscure obliquely longitudinal sulcus on the outer surface; merus small, distally truncated, with the antero-external angle rounded and very slightly prominent, and bearing the next joint at the antero-internal angle ; exognath very slender, narrow, and distally acute.   Chelipedes in the male moderately developed ; merus very finely denticulated on its anterior margin; carpus with a small tooth on its inner margin, anterior to which the margin is granulated ; palm slightly inflated, and obscurely granulated on its outer surface and more distinctly on its upper margin; fingers about as long as the palm, denticulated on the inner margins, slightly incurved, and distally acute ; ambulatory legs rather elongated, with the merus-joints dilated and compressed, and armed with a spiniform tooth near to the distal extremity of the upper margin; the following joints slender and very hairy ; dactyli armed below with a series of short spines, and with a small terminal claw.   Colour (in spirit) reddish-brown.

The largest male presents the following dimensions :—

Adult ♂.						Lines.	Millims.
Length of carapace, rather over,	.		.		.	5	11
Breadth of carapace, nearly	.	.	.		.	6	12
Length of a chelipede,	.	.	.		.	8½	18
Length of second ambulatory leg, nearly,			.		.	12	25

The smallest ova-bearing female measures as follows :—

Adult ♀.				Lines.	Millims.
Length and breadth of carapace, nearly,	.	.	.	4½	9
Length of a chelipede, about	.	.	.	.	6	13
Length of second ambulatory leg,	.	.	.	.	8½	18

A good series of specimens (mostly females) were dredged on the Agulhas Bank, south of Cape Agulhas, South Africa, in lat. 35° 4′ 0″ S., long. 18° 37′ 0″ E., in 150 fathoms (Station 142).

In this species the male verges are exserted near the margins of the sternum and the bases of the fifth ambulatory legs.

The pubescent body and limbs and the form of the front distinguish *Litocheira kingsleyi* from the other species referred to this genus.

Rhizopinæ.

Rhizopinæ, Stimpson, Proc. Acad. Nat. Sci. Philad., p. 95, 1858.

This subfamily was established by Dr. Stimpson for certain small genera which resemble the Carcinoplacidæ in the form of the carapace, whose antero-lateral margins are arcuated, and in the characters drawn from the front, orbits and outer maxillipedes, but the antero-lateral margins of the carapace are usually entire, rarely dentated, and the post-abdomen of the male does not cover the whole width of the sternum at the base (except perhaps sometimes in *Typhlocarcinus*).

The characters distinguishing this group from the Carcinoplacidæ are not invariably constant,[1] but the genera referred to it are usually at once recognisable by their small size, small orbits, small and deflexed front, and by the arcuated, subentire, antero-lateral margins and subparallel postero-lateral margins of the carapace.

Ceratoplax, Stimpson.

Ceratoplax, Stimpson, Proc. Acad. Nat. Sci. Philad., p. 96, 1858.

Carapace transverse, longitudinally convex and nearly smooth on the dorsal surface, with the antero-lateral margins arcuated and entire or subentire ; the postero-lateral

[1] I have myself described a species of Carcinoplax (*Carcinoplax integra*) with entire antero-lateral margins.

margins nearly parallel. Front deflexed and rather narrow, with the anterior margin slightly arcuated and entire, or with an obscure median notch. Epistoma transverse. The ridges of the endostome are obsolete or imperfectly defined. The post-abdomen of the male is distinctly seven-jointed and it does not occupy nearly the whole width of the sternum at the base. The orbits are rather large, normally excavated, their superior margins are not (as in *Scalopidia*) nearly continuous with the front and antero-lateral margins; the eyes have the corneæ normally developed, the margins acute, ciliated. The antennules are transversely plicated. The basal joint of the antennæ is rectangular and usually does not reach the front; the flagellum is somewhat elongated. The exterior maxillipedes, when closed, have no interspace between them; the ischium-joint is longitudinally sulcated and is not produced at its antero-internal angle; the merus is truncated, shorter than the ischium, and is considerably produced at its antero-external angle; the following joint is articulated at the antero-internal angle of the merus, which angle is usually slightly emarginated. The chelipedes are subequal and of moderate size; merus trigonous and short; carpus with a spine on its interior margin; palm short and compressed, fingers distally acute. The ambulatory legs are moderately elongated, with the joints slender and unarmed; dactyli styliform.

From *Scalopidia* this genus is at once distinguished by the normally excavated orbits, and from *Typhlocarcinus* by the larger orbits, the form of the eye-peduncles and of the merus of the exterior maxillipedes, whose merus-joint is produced at the antero-external angle. It is much more nearly allied to *Rhizopa*, but, if I have rightly identified specimens in the British Museum collection with this genus, the form of the merus of the exterior maxillipedes will also, perhaps, suffice to distinguish it generically.

The following species have been referred to *Ceratoplax:*—

> *Ceratoplax ciliata*, Stimpson.   North China Sea (20 fathoms); Torres Strait.
> *Ceratoplax arcuata*, Miers.   North Australia, Port Darwin (12 fathoms); South of New Guinea (28 fathoms).
> *Ceratoplax* (?) *lævis*, Miers.   Arafura Sea (32 to 36 fathoms).

Ceratoplax ciliata, Stimpson (Pl. XIX. fig. 3).

Ceratoplax ciliata, Stimpson, Proc. Acad. Nat. Sci. Philad., p. 96, 1858.

Torres Strait, August 1874 (a female). This specimen has the dorsal surface of the carapace clothed with a few hairs; the ambulatory legs, although compressed, are rather slender.

Adult ♀.		Lines.	Millims.
Length of carapace,	· ·	3	6·5
Breadth of carapace,	· ·	4	8·5

It is unfortunately in very imperfect condition, having lost the right chelipede and several of the ambulatory legs, which are rather more slender than in the description of Dr. Stimpson, and the front is subentire, not distinctly notched.

Ceratoplax arcuata, Miers.

Ceratoplax arcuata, Miers, Crust. in Rep. Zool. Coll. H.M.S. "Alert," p. 243, pl. xxv. fig. B, 1884.

A female specimen dredged south of New Guinea, in 28 fathoms, in lat. 9° 59′ 0″ S., long. 139° 42′ 0″ E. (Station 188) is referred, but doubtfully, to this species. This example is of much larger size than the small male described in the Report cited; the whole animal is more pubescent and the carapace proportionately broader; the subdistal tooth or prominence on the upper margin of the merus of the chelipedes, which is very obscurely indicated in the original type, is more distinctly developed.[1] It has the following dimensions :—

Adult ♀.	Lines.	Millims.
Length of carapace, nearly	4½	9·6
Breadth of carapace, .	5¼	11·5

Notonyx, A. Milne Edwards.

Notonyx, A. Milne Edwards, Nouv. Archiv. Mus. Hist. Nat., vol. ix. p. 268, 1873.

Carapace nearly quadrilateral, with the antero-lateral angles rounded, subcristated, and the lateral margins straight, entire, and subparallel; longitudinally it is slightly convex, and the dorsal surface smooth and polished. Front deflexed, about one-third the width of the carapace, with the anterior margin straight and entire. The orbits, antennæ and post-abdomen are nearly as in *Ceratoplax*. The eye-peduncles have the corneæ normally developed. The exterior maxillipedes are nearly as in *Ceratoplax*, but the merus-joint, in the specimens I have examined, is subquadrilateral, distally truncated, and is not produced at its antero-external angle. The chelipedes are subequal and moderately developed; merus trigonous; carpus with a tubercle or prominence, not a spine, on its interior surface; palm short and compressed, cristated below; fingers distally acute. The ambulatory legs, as in *Ceratoplax*, are slender and moderately elongated, with the joints unarmed; dactyli nearly straight.

[1] I very much doubt the generic distinctness of *Rhizopa gracilipes*, Stimpson (from Hong-Kong), from this species. In specimens, probably from the Chinese Seas, referred doubtfully to *Rhizopa gracilipes* in the British (Natural History) Museum, the ocular corneæ are minute and inferior as in *Ceratoplax*, but the merus of the exterior maxillipedes is not produced at its antero-external angle; the basal antennal joint is more robust and quadrate. In the fully grown specimen the frontal margin is entire, and the palms of the chelipedes are cristate and externally glabrous, as in Stimpson's description.

The single species, *Notonyx nitidus*, which has been referred to this genus, occurs, but rarely, at New Caledonia and the Fiji Islands, and has been taken by the Challenger off the south coast of New Guinea in 28 fathoms.

Notonyx is very nearly allied to *Ceratoplax*, and were it not for a slight difference in the form of the merus of the exterior maxillipedes, could, I think, hardly be distinguished from that genus.

Notonyx nitidus, A. Milne Edwards.

> *Notonyx nitidus*, A. Milne Edwards, Nouv. Archiv. Mus. Hist. Nat., vol. ix. p. 269, pl. xli. fig. 3, 1873.

South of New Guinea, 28 fathoms, in lat. 9° 59′ S., long. 139° 42′ E. (Station 188). An adult but small male and an adult female.

	Lines.	Millims.
Adult ♀.		
Length of carapace, rather over	3	7
Breadth of carapace, rather over	4	9

In these specimens, the slender basal antennal joint does not quite reach the subfrontal process; the merus of the exterior maxillipedes, as in the description of Milne Edwards, is not produced at the antero-external angle, and herein differs from the figure cited, where it is shown as slightly produced.

Subfamily 2. OCYPODINÆ.

> *Ocypodiens*, Milne Edwards, Hist. Nat. Crust. vol. ii. p. 39, 1837.
> *Gonoplaciens* (pt.), Milne Edwards, tom. cit., p. 56, 1837.

Carapace transverse, trapezoidal or quadrate, with the antero-lateral angles frequently produced and acute, and the lateral margins straight, entire or incised; front narrow or of moderate width. Orbits usually very considerably developed and occupying nearly the whole or a great part of the anterior face of the body below the front. The fifth joint of the exterior maxillipedes usually articulates at the antero-external angle of the merus. Chelipedes in the adult male subequal, or sometimes very unequal. Ambulatory legs moderately elongated, with the dactyli styliform, compressed, not spinose.

I divide the genera of this group into three sections, as follows:—

1. Ocypodinæ.

Carapace trapezoidal with the antero-lateral margins not greatly produced, and entire; the fifth joint of the endognath of the exterior maxillipedes articulated at its apex, or at its antero-external angle ; the post-abdomen in the male usually much narrower than the sternum at the base.

Genera:—*Ocypoda*, Fabricius; *Gelasimus*, Latreille; *Acanthoplax*, Milne Edwards (perhaps to be regarded as a subgenus of *Gelasimus*); *Cleistostoma*, de Haan ; *Helœcius*, Dana.

2. Gonoplacinæ.

Gonoplaces cancéroïdes, Milne Edwards (pt.), Ann. d. Sci. Nat., ser. 3, vol. xviii. p. 162, 1852.

Carapace trapezoidal, with the antero-lateral angles produced and spiniform ; the fifth joint of the endognath of the exterior maxillipedes articulated with the merus at its antero-internal angle ; post-abdomen in the male occupying the whole width of the sternum, between the bases of the fifth ambulatory legs.

Genera :—*Gonoplax*, Leach, and *Ommatocarcinus*, White.[1]

3. Macrophthalminæ.

Gonoplaces typus, Milne Edwards, Ann. d. Sci. Nat., ser. 3, Zool. xviii. p. 155, 1852.
Macrophthalminæ, Dana, U.S. Explor. Exped., vol. xiii., Crust., p. 312, 1852.

Carapace transverse and quadrate, depressed, and dentated behind the antero-lateral angles, which are not greatly produced. Front narrow or of moderate width. The post-abdomen in the male does not cover the whole width of the sternum at the base, between the coxal joints of the fifth ambulatory legs. The fifth joint of the endognath of the exterior maxillipedes is articulated with the merus at or near its antero-external angle.

Genera :—*Macrophthalmus*, Latreille; *Euplax*, Milne Edwards (= *Chænostoma*, Stimpson, subgenus); *Hemiplax*, Heller; *Rhaconotus*, Gerstæcker; *Ilyoplax*, Stimpson.

Ocypodinæ.

Ocypoda, Fabricius.

Ocypoda, Fabricius (pt.), Entom. Syst. Suppl. p. 347, 1798.
    „   Milne Edwards, Hist. Nat. Crust., vol. ii. p. 41, 1837; Ann. d. Sci. Nat., ser. 3, Zool. xviii. p. 141, 1852.
    „   Kingsley, Proc. Acad. Nat. Sci. Philad., p. 179, 1880.
    „   Miers, Ann. and Mag. Nat. Hist., ser. 5, vol. x. p. 376, 1882.

Carapace quadrate, moderately transverse and convex, with the dorsal surface closely granulated, the lateral margins entire, the cervical and cardiaco-branchial sutures in part

[1] The genus *Prionoplax*, Milne Edwards, which that author arranges with these genera, is, I think, better placed in the family Carcinoplacidæ ; it is intermediate in structure between *Pseudorhombila* and *Macrophthalmus*.

distinct, the antero-lateral angles moderately prominent. The front (as in *Macrophthal-mus* and *Gelasimus*) is very narrow and deflexed; the orbits are very large and open, and extend along the whole anterior face of the body, between the front and antero-lateral angles (which are not very prominent), and their inferior margins are usually divided by a hiatus or fissure. The ridges of the endostome are usually not developed. The post-abdomen in the male is narrow and distinctly seven-jointed, with the terminal segment small and triangulate. The eye-peduncles are very large, and are jointed near the base, the basal part is short, and the terminal portion is often prolonged at its distal extremity as a spine or tubercle; the corneæ, which are of great size, cover a great part of the inferior surface of the mobile portion of the eye-peduncles. The antennules are partially concealed by the front; the antennæ are very small, and are placed beneath the eye-peduncles in the narrow hiatus between the bases of the antennules and the interior subocular lobe of the orbit; their basal joints are very short, and the flagella scarcely exceed the peduncles in length. The exterior maxillipedes are closely applied to the buccal cavity; the ischium-joints are longer than the merus-joints, and are distally truncated; the merus-joints are longer than broad, distally truncated, not emarginated at the antero-internal angles, and the next joint is articulated at the antero-external angle of the merus. The chelipedes in the adult male are unequal and well developed; the merus-joint in the larger chelipede is trigonous, with the superior and inferior margins denticulated; carpus short, with usually a lobe or tooth on the inner margin; hand vertically deep and compressed, the palm with a stridulating ridge on its inner surface (except in *Ocypoda cordimana*), composed of a vertical series of short raised lines or tubercles; fingers either distally acute or truncated, and denticulated on the inner margins. The ambulatory legs are somewhat elongated, with the joints usually granulated and the dactyli styliform.

The species are found on the shores of nearly all the warmer temperate and tropical regions of the globe.

Ocypoda ceratophthalma (Pallas).

Cancer ceratophthalmus, Pallas, Spicilegia Zoologica, p. 83, pl. v. figs. 7–8, 1772.
Ocypoda ceratophthalma, Fabricius, Entom. Syst. Suppl., p. 347, 1798.
"   "   Milne Edwards, Hist. Nat. Crust., vol. ii. p. 48, 1834; Crust. in Cuvier, Règne Anim., ed. 3, Atlas, pl. xvii. fig. 1.
"   "   Kingsley, Proc. Acad. Nat. Sci. Philad., p. 179, 1880, and references to synonyma, except *Ocypoda ægyptiaca*.
"   "   Miers, Ann. and Mag. Nat. Hist., ser. 5, vol. x. p. 379, pl. xvii. fig. 1, 1882.

North Australia, Raine Island, August 1874 (an adult male); Fiji Islands, Kandavu (a series of specimens, male, female, and young).

The adult male from Raine Island measures as follows:—

	Adult ♂.	Lines.	Millims.
Length of carapace, about	. . .	17½	37·5
Breadth of carapace, nearly	. . .	20	42

In this specimen the terminal styles of the ocular peduncles are normally developed, and project considerably beyond the antero-lateral angles of the carapace, and the tubercles and striæ of the stridulating ridge of the larger chelipede are distinct and well defined.

In the larger specimens from Kandavu (length of carapace 12½ lines, 26 mm.) the terminal styles of the ocular peduncles are very short and tuberculiform, as in the variety of this species, designated by Milne Edwards *Ocypoda brevicornis*;[1] in the smaller specimens (length of carapace 9 lines, 19 mm., or under), the ocular styles are obsolete and the eyes distally rounded. An adult female from Kandavu, which is probably an abnormal variety of this species, has short terminal ocular styles, but scarcely any trace of a stridulating ridge on the inner margin of the palm of the larger chelipede, and the fingers of the smaller chelipede are slightly dilated at the distal extremity, in which character this specimen exhibits some approach to *Ocypoda macrocera*. There is also in the Challenger collection a series of specimens (not fully grown) from the Arrou Islands, which I suppose belong to *Ocypoda ceratophthalma* rather than to *Ocypoda kuhlii*, because (although the terminal styles of the ocular peduncles are not developed) the stridulating ridge of the chelipedes (in the larger specimens in the series) is coarsely striated above, finely striated below, and the antero-lateral angles of the carapace are but slightly prominent.

Young specimens from the Philippine Islands (Samboangan, 10 fathoms), may belong either to this species or to *Ocypoda kuhlii*; they have no trace of the terminal ocular styles, but the stridulating ridge of the chelipedes is striated rather than tuberculated, as in *Ocypoda kuhlii*.

A young male from Hilo, Sandwich Islands (beach) has no indications of ocular styles, but the stridulating ridge is coarsely and evenly striated, as in the adult examples from these islands referred to in my paper on the genus (*tom. cit.*, p. 380).

Small specimens of *Ocypoda* (too young to be assigned to any species with certainty) are in the Challenger collection, from the South Australian Coast, 2 to 10 fathoms (April 1874), from the beach at Botany Bay, and from Matuku, Fiji Islands, "freshwater" (July 24, 1874). The length of the carapace in the largest of these specimens (that from Matuku) does not exceed 4½ lines (9 mm.). None present any indications of terminal ocular styles. In the specimen from Matuku, and in one from the South Australian Coast, a striated stridulating ridge is very obscurely indicated, and these specimens may also presumably be young examples of *Ocypoda ceratophthalma*.

[1] Hist. Nat. Crust., vol. ii. p. 48, 1837.

Ocypoda cursor (Linné).

Cancer cursor, Linné (pt.), Syst. Nat., ed. xii. p. 1039, 1766.
Ocypoda ippeus, Olivier, Voy. dans l'Empire Ottoman, Atlas, pl. xxx. fig. 1.
    ,,   ,,   Audouin, Explic. des planches in Savigny, Descr. de l'Egypte, Atlas, pl. i. fig. 1.
    ,,   ,,   Milne Edwards, Hist. Nat. Crust., vol. ii. p. 47, 1837.
    ,,   ,,   Moseley, Notes by a Naturalist on the Challenger, pp. 48, 49, woodcut, 1879.
    ,,   *cursor*, de Haan, Crust. in v. Siebold, Fauna Japonica, p. 29, 1835.
    ,,   ,,   Milne Edwards, Ann. d. Sci. Nat., ser. 3, Zool. xviii. p. 142, 1852.
    ,,   ,,   Kingsley, Proc. Acad. Nat. Sci. Philad., p. 182, 1880.
    ,,   ,,   de Man, Notes Leyden Mus., p. 248, 1881.
    ,,   ,,   Miers, Ann. and Mag. Nat. Hist., ser. 5, vol. x. p. 376, 1882.
    ,,   ,,   Studer, Abhandl. d. k. Akad. d. Wiss. Berlin, Abh. ii. p. 13, 1882.

Cape Verde Islands, St. Vincent, July 1873 (a large series of specimens, adult and young); St. Jago, Cape Verde Islands, August 1873 (an adult male).

The largest male in the series from St. Vincent has the following dimensions :—

	Lines.	Millims.
Adult ♂.		
Length of carapace, .	17	36
Greatest breadth of carapace,	20½	43·5

In the adult examples of large size, the pencil of setæ borne at the end of the eye-peduncles, characteristic of this species, is long and thick ; in the smaller specimens, however, it is little developed, and in immature specimens of small size no trace of it exists, and the eye-peduncles are rounded at the distal extremity. The stridulating ridge of the chelipedes, which in adult examples of both sexes is finely and evenly striated throughout its length, is absent from the very smallest examples in the Challenger series.

Ocypoda arenaria (Catesby).

Cancer arenarius, Catesby, Hist. of the Carolinas, vol. ii. p. 35, pl. xxxv., 1731, 1771.
Ocypoda arenaria, Say, Journ. Acad. Nat. Sci. Philad., vol. i. p. 69, 1817.
    ,,   ,,   Milne Edwards, Hist. Nat. Crust., vol. ii. p. 44, pl. xix. figs. 13, 14, 1837.
    ,,   ,,   Kingsley, Proc. Acad. Nat. Sci. Philad., p. 184, 1880, and references to synonyma.
    ,,   ,,   Miers, Ann. and Mag. Nat. Hist., ser. 5, vol. x. p. 384, pl. xvii. fig. 7, 1882.
    ,,   *rhombea*, Milne Edwards, Hist. Nat. Crust., vol. ii. p. 46, 1837; Ann. d. Sci. Nat., ser. 3, Zool. xviii. p. 143, 1852.
    ,,   ,,   Dana, U.S. Explor. Exped., vol. xiii., Crust. I, p. 322, pl. xix. fig. 8, 1852.

Bermuda (an adult female), Bahia; shallow water (an adult male).

	Lines.	Millims.
Adult ♂.		
Length of carapace, about	14½	31
Breadth of carapace,	17½	37

Gelasimus, Latreille.

Gelasimus, Latreille, Nouv. Dict. d. Hist. Nat., ed. 2, vol. xviii. p. 286, 1880.

,, Milne Edwards, Hist. Nat. Crust., vol. ii. p. 49, 1837; Ann. d. Sci. Nat., ser. 3, vol. xviii. p. 144, 1852.

,, Kingsley, Proc. Acad. Nat. Sci. Philad., p. 133, 1880.

Carapace transverse, longitudinally convex, usually smooth on the dorsal surface, but sometimes granulated, with the cervical and cardiaco-branchial sutures usually more or less distinctly defined, with the antero-lateral angles usually prominent and acute, and the lateral margins nearly straight, and convergent to the posterior margin. The front is deflexed, usually very narrow, almost linear between the bases of the eye-peduncles, but sometimes much broader, and at the base nearly equalling one-third the width of the carapace at the anterior margin. The orbits are very large and, as in *Macrophthalmus*, extend along the whole anterior surface of the carapace, between the front and antero-lateral margins. The longitudinal ridges of the endostome are usually more or less distinctly developed. The post-abdomen in the male is narrow and distinctly seven-jointed, and its base does not occupy the whole width of the sternum between the bases of the ambulatory legs. The eye-peduncles (as in *Macrophthalmus*) are very slender and elongated, reaching, or nearly reaching, the antero-lateral angles of the carapace. The antennulary flagella are usually somewhat obliquely plicated. The basal joint of the antennæ is small, and placed beneath the bases of the eye-peduncles; the flagella are of moderate length. The ischium of the exterior maxillipedes is much larger than the merus, and is not produced at its antero-internal angle; the merus is small, usually transverse, distally truncated, and not emarginated at the antero-internal angle, and the following joint is articulated at the antero-external angle of the merus. The chelipedes are very unequally developed, either the right or left may be the larger in the same species; the merus in the larger chelipede is usually trigonous and prolonged beyond the antero-lateral angles of the carapace; the carpus is moderately elongated, and has usually no spine on its interior surface; the hand is compressed and enormously developed, usually greatly exceeding in length the three preceding joints; the palm is much shorter than the fingers, and is usually obliquely cristated on the inner surface; the fingers are distally acute or subacute, granulated, and usually lobated on the inner margins; in the smaller chelipede (and in both chelipedes in the female) the joints are slender and feeble. The ambulatory legs are of moderate length and present nothing remarkable; the merus-joints are compressed, and the dactyli styliform.

The described species are extremely numerous, and occur in all the warmer temperate, subtropical, and tropical regions of the globe. The two which are mentioned below have been described since the publication of Mr. Kingsley's list.

Gelasimus thomsoni, Kirk. New Zealand (Wellington).

Gelasimus cimatodus, Rochebrune. Senegambia.

The nearest ally to this genus is *Helœcius*, Dana, which is distinguished by the equal chelipedes and somewhat differently shaped post-abdomen of the male, which is broader at the base, &c.

The genus *Acanthoplax*, Milne Edwards, is united by Kingsley with *Gelasimus*, and cannot, I think, be regarded as more than a subgenus; it is distinguished, according to Milne Edwards, merely by having the branchial regions of the carapace armed with a marginal series of large spiniform tubercles.

Gelasimus vocans (Linné).

> †† *Cancer vocans*, Linné, Syst. Nat., ed. xii. p. 1041, 1766.
> *Gelasimus vocans*, Milne Edwards, Ann. d. Sci. Nat., ser. 3, Zool. xviii. p. 145, pl. iii. fig. 4, 1852, *nec* Hist. Nat. Crust.
> *Gelasimus nitidus*, Dana, U.S. Explor. Exped., vol. xiii., Crust., p. 316, pl. xix. fig. 5, 1852.

Fiji Islands, Kandavu (an adult male); Arrou Islands (an adult male).

The form thus designated is distinguished by the form of the larger chelipede in the male, whose palm is coarsely and strongly granulated externally, with the pollex or lower finger externally concave at the base, and bearing two strong triangular lobes, situated, one at about the middle of its inner margin, and one near to the distal extremity (besides a smaller sub-basal tooth on the inner margin, which is not always present). The figures cited give an excellent representation of this species.

Mr. Kingsley, in his recent Revision of the Gelasimi,[1] refers to this species as *Gelasimus cultrimanus*, Adams and White, but in the specimens designated by White *Gelasimus cultrimanus*, in the collection of the British Museum, the larger chelipede in the male is much more elongated, the proximal tooth of the two large triangular teeth of the pollex is always wanting, and the distal one is much less prominent and triangulate, and these specimens certainly belong to a distinct species or a very distinct and well marked variety. It may become necessary, for the reasons urged by Kingsley, to abandon the designation *Gelasimus vocans* for this species altogether.[2]

Adult ♂.				Lines.	Millims.
Length of carapace, rather over	.	.	.	6	13
Breadth of carapace,	.	.	.	9	19

---

[1] *Proc. Acad. Nat. Sci. Philad.*, p. 140, pl. ix. fig. 7, 1880.

[2] I retain, for the present, the Linnean name for this species, because it is so used by Milne Edwards in his later monograph of the group, and by other authors. If, however, the Linnean designation be properly referable to *Gelasimus tetragonon*, or to any other species, it will be necessary to use Dana's designation, *Gelasimus nitidus*, for the present form, since his figure (*tom. cit.*, pl. xix. fig. 5c) certainly belongs to it.

Gelasimus rubripes, Jacquinot and Lucas.

> *Gelasimus rubripes*, Jacquinot and Lucas, Crust. in Voy. au Pôle Sud. Zool., iii. p. 66, pl. vi. fig. 2, 1853.
>
> „ „ Milne Edwards, Ann. d. Sci. Nat., ser. 3, Zool., vol. xviii. p. 148, pl. iv. fig. 12, 1852.
>
> „ „ Kingsley, *tom. cit.*, p. 145, pl. x. fig. 17, 1880.

Philippines, Samboangan, 10 fathoms (an adult male). This specimen agrees well with the figure of Jacquinot and Lucas.

Specimens of this species, which have two triangular teeth on the inner margin of the palm, might at first sight be mistaken for *Gelasimus vocans*, but the series referred to *Gelasimus rubripes*, in the British Museum collection, may always be distinguished by having the dactylus granulated at and near the base and longitudinally sulcated on its outer surface.

Adult ♂.		Lines.	Millims.
Length of carapace, about	.	8½	18·5
Breadth of carapace, nearly	.	13	27·5

Gelasimus tetragonon (Herbst).

> ? *Cancer tetragonon*, Herbst, Naturgesch. der Krabben u. Krebse, Heft 1, p. 257, pl. xx. fig. 110, 1782, ♀.
>
> *Gelasimus tetragonon*, Rüppell, Beschreib. 24 Arten kurzschwänzigen Krabben d. rothen Meeres, p. 25, pl. v. fig. 5, 1830.
>
> „ „ Milne Edwards, Hist. Nat. Crust., vol. ii. p. 52, 1837; Ann. d. Sci. Nat., *tom. cit.*, p. 147, pl. iii. fig. 9, 1852.
>
> „ „ Kingsley, Proc. Acad. Nat. Sci. Philad., p. 143, pl. ix. fig. 11, 1880, where references to literature.

Two males were obtained at Tahiti, near the reefs, and three at Papiete, in the same island, on the shore and at the river mouths; also a male and two females at the Arrou Islands.

This species varies much in the coloration and in the tuberculation of the larger chelipede, whose fingers are granulated on the inner margins and bear ordinarily two or three larger tubercles, but these latter are sometimes absent from the inner margins of one or both fingers. The coloration in the Challenger spirit specimens is purplish-brown, with irregular pale markings; the chelipedes yellowish, with a large red patch at the base of the pollex, which extends sometimes over the whole outer surface of the palm.

Adult ♂.		Lines.	Millims.
Length of carapace,	.	10	21
Breadth of carapace, nearly	.	15	31·5

Gelasimus annulipes, Milne Edwards.

Gelasimus annulipes, Milne Edwards, Hist. Nat. Crust., vol. ii. p. 55, pl. xviii. figs. 10–13, 1837; Ann. d. Sci. Nat., ser. 3, Zool., vol. xviii. p. 149, pl. iv. fig. 15, 1852.

    „   „   Hilgendorf, Monatsber. d. k. preuss. Akad. d. Wiss. Berlin, p. 803–805, 1878.

    „   „   Kingsley, Proc. Acad. Nat. Sci. Philad., p. 148, pl. x. fig. 22, 1880, and references to literature, in part only.

    „   *porcellanus*, White, Proc. Zool. Soc. Lond., p. 85, 1847; Crust. in Zool. H.M.S. "Samarang," p. 50, 1848.

    „   „   Milne Edwards, tom. cit., p. 151, 1852.

    „   „   Kingsley, tom. cit., p. 155, 1880.

    „   *perplexus*, Milne Edwards, Ann. d. Sci. Nat., tom. cit., p. 150, pl. iv. fig. 18, 1852; fide Hilgendorf, tom. cit., p. 806, 1878.

    „   *marionis*, Milne Edwards, Hist. Nat. Crust., vol. ii. p. 53, 1837, not of Desmarest (?)

? „ *splendidus*, Stimpson, Proc. Acad. Nat. Sci. Philad., p. 99, 1858.

    „   „   Kingsley, tom. cit., p. 149, 1880.

? „ *pulchellus*, Stimpson, tom. cit., p. 100, 1858.

    „   *rectilatus*, Lockington, Proc. Calif. Acad. Nat. Sci., pt. 1, p. 148, 1876; fide Kingsley.

    „   *annulipes*, var. *albimana*, Kossmann, Zool. Ergebn. einer Reise Küstengeb. d. rothen Meeres, Brachyura, p. 53, 1877.

Philippines, Samboangan, 10 fathoms (three males); Fiji Islands, Kandavu (a good series of specimens, mostly males); Matuku (three males and two females).

Adult ♂.	Lines.	Millims.
Length of carapace, about	3½	7·5
Breadth of carapace,	6	13

In the specimens I refer to this species the carapace is moderately convex, smooth, and shining; the front ordinarily subtruncated at the distal extremity, the posteriorly convergent lateral margins defined by a straight line, which proceeds from the rather prominent acute antero-lateral angles of the carapace nearly to the rounded postero-lateral angles. The larger chelipede in the adult male has the hand elongated, externally granulated, the palm about once and a half as long as broad, with a vertical impressed line near the base of the fingers, internally armed with an angulated, coarsely granulated prominence near the base, and with two short granulated ridges near the bases of the fingers, the fingers are granulated on their inner margins; the pollex has two or three larger teeth or prominences, of which one is usually situated at about the middle of the inner margin, and one (triangulate) near the distal extremity; the dactylus is flat externally, with the margins subparallel nearly to the distal extremity, which is strongly incurved, and has usually two or three larger granules on the inner margin.

In the smaller males the tubercles and prominences of the inner margin of the fingers are often deficient, the subdistal triangulate tooth of the pollex or lower finger being that most permanent and characteristic of the species.[1]

Gonoplacinæ.

Gonoplax, Leach.

Gonoplax, Leach, Trans. Linn. Soc. Lond., vol. xi. p. 323, 1815.

„ Milne Edwards, Hist. Nat. Crust., vol. ii. p. 60, 1837; Ann. d. Sci. Nat, Zool., vol. xviii. p. 162, 1852.

Carapace transverse, longitudinally rather convex, dorsally smooth, with the antero-lateral angles terminating in spines, which are not greatly produced, as in *Ommatocar-cinus*. The front is deflexed, and rather less than one-third of the width of the carapace, and its anterior margin is straight ; the orbits are well defined, and extend along the whole anterior margin of the body between the front and antero-lateral angles. The epistoma is transverse; the ridges of the endostome or palate are nearly obsolete ; the post-abdomen in the male is rather broad, distinctly seven-jointed, and at the base covers the whole width of the sternum. The eye-peduncles are elongated. The antennules are transversely plicated. The basal antennal joint is small, and does not reach the infero-lateral process of the front, and the flagellum is somewhat elongated. The exterior maxillipedes present nothing remarkable; their ischium-joints are longitudinally sulcated, and are not produced at the antero-internal angles ; the merus-joints are distally truncated, the antero-external angles not produced, and the antero-internal angles, where the next joint articulates, usually slightly emarginated. The chelipedes in the adult male are subequal and very considerably elongated ; the merus-joints subcylindrical, and often exceeding in length the width of the carapace ; carpus short, without a spine on the interior surface ; palm about as long as the merus and compressed, fingers rather robust and compressed, dentated or tuberculated on the inner margins, and distally acute. The ambulatory legs are moderately elongated and slender, with the dactyli styliform.

To the long known European and North Atlantic species of this genus is to be added an apparently new form, which I propose to designate *Gonoplax sinuatifrons*.

[1] I have thought it useful to give the leading references to the synonyms of this species, so far as I am acquainted with them, since Mr. Kingsley's are incorrect in some particulars, or incomplete. Thus the West American *Gelasimus macrodactylus*, Milne Edwards and Lucas, cannot, in my opinion, be regarded as synonymous with *Gelasimus annulipes*, but, as the description and figure and specimens in the British (Natural History) Museum show, is a species with much more convex carapace and less prominent antero-lateral angles, and with differently shaped chelipedes and fingers. *Gelasimus perplexus* is stated by Hilgendorf to be synonymous, not with *Gelasimus chlorophthalmus*, Milne Edwards, as stated by Kingsley, but with *Gelasimus annulipes*. *Gelasimus grimaldi*, Milne Edwards, may also possibly be a variety of this species, but specimens referred to it in the British Museum collection may be distinguished by the rounder front and more tapering and slender dactylus of the larger chelipede, while the palm has a cicatrice on its outer surface. *Gelasimus pulchellus*, Stimpson, from Tahiti, is, to judge from the short diagnosis, a species or variety intermediate between *Gelasimus annulipes* and *Gelasimus grimaldi*.

Gonoplax sinuatifrons, n. sp. (Pl. XX. fig. 2).

The carapace is shaped nearly as in *Gonoplax rhomboides* (Linné), *i.e.*, it is moderately convex in a longitudinal direction, transverse, with the antero-lateral angles prominent and spiniform, the lateral margins slightly tumid at the hepatic regions, and thence slightly convergent to the postero-lateral angles; the dorsal surface is smooth and glabrous. The front is rather less than one-third the width of the carapace, its anterior margin not straight, as in *Gonoplax rhomboides*, but with a wide, shallow, median sinus; the lateral angles are rounded. The upper and lower orbital margins are undulated nearly as in *Gonoplax rhomboides*, which this species closely resembles also in the form of the ocular peduncles, antennæ, and outer maxillipedes. Chelipedes, in the small female, of moderate length; merus with a small tooth near the middle of its upper margin, as in *Gonoplax rhomboides*, and with another on the inferior margin; carpus with a tooth on its inner margin, but with the small tooth of the outer surface (present in *Gonoplax rhomboides*) nearly obsolete; chela less elongated than in *Gonoplax rhomboides*, and deeper in proportion to its length; the palm is rounded above and carinated below as in the European species, and the fingers somewhat more robust. The ambulatory legs as in *Gonoplax rhomboides*. Colour (in spirit) yellowish brown.

Young ♀.				Lines.	Millims.
Length of carapace, rather over	.	.	.	3	7
Breadth of carapace,	.	.	.	4½	9·5
Length of a chelipede,	.	.	.	6½	13·5
Length of second ambulatory leg,	.	.	.	8½	18·5

The unique specimen (a female, probably not fully grown) was dredged at Amboyna, in 15 to 25 fathoms.

This specimen is only distinguishable from the well-known European *Gonoplax rhomboides* by slight differences in the form of the front and chelipedes, which are referred to above, but probably additional specific characters would be derived from adult male examples.

Ommatocarcinus, White.

Ommatocarcinus, White, Append. in Stanley, Voy. H.M.S. "Rattlesnake," vol. ii. p. 393, pl. v. fig. 1, 1852.

" Milne Edwards, Ann. d. Sci. Nat., ser. 3, Zool., vol. xviii. p. 163, 1852.

Carapace transverse, longitudinally rather convex, with the dorsal surface smooth, the sides converging slightly to the posterior margins; the antero-lateral angles prolonged as long spines. The front is very narrow, less than one-sixth the anterior width of the carapace in the adult male, it is deflexed, and constricted between the bases of the eye-peduncles as in *Macrophthalmus*. The epistoma is linear-transverse. The endostome is without distinct longitudinal ridges; these are only very faintly indicated in their basal portion. The post-abdomen in the male is distinctly seven-jointed and resembles that of *Gonoplax*, but is distally rather broader; as in that genus, the two basal segments occupy the whole of the interspace between the bases of the ambulatory legs. The eye-pedicels are slender and greatly elongated, reaching the extremities of the antero-lateral angles of the carapace or even prolonged slightly beyond them. The antennules are transversely plicated. The antennæ occupy the narrow interior hiatus of the orbits; the basal joints are very short and do not nearly reach the lateral angles of the front; the flagella are moderately elongated. The ischium-joint of the exterior maxillipedes is not produced at its antero-internal angle; the merus is short, distally truncated, with the antero-external angles rounded. The chelipedes (in the adult males) are subequal and very greatly developed, sometimes exceeding in length three and a half times the greatest width of the carapace; the merus is subcylindrical and projects far beyond the antero-lateral angles of the carapace; the carpus is short and unarmed; palm slightly compressed, and nearly as long as the merus; fingers enclosing a wide basal interspace, compressed, distally acute, and irregularly dentated on the anterior margins. The ambulatory legs are elongated and compressed, without spines, and the dactyli are styliform and about as long as the penultimate joints.

Perhaps the nearest ally of this genus is *Gonoplax*, which *Ommatocarcinus* resembles in general form, in the great development of the chelipedes, in that the abdomen covers the whole width of the sternum at the base, and especially in the articulation of the fourth joint of the exterior maxillipedes with the merus at or near the antero-internal angle of the latter joint. *Ommatocarcinus* is distinguished from *Gonoplax* by the narrower front, the greater development of the eye-peduncles, and the greatly developed antero-lateral spine of the carapace.

I believe the only described species of this genus is the following :—

Ommatocarcinus macgillivrayi, White.

> *Ommatocarcinus macgillivrayi*, White, Append. in Stanley, Voy. of the "Rattlesnake," vol. ii. p. 393, pl. v. fig. 1, 1852.
>
> „   „   Milne Edwards, Ann. d. Sci. Nat., ser. 3, Zool., vol. xviii. p. 163, 1852.

Queen Charlotte Sound, near Long Island, New Zealand, 10 fathoms, Station 167A (a young male and two small but ova-bearing females).

The smallest female has the following dimensions :—

♀.		Lines.	Millims.
Length of carapace, a little over		4	9
Breadth of carapace (to base of the spines at the antero-lateral angles),		7	15

In all these specimens the chelipedes are of moderate length, not greatly elongated, as in the large Australian males, the types of the species, in the British (Natural History) Museum; the front, in the females especially, is somewhat broader, and the chelipedes relatively much shorter, with the merus-joints angulated, and armed with two or three spinules on the anterior margin, near the distal extremity, besides the spinule on the middle of the posterior margin, which is found also in the large adult males. These differences will perhaps be found to be of specific importance.

The largest female measures as follows :—

Adult ♀.		Lines.	Millims.
Length of carapace,	.	5	10·5
Breadth to base of antero-lateral spine,	.	9	19
Length of a chelipede,	.	10	21

Macrophthalminæ.

Macrophthalmus, Latreille.

Macrophthalmus, Latreille, in Cuvier, Règne Animal, ed. 2, vol. iv. p. 44, 1829.
　　　　,,　　　　Milne Edwards, Hist. Nat. Crust., vol. ii. p. 63, 1837; Ann. d. Sci. Nat., ser. 3, Zool. xviii. p. 155, 1852.

Carapace usually quadrate, depressed and broadly transverse, with the cervical and cardiaco-branchial sutures distinct on the lateral margins, nearly straight, and armed with one or more teeth behind that at the exterior angle of the orbit; front very narrow and deflexed, so as in great measure to conceal the antennules, and distally truncated. The orbits are very large and occupy the whole width of the anterior face of the carapace, between its antero-lateral angles and the front, and they are not defined externally, beneath the antero-lateral or exterior orbital angle. The epistoma is very narrow, usually linear-transverse. The longitudinal ridges of the endostome or palate are either absent or very obscurely indicated. The post-abdomen (in the male) is distinctly seven-jointed, and does not occupy the whole width of the sternum, between the bases of the fifth ambulatory legs. The ocular pedicels are slender and considerably elongated, and in certain species are even prolonged laterally beyond the exterior angle of the orbit. The antennules are transversely plicated. The antennæ are not excluded from the orbital hiatus, their basal portion is very short; the basal joint is usually transverse and does not nearly reach the frontal margin; the flagellum is somewhat elongated. The exterior maxillipedes do not meet along their inner margins; the ischium and merus-joints of the endognathi are distally truncated, or their anterior margins are even slightly concave; the carpus is articulated with the merus at its antero-external angle. The chelipedes (in the adult males) are usually subequal and somewhat elongated, with the merus trigonous; carpus without or with only a small spine on the interior margin; palm elongated, compressed, and rounded or subcarinated on the superior margin; fingers compressed and minutely denticulated on the interior margins, and usually armed with one or two larger lobes or teeth, and distally acute or somewhat excavated. Ambulatory legs somewhat elongated, robust, with the merus-joints compressed, and usually armed with a subdistal spine on the superior margin; dactyli slender, nearly straight, and styliform.

The described species of this genus are numerous and occur in the littoral or shallow waters of all parts of the Indo-Pacific region.

The following are recent forms which have been described since the publication of Milne Edwards's memoir in 1852. They apparently all belong to his second section (§ 2) of the genus.

Macrophthalmus dentatus, Stimpson. Hong Kong.
Macrophthalmus convexus, Stimpson. Loo-Choo.
Macrophthalmus bicarinatus, Heller. Nicobars.
Macrophthalmus grandidieri, A. Milne Edwards. Zanzibar.
Macrophthalmus inermis, A. Milne Edwards. Sandwich Islands; New Caledonia.
Macrophthalmus lævis, A. Milne Edwards. Indian Ocean.
Macrophthalmus græffei, A. Milne Edwards. Upolu, Samoan Islands.
Macrophthalmus quadratus, A. Milne Edwards. New Caledonia.
Macrophthalmus punctulatus, Miers. New South Wales, Port Jackson.
Macrophthalmus latifrons, Haswell. Victoria, Port Philip.[1]

Macrophthalmus podophthalmus, Eydoux and Souleyet.

Macrophthalmus podophthalmus, Eydoux and Souleyet, Crust. in Voy. d. "Bonite," Zool., vol. i. pl. iii. fig. 6, 1841.
,, ,, Milne Edwards, Ann. d. Sci. Nat., ser. 3, Zool., vol. xviii. p. 155, 1852.

Torres Strait, August 1874 (a small male).

A young female of very small size from the Arafura Sea (depth not stated), may perhaps be regarded as the young of this species, although in the length of the ambulatory legs it approaches *Macrophthalmus sulcatus*, as described by Milne Edwards,[2] from the Mauritius. In this specimen the ocular peduncles project beyond the antero-lateral angles of the carapace by little more than the length of their corneæ; the carapace is less widely transverse than in the adult *Macrophthalmus podophthalmus*,

[1] *Macrophthalmus brevis* (Herbst), from the Red Sea and Zanzibar, which Milne Edwards and Hilgendorf placed (the former doubtfully) as synonymous with *Macrophthalmus carinimanus*, Latreille, is regarded by M. de Man (Notes Leyden Mus., vol. ii. p. 70, 1879) as distinct from that species; *Macrophthalmus pollewi*, Hoffmann, from Sakatia Island, is thought by the same author to be synonymous with *Macrophthalmus latreillei*, Desmarest, which has been recorded both as a recent and fossil species.
[2] Tom. cit., p. 156, 1852.

and the transverse sulci of the carapace are not more deeply indicated than in that species. The ambulatory legs are relatively more elongated and very slender (one only remains perfect, and is detached in this specimen). Its dimensions are as follows :—

	Lines.	Millims.
Young ♀.		
Length of carapace, nearly	2	4
Breadth of carapace,	2½	5·3

Macrophthalmus serratus, Adams and White (Pl. XX. fig. 1).

Macrophthalmus serratus, Adams and White, Crust. in Zool. H.M.S. "Samarang." p. 51, 1848.

"   "   Milne Edwards, Ann. d. Sci. Nat., ser. 3, Zool., vol. xviii. p. 159, 1852.

Japan, Kobé, 8 to 10 fathoms (a small male and female).

These specimens are referred with much doubt to this species, on account of their very small size. One of the chelipedes in the male is deficient. The carapace is more finely granulated than in adult examples of *Macrophthalmus serratus*, the front perhaps somewhat broader, and the antero-teeth much less prominent. As in adult specimens there are obscure indications of a fourth lateral marginal tooth. In the female the first lateral tooth is less prominent than the following. I have observed a similar variation in adult males.

The figure of the adult is from a male of large size in the collection of the British (Natural History) Museum. In an adult female, also from the Philippines, in the collection, the antero-lateral teeth are less prominent, and the small Challenger specimens in these particulars more nearly approach this example.

Hemiplax, Heller.

Hemiplax, Heller, Crust. in Reise der "Novara," p. 40, 1865.

"   Miers, Cat. New Zeal. Crust., p. 33, 1876.

This genus was established by Dr. Heller for a species very nearly allied to *Macrophthalmus*, but differing from the typical species of that genus in the less transverse carapace and broader front, which is at least one-third of the width of the carapace at the antero-lateral angles. In all other characters it nearly resembles *Macrophthalmus*. As in that genus the exterior maxillipedes do not meet along the inner margins, and their merus-joints are short, distally truncated, and are not, as in *Metaplax* (to which *Hemiplax* is also very nearly related), traversed externally by an oblique piliferous crest. The chelipedes (as in *Macrophthalmus*) are subequal, and shorter than the ambulatory legs, and the fingers are finely denticulated on the inner margins.

I think that this genus would perhaps be better regarded as a subgenus of *Macrophthalmus*, and might perhaps include those forms (*Macrophthalmus quadratus*, A. Milne Edwards, *Macrophthalmus punctulatus*, Miers, &c.) which differ from the typical *Macrophthalmi* in their narrower carapace and broader front, and from *Euplax* in that the eye-peduncles reach, or nearly reach, the antero-lateral angles of the carapace.

Hemiplax hirtipes, Heller.

Metaplax hirtipes, Heller, Verhandl. d. k. k. zool.-bot. Gesellsch. Wien, vol. xii. p. 521, 1862.
Hemiplax hirtipes, Heller, Crust. in Reise der "Novara," p. 40, pl. iv. fig. 3, 1865.
    „   „   Miers, Cat. New Zeal. Crust., p. 34, 1876.

New Zealand, Queen Charlotte Sound, 10 fathoms (Station 167A). Five small specimens, males and females.

The chelipedes and ambulatory legs are more slender than in the fully grown adult, the carapace perhaps narrower, and the ambulatory legs less hairy. The largest male measures as follows :—

Adult ♂.	Lines.	Millims.
Length of carapace, about	3	6·5
Breadth of carapace, about	4	8·5

Euplax, Milne Edwards.

Euplax, Milne Edwards, Ann. d. Sci. Nat., ser. 3, Zool. xviii. p. 160, 1852.
    „   A. Milne Edwards, Nouv. Archiv. Mus. Hist. Nat., vol. ix. p. 281, 1873.
Chaenostoma, Stimpson, Proc. Acad. Nat. Sci. Philad., p. 97, 1858, subgenus (?).

This genus was established by Milne Edwards for two species which very nearly approach *Macrophthalmus* in all structural characteristics, and are distinguished merely by the shorter eye-peduncles, which do not nearly reach the exterior orbital angle (which is less developed than in the typical *Macrophthalmi*) ; the carapace is narrower than in the typical forms of *Macrophthalmus* and is uniformly granulated, and the front broader in width, nearly equalling the length of the eye-peduncles. It is very nearly connected with *Macrophthalmus*, through such forms as *Macrophthalmus quadratus*, A. Milne Edwards, and *Macrophthalmus punctulatus*, Miers.

Of the two species assigned by Milne Edwards to this genus, one, *Euplax boscii*, is distributed throughout the Indo-Pacific region, the other, *Euplax leptophthalmus*, Milne Edwards, occurs on the coast of Chili.

Euplax (Chænostoma) boscii (Audouin).

Macrophthalmus boscii, Audouin, Explic. des planches in Savigny, Crust. de l'Egypte, pl. ii. fig. 1.
Euplax boscii, Milne Edwards, Ann. d. Sci. Nat., ser. 3, Zool., vol. xviii. p. 160, 1852.
    „   (*Chænostoma*) *boscii*, A. Milne Edwards, Nouv. Archiv. Mus. Hist. Nat., vol. ix. p. 281,
                    1873, and reference to synonyms.
    „   „   „   Miers, Crust. in Rep. Zool. Coll. H.M.S. "Alert," pp. 238, 542,
                    1884.

Fiji Islands, Kandavu, July 1874 (an adult female, bearing ova).

Adult ♀ .					Lines.	Millims.
Length of carapace, nearly	.	.	.	.	5	10·5
Breadth of carapace, nearly	.	.	.	.	6	12·5

Family III. GRAPSIDÆ.

Grapsoidiens, Milne Edwards, Hist. Nat. Crust. vol. ii. p. 68, 1837.
Grapsidæ, Dana, U.S. Explor. Exped., vol. xiii., Crust., p. 329, 1852.
    „   Kingsley, Proc. Acad. Nat. Sci. Philad., p. 187, 1880.
Grapsinæ (pt.), Milne Edwards, Ann. d. Sci. Nat., ser. 3, Zool. xx. p. 163, 1853.

Carapace depressed or moderately convex, more or less quadrilateral, with the lateral margins straight or slightly arcuated. Front usually broad, more rarely of moderate width, never very narrow. Orbits and eye-peduncles of moderate size. The post-abdomen usually covers the whole width of the sternum at the base, between the coxæ of the fifth ambulatory legs. The carpal joint of the endognath of the exterior maxillipedes articulates at the summit or at the antero-external angle of the merus, not at the antero-internal angle. The chelipedes (in the adult males) are usually subequal and are moderately developed. The dactyli of the ambulatory legs are styliform, compressed, and are sometimes smooth, sometimes armed with strong spines.

The species are nearly always littoral or shallow-water forms, but rarely inhabit deep water (*e.g.*, *Euchirograpsus*). Both genera and species have been recently enumerated by Kingsley in the memoir cited above.

Subfamily 1. GRAPSINÆ.

Grapsinæ, Kingsley, tom. cit., p. 189.

Antennules more or less transverse and covered by the front, which is entire, not longitudinally cleft.

This subfamily is divided by Mr. Kingsley (after Dana) into the following sections (or tribes) :—

1. Grapsini.

Grapsinæ, Dana, *tom. cit.*, p. 331.

Exterior maxillipedes without an oblique piliferous crest on the ischial and meral joints.

2. Sesarmini.

Sesarminæ, Dana, *tom. cit.*, p. 333.

Merus and ischium of the exterior maxillipedes crossed obliquely by a piliferous ridge.

Nautilograpsus, Milne Edwards.

Nautilograpsus, Milne Edwards, Hist. Nat. Crust., vol. ii. p. 89, 1837; Ann. d. Sci. Nat., ser. 3, Zool. xx. p. 173, 1853.

,, Kingsley, Proc. Acad. Nat. Sci. Philad., p. 201, 1880.

Planes (Leach, MS.), Bowdich, Excursion to Madeira and Porto Santo, p. 15, fig. 2, 1825 (description insufficient).

Carapace subquadrate, with the postero-lateral margins somewhat convergent, dorsally smooth and slightly convex, and usually with a slightly indicated lateral post-orbital tooth. The front is broad, usually about half the width of the carapace; its anterior margin projecting slightly and nearly straight. The orbits are small, and the margins entire or with only a very small notch beneath the eye-peduncles, near the exterior orbital tooth. The epistoma is short and broadly transverse. The longitudinal ridges of the endostome are distinct and well-defined. The post-abdomen in the male is distinctly seven-jointed, and its basal segments occupy the whole width of the sternum, between the coxæ of the ambulatory legs. The eye-peduncles are short and thick. The antennules are transversely plicated. The basal joint of the antennæ is short and robust, and is produced at its antero-external angle, which forms a lobe or tooth, and lies within the interior hiatus of the orbit; the flagellum is short. The exterior maxillipedes have a rhomboidal gape; the merus-joints are distally truncated, and their anterior margins are even slightly concave; the carpal joints are articulated near the rounded antero-external angles of the merus-joints. The chelipedes (in the adult males) are rather robust, subequal, of moderate size; the merus-joints trigonous, and the anterior margins dentated; carpi with a tooth or spine on the interior margins, palms somewhat turgid, rounded above; fingers dentated on the interior margins and distally acute. The ambulatory legs are short, with the joints compressed, and (the merus-joints especially) somewhat dilated; the penultimate joints are ciliated on the superior margins, and the inferior margins are spinuliferous; dactyli short, compressed and spinuliferous.

Nautilograpsus is allied in some particulars to *Trapezia* in the Cancroidea, and to *Litocheira*, Kinahan, in the Carcinoplacidæ, from both of which it is distinguished by the broader basal antennal joint and the compressed and robust ambulatory legs.

There is probably but a single species of this genus (the common Gulf-Weed Crab), which occurs nearly everywhere on floating weed in the temperate and tropical seas of the globe, and has been referred to under many different specific names.

Nautilograpsus minutus (Linné).

> *Cancer minutus*, Linné, Syst. Nat., ed. xii., p. 1040, 1766.
> *Grapsus pusillus*, de Haan, Crust. in v. Siebold, Fauna Japonica, pp. 32, 59, pl. xvi. fig. 2, 1835.
> *Nautilograpsus minutus*, Milne Edwards, Hist. Nat. Crust., vol. ii. p. 90, 1837; Ann. 4. Sci.
> Nat., tom. cit., p. 174, 1853.
> „   „   Kingsley, tom. cit., p. 202, 1880, et synonyma.

The specimens in the collection are from the following localities :—

Gomera, Canary Islands, February 1, 1873 (an adult male); between Bermuda and the Azores, June 20, 1873 (a large series of specimens, attached to *Ianthina* and other floating objects); off Sombrero Island, West Indies, March 15, 1873 (a small female); gulf-weed in the North-West Atlantic, April 1873 (numerous specimens); from *Fucus* in the North Atlantic, June 26, 1873 (an adult male and two females); South Pacific, near the Kermadec Islands, on the surface, among seaweed, July 15, 1874 (numerous specimens); North Pacific, off Volcano Island, April 4, 1875, from tube containing surface dredgings (a female); coast of Japan, June 1875 (an adult female); North-West Pacific, surface, June 1875 (two adult females).

Specimens of this genus show a considerable degree of variation in the convexity of the carapace, the development of the antero-lateral marginal tooth, which is sometimes obsolete, in the coloration of the body and limbs, &c., but I cannot find any valid characters for the distinction of the numerous supposed species which have been described by authors.

	Lines.	Millims.
Adult ♂.		
Length and breadth of carapace,[1] .	9	19

Grapsus, Lamarck.

> *Grapsus*, Lamarck (pt.), Syst. Anim. sans Vert., v. p. 247, 1818.
> „   Milne Edwards (pt.), Hist. Nat. Crust., vol. ii. p. 83, 1837; Ann. d. Sci. Nat.,
> ser. 3, Zool., vol. xx. p. 166, 1853.
> „   Kingsley, Proc. Acad. Nat. Sci. Philad., p. 192, 1880.

Carapace depressed, with the cervical sutures strongly defined; the lateral margins regularly arcuated, and armed with a single tooth behind the exterior orbital angle, the dorsal surface marked with transverse raised lines, which are strongest on the branchial regions; the front is of moderate width, strongly deflexed, and its anterior margin is entire and slightly arcuated; the orbits of moderate size, rather deep, and their inferior

---

[1] This is rather more than the average size of adult examples, but I have examined yet larger specimens in the collection of the British (Natural History) Museum.

margins have a fissure or notch, near the exterior orbital tooth; the epistoma is transverse and rather large; the buccal cavity small, and the ridges of the endostome or palate distinctly defined. The post-abdomen (in the male) is distinctly seven-jointed, and its basal segments cover the whole width of the sternum between the bases of the fifth ambulatory legs. The eye-peduncles are robust and short; the antennules are transversely plicated in very narrow fossetts. The basal antennal joint is very short and is produced at its antero-external angle; it lies within the interior orbital hiatus, between the front and the interior subocular lobe of the orbit, which is dentiform and acute; the flagellum is short. The exterior maxillipedes have a rhomboidal gape, and their endognathi are narrow (the ischium-joints are not in contact at the base); the merus-joints are truncated or slightly concave at the distal extremity, the carpi are articulated at the distal extremity of the merus-joints near the antero-external angle.

The chelipedes (in the adult male) are robust and rather short; merus-joints trigonous, with the anterior margins dentated; carpus with a strong lobe or tooth on the inner margin; palm short, granulated above; dactyli denticulated on the inner margin and excavated at the distal extremity. Ambulatory legs large and robust, with the merus-joints dilated and compressed; dactyli strongly spinulose.

There are probably but two distinct species of this long-known genus, to one or other of which many of the forms briefly characterised by M. H. Milne Edwards and other authors are to be referred as synonymous, or at most, as varieties.

> *Grapsus maculatus* (Catesby). Common on all the warmer temperate and tropical coasts and islands both of the Indo-Pacific and Atlantic regions.
>
> *Grapsus trigosus* (Herbst). Common on the shores and islands of the Indo-Pacific region.[1]

Grapsus maculatus (Catesby).

Pagurus maculatus, Catesby, Nat. Hist. of the Carolinas, vol. ii. p. 36, pl. xxxvi. fig. 1, 1743 and 1771.

Cancer grapsus, Linné, Syst. Nat., ed. xii. p. 1048, 1766.

Grapsus pictus, Latreille, Hist. Nat. Crust. et Ins., vol. vi. p. 69, pl. xlvii. fig. 2, 1803–1804.

   „   „   Milne Edwards, Hist. Nat. Crust., vol. ii. p. 86, 1837; Crust. in Cuvier, Règne Animal, ed. 3, pl. xxii. fig. 1.

   „   *maculatus*, Milne Edwards, Ann. d. Sci. Nat., ser. 3, Zool., vol. xx. p. 167, pl. vi. fig. 1, 1853.

   „   „   Kingsley, Proc. Acad. Nat. Sci. Philad., p. 192, 1880, et synonyma (?).

   „   *pictus*, var. *ocellatus*, Studer, Abhandl. d. k. Akad. d. Wiss. Berlin, Abh. ii. p. 14, 1882.

Of this common and widely distributed species, specimens are in the collection from the following localities:—Bermuda, an adult male (in spirits), and an adult male and

---

[1] *Grapsus gracilipes*, Milne Edwards, is retained as distinct by Kingsley, but is regarded by M. de Man as a variety of *Grapsus maculatus*. *Grapsus simplex*, Herklots, referred to by de Man, may be a distinct species.

female in imperfect condition (preserved dry); St. Paul's Rocks, August 1873 (a good series of specimens); St. Michael's, Fernando Noronha, on the rocks, September 1873 (two males and two females); Ascension Island, on the shore (a small male); and St. Vincent, Cape Verde Islands, July 1873 (two females and two young males).

Grapsus maculatus, as has been repeatedly observed, varies much in the coloration of the carapace and limbs. The specimens in the Challenger series from Bermuda and St. Paul's Rocks, which are the largest and best preserved in the collection, nearly resemble in colour the excellent figure given by M. H. Milne Edwards in the large illustrated edition of the Règne Animal de Cuvier, which is reproduced, uncoloured, in his monograph of the group in the Ann. d. Sci. Nat. (*tom. cit.*), and which may, I presume, be regarded as typical of the species. To this (the typical) variety also probably belong the specimens from Fernando Noronha and Ascension Island.

The specimens from the Cape Verde Islands are perhaps referable to the variety described by M. Milne Edwards as *Grapsus webbi*, in which the spots of the carapace are smaller and more numerous, but the two are connected by imperceptible gradations.

The variety designated by Studer, *Grapsus pictus*, var. *ocellatus*, ought not to be separated, I think, from the typical form.

Adult ♂.	Lines.	Millims.
Length of carapace, about	30	63·5
Breadth of carapace, .	33	70

Grapsus strigosus (Herbst).

> *Cancer strigosus*, Herbst, Naturgesch. der Krabben u. Krebse, vol. iii. Heft 1, p. 55, pl. xlvii. fig. 7, 1799.
> *Goniopsis flavipes*, MacLeay, in Smith, Illustr. Zool., South Africa, p. 66, 1838.
> *Grapsus strigosus*, Latreille, Hist. Nat. Crust. et Ins., vol. vi. p. 70, 1803–1804.
> ,,   ,,   Milne Edwards, Hist. Nat. Crust., vol. ii. p. 87, 1837; Ann. d. Sci. Nat., *tom. cit.*, p. 169, 1853.
> ,,   ,,   A. Milne Edwards, Nouv. Archiv. Mus. Hist. Nat., vol. ix. p. 285, 1873.
> ,,   ,,   Kingsley, Proc. Acad. Nat. Sci. Philad., p. 194, 1880, et synonyms.

Arrou Islands (a small female).

The front is transversely wider and less abruptly deflexed; the postfrontal lobes less prominent, and the epistoma is wider and shorter in this species than in *Grapsus maculatus*.

The Challenger specimen is a small one and faded, presenting but slight traces of the original coloration.

♀.	Lines.	Millims.
Length of carapace, .	10	21
Breadth of carapace, about .	10½	22·5

Leptograpsus, Milne Edwards.

Leptograpsus, Milne Edwards (pt.), Ann. d. Sci. Nat., ser. 3, Zool., vol. xx. p. 171, 1853.
,,   Stimpson, Proc. Acad. Nat. Sci. Philad., p. 101, 1858.
,,   Kingsley, Proc. Acad. Nat. Sci. Philad., p. 196, 1880.

This genus, as restricted by Stimpson and Mr. Kingsley, is very nearly allied to *Grapsus*, but distinguished as follows :—The front is nearly horizontal, but very slightly deflexed, and the postfrontal or protogastric lobes of the carapace are but very obscurely developed. There are two teeth behind the exterior orbital tooth on the lateral margins, which are arcuated as in *Grapsus*. The basal antennal joint is broader, and the lobe at the antero-external angle fully reaches the apex of the interior subocular lobe of the orbit, which is shorter than in *Grapsus*; the exterior maxillipedes (the merus-joints of the endognaths especially) are broader than in that genus.

I doubt the advisability of regarding *Leptograpsus* as a distinct genus; as restricted above, it will contain only the following species :—

Leptograpsus variegatus (Fabricius). Australian Coasts; Norfolk Island; New Zealand; Marianne Islands; Shanghai; Chili; St. Ambrose Islands; Juan Fernandez; Pernambuco; and perhaps the Canary Islands (Milne Edwards, as *Leptograpsus bertheloti*).[1]

Leptograpsus variegatus (Fabricius).

Cancer variegatus, Fabricius, Entom. Syst. Suppl., vol. 2, p. 450, 1793.
Grapsus variegatus, Latreille, Hist. Nat. Crust. et Ins., vol. vi. p. 71, 1803–1804.
,,   ,,   Guérin, Icon. du Règne Animal, Crust., pl. vi. fig. 1.
,,   ,,   Milne Edwards, Hist. Nat. Crust., vol. ii. p. 87, 1837.
Leptograpsus variegatus, Milne Edwards, Ann. d. Sci. Nat., tom. cit., p. 171, 1853.
,,   ,,   Kingsley, Proc. Acad. Nat. Sci. Philad., p. 196, 1880, et synonyma (?).

Valparaiso, on the shore (an adult female).

Adult ♀.		Lines.	Millims.
Length of carapace,	.	19½	41
Breadth of carapace,	.	22	46·5

Metopograpsus, Milne Edwards.

Metopograpsus, Milne Edwards (§ 1), Ann. d. Sci. Nat., ser. 3, Zool., vol. xx. p. 164, 1853.
,,   Kingsley, Proc. Acad. Nat. Sci. Philad., p. 190, 1880.

Carapace depressed, subtrapezoidal, broader than long, with the antero-lateral margins straight or nearly straight, entire or unidentated, and slightly convergent to the posterior

[1] *Grapsus incrustus*, Hess, from Sydney, may belong to this genus, but, on account of the long curved dactyli, I think it more probably belongs to *Cyrtograpsus* or *Pseudograpsus*.

margins. The dorsal surface plicated only on the lateral margins. Front very broad, deflexed, with the anterior margin straight or slightly sinuated. Orbits small, with the interior subocular lobe usually very broad, reaching the front, and completely excluding the antennæ from the orbit. Epistoma very short, almost linear-transverse. Endostome with the longitudinal ridges usually (but not invariably) distinctly developed. Post-abdomen (in the male) distinctly seven-jointed, and covering the whole width of the sternum at the base. Eye-peduncles short, robust. Antennules transversely plicated in short, wide fossettes. Antennæ with the basal joint very short, completely excluded from the orbit, and more or less produced at its antero-external angle; flagellum short. Exterior maxillipedes with a rhomboidal gape; the merus-joints short, and broader than in *Grapsus*, distally truncated, and bearing the next joint at their summits, near the antero-external angle. Chelipedes subequal, and moderately developed, with the merus-joints trigonous and their anterior margins distally dentated; palms somewhat turgid, rounded above and below; fingers excavated at the distal extremity. Ambulatory legs of moderate length, with the merus-joints dilated and compressed, their anterior margins with a subterminal spine, and their posterior margins with several spines at the distal extremity; the dactyli rather short and spinuliferous.

This genus is distinguished from *Pachygrapsus*, its nearest ally, by the less convex carapace, which is usually less distinctly plicated on the dorsal surface, and by the great development of the interior subocular lobe of the orbit. The species, which with their synonyms have been enumerated by Mr. Kingsley, are apparently confined to the Indo-Pacific region.

Metopograpsus messor (Forskål).

Cancer messor, Forskål, Descript. animalium quæ in itinere orientali observavit, P. Forskål, p. 88, 1775.

Grapsus guimardii, Audouin, Explic. des planches in Savigny, Descr. de l'Egypte, Crust. Atlas, pl. ii. fig. 3.

   ,,   *messor*, Milne Edwards, Hist. Nat. Crust., vol. ii. p. 88, 1837.

Metopograpsus messor, Milne Edwards, Ann. d. Sci. Nat., tom. cit., p. 165, 1853.

   ,,   ,,   Kingsley, tom. cit., p. 190, 1880, and references to synonyms.

Grapsus (Pachygrapsus) æthiopicus, Hilgendorf, Crust. in Van der Decken, Reise in Ost-Afrika, vol. iii. p. 88, pl. iv. fig. 2, 1869; teste, Hilgendorf, Monatsber. d. k. preuss. Akad. d. Wiss. Berlin, p. 808, 1878.

Metopograpsus messor, var. *frontalis*, Miers.

Metopograpsus messor, var. *frontalis*, Miers, Ann. and Mag. Nat. Hist., ser. 5, vol. v. p. 311, 1880.

Kandavu, Fiji (a small male); Tahiti, near the reefs (two males and two females); Papiete, Tahiti, in brackish water at the mouth of a stream (several specimens, mostly females); Sandwich Islands, Hilo, on the beach (several males and females).

The specimens collected show considerable variation in the coloration and markings of the carapace.

Adult ♂.	Lines.	Millims.
Length of carapace, about	10½	22·5
Breadth of carapace, nearly	13	27

Pachygrapsus, Randall.

Pachygrapsus, Randall, Journ. Acad. Nat. Sci. Philad., vol. viii. p. 126, 1839.
,, Milne Edwards, Ann. d. Sci. Nat., ser. 3, Zool., vol. xx. p. 166, 1853.
,, Kingsley (pt.), Proc. Acad. Nat. Sci. Philad., p. 198, 1880.

This genus in all of its characters is very nearly allied to Metopograpsus, but is distinguished by the somewhat more convex carapace, which is usually very distinctly plicated over the whole of the dorsal surface, and by the lesser development of the interior subocular lobe of the orbit, which does not reach the front, so that the produced antero-external lobe of the basal antennal joint usually enters slightly within the orbital hiatus.

The species are widely distributed over the warmer, temperate, and tropical shores and islands, both of the Indo-Pacific and Atlantic regions. The single species occurring in the Challenger collection has (as the specimens collected show) a very extended geographical range. For further details upon the distribution of this genus I may refer to Mr. Kingsley's memoir.[1]

Pachygrapsus transversus, Gibbes.

Pachygrapsus transversus, Gibbes, Proc. Amer. Assoc. Adv. Sci., p. 181, 1850.
,, ,, Kingsley, Proc. Acad. Nat. Sci. Philad., p. 199, 1880, where references to synonyms.
Goniograpsus innotatus, Dana, Proc. Acad. Nat. Sci. Philad., p. 249, 1851; U.S. Explor. Exped., vol. xiii., Crust., p. 345, pl. xxi. fig. 9, 1852.

Bermuda, on the shore (an adult male and two females); St. Vincent, Cape Verde Islands, July 1873 (four males, of which two are adult, and a young female); Australia, Port Jackson, Sow and Pig's Bank, in 6 fathoms (an adult male).

I can see no distinctions of specific importance in the specimens from these widely-distant localities.[2]

[1] I would suggest here, that the genus Goniograpsus, Dana, which is not retained as distinct either by Stimpson or Kingsley, may be conveniently restricted to and used as a generic designation for the Mediterranean Pachygrapsus marmoratus, and (probably) the Chilian Pachygrapsus pubescens, Heller, and Pachygrapsus latipes, in which the carapace has two teeth behind the tooth at the exterior orbital angle. Goniograpsus marmoratus, which is the only one of these species I have examined, is further distinguished from the typical Pachygrapsi by the nearly horizontal front and smoother quadrate carapace.

[2] It is very probable that this species may be identical with the earlier described Pachygrapsus maurus (Lucas), from the Mediterranean, but as I have examined no Mediterranean examples of the genus, I do not venture to unite the two forms. Of the numerous synonymical citations admitted by Kingsley, there are one or two which I have not personally verified.

An adult male from Port Jackson measures :—

Adult ♂.	Lines.	Millims.
Length of carapace, . .	6	13
Breadth of carapace, nearly, .	8½	17·5

Geograpsus, Stimpson.

Geograpsus, Stimpson, Proc. Acad. Nat. Sci. Philad., p. 101, 1858.
    „   Kingsley, Proc. Acad. Nat. Sci. Philad., p. 195, 1880.
Discoplax, A. Milne Edwards, Ann. Soc. Entom. France, ser. 4, vol. vii. p. 284, 1867; Nouv. Archiv. Mus. Hist. Nat., vol. ix. p. 293, 1873.
Orthograpsus, Kingsley, Proc. Acad. Nat. Sci. Philad, p. 194, 1880.

This genus is allied to *Leptograpsus* and to *Pachygrapsus*, but is distinguished from both by the form of the carapace, which is depressed, plicated only near the lateral margins, with the antero-lateral margins straight or arcuated only at the hepatic regions, posteriorly nearly straight ; they are armed with a single tooth behind the exterior orbital tooth. Front of moderate width and deflexed. The orbits, epistoma, buccal cavity, post-abdomen, and eye-peduncles present nothing remarkable. The basal antennal joint is short and but slightly produced at its antero-external angle. As in *Grapsus*, the endognathi of the exterior maxillipedes are narrow ; the merus-joints, in particular, slender and elongated, distally truncated, and bearing the next joint at or near the antero-external angle. The chelipedes and ambulatory legs resemble those of *Leptograpsus* and *Pachygrapsus*, but the fingers of the palms of the chelipedes are acute or subacute, not excavated at the distal extremities.

The species occur both in the Indo-Pacific and Atlantic regions, and are all, I believe, littoral or shallow-water forms.

The genus *Discoplax* is united by Kingsley with *Geograpsus*, but may prove to be distinct in the figure of the type (*Discoplax longipes*, A. Milne Edwards), the carapace is represented as strongly arcuated and granulated anteriorly, and the merus of the exterior maxillipedes is shorter and broader than in the typical *Geograpsi*.

The species of *Geograpsus* as restricted above are :—

Geograpsus lividus, Milne Edwards (= *Geograpsus brevipes*, Milne Edwards; *Geograpsus occidentalis*, Stimpson). West Indies; California; Chili.

Geograpsus crinipes, Dana (= *Geograpsus depressus*, Heller, *fide* Kingsley). Polynesian Islands.

Geograpsus longitarsis (Dana). Paumotu Archipelago.

Geograpsus grayi, Milne Edwards (= *Geograpsus rubidus*, Stimpson). Indo-Pacific Region.

Geograpsus eydouxi (Milne Edwards). Chili. (This species is regarded by
Kingsley as identical with *Pachygrapsus crassipes*, Randall, but I believe
its true place to be in the genus *Geograpsus*, since Milne Edwards says of
the chelipedes "*pinces aigües*.")

Geograpsus (?) *longipes* (A. Milne Edwards). New Caledonia.

Geograpsus hillii (Kingsley). West Indies ; Florida.

Geograpsus grayi [1] (Milne Edwards).

Grapsus grayi, Milne Edwards, Ann. d. Sci. Nat., *tom. cit.*, p. 170, 1853.

Geograpsus grayi, A. Milne Edwards, Nouv. Archiv. Mus. Hist. Nat., vol. ix. p. 288, pl. xvi.
     fig. 1, 1873.

,,   ,,   Kingsley, Proc. Acad. Nat. Sci. Philad., p. 196, 1880, et synonyma.

Fiji Islands, Kandavu (two small males).

♂.	Lines.	Millims.
Length of carapace, about	7	15
Breadth of carapace, about	8	17

Pseudograpsus, Milne Edwards.

Pseudograpsus (pt.), Milne Edwards, Hist. Nat. Crust., vol. ii. p. 81, 1837 ; Ann. d. Sci. Nat.,
     ser. 3, Zool., vol. xx. p. 191, 1853.

,,   A. Milne Edwards, Nouv. Archiv. Mus. Hist. Nat., vol. iv. p. 176, 1868.

,,   Kingsley, Proc. Acad. Nat. Sci. Philad., p. 204, 1880.

Carapace subquadrate, depressed or nearly flat on the dorsal surface, with the antero-
lateral margins arcuated and dentated ; the branchial regions not defined as in *Varuna* ;
in the middle of the carapace is an H-shaped impression formed by part of the cervical
and cardinco-branchial sutures ; the epigastric lobes are well-defined ; the front is but
slightly deflexed, and its anterior margin is sinuated or nearly straight. Orbits small,
with the interior subocular lobe acute or subacute. Epistoma very short and widely
transverse. Buccal cavity large ; the ridges of the endostome or palate not very distinct.
Post-abdomen of the male distinctly seven-jointed ; the basal segments do not cover the
whole width of the sternum, between the coxæ of the fifth ambulatory legs. Eye-
peduncles short. Antennules transversely plicated. Antennæ with the basal joint short
and not markedly dilated at the antero-external angle, it sometimes does not reach the
infero-lateral frontal process. Exterior maxillipedes with the exognath much dilated,
as broad or nearly as broad as the ischium of the endognath ; the merus of the endognath

---

[1] It is possible, as Mr. Kingsley has pointed out, that the species described by Dana as *Grapsus crassipes* is identical
with *Geograpsus grayi*, and if so, his name having priority, must be used for the species, but I prefer to refer to it for the
present under the designation which has been more generally adopted for it, since Dana's description is somewhat insufficient, and may belong to a species of *Pachygrapsus*.

is much enlarged, with the antero-external angle much produced and rounded ; the next joint is articulated at or near the middle of the distal margin of the merus. The chelipedes (in the male) are often considerably developed; the merus is trigonous; carpus without a spine on its interior margin ; palm sometimes compressed but not flattened or concave, on the exterior surface, and usually with a lanate patch of hair on the exterior surface at the base of the fingers, which are distally acute or subacute. Ambulatory legs with the joints not dilated ; dactyli styliform, without marginal spines.

Thus characterised the genus *Pseudograpsus* is nearly allied to *Varuna* (see below), and to a genus represented by two or three species in the collection of the British (Natural History) Museum, which I identify, somewhat doubtfully, with *Ptychognathus*, Stimpson (*Gnathograpsus*, A. Milne Edwards), and which is distinguished from *Pseudograpsus* by the flatter carapace, with nearly horizontal front, and without distinct epigastric lobes, and the enormous development of the exognathi of the exterior maxillipedes. Another genus, represented by a single species from the Fiji Islands and New Hebrides in the collection of the museum, which I will designate *Macrograpsus*, is characterised by the form of the palms and dactyli of the chelipedes in the male, which are greatly dilated and flattened on their exterior surfaces. The species may be designated *Macrograpsus orientalis*. The species of *Pseudograpsus* have been enumerated by Mr. Kingsley.[1]

The genera *Cœlochirus*, Nauck, and *Pachystomum*, Nauck (Zeitschr. f. wiss. Zool., vol. xxxiv. pp. 66, 67, 1880), based on types from the Philippines, and which are allied to *Pseudograpsus*, are too briefly described for certain identification with any of the above-mentioned genera.

Pseudograpsus albus, Stimpson.

Pseudograpsus albus, Stimpson, Proc. Acad. Nat. Sci. Philad., p. 104, 1858.
　　　　　　 „　　　 „　 A. Milne Edwards, Nouv. Archiv. Mus. Hist. Nat., vol. ix. p. 314, pl. xviii. fig. 2, 1873.
　　　　　　 „　　　 „　 Kingsley, tom. cit., p. 205, 1880.

Kandavu, Fiji Islands, July 1874. (Two males).

In the specimens I refer to this species, the front curves slightly downwards, and the epigastric lobes are distinctly defined. The antero-lateral margins also are slightly arcuated.

♂.	Lines.	Millims.
Length of carapace, nearly	4½	9
Breadth of carapace, nearly	5	10

These characters are perhaps the best that can be cited to distinguish this genus from *Ptychognathus*, Stimpson (*Gnathograpsus*, A. Milne Edwards), since in the

[1] *Gnathograpsus pilipes*, A. Milne Edwards, is, for reasons I have indicated below, to be referred, I think, to the genus *Pseudograpsus*.

produced antero-external angle of the merus of the outer maxillipedes, and in the dilatation of their exognath, certain species of the former genus nearly approach *Ptychognathus*. If the genera be thus defined, *Gnathograpsus pilipes*, A. Milne Edwards, will be better placed in *Pseudograpsus* than in *Ptychognathus*, where both Kingsley and myself have placed it.

Platygrapsus, Stimpson.

Platynotus, de Haan, Crust. in v. Siebold, Fauna Japonica, p. 34, pl. D., 1835 ; name previously used.
Platygrapsus, Stimpson, Proc. Acad. Nat. Sci. Philad., p. 104, 1858.

This genus is very nearly allied in external appearance and in nearly all structural points to *Pseudograpsus*, but is distinguished by the form of the exterior maxillipedes, whose ischium-joint is short, very obliquely, not squarely truncated at the distal extremity. The merus-joint is large, distally truncated, and the next joint articulates with it near its antero-external angle, which is not produced as in *Pseudograpsus*; the exognath also is narrower than in that genus. As in *Pseudograpsus*, the carapace is depressed, with the antero-lateral margins slightly arcuated and dentated ; the epigastric lobes are developed ; the post-abdomen of the male does not cover the whole width of the sternum at the base. The chelipedes are well developed, and the dactyli of the ambulatory legs are styliform and without marginal spines.

The single species, *Platygrapsus depressus*, occurs on the shores and islands of China and Japan.

Platygrapsus depressus (de Haan).

Platynotus depressus, de Haan, Crust. in v. Siebold, Fauna Japonica, pp. 34, 63, pl. viii. fig. 2, 1835.
     "   "   Milne Edwards, Ann. d. Sci. Nat., tom. cit., p. 199, 1853.
Platygrapsus depressus, Stimpson, Proc. Acad. Nat. Sci. Philad., p. 104, 1858.
     "   "   Miers, Proc. Zool. Soc. Lond., p. 37, 1879.
     "   "   Kingsley, tom. cit., p. 211, 1880, et synonyma.

Japan, Kobé, 50 fathoms, lat. 34° 38' 0" N., long. 135° 1' 0" E., Station 233a. (A small female bearing ova).

♀.	Lines.	Millims.
Length of carapace,	3	6·5
Breadth of carapace,	3½	7

Brachynotus, de Haan.

Brachynotus, de Haan, Crust. in v. Siebold, Fauna Japonica, decas ii. p. 34, 1835.
„   Hilgendorf, Sitz. Gesellsch. Freunde zu Berlin, p. 68, 1882.
Heterograpsus, Lucas, Anim. articulés in Explor. Sci. de l'Algérie, vol. i. p. 18, 1849.
„   Milne Edwards, Ann. d. Sci. Nat., ser. 3, Zool., vol. xx. p. 192, 1853.
„   Kingsley, Proc. Acad. Nat. Sci. Philad., p. 207, 1880, et synonyms.

This genus is very nearly allied to *Pseudograpsus*, but may be distinguished by the form of the exterior maxillipedes, whose merus-joint is not produced at its antero-external angle as in that genus; the exognath also is much narrower. As in *Pseudograpsus*, there are normally two teeth behind the orbital angle. From *Platygrapsus*, it is distinguished by the shorter merus-joint and the form of the ischium of the exterior maxillipedes, which is squarely, not obliquely truncated, at the distal extremity; *Cyrtograpsus*, Dana, has a differently shaped carapace, with normally three teeth behind the exterior orbital angle, a narrower front, and widely gaping maxillipedes.[1]

The species, which are enumerated by Kingsley, are distributed throughout the Indo-Pacific region, and one, *Brachynotus sexdentatus* occurs in the Mediterranean. To those he mentions, the following is to be added :—

Brachynotus longitarsis (Miers).   Japanese and Corean Seas.

The Oriental forms (true *Heterograpsi*) may perhaps be still regarded as subgenerically distinct from the Mediterranean type of the genus *Brachynotus*, on account of their broader, more depressed carapace, which has a median dorsal H-shaped impression (which does not exist in *Brachynotus sexdentatus*, Risso), by the less prominent and less accentuated lateral marginal teeth, and by the more prominent front.[2]

Brachynotus (*Heterograpsus*) *penicillatus* (de Haan).

Eriocheir penicillatus, de Haan, Crust. in v. Siebold, Fauna Japonica, pp. 32, 60, pl. xi. fig. 6, 1835.
Heterograpsus penicillatus, Stimpson, Proc. Acad. Nat. Sci. Philad., p. 104, 1858.
„   „   Kingsley, tom. cit., p. 209, 1880.

Japan, Oosima, from pools (three adult males).

Adult ♂ .			Lines.	Millims.
Length of carapace, nearly	.	.	9	18·5
Breadth of carapace,	.	.	10	21

---

[1] The species with three antero-lateral marginal teeth, constituting Kingsley's Section B of the genus *Heterograpsus*, would perhaps be better referred to *Cyrtograpsus*.

[2] Dr. Hilgendorf (tom. cit., p. 70) has proposed the name *Brachynotus edwardsii* for *Heterograpsus sexdentatus*, Milne Edwards, not Risso.

Varuna, Milne Edwards.

Varuna, Milne Edwards, Dict. Class. d'Hist. Nat., vol. xvi. p. 511, 1830; Hist. Nat. Crust., vol. ii. p. 94, 1837; Ann. d. Sci. Nat., ser. 3, Zool., vol. xx. p. 176, 1853.

,, Kingsley, Proc. Acad. Nat. Sci. Philad., p. 205, 1880.

Trichopus, de Haan (subgenus), Crust. in v. Siebold, Fauna Japonica, decas ii. p. 32, 1835.

This genus is nearly allied to *Pseudograpsus*, but may be distinguished as follows :— The front and the antero-lateral teeth of the carapace are more prominent ; the front nearly horizontal. The branchial regions of the carapace are laterally defined by a granulated line. The orbits are much deeper, and are widely open below. The exognath of the exterior maxillipedes (although rather broad) is much less developed. The chelipedes (in the male) are subequal and somewhat more slender ; carpus with a spine on its interior surface. The fifth to the seventh joints of the ambulatory legs are ciliated, compressed, and somewhat dilated and adapted for swimming.

This genus contains but a single species, *Varuna litterata*, which occurs commonly throughout the Indo-Pacific region.

Varuna litterata (Fabricius).

Cancer litteratus, Fabricius, Entem. Syst. Suppl., p. 342, 1798.

Varuna litterata, Milne Edwards, Dict. Class. d'Hist. Nat., vol. xvi. p. 511, 1830; Hist. Nat. Crust., vol. ii. p. 95, pl. xiv. *bis*, fig. 8, 1837; Ann. d. Sci. Nat, *tom. cit.*, p. 176, 1853.

,, ,, Dana, U.S. Explor. Exped., vol. xiii., Crust , p. 336, pl. xx. fig. 8, 1852.

,, ,, Kingsley, *tom. cit.*, p. 205, 1880, and references to literature.

Philippines, Pasuanaca, Mindanao, in the rivers (a male); north coast of New Guinea (three males and three females); Banda, fresh water (a young male).

In these specimens the frontal margin is perfectly straight, and is less rounded at the lateral angles than is usual in the genus.

The colour (in spirit) is a deep chocolate-brown ; on either side of the gastric region and in front of the branchial regions are three or four yellow spots.

	Lines.	Millims.
Length of carapace, nearly	21	44
Breadth of carapace, .	23½	50

Epigrapsus, Heller.

Epigrapsus, Heller, Verhandl. d. k. k. zool.-bot. Gesellsch. Wien, vol. xii. p. 522, 1862.

,, Kingsley, Proc. Acad. Nat. Sci. Philad., p. 192, 1880.

Nectograpsus, Heller, Crust. in Reise der "Novara," p. 56, 1865.

Carapace depressed, smooth and polished on the dorsal surface, nearly quadrate, with the antero-lateral angles rounded and entire. Front of moderate width, deflexed ; no postfrontal crest. Orbits small, the superior margin entire, the inferior margin scarcely defined, the interior subocular lobe small and dentiform; from the base of this lobe a raised line or crest extends beneath the orbits in a straight line continuous with the anterior margin of the epistoma nearly to the subhepatic regions. Epistoma transverse and very broad. Ridges of the endostome very indistinct. Post-abdomen in the male distinctly seven-jointed, the terminal segment narrower than the preceding; it covers the whole width of the sternum at the base. Eye-peduncles small. Antennules transversely plicated and partially concealed by the deflexed front. Antennæ very small; the basal joint short, quadrate, and occupying the interior orbital hiatus. Exterior maxillipedes rather short and broad, with a rhomboidal gape; the merus of the endognath oblique, nearly as large as the ischium, distally truncated, and bearing the next joint near its antero-external angle. Chelipedes (in the adult male) well developed; merus trigonous, unarmed; carpus without a spine on its interior surface; palm smooth, rounded above; fingers subacute. Ambulatory legs rather slender and smooth; dactyli styliform and spinulose.

The nearest ally to this genus is probably *Acmæopleura*, very briefly characterised by Stimpson, in which form, however, the orbits are nearly complete inferiorly, and the maxillipedes have not a rhomboidal gape.

The single described species of *Epigrapsus* (*Epigrapsus politus*) occurs at the Nicobars and various islands of the Polynesian group (New Hebrides, Fiji Islands, Tahiti).

Epigrapsus politus, Heller.

Epigrapsus politus, Heller, Verhandl. d. k. k. zool.-bot. Gesellsch. Wien, xii. p. 522, 1862.
„ „ Kingsley, tom. cit. p. 192, 1880.
Nectograpsus politus, Heller, Crust. in Reise der " Novara," p. 57, pl. v. fig. 3, 1865.

New Hebrides, Api (an adult male).

In this specimen, and in others from the Fiji Islands in the collection of the British (Natural History) Museum, the carapace is slightly broader in front and narrower behind than in the figure cited, and all the post-abdominal segments in the male are distinct.

Adult ♂.	Lines.	Millims.
Length of carapace,	6½	14
Greatest breadth of carapace,	8	17

Goniopsis, de Haan.

Goniopsis, de Haan (pt.), Crust. in v. Siebold, Fauna Japonica, decas ii. p. 33, 1835.
„ Milne Edwards, Ann. d. Sci. Nat., ser. 3, Zool. xx. p. 164, 1853.
„ Kingsley, Proc. Acad. Nat. Sci. Philad., p. 189, 1880.

Carapace vertically thick, subquadrate, flat above, with the cervical suture strongly defined, the lateral margins armed with a single tooth behind the exterior orbital angle, the anterior margin abruptly four-lobed above the front, which is broad and vertically deflexed. Orbits rather large, with two hiatuses below, and with the interior subocular lobe well-developed. Epistoma very short, transverse. The ridges of the endostome distinctly developed. Post-abdomen (in the male) rather broadly triangulate, distinctly seven-jointed, and occupying the whole width of the sternum at the base between the coxæ of the fifth ambulatory legs. Eye-peduncles short, robust. Antennules transversely plicated in the narrow, wide fossettes. Antennæ with the basal joint broad and very short; they are completely excluded from the orbit by the intervening subocular orbital lobe. Exterior maxillipedes narrow, widely gaping and remote from one another at the base (as in *Grapsus*); the merus of the endognath nearly as long as the ischium, widening to the distal extremity, which is truncated; the next joint articulated near the antero-external angle of the merus; exognath very slender. Chelipedes (in the adult male) robust; merus trigonous, with the anterior margin dentated; carpus and hand strongly granulated; palm somewhat compressed; fingers but slightly excavated or subacute. Ambulatory legs robust, with the merus-joints broadly dilated and compressed; penultimate joints spinuliferous and dactyli armed with strong spines.

The genus is somewhat intermediate between *Grapsus*, from which it is distinguished by the great development of the interior subocular lobe and form of the carapace, and *Sesarma*.

There are but two species:—*Goniopsis cruentatus*, whose synonyms and wide distribution are referred to by Kingsley and S. J. Smith, and *Goniopsis pulcher* (Lockington), which is apparently distinguished by the coloration, and occurs on the coasts of Lower California and Panama (Coll. Brit. Mus.).

Goniopsis cruentatus (Latreille).

Grapsus cruentatus, Latreille, Hist. Nat. Crust. et Ins., vol. vi. p. 70, 1803–1804.
    „   „   Milne Edwards, Hist. Nat. Crust., vol. ii. p. 85, 1837.
Goniopsis cruentatus, de Haan, Crust. in v. Siebold, Fauna Japonica, p. 33, 1835.
    „   „   Milne Edwards, Ann. d. Sci. Nat., tom. cit., p. 164, pl. vii. figs. 2, 2a, b, 1853.
    „   „   Kingsley, tom. cit., p. 190, 1880, et synonyma.
Goniograpsus cruentatus, Dana, U.S. Explor. Exped., vol. xiii., Crust., p. 342, pl. xxi. fig. 7, 1852.

Bermuda (a large series of specimens, males and females); some of these are labelled as from Hungry Bay, in the mangrove swamps.

Adult ♂.		Lines.	Millims.
Length of carapace,	.	18	40·5
Breadth of carapace,	.	22½	47·5

Helice, de Haan.

Helice, de Haan (subgenus), Crust. in v. Siebold, Fauna Japonica, decas ii. p. 28, 1835.
„ Milne Edwards, Ann. d. Sci. Nat., ser. 3, Zool., vol. xx. p. 189, 1853.
„ Kingsley, Proc. Acad. Nat. Sci. Philad., p. 219, 1880.

Carapace dorsally thick and convex, transverse, quadrate, with the antero-lateral margins straight and dentated. Front of moderate width and curving downwards, but not abruptly deflexed ; the anterior margin not projecting as in the genus or subgenus *Paragrapsus*. The orbits are large. The epistoma is transverse and very short. The endostomian ridges are distinctly defined. The post-abdomen in the male is distinctly seven-jointed, and at the base does not quite cover the whole width of the sternum between the bases of the fifth ambulatory legs. The eye-peduncles are rather short, and do not reach the antero-external angle of the orbit. The antennules are transversely plicated. The basal antennal joint is very small, and does not reach the extero-lateral angle of the front, and it lies within the interior hiatus of the orbit. The exterior maxillipedes have between them, when closed, a rhomboidal gape ; the merus of the endognath is externally concave, as long or nearly as long as the ischium, and widens to the distal extremity, which is truncated or slightly concave ; the following joint is articulated at or near the antero-external angle of the merus ; the exognath is very narrow. The chelipedes (in the adult male) are subequal and moderately developed ; merus trigonous ; carpus angulated, but usually without a spine at the interior angle ; palm rounded or subcarinated above, fingers subexcavated at the distal extremity. Ambulatory legs rather slender, with the merus-joints not markedly dilated ; the dactyli slender, styliform, and unarmed.

This genus is very nearly allied to *Chasmagnathus*, de Haan, of which it should, perhaps, be considered a subgenus. It is scarcely distinguished from it by any constant character, except the nearly straight, not arcuated, antero-lateral margins of the carapace in the male. From *Paragrapsus*, Milne Edwards, which Mr. Kingsley unites with *Chasmagnathus*, it is further distinguished by the thicker body and the non-projecting anterior margin of the front.

Besides the species enumerated by Kingsley, the following should perhaps be referred to *Helice* :—

  Helice (?) *longipes* (*Sesarma longipes*, Krauss). South Africa, Umlaas (? Umlazi) River.

Helice latreillei, Milne Edwards, var. (Pl. XXI. fig. 2).

Cyclograpsus latreillei, Milne Edwards, Hist. Nat. Crust., vol. ii. p. 80, 1837.
Helice latreillei, Milne Edwards, Ann. d. Sci. Nat., tom. cit., p. 190, 1852.
„ „ Kingsley, tom. cit., p. 220, 1880.

Philippine Islands, Samboangan, 10 fathoms, on the reefs (a small male); Fiji Islands, Kandavu (an adult female).

In the specimens, referred somewhat doubtfully to this species,[1] the carapace and chelipedes are punctulated and the frontal and gastric regions marked with a median posteriorly bifurcating suture. In the larger (female) specimen there is no trace of the posterior (fourth) rudimentary tooth of the lateral margins, which is very faintly indicated on one side only in the smaller male. The female specimen has also a relatively somewhat broader carapace, with the sides anteriorly slightly arcuated, and the antero-lateral teeth less prominent, as in the species of *Chasmagnathus*, and may perhaps belong to a distinct species, on which account I think it useful to figure it. Its dimensions are as follows :—

	Lines.	Millims.
Adult ♀.		
Length of carapace, .	6	12·5
Breadth of carapace, nearly	8	16·5

Helice crassa, Dana.

> *Helice crassa*, Dana, Proc. Acad. Nat. Sci. Philad., p. 252, 1851 ; U.S. Explor. Exped., vol. xiii., Crust., p. 367, pl. xxxii. fig. 8, 1852.
> „ „ Kingsley, *tom. cit.*, p. 220, 1880, et synonyma (?).

Port Jackson, Sow and Pig's Bank, 6 fathoms (a small male).

	Lines.	Millims.
♂.		
Length of carapace, nearly	4	8
Breadth of carapace, .	4½	9

The carapace in this specimen, which is not fully grown, is flatter, and the body more depressed than is usual in the genus.

Sesarma, Say.

> *Sesarma*, Say, Journ. Acad. Nat. Sci. Philad., vol. i. p. 76, 1817.
> „ Milne Edwards, Ann. d. Sci. Nat., ser. 3, Zool., vol. xx. p. 181, 1853.
> „ Kingsley, Proc. Acad. Nat. Sci. Philad., p. 213, 1880, et synonyma.

Body thick. Carapace quadrate, and normally broader than long, with the sides straight, not arcuated, entire or dentated. Front rather broad, and abruptly deflexed, with the anterior margin straight or sinuated ; the epigastric lobes prominent, and defined externally by a suture (except in the forms separated by Milne Edwards as *Holometopus*). Orbits large, with a hiatus below the exterior angle; the interior subocular lobe is moderately developed and does not exclude the antenna from the orbital cavity. Epistoma transverse and very short. Endostomian ridges usually distinctly developed. Post-abdomen (in the male) distinctly seven-jointed, and covering the whole width of the

[1] They agree closely with specimens from the Mauritius, whence were obtained the types in the British (Natural History) Museum.

sternum between the bases of the fifth ambulatory legs. Eye-peduncles short, robust.
Antennules transversely or almost transversely plicated. Antennæ situated in the interior
orbital hiatus ; the basal joint is very short and does not reach the front, and is slightly
produced at its antero-lateral angle ; flagellum of moderate length. Exterior maxillipedes
narrow, with a rhomboidal gape ; the merus of the endognath is somewhat elongated and
is rounded at the distal extremity, where the following joint is articulated ; exognath
small and slender. Chelipedes (in the adult male) usually subequal and moderately
developed ; merus trigonous ; carpus usually angulated, without a spine on the interior
margin ; palm slightly compressed ; fingers distally subacute, or but slightly excavated.
Ambulatory legs moderately elongated ; merus-joints compressed, and usually armed with
a subdistal spine or tooth on their superior margins ; dactyli styliform, slightly
compressed, and without spines.

The described species are very numerous, and are of common occurrence on the shores
and in the shallow waters of all the tropical and subtropical regions of the globe. They
occur on the African coast southward at least to Natal, and one species (*Sesarma
pentagona*, Hutton) has been described from New Zealand.

The following are forms which are not referred to by Mr. Kingsley in his list of the
species of this genus. Except *Sesarma miniata*, they are all too recently described to
be inserted in it :—

     Sesarma miniata, de Saussure.   West Indies.
     Sesarma granosimana, Miers.   Indo-Malaysian Seas.
     Sesarma büttikoferi, de Man.   Liberia.
     Sesarma kamermani, de Man.   Congo Coast.[1]

Sesarma mülleri, A. Milne Edwards (Pl. XXI. fig. 3).

Sesarma mülleri, A. Milne Edwards, Nouv. Archiv. Mus. Hist. Nat., vol. v., Bulletin, p. 29, 1869.
     „   „   Kingsley, *tom. cit.*, p. 216, 1880.

Bahia, shallow water (a small male).

The identification is somewhat uncertain, since the diagnosis of the species is brief.[2]

♂ .					Lines.	Millims.
Length of carapace, rather over .		.	.	.	5	11
Breadth of carapace,	.	.	.	.	6½	13·5

[1] I have proposed (*Proc. Zool. Soc. Lond.*, p. 70, 1881) *Sesarma stimpsonii* as an alternative name for specimens
from Monte Video, doubtfully referred to *Sesarma angustipes*.

[2] This specimen differs from much larger examples in the collection of the British Museum, which have also been
referred to *Sesarma mülleri*, in having the frontal margin perfectly straight without a median sinus, the sutures defining
the postfrontal or epigastric lobes less deep and distinct, the sides of the carapace behind the exterior angle of the orbit
more distinctly convergent ; the superior margin of the dactyl of the chelipedes is less distinctly granulated, but the
palm is more distinctly granulated. The carapace in the largest specimen in the British Museum measures nearly 16
lines (33·5 mm.) in length, and about 17½ lines (37 mm.) in width.

Sesarma schüttei, Hess.

Sesarma schüttei, Hess, Archiv f. Naturgesch., vol. xxxi. p. 150, pl. vi. fig. 11*, 1865.
„   „   Kingsley, *tom. cit.*, p. 217, 1880.
„   „   Haswell, Cat. Aust. Crust., p. 109, 1882.

New South Wales, Hawkesbury, brackish water (an adult male).

This specimen is without the chelipedes, but it agrees in all particulars with the description of Hess. The colour (in spirit) is purplish or blackish-brown.

Adult ♂.	Lines.	Millims.
Length of carapace, nearly	8½	17·5
Breadth of carapace, nearly	10	20·5

Sesarma (Holometopus) aubryi, A. Milne Edwards.

Sesarma (Holometopus) aubryi, A. Milne Edwards, Nouv. Archiv. Mus. Hist. Nat., vol. v., Bulletin, p. 29, 1869 ; vol. ix. p. 307, pl. xvi. fig. 3, 1873.
„   *aubryi*, Kingsley, *tom. cit.*, p. 214, 1880.

New Hebrides, Api (in sea-water), a series of specimens (males and females); Arrou Islands (a female); Wild Islands, Admiralty Island (on the beach), several specimens, males and females.

Adult ♂.	Lines.	Millims.
Length of carapace, about	4½	10
Breadth of carapace, about	5	11
Length of a chelipede, .	5½	11·5
Length of first ambulatory leg,	6½	13·5

Subfamily 2. PLAGUSIINÆ.

Plagusiinæ, Dana, U.S. Explor. Exped., vol. xiii., Crust. 1, p. 333, 1852.
„   Miers, Ann. and Mag. Nat. Hist., ser. 5, vol. i. p. 147, 1878.
„   Kingsley, *tom. cit.*, p. 223, 1880.

Carapace much depressed ; antennules transverse, and exposed and visible in a dorsal view in deep longitudinal clefts or sinuses of the front.

This subfamily includes only the genera *Plagusia*, Latreille, and *Leiolophus*, Miers (=*Acanthopus*, de Haan, *nom. præoc.*).

Plagusia, Latreille.

Plagusia, Latreille (pt.), Gen. Crust. et Ins., vol. i. p. 33, 1806.
„   Milne Edwards (pt.), Hist. Nat. Crust, vol. ii. p. 90, 1837 ; Ann. d. Sci. Nat., ser. 3, Zool., vol. xx. p. 178, 1853.
„   Miers, Ann. and Mag. Nat. Hist., ser. 5, vol. i. p. 148, 1878, et synonyma.

Carapace depressed, and little, if at all, broader than long ; with the antero-lateral margins dentated, and slightly divergent to the bases of the third pair of legs. The front is not deflexed, is less than half the width of the carapace, and is deeply cleft by the antennulary fossæ; these fossæ, in which the antennules are vertically plicated, are open in a dorsal view.

The orbits are deep, and their interior hiatuses are large and open. The epistoma is very short and transverse. The anterior margin of the buccal cavity is cristated. The endostomian ridges are usually distinctly developed. The post-abdomen (in the male) is broad and triangulate, and occupies at the base the whole width of the sternum between the coxæ of the fifth ambulatory legs, and is distinctly seven-jointed. The eye-peduncles are very short and robust. The basal joint of the antennæ is considerably dilated and is slightly produced at its anterior angles; the antero-internal angle reaches the front, and the following joint is articulated in the middle of the distal margin of the basal antennal joint.

The exterior maxillipedes are rather small; the merus of the endognath is shorter than the ischium, but not narrower at the base than that joint, distally truncated, and bears the next joint at its summit; the exognath is slender. Chelipedes (in the male) subequal, and moderately developed; merus trigonous, carpus and palm usually granulated; palm rounded above and not dilated; fingers distally excavated, with corneous tips. Ambulatory legs large and robust; the merus-joints compressed, and with one or more spines on the anterior margin; dactyli strongly spinuliferous.

In the nearly allied genus *Leiolophus*, the merus-joint of the exterior maxillipedes is very small; much narrower at the base than the distal extremity of the ischium, and the dorsal surface of the carapace is marked with smooth naked lines or ridges.

Plagusia depressa (Fabricius).

Cancer depressus, Fabricius, Entom. Syst. Suppl., p. 406, 1775.
Cancer squamosus, Herbst, Naturgesch. der Krabben u. Krebse, vol. i. p. 260, pl. xx. fig. 113, *bad*, 1790.
Plagusia depressa, Say, Journ. Acad. Nat. Sci. Philad., vol. i. p. 100, 1815.
,,   ,,   Miers, Ann. and Mag. Nat. Hist., ser. 5, vol. i. p. 149, 1878, and references to synonyms.
,,   ,,   de Man, Notes Leyden Mus., vol. v. p. 168, 1883.
,,   *tuberculata*, Lamarck, Hist. Anim. sans Vert., p. 247, 1818.
,,   ,,   (var.), Miers, *tom. cit.*, p. 148, 1878, et synonyms.

St. Vincent, Cape Verde Islands, July 1873 (an adult male).

In this specimen the lobes above the bases of the second and third ambulatory legs are either not at all or very obscurely dentated. There can be little doubt that the Indo-Pacific form (*Plagusia tuberculata*, Lamarck), which I, following the late Dr. Stimpson, regarded as a distinct species, on material which further research has shown to be insufficient, is to be regarded as merely a variety of the Atlantic *Plagusia depressa*.

Adult ♂.	Lines.	Millims.
Length of carapace, nearly	21	44
Breadth of carapace, nearly	22½	47·5

Plagusia immaculata, Lamarck (Pl. XXII. fig. 1).

> *Plagusia immaculata*, Lamarck, Hist. Anim. sans Vert., v. p. 247, 1818.
> ,,   ,,   Miers, Ann. and Mag. Nat. Hist., tom. cit., p. 150, 1878.
> ,,   *depressa*, Latreille, Encycl. Méth., x. p. 145, 1825.
> ,,   ,,   Milne Edwards, Hist. Nat. Crust., vol. ii. p. 93, 1837; Ann. d. Sci. Nat., ser. 3, Zool., vol. xx. p. 179, 1853, *nec Cancer depressus*, Fabricius.

North of New Guinea, from driftwood, February 21, 22, 1875 (a female and several small males).

The largest male measures as follows:—

♂.	Lines.	Millims.
Length of carapace, nearly	8½	17·5
Breadth of carapace,	9	19

It is possible that this form, like the preceding, may prove to be merely a small convex, smoother and glabrous variety of *Plagusia depressa*, but I hesitate to unite it with that species on the authority of the series at present contained in the collection of the British (Natural History) Museum and in the Challenger collection.[1]

Plagusia chabrus (Linné) (Pl. XXII. fig. 1d).

> *Cancer chabrus*, Linné, Mus. Lud. Ulrici, p. 438, 1764; Syst. Nat., ed. xii. p. 1044, 1766.
> *Plagusia chabrus*, White, Ann. and Mag. Nat. Hist., vol. xvii. p. 497, 1846.
> ,,   Miers, tom. cit., p. 152, 1878.
> ,,   *tomentosa*, Milne Edwards, Hist. Nat. Crust., vol. ii. p. 92, 1837; Ann. d. Sci. Nat., ser. 3, Zool., vol. xx. p. 178, 1853.
> ,,   ,,   Krauss, Die Süd-Afrik. Crust., p. 42, pl. ii. fig. 6, 1843.

South Africa, Simon's Bay, 10 to 20 fathoms (an adult male and a young example).[2]

Adult ♂.	Lines.	Millims.
Length of carapace,	19	40·5
Breadth of carapace, about	21	44

[1] The British Museum has lately received specimens of the very distinct *Plagusia speciosa*, Dana (of which I had seen no specimens when I revised the species of this subfamily in 1877), from Paumotu.

[2] In the South African specimens of this species, which I have examined, the branchial regions are smooth, or very nearly so; in specimens from New Zealand they bear several small ridges or tubercles near the posterior lateral angles, but I do not think this to be a character of specific importance, though a larger series might show it to be so.

Family IV. PINNOTHERIDÆ.

Pinnotheriens, Milne Edwards, Hist. Nat. Crust., vol. ii. p. 28, 1837.

Carapace usually more or less membranaceous, convex or depressed, with the antero-lateral margins entire or but very slightly dentated ; front, orbits, and eye-peduncles very small. The post-abdomen of the male usually does not cover the sternum between the bases of the fifth ambulatory legs. The buccal cavity is usually arcuated anteriorly. The merus (and usually the ischium) of the exterior maxillipedes is well developed, and the carpal joint is articulated at the summit or at the antero-internal angle, or more rarely at the antero-external angle of the merus. Chelipedes (in the adult males) small, or moderately developed ; ambulatory legs slender and usually naked ; dactyli styliform, unarmed.

The species are of small size, and many are found living within the shells of bivalve Mollusca, *Echini*, tubicolous Annelida, &c.

However characterised, the Pinnotheridæ constitute at present a somewhat hetero-geneous group, which I think it most convenient to define in the sense originally indicated by Milne Edwards in 1834. In his later monograph (1852, 1853) Milne Edwards separated *Myctiris, Scopimera*, and *Doto (Dotilla)* from the Pinnotheridæ. By Dana (1852) *Myctiris* is constituted the type of a distinct family and arranged next to the Pinnotheridæ, but *Scopimera* and *Dotilla* are widely separated from *Myctiris* and placed in the Macrophthalmidæ.

The few forms occurring in the collection of H.M.S. Challenger must be arranged in three distinct subfamilies, as will be seen in the following synoptical arrangement of the genera of this group, wherein I have recharacterised the subdivisions established by Milne Edwards, and have added a new subfamily to those previously established, which includes a few forms not represented in the Challenger collection.

Subfamily 1. PINNOTHERINÆ.

Pinnotherinæ, Milne Edwards, Ann. d. Sci. Nat., ser. 3, Zool., vol. xx. p. 216, 1853.
    „   Dana, U.S. Explor. Exped., vol. xiii., Crust. 1, p. 378, 1852.

Carapace usually convex, subglobose or transverse. Front not rostrated. Ischium joint of the exterior maxillipedes usually rudimentary or very short ; dactyli of the ambulatory legs very short. The ambulatory legs all well developed.

Genera :—*Pinnotheres*, Latreille ; *Ostracotheres*, Milne Edwards ; *Pinnixa*, White (= *Tubicola*, Lockington) ; *Xanthasia*, White ; *Pinnotherelia*, Lucas ; *Fabia*, Dana ; *Pinnaxodes*, Heller (?) ; *Dissodactylus*, Smith (?) ;[1] *Mala-cosoma*, de Man ; *Holothuriophilus*, Nauck.

[1] This genus is regarded by S. I. Smith as typical of a distinct subfamily.

Subfamily 2. HEXAPODINÆ.

† *Asthenognathidæ*, Stimpson, Proc. Acad. Nat. Sci. Philad., p. 107, 1858, *descript. nulli.*

Characters of Pinnotherinæ, but the ischium of the exterior maxillipedes and the dactyli of the ambulatory legs are longer and more distinctly developed, and the ambulatory legs of the fifth pair, which are often short in *Pinnixa*, are altogether aborted or represented by a mere rudiment.

Genera :—*Hexapus*, de Haan ; *Amorphopus*, Bell ; *Thaumastoplax*, Miers ; *Asthenognathus*, Stimpson, may belong to this group, but nothing is said with regard to the form of the fifth ambulatory legs, if developed ; *Xenophthalmus*, White, which is somewhat intermediate between this group and the Pinnotherinæ, differs from the genera of both groups in the form of the orbits, which are narrow, longitudinal and open above.

Subfamily 3. MYCTIRINÆ.

Carapace very convex, usually subglobose. Front rostrated ; rostrum deflexed. The exterior maxillipedes bulge out beyond the buccal cavity, and the merus and ischium joints are largely developed. Dactyli of the ambulatory legs styliform, well developed. This subfamily establishes the transition to the Ocypodidæ.

Genera :—*Myctiris*, Latreille ; *Dotilla*, Stimpson (*Doto*, de Haan) ; *Scopimera*, de Haan.

Subfamily 4. HYMENOSOMINÆ.

Hymenosominæ, Milne Edwards, *tom. cit.*, p. 221, 1853.
Hymenicinæ, Dana, *tom. cit.*, p. 379, 1852.

Carapace flattened on the dorsal surface, more or less, triangulate. Front rostrated. Ischium of the exterior maxillipedes and dactyli of the ambulatory legs well developed.

Genera :—*Hymenosoma*, Desmarest ; *Halicarcinus*, White (= *Liriopea*, Nicolet) ; *Hymenicus*, Dana ; *Elamena*, Milne Edwards ; *Elamenopsis*, A. Milne Edwards ; *Trigonoplax*, Milne Edwards ; *Rhynchoplax*, Stimpson.

Subfamily 1. PINNOTHERINÆ.

Pinnotheres, Latreille.

Pinnotheres, Latreille, Hist. Nat. Crust. et Ins., vol. vi. p. 78, 1803–1804.
„ Milne Edwards, Hist. Nat. Crust., vol. ii. p. 30, 1837; Ann. d. Sci. Nat., ser. 3, Zool., vol. xx. p. 216, 1853.
Pinnotheres, Dana, U.S. Explor. Exped., vol. xiii., Crust. i, p. 378, 1852.

Body soft and membranaceous; carapace smooth, subglobose, the regions not defined, the lateral margins regularly arcuated; orbits very small, usually nearly circular, with the margins subentire and very little prominent. Front narrow, with the anterior margin nearly straight. Epistoma very short and transverse. Buccal cavity wide and transverse, and anteriorly arcuated, the endostomian ridges obsolete or nearly so. Post-abdomen narrow and distinctly seven-jointed, its base does not cover the whole width of the sternum between the coxæ of the fifth ambulatory legs. Eye-peduncles very short. Antennules transversely plicated within the small fossettes, which are little broader than long. Antennæ very small and placed within the anterior hiatus of the orbit; basal antennal joint slender and short, barely attaining the infero-lateral angle of the front. Exterior maxillipedes oblique, with the ischium-joints of the endognath short or rudimentary, the merus very large, usually curved, and widening more or less to the distal extremity, which is rounded or subtruncated, the small dactyl articulated on the interior margin of the penultimate joint. Chelipedes in the male small, with the joints smooth, merus trigonous, carpus without a spine on its anterior margin, palm slightly compressed and rounded above and below, fingers acute. Ambulatory legs of moderate length and very slender, with the joints smooth; dactyli slightly arcuated and usually shorter than the penultimate joints.

The species are all of small size, and (as is well known), inhabit the shells of *Mytilus*, *Pinna*, and other bivalves, and are found probably in all the temperate and tropical regions of the globe.

The following species are either not referred to or have been added since the publication of Milne Edwards's Memoir in 1853. As the discrimination of the species is difficult, and a revision of the genus is much needed, it is possible that some are not really specifically distinct.

Pinnotheres maculatum, Say. ⎫
Pinnotheres byssomiæ, Say. ⎬ All from the Eastern and Southern Coasts of
Pinnotheres depressum, Say. ⎪ the United States.
Pinnotheres monodactylum, Say. ⎭
Pinnotheres pholadis, de Haan. Japan.
Pinnotheres obscurus, Stimpson. Hong-Kong.
Pinnotheres boninensis, Stimpson. Bonin Island.
Pinnotheres parvulus, Stimpson. China Seas.
Pinnotheres margarita, Smith. Bay of Panama, Pearl Islands, and Lower California.
Pinnotheres lithodomi, Smith. Pearl Islands and Lower California.
Pinnotheres ascidiicola, Hesse. Coast of France.
Pinnotheres pectunculi, Hesse. France.
Pinnotheres angelicus, Lockington. Gulf of California.
Pinnotheres flavus, Nauck. Philippines.

Pinnotheres villosulus, Guérin-Méneville (Pl. XXII. fig. 2).

Pinnotheres villosulus, Guérin-Méneville, Crust. in Voy. de la "Coquille," vol. ii. p. 13, 1830; Icon. du Règne Animal, Crust., p. 7, pl. iv. fig. 6.

„   „   Milne Edwards, Ann. d. Sci. Nat., ser. 3, Zool., vol. xx. p. 218, pl. xi. fig. 8, 1853.

?   „   „   Miers, Crust. in Rep. Zool. Coll. H.M.S. "Alert," p. 247, 1884.

Torres Strait (found within the pearl-oyster). An adult female was presented to the Challenger staff at Sydney, in June 1874.

As this specimen differs somewhat from the descriptions and figures in the form of the exterior maxillipedes, &c., I append the following description. I had at first regarded it as specifically distinct.

The body is everywhere clothed with a short close pubescence. The carapace (in the single specimen examined) is somewhat indurated for a species of this genus and rather flat, with an obscure prominence on the gastric and cardiac region; the general outline is nearly circular, the lateral margin is thin-edged and acute on the sides of the branchial regions. The front is deflexed and trilobate, the median lobe broad, triangulate, the lateral lobes small and dentiform, and constituting the inner angles of the orbits, which are small, nearly circular, with a wide inner hiatus. The post-abdomen (in the female) covers the whole sternal surface, and all the segments are distinct. The eyes are very small, with short thick peduncles and minute terminal cornea. The antennules lie somewhat obliquely in very large fossettes. The joints of the peduncles of the antennæ are large and robust, and are apparently partly coalescent with one another and with the inferior wall of the orbit; the flagellum minute and rudimentary. The outer maxillipedes are of moderate size; merus elongated, and widening somewhat to the distal extremity; carpus and penultimate joints robust and rather short, the penultimate joint but slightly rounded at the distal extremity, the dactylus minute, articulated at the middle of the inner margin of the penultimate joint, and not nearly reaching to its distal extremity; exognath rudimentary. The chelipedes (in the female) are equal and moderately robust; merus and carpus subequal, without spines or teeth; palm little longer than the carpus, rounded above and smooth on its outer surface; fingers meeting along their inner edges and acute at the apices. Ambulatory legs moderately robust and slightly compressed; dactyli slender, acute, half as long as the penultimate joints. Colour (in spirit) greyish-brown. The single specimen (an adult female) has the following dimensions :—

Adult ♀.							Lines.	Millims.
Length of carapace,	.	.	.	.	.	.	6½	13·5
Breadth of carapace, nearly	.	.	.	.	.	.	7	14
Length of a chelipede, rather over	.	.	.	.	.	6	13
Length of second ambulatory leg, about		.	.	.	.	7	15

Subfamily 2. Myctirinæ.

Myctiris, Latreille.

Myctiris, Latreille (Mictyris), Crust. in Cuvier, Règne Anim., ed. 1, vol. iii. p. 21, 1817.
„   Milne Edwards (Myctiris), Hist. Nat. Crust., vol. ii. p. 36, 1837 ; Ann. d. Sci. Nat.,
      ser. 3, Zool., vol. xviii. p. 154, 1852.

Body very thick and dorsally convex, nearly smooth, with the cervical and cardiaco-branchial sutures strongly defined. There are no distinct orbits, but a small postocular spine. The front is very narrow and deflexed, and is triangulate or subacute at the distal extremity. The epistoma is very short and transverse. The post-abdomen in the male, as in the female, is broad, and widens from the base to the penultimate segment ; the terminal segment is narrow, and rounded at the distal extremity. The eyes are very short and exposed in a dorsal view. The antennules and antennæ are very small ; the antennæ in particular very short and slender, and inserted beneath the bases of the eye-peduncles. The exterior maxillipedes project very markedly from the buccal cavity, and are closely applied to one another along their inner margins ; the ischium is largely developed and truncated at the distal extremity ; the merus is somewhat narrowed and rounded at the distal extremity, where it bears the following joint. The chelipedes (in the male) are slender and moderately elongated ; carpus and palm compressed, and usually carinated above, carpus usually longitudinally sulcated on the exterior surface ; palm very short ; fingers straight and acute ; dactylus with only one large tooth on the inner margin near the base. Ambulatory legs slender and somewhat elongated ; dactyli nearly straight, and not denticulated on the margins.

The following species are referable to this genus, Myctiris longicarpus, White (= Myctiris brevidactylus, Stimpson ?), common on the shores of Australia, and occurring in the Malaysian subregion and at New Caledonia, and Myctiris platycheles, Milne Edwards (= Myctiris subverrucatus, White), common in Australia.

Myctiris deflexifrons, a species mentioned by de Haan, has apparently never been described.

Myctiris longicarpus, Latreille.

Mictyris longicarpus, Latreille, Gen. Crust. et Ins., vol. i. p. 41, 1806.
Myctiris longicarpus(is), Milne Edwards, Hist. Nat. Crust., vol. ii. p. 37, 1837 ; Crust. in Cuvier,
      Règne Anim., pl. xviii. fig. 2 ; Ann. d. Sci. Nat., ser. 3, Zool., vol. xviii. p. 154,
      1852.
Mycteris longicarpus, Haswell, Cat. Australian Crust., p. 116, 1882.
Myctiris brevidactylus, Stimpson, Proc. Acad. Nat. Sci. Philad., p. 99, 1858, var.

Albany Island, near Cape York, North Australia. Numerous specimens, none of large size, were collected.

An adult male measures as follows :—

Adult ♂.	Lines.	Millims.
Length of carapace, .	6½	13·5
Breadth of carapace, .	5½	11·5

This species nearly resembles the following, but may always be distinguished by the smooth, not granulated, dorsal surface of the carapace.

Myctiris platycheles, Milne Edwards.

> *Myctiris platycheles*, Milne Edwards, Ann. d. Sci. Nat., *tom. cit.*, p. 154, 1852.
> " Targioni-Tozzetti, Crostacei del Viaggio della Magenta, p. 186, pl. xi. fig. 6, *a–e*, 1877.
> " " Haswell, *tom. cit.*, p. 117, 1882 (*Myctiris*).
> = *subverrucatus*, White, List Crust. Brit. Mus. p. 34, 1847; *descript. nullâ.*
> " " Kinahan, Journ. Roy. Dublin Soc., vol. i. p. 123, 1858.

New South Wales, Botany Bay (beach).　A good series of specimens.

Adult ♂.	Lines.	Millims.
Length of carapace, .	8	17
Breadth of carapace, nearly	7	14

Subfamily 3. HYMENOSOMINÆ.

Hymenosoma, Leach.

> *Hymenosoma*, Leach (*ined.* ?), Desmarest, Consid. sur les Crust., p. 163, 1825.
> " Milne Edwards, Hist. Nat. Crust., vol. ii. p. 35, 1837; Ann. d. Sci. Nat., ser. 3, Zool., vol. xx. p. 222, 1853.

Carapace dorsally flattened, and rounded on the posterior and lateral margins, which are defined by a granulated line or ridge, which is not dentated, as in *Halicarcinus;* beneath this line the carapace usually slopes outwards obliquely to the bases of the ambulatory legs. The front is simple, triangulate, and nearly horizontal ; the hepatic region is tuberculated. There is a distinct postocular spine or tooth. Scarcely any trace exists of an epistoma, or of longitudinal ridges on the endostome. The post-abdomen of the male is very narrow, and is six-jointed, and its base barely occupies the whole width of the sternum between the last ambulatory legs; the sternum is large and circular. The eye-peduncles are short ; the antennulary fossettes are not separated from the orbits ; the antennæ are somewhat elongated, and their basal joint is very short. The exterior maxillipedes are not arcuated, and have the ischium and merus-joints well developed, the merus usually obliquely truncated, and bearing the following joint near its antero-external angle, at the distal extremity. The chelipedes in the adult male are subequal and rather slender, with the merus (in the species I have examined) short and rounded, or obscurely trigonous; carpus without a spine or tooth on the inner margin; palm small, compressed; fingers straight and distally acute. Ambulatory legs very slender and of moderate length; dactyli nearly straight and as long as or longer than the penultimate joints.

Two of the described species of this genus (which are possibly not distinct the one from the other) *Hymenosoma orbiculare*, Latreille, and *Hymenosoma geometricum*, Stimpson, occur at the Cape of Good Hope in rather shallow water; a third species, referred by Milne Edwards to this genus, *Hymenosoma gaudichaudii*, Guérin-Méneville, occurs on the Australian coasts.[1]

Hymenosoma orbiculare, Desmarest.

Hymenosoma orbiculare, Desmarest, Consid. sur les Crust., p. 163, pl. xxvi. fig. 1, 1825.
    „   „   Milne Edwards, Hist. Nat. Crust., vol. ii. p. 36, 1837; Crust. in Cuvier, Règne Anim., ed. 3, Atlas, pl. xxxv. fig. 1; Ann. d. Sci. Nat., ser. 3, Zool., vol. xx. p. 222, 1853.
Leuchium orbiculare, MacLeay, Invert. in Smith, Zool. South Africa, p. 68, 1849.
Hymenosoma geometricum, Stimpson, Proc. Acad. Nat. Sci. Philad., p. 108, 1858, var. (?)

Numerous specimens were collected in Simon's Bay, South Africa, in 5 to 20 fathoms (mostly of small size).

The largest male in the series has the following dimensions :—

	Lines.	Millims.
Length of carapace, ...	5½	11·5
Greatest breadth of carapace, rather over	5	11
Length of a chelipede, ...	6	12·5
Length of second ambulatory leg, rather over . . .	10	21·5

Halicarcinus, White.

Halicarcinus, White, Ann. and Mag. Nat. Hist., ser. 1, vol. xviii. p. 178, 1846.
    „   Milne Edwards, Ann. d. Sci. Nat. ser. 3, Zool., vol. xx. p. 222, 1853.
Liriopea, Nicolet, in Gay, Hist. de Chile, Zool., vol. iii., Crust., p. 160, pl. i. fig. 1, 1849.

This genus is very nearly allied to the preceding (*Hymenosoma*), but the typical species may be distinguished by the following characters :—The carapace is more transverse and rounded, less distinctly triangulate. The front is distinctly trilobate. The epistoma is transverse and distinctly developed. The basal segment of the post-abdomen in the male occupies the whole width of the sternum, between the bases of the fifth ambulatory legs. The exterior maxillipedes (in the typical species at least) are more robust, the ischium

[1] Mr. W. A. Haswell, in his recent Catalogue of the Australian Crustacea, would unite with *Hymenosoma* all the species referred by authors to *Hymenicus*, Dana, and *Halicarcinus*, White.

and merus-joints broader and shorter. The chelipedes and ambulatory legs resemble those of *Hymenosoma* but the chelipedes are more robust and more strongly developed, with the palms more turgid, and the dactyli of the ambulatory legs are more distinctly falcated.

The following forms seem to be referable to this genus, which is distinguishable from *Hymenosoma* and *Hymenicus* by the distinctly trilobated front, whose lobes are separated to the base :—

Halicarcinus planatus. Common throughout the whole Antarctic Region.
Halicarcinus ovatus, Stimpson. The representative on the Australian Coasts of *Halicarcinus planatus.*
Halicarcinus leachii, Nicolet. Chili.
Halicarcinus larasii, Nicolet. Chili.

Halicarcinus planatus (Fabricius).

Cancer planatus, Fabricius, Entom. Syst., vol. ii. p. 443, 1793.
Halicarcinus planatus, White, Ann. and Mag. Nat. Hist., vol. xviii. p. 178, pl. ii. fig. 1, 1846.
  ,,   ,,   Milne Edwards, Ann. d. Sci. Nat., ser. 3, vol. xx. p. 222, 1853.
  ,,   ,,   Miers, Phil. Trans., vol. clxviii. p. 201, 1879, et synonyma.
Hymenosoma planatum, Haswell, Cat. Australian Crust., p. 114, 1882.

Specimens of this common and widely distributed inhabitant of the Antarctic or Austral region, and which is the only Brachyurous Decapod proper to that wide area of distribution, occurred at the following localities :—

Between the Cape of Good Hope and Kerguelen (near Marion Island) on December 26, 1873, in 50 to 75 fathoms (an adult female); off Prince Edward Island, 85 to 150 fathoms (an adult female); at Kerguelen Island, in rock-pools (a male and a female); New Zealand, Cape Campbell (an adult male, from the Wellington Museum); Falkland Islands, 4 fathoms, in lat. 51° 32′ 0″ S., long. 58° 6′ 0″ W., Station 316 (five males and females); Port William (an adult male).

One of the specimens from the Falkland Islands (Station 316) has the carapace somewhat more hairy and convex than is usual in the species.

An adult male measures as follows :—

Adult ♂.	Lines.	Millims.
Length of carapace, about	5	11
Breadth of carapace, about	6¼	13·5

Halicarcinus ovatus, Stimpson.

Halicarcinus ovatus, Stimpson, Proc. Acad. Nat. Sci. Philad., p. 109, 1858; Targioni-Tozzetti, Crostacei brachiuri e anomouri del viaggio della "Magenta," p. 173, pl. x. fig. 5, *a-d;* pl. xi. figs. 3, 3a, 1877.

An adult, but small female was obtained off the South Australian Coast, in 2 to 10 fathoms, in April 1874, and two young specimens off the entrance to Port Philip, in 33 fathoms (Station 161).

This species, which I formerly regarded as synonymous with *Halicarcinus planatus*,[1] can apparently always be distinguished by the more triangulate form of the carapace and the flattened triangulate frontal teeth, which are closely approximated at their bases. The species, so far as I am aware, does not occur elsewhere than in the Australian Seas.

Adult ♀.	Lines.	Millims.
Length of carapace,	2½	5
Breadth of carapace, nearly	3	6

OXYSTOMATA or LEUCOSIIDEA.

Orbiculata, Latreille (pt.), Fam. Nat., p. 271, 1825.
Oxystomes, Milne Edwards (pt.), Hist. Nat. Crust., vol. i. p. 265, 1834 ; vol. ii. p. 96, 1837.
Leucosoidea vel *Oxystomata*, Dana, U.S. Explor. Exped., vol. xiii., Crust. 1, pp. 68, 389, 1825.
Oxystomata, Miers, Cat. New Zeal. Crust., p. 54, 1876.

Family I. CALAPPIDÆ.

Calappiens (pt.), Milne Edwards, Hist. Nat. Crust., vol. ii. p. 100, 1837.
Calappidæ, Dana, U.S. Explor. Exped., vol. xiii., Crust. 1, p. 390, 1852.
    „   Miers, Cat. New Zeal. Crust., p. 54, 1876.

Afferent canals to the branchiæ opening behind the pterygostomian regions, and in front of the chelipedes ; the carpal and following joints of the merus of the endognath of the exterior maxillipedes are not wholly concealed by the merus-joint. (The intromittent sexual organs in the male are exserted from the bases of the legs of the fifth pair.)

[1] Cat. New Zeal. Crust., p. 49, 1876.

Subfamily 1. CALAPPINÆ.

Calappinæ, Dana, *tom. cit.*, p. 390, 1852.

The eight posterior legs are gressorial, not natatorial, *i.e.*, the dactyli not dilated and expanded.

The genera referred to this subfamily are :—*Camara*, de Haan ; *Calappa*, Fabricius (= *Lophos*, de Haan, subgenus; *Gallus*, de Haan, subgenus) ; *Paracycloïs*, n. gen.; *Mursia*, Desmarest (= *Theolia*, Lucas) ; *Cryptosoma*, Brullé (= *Cycloës*, de Haan) ; *Platymera*, Milne Edwards ; *Acanthocarpus*, Stimpson.

Subfamily 2. ORITHYINÆ.

Orithyinæ, Dana, *tom. cit.*, p. 391, 1852.

The eight posterior legs are natatorial, *i.e.*, with the dactyli more or less dilated and compressed, those of the fifth pair lanceolate, ovate.

Genus :—*Orithyia*, Fabricius.

This subfamily is not represented in the Challenger collection.

Calappa, Fabricius.

Calappa, Fabricius, Entom. Syst. Suppl., p. 345, 1798.
      „   Milne Edwards, Hist. Nat. Crust., vol. ii. p. 102, 1837.
Lophos, de Haan, subgenus, Crust. in v. Siebold, Fauna Japonica, decas iii. p. 69, 1837.
Gallus, de Haan, subgenus, *tom. cit.*, p. 70, 1837.

Carapace very convex and tuberculated, rounded in front, with the antero-lateral margins regularly arcuated and granulated or toothed ; and the postero-lateral margins prolonged into two large rounded wings or clypeiform expansions, which partly cover the ambulatory legs, but are not developed to so great a degree as in *Camara*, de Haan, and are dentated on the margins, not entire, as in that genus. The front is very small, triangulate or concave in the median line. Orbits very small, circular. The post-abdomen (in the male) covers the sternum at the base, between the bases of the fifth ambulatory legs, and is usually five-jointed. Eye-peduncles short, robust. Antennules nearly vertically plicated. Antennæ with the basal joint usually more or less dilated, and occupying the wide inferior hiatus of the orbit ; flagellum short. The exterior maxillipedes do not entirely cover the buccal cavity (which is narrowed and prolonged in its anterior part, and divided by a median septum); their ischium-joint is not produced at its antero-internal angle ; the merus is subacute or distally truncated, and emarginate at the antero-internal angle, where the next joint is articulated. The chelipodes are equal and very large, and can be closely applied to the body ; merus and carpus trigonous, palm vertically very deep, and laterally compressed, and armed above with a strongly dentated crest ; the fingers of the right and left chelipedes are usually somewhat dissimilar, and are acute at the distal extremity. Ambulatory legs slender and of moderate length, with the dactyli styliform.

The species are rarely found in deep water.

The following are species of this genus which have been described since the publication of Milne Edwards's work in 1837.

> *Calappa convexa*, Saussure (= *Calappa xantusiana*, Stimpson). West Coast of Mexico; California. (This species is the western representative of, or perhaps identical with, *Calappa gallus*).
>
> *Calappa rubroguttata*, Herklots (= *Calappa bocagei*, B. Capello). West Africa.
>
> *Calappa pelii*, Herklots. West Africa.
>
> *Calappa guérinii*, B. Capello. India, Yannou.
>
> *Calappa moniziana*, B. Capello. Cape of Good Hope.
>
> *Calappa angustata*, A. Milne Edwards. West Indian Seas (to 115 fathoms).
>
> *Calappa depressa*, n. sp. South Australia.

I have described a new variety of *Calappa gallus* (var. *bicornis*) from the Providence Islands and Indian Ocean. M. de Haan also refers to a species, *Calappa gallina* (Herbst), not mentioned by Milne Edwards.

Calappa flammea (Herbst) (Pl. XXIII. fig. 1).

> 1 *Cancer flammea*, Herbst, Naturgesch. der Krabben u. Krebse, vol. ii. p. 161, pl. xl. fig. 2, 1793.
> *Calappa flammea*, Bosc, Hist. Nat. Crust., vol. i. p. 185 (?).
>
> „   „   White, List Crust. Brit. Mus., p. 44, 1847.
> „   „   *marmorata*, Desmarest, Consid. sur les Crust., p. 109, 1825 (?).
> „   „   Milne Edwards, Hist. Nat. Crust., vol. ii. p. 104, 1837.
> „   „   Gibbes, Proc. Amer. Assoc. Adv. Sci., p. 183, 1850.
> „   „   Kingsley, Proc. Acad. Nat. Sci. Philad., p. 402, 1879 ; not *Calappa marmorata*, Fabricius,
> „   *granulata*, de Haan, Crust. in v. Siebold, Fauna Japonica, dec. iii. p. 70, 1837 ; not *Cancer granulatus*, Linnæus (*type*) (?).

An adult female and two smaller females are in the collection from Bermuda, and an adult male labelled as from Simon's Bay, Cape of Good Hope, 10 to 20 fathoms.[1]

		Lines.	Millims.
Adult ♀.			
Length of carapace,	. . . . .	40	84·5
Breadth of carapace,	. . . .	58½	124

The adult male from the Cape is rather smaller.

[1] Herbst's name is cited by Milne Edwards for this species, and must be adopted, as having priority, if the quotation be correct, but it is possible that Herbst's species is not identical with this Atlantic and South African forms. The brief description, however, and rude figure apply fairly well. In the adult female in the Challenger collection, from Bermuda, the faint brownish-pink markings on the carapace are scarcely reticulated; in the smaller specimens from Bermuda, and in the large male from the Cape of Good Hope, they are almost entirely absent; in an adult male from St. Croix, they form more distinct but still irregular reticulations, but in one from the collection of the late General Hardwicke, and presumably from the Indian Ocean, the reticulations on the anterior half of the carapace are very regular and distinct. Except in coloration the specimens from these widely distant localities do not present any marked distinctions. Herbst gives "Ostindien" as the habitat of his type.

Calappa hepatica (Linné).

Cancer hepaticus, Linné, Mus. Lud. Ulrici, p. 448, 1764 ; Syst. Nat., ed. xii. p. 1048, 1766.
„   tuberculatus, Herbst, Naturgesch. der Krabben u. Krebse, vol. i. p. 204, pl. xiii. fig. 78, 1790.
Calappa tuberculata, Fabricius, Entom. Syst. Suppl., p. 345, 1798.
„   „   Guérin, Icon. du Règne Animal, Atlas, Crust., pl. xii. fig. 2.
„   „   Milne Edwards, Hist. Nat. Crust., vol. ii. p. 106, 1837.
„   „   Dana, U.S. Explor. Exped., vol. xiii., Crust., p. 393, 1852.
„   „   A. Milne Edwards, Nouv. Archiv. Mus. Hist. Nat., vol. x. p. 55, 1874.
„ hepatica, de Haan, Crust. in v. Siebold, Fauna Japonica, dec. iii. p. 70, 1837.
„   „   White, List Crust. Brit. Mus., p. 44, 1847.
„   „   Miers, Cat. New Zealand Crust., p. 55, 1876.
„   „   Haswell, Cat. Australian Crust., p. 136, 1882.
„ sandwichica, Eydoux and Souleyet, Crust. in Voy. de la "Bonite," Atlas, pl. iii. fig. 10, 1841-1852.

Specimens of this common and widely distributed species are in the collection from Tongatabu, reefs (an adult male); Amboina, 10 fathoms (an adult female); and Honolulu, reefs (an adult male and female).

The largest example (female from Honolulu) measures as follows:—

Adult ♀.	Lines.	Millims.
Length of carapace, about, .	20¼	43·5
Breadth of carapace, about .	33	70

Calappa granulata (Linné).

Cancer granulatus, Linné, Syst. Nat., ed. xii. p. 1013, 1766, type, not synonym.
„   „   Herbst, Naturgesch. der Krabben u. Krebse, vol. i. p. 200, pl. xii. figs. 75, 76, 1782.
Calappa granulata, Fabricius, Entom. Syst. Suppl., p. 346, 1798.
„   „   Latreille, Hist. Nat. Crust. et Ins., vol. v. p. 392, pl. xliii. figs 1, 2, 1803-1804, after Herbst.
„   „   Risso, Hist. Nat. de l'Europe méridionale, vol. v. p. 30, 1826.
„   „   Roux, Crust. de la Méditerranée, pls. ii. and xvi. 1828-1830.
„   „   Milne Edwards, Hist. Nat. Crust., vol. ii. p. 103, 1837 ; Crust. in Cuvier, Règne Animal, ed. 3, pl. xxxviii. fig. 1.
„   „   Heller, Crust. des südlichen Europa, p. 130, pl. iv. fig. 3, 1863.
Cancer sanguineo-guttatus, Herbst, op. cit., vol. iii. pp. 20, 48, 1803, var. (?).
Calappa sanguineo-guttata, de Haan, Crust. in v. Siebold, Fauna Japonica, dec. iii. p. 70, 1837, var. (?).

Two small and immature specimens dredged off Fayal, in 50 to 90 fathoms, are referred to this species.

They scarcely differ from adult examples except in the more distinct tuberculation of the carapace. The colour is a nearly uniform yellowish-white.

Young.	Lines.	Millims.
Length of carapace, about	7	15
Breadth of carapace at lateral expansions,	8	17
Length of a cheliped, about	8	17

Calappa gallus (Herbst).

Cancer gallus, Herbst, Naturgesch. der Krabben u. Krebse, vol. iii. p. 46, pl. lviii. fig. 1, 1803.
Calappa gallus, Milne Edwards, Hist. Nat. Crust., vol. ii. p. 105, 1837.
   ,,   ,,   Dana, U.S. Explor. Exped., vol. xiii., Crust., p. 393, 1852.
   ,,   ,,   F. de R. Capello, Journ. de Sci. Math. Phys. Nat. Lisboa, vol. iii. p. 133, pl. ii. fig. 4, 1871.
Gallus gallus, de Haan, Crust. in v. Siebold, Fauna Japonica, p. 70, 1837.
Calappa galloides, Stimpson, Ann. Lyc. Nat. Hist. New York, vol. vii. p. 71, 1859, var.

Bermuda, April 1873 (an adult female); Cape Verde Islands, St. Vincent, July 1873 (an adult male of large size).

	Lines	Millims.
Adult ♂.		
Length of carapace, .	27½	58·5
Breadth of carapace,	40½	85·5

Two very small and young specimens, dredged at the island of Fernando Noronha, in 7 to 20 fathoms, in September 1872, may perhaps belong to this species, but are too small for certain identification, the length of the carapace of the largest being only four lines (8·5 mm.).

To *Calappa gallus*, also, must, I think, be referred a small male dredged at Amboina in 15 to 25 fathoms, in October 1874, length of carapace nearly 5 lines (10 mm.), and breadth at lateral expansions 5½ lines (11·5 mm.), which has the carapace covered with smoother, more regularly disposed, rounded bosses or tubercles. Specimens of the same variety are in the collection of the British (Natural History) Museum from the Philippines (Cuming) and Oriental Seas (H.M.S. "Samarang"). The rostrum is not deeply notched as in the variety *bicornis*, Miers.

Calappa lophos (Herbst).

Cancer lophos, Herbst, Naturgesch. der Krabben u. Krebse, vol. i. (Heft. 1), p. 201, pl. xiii. fig. 77, 1782.
Calappa lophos, Fabricius, Entom. Syst. Suppl., p. 346, 1798.
   ,,   ,,   Milne Edwards, Hist. Nat. Crust., vol. ii. p. 104, 1837.
   ,,   ,,   White, List Crust. Brit. Mus., p. 45, 1847.
Lophos lophos, de Haan, Crust. in v. Siebold, Fauna Japonica, dec. iii. p. 72, pl. xx. fig. 1, 1837.

Off Port Jackson, June 3, 1874, in 30 to 35 fathoms (a very small male).

	Lines	Millims.
Length of carapace, rather over	3	7
Breadth of carapace, .	3½	7·5

This specimen is very doubtfully referred to *Calappa lophos*, on account of its very small size. The carapace is distinctly tuberculated (as in some varieties of *Calappa lophos*), but is not granulated and depressed as in *Calappa depressa*. Scarcely any traces exist of the tubercles on the outer surface of the palms of the chelipedes, which are discernible in specimens of *Calappa lophos* of larger size but not fully grown.

Calappa depressa, n. sp. (Pl. XXIII. fig. 2).

Carapace much more depressed than is usual in the genus and comparatively narrow in proportion to its length; its surface is tuberculated and finely granulated. Of the tubercles (which are not large), there is a longitudinal median series on the gastric and cardiac regions, and several on either side of this series on the hepatic and front of the branchial regions; the postfrontal region is nearly smooth; the lateral clypeiform prolongations of the carapace are little developed, and are covered above with transverse, granulated, piliferous lines. The front projects but little, and is divided by a deep median notch. The antero-lateral margins of the carapace are very obscurely toothed; its clypeiform prolongations are armed on the sides with distant serratures and are posteriorly entire, as is also the posterior margin of the carapace. The ridge upon the pterygostomian region is marked by a notch or fissure, where it unites with the inferior wall of the orbit. All the post-abdominal segments (in the male) are distinct; the penultimate and terminal segments are the longest, the terminal longer than the penultimate segment, triangulate, and distally acute; the anterior margin of the third segment is a little broader than the base of the following (fourth) segment. The eyepeduncles are slender and elongated. The orbits, antennæ, and outer maxillipedes are very similar to the same parts in *Calappa lophos* (Herbst). The chelipedes also nearly resemble those of *Calappa lophos*; the merus having, as in that species, a strong transverse subdistal carina on its outer surface, which is not notched or spinose, but the carpus and the palms of the chelipedes are more granulated on their outer surface; the dactyl, also, is granulated externally near the base, and has (as in *Calappa lophos*), a rounded lobe on its upper margin close to the articulation with the palm. The ambulatory legs, as in other species of the genus, are small, smooth, slender, and compressed.

The colour (in spirit) is light yellowish-brown; the lateral expansions of the carapace, outer surface of the chelipedes, and the ambulatory legs of a pinker hue; the inner surface of the palms are brownish-pink upon a yellowish ground; the coloration being disposed, in places, in wavy lines.

Adult (♂) ♀.	Lines.	Millims.
Length of carapace, nearly	9	18·5
Breadth of carapace at lateral expansions,	10½	22
Length of a chelipede,	10	21
Length of second ambulatory leg,	9	19

A female, probably adult, was obtained off the South Australian Coast, in 2 to 10 fathoms, in April 1874.

The nearest ally to this species with which I am acquainted is *Calappa lophos* (Herbst), from which *Calappa depressa* is distinguished by the more depressed and tuberculated carapace, with less strongly-toothed lateral expansions, the more distinctly granulated chelipedes, &c.[1]

Young specimens of *Calappa hepatica* may be distinguished from *Calappa depressa* by their broader carapace, with less deeply emarginate front.

Paracyclois, n. gen.

Carapace about as long as broad, and moderately convex; front narrow and trilobated; the median lobe rounded and much broader than the lateral lobes. No lateral epi-branchial spine or tooth; the antero-lateral margins are regularly arcuated and entire; there is in the middle of each of the postero-lateral margins a strongly spiniferous lobe (the rudiment of the postero-lateral clypeiform prolongations of the carapace which are fully developed in *Calappa*). The subhepatic regions of the carapace are concave; the channel thus formed communicating with the antennary region (and thereby with the buccal cavity) by a notch situated between it and the inferior wall of the orbit. Post-abdominal segments distinct. Eye-peduncles short, robust, closely encased in the oval orbits. Antennules obliquely plicate. Antennæ with a quadrate basal joint, which does not reach the frontal margin, and a very short flagellum. Outer maxillipedes with the ischium longer than broad and longer than the merus, which is distally truncated, with the antero-internal angle very distinctly notched; the following joints are exposed as in *Calappa*; the exognath is slender, straight, and narrows slightly to its distal extremity, which does not reach the antero-external angle of the merus of the endognath. Cheli-pedes and ambulatory legs as in *Calappa* and *Cryptosoma*, and the description of these limbs in *Calappa* applies to *Paracyclois*.

This remarkable type apparently connects the genera *Cryptosoma* and *Platymera* with *Calappa* through such forms as *Calappa* (*Gallus*) *gallus* (Herbst). As in these genera the merus of the outer maxillipedes is distally truncated, and bears the next joint at its antero-internal angle, which is prolonged in the form of a lobe or tooth, but *Para-cyclois* is distinguished from the first two of the above mentioned genera by the absence of any lateral spine on the margin of the carapace, and the broader basal antennal joint, and from *Calappa* by the absence of the clypeiform prolongations of the carapace, which are represented by a slight protuberance of the postero-lateral margins in *Paracyclois*, which protuberance bears several strong spines.

[1] There are in the British Museum collection two small specimens of this species without special indication of locality, collected in the Voyage of H.M.S. "Samarang." All the specimens I have examined are of small size.

As in the genus *Mursia*, the posterior margin of the carapace bears distinct lobes or teeth; but *Mursia* is at once distinguished by the form of the merus of the exterior maxillipedes, and the strongly developed lateral spine of the carapace.

Acanthocarpus, Stimpson,[1] is (as its name imports), distinguished by the extraordinarily developed carpal (or meral) spine of the chelipedes.

Paracyclois milne-edwardsii, n. sp. (Pl. XXIV. fig. 1).

The carapace is irregularly orbiculate, convex and broadest at a point situated a little in advance of the middle of the lateral margins; its dorsal surface, except on the intestinal region, and the postero-lateral parts of the branchial regions, is coarsely granulated and covered with low, smooth, rounded tubercles, which diminish in size towards the lateral and postero-lateral margins. The median frontal lobe is broadly rounded, with three low tubercles on its upper surface, the lateral margins of the carapace sweep round in a regular curve to the protuberances of the postero-lateral margins, which bear four unequal spines; the lateral and postero-lateral margins, and the parts of the carapace immediately adjoining the three tubercles of the posterior margin, are granulated. The pterygostomian regions are smooth. The post-abdomen (in the female) is rather narrow, with subparallel sides; the five first segments are transverse and short (the second segment with two small lateral protuberances), the sixth segment is quadrate, and slightly broader than long; the terminal segment is triangulate, somewhat longer than broad, and distally acute. The eye-peduncles are short and thick, and granulated above, the corneæ occupying a great part of their inferior surface. The quadrate basal joint of the antennæ lies loosely within the orbital hiatus, the following peduncular joints are slender and very short. The exterior maxillipedes have been already described; their ischium-joints are denticulated on the inner margins, and the merus-joints are slightly concave on their exterior surface. The chelipedes and ambulatory limbs are nearly as in *Calappa*, e.g., *Calappa gallus*; as in that species the merus of the chelipedes has a subdistal crest on its outer surface, but this crest is armed with short spines (not dentated as in *Calappa gallus*); the merus, in front of this ridge, and the carpus and palm are externally granulated, and the carpus and palm also tuberculated, and the palm dentated on its upper margin, as in *Calappa gallus*, but the tubercles and spines are less prominent than in that species; the granules of the lower part of the exterior surface of the palm are very numerous and regular. As in specimens of *Calappa gallus* I have examined, the dactyli of the chelipedes are dissimilar; that of the left chelipede being much more slender than the right, and sinuated. The ambulatory legs are compressed, with the carpus-joints obscurely bicarinated above, and the carinæ (in the two last pairs) granulated; the merus in the fifth pair is denticulated on its inferior margin. Colour (in spirit) yellowish-white.

[1] *Bull. Mus. Comp. Zool.*, vol. ii. p. 153, 1871.

The unique specimen presents the following dimensions :—

Adult ♀.							Lines.	Millims.
Length of carapace, about	.	.	.	.	.	.	23	49
Breadth of carapace,	.	.	.	.	.	.	23½	50
Length of a chelipede, about	.	.	.	.	.	24½	52
Length of first ambulatory leg,	.	.	.	.	.	25½	54

It was dredged north of the Admiralty Islands, in 150 fathoms, in lat. 1° 54′ 0″ S., long. 146° 39′ 40″ E. (Station 219).

Mursia, Desmarest.

Mursia, Desmarest, Consid. sur les Crust., p. 108, footnote, 1825.

   „   Latreille, Crust. in Cuvier, Règne Animal, ed. 2, p. 39, 1829.

   „   Milne Edwards, Hist. Nat. Crust., vol. ii. p. 109, 1837.

   „   Dana, U.S. Explor. Exped., vol. xiii., Crust. 1, p. 391, 1852 ; not *Mursia* of Leach, MSS.

Thealia, Lucas, Ann. Soc. Entom. France, ser. 1, vol. viii. p. 577, 1839.

   „   Dana, tom. cit., p. 391, 1852.

Carapace transverse and moderately convex, with the dorsal surface tuberculated, some of the tubercles disposed in five longitudinal series ; the antero-lateral margins regularly arcuated and terminating in a well-developed lateral epibranchial spine, the postero-lateral margins straight, and without any trace of the clypeiform expansions characteristic of *Calappa*. Front small, with a median tubercle or tooth. Orbits with usually one or two closed fissures in the superior margin and with a wider hiatus in the inferior margin. Subhepatic channels well developed, as in *Paracyclois*. Post-abdomen in the male with two or three of the intermediate segments coalescent ; it covers at the base the whole width of the sternum between the coxæ of the fifth ambulatory legs. The robust eye-peduncles fill or nearly fill the orbital cavities. The antennules are somewhat obliquely plicated. The basal antennal joint is rather slender, and occupies the interior hiatus of the orbit ; the flagellum is well developed. The exterior maxillipedes (as in *Calappa*) do not cover the anterior part of the buccal cavity ; the ischium of the endognath is not produced at its antero-internal angle ; the merus is obliquely truncated at the distal extremity, and the carpal joint is articulated at the antero-external angle of the merus, the exognath is slender and straight: The chelipedes and ambulatory legs are nearly as in *Calappa*, but the chelipedes are much less developed, the palm not so deep, and the merus bears one or two distal spines, not a dentated crest, on the outer surface, and the ambulatory legs are relatively longer.

Besides the two species referred to below, *Mursia acanthophora* (Lucas) = *Mursia armata*, de Haan, is to be included in this genus. This species occurs in China and Japan.

In the nearly allied genus *Platymera*, Milne Edwards, the carapace is much more transverse, and the lateral spine very greatly developed, but in this genus the merus of the exterior maxillipedes is distally truncated, deeply notched on the inner margin, with a tooth or lobe at the antero-internal angle, above the point of articulation with the next joint, somewhat as in *Cryptosoma*.

Mursia cristimana.

Mursia maina en colte, Desmarest, Consid. sur les Crust., pl. ix. fig. 3, 1825.
Mursia cristata, Milne Edwards, Hist. Nat. Crust., vol. ii. p. 109, 1837 ; Crust. in Cuvier, Règne Animal, ed. 3, pl. xiii. fig. 1.
„   „   Studer, Abhandl. d. k. Akad. d. Wiss. Berlin, p. 15, 1882 ; not Mursia cristata, Leach, in Coll. Brit. Mus.
" Mursia cristimana, Latreille," de Haan, Crust. in v. Siebold, Fauna Japonica, pp. 70, 73, 1837.
„   „   Krauss, Die süd-afrikanischen Crust., p. 52, 1843.
Cryptosoma orientis, Adams and White, Crust. in Zool. H.M.S. "Samarang," p. 62, pl. xiii. fig. 4, 1848 var. (?).

Cape of Good Hope, Simon's Bay (an adult male); Sea Point, near Cape Town (an adult female); Agulhas Bank, in 150 fathoms, in lat. 35° 4′ 0″ S., long. 18° 37′ 0″ E., Station 142 (a male and two females).

The adult male from Simon's Bay is without chelipedes, and presents the following dimensions :—

Adult ♂.	Lines.	Millims.
Length of carapace,	13	27·5
Breadth of carapace, at base of lateral spines,	14½	31
Length of second ambulatory leg,	22½	47·5

Cryptosoma orientis of Adams and White differs in nothing but in the slightly broader front, with somewhat less prominent median tooth, and the somewhat straighter posterior margin of the carapace, and is probably only a variety of *Mursia cristimana*.

Mursia curtispina, n. sp. (Pl. XXIV. fig. 2).

This new species so nearly resembles in all its characters *Mursia armata*, de Haan (*Thealia acanthopora*, Lucas), that the detailed description which follows is scarcely needed ; it is distinguished, however, by the somewhat narrower carapace, which has the antero-lateral margins more arcuated, and is armed with proportionately shorter lateral spines, and with three very small equal tubercles on the posterior margin, in place of the two larger prominences of *Mursia armata*. The front also is somewhat broader, with a smaller, less prominent median cusp.

The carapace, as in *Mursia armata*, is transverse, convex, and granulated ; the granules more regular and even than in that species; as in *Mursia armata*, its dorsal surface has five longitudinal series of larger granulated prominences, one of which is in the median line; the antero-lateral margins are minutely granulated; the lateral spine is short, about one-seventh the width of the carapace; the granuliform prominences of the posterior margin are scarcely larger than the granules of the antero-lateral margins. The frontal margin projects but little beyond the inner angles of the orbits and is armed with three small obtuse teeth, the median is scarcely more prominent than the lateral teeth. As in *Mursia armata*, the orbits have a closed fissure in their superior margins and a deep and well-defined hiatus in the inferior margins; the inner subocular lobe of the orbit is prominent and subacute; the subhepatic regions of the carapace are deeply channelled, as in *Mursia armata*. All the segments of the post-abdomen (in the female) are distinct; the second segment is armed with a trilobate crest, as in *Mursia armata*, but the lobes are broader, more prominent, and entire. The eyes, antennæ, and outer maxillipedes, are disposed as in *Mursia armata*; the chelipedes and ambulatory legs, also, are nearly as in that species, but the palms of the chelipedes are externally more regularly granulated; and there are three subdistal spines on the outer surface of the merus-joint; the one nearest the upper margin of this joint is very small, and is usually obsolete in *Mursia armata*. Colour (in spirit) yellowish-brown, inclining to pink on the chelipedes; the apices of the dactyli of the ambulatory legs are brown-pink, and a patch of the same colour ornaments the inner surface of the palms of the chelipedes.

	Lines.	Millims.
Adult ♀.		
Length of carapace, ...	13¼	29
Breadth of carapace, ...	16	34
Length of a chelipede, rather over	19	40·5
Length of second ambulatory leg, ...	22½	47·5

Fiji Islands, 315 fathoms, lat. 19° 9′ 35″ S., long. 179° 41′ 50″ E., Station 173 (an adult female).

Cryptosoma, Brullé.

Cryptosoma, Brullé, Crust. in Webb et Berthelot, Hist. Nat. des Iles Canaries, p. 16, 1836–1844.
„   Milne Edwards, Hist. Nat. Crust., vol. ii. p. 110, 1837.
Cyclöes, de Haan, Crust. in v. Siebold, Fauna Japonica, p. 68, 1837.
„   Dana, U.S. Explor. Exped., vol. xiii., Crust. 1, p. 391, 1852.

In this genus, which is nearly allied to the preceding, the carapace is always suborbiculate, with the postero-lateral margins slightly concave, the front in the middle line emarginate, the lateral spine very small or reduced to a mere tubercle. The subhepatic channels are less distinctly defined. The merus of the exterior maxillipedes is produced at the antero-internal angle into a small tooth or lobe above the carpal joint, which is articulated near the antero-internal angle. The chelipedes (as in *Calappa*) are strongly developed; the ambulatory legs robust, with the joints dilated and compressed.

In the nearly allied genus, *Acanthocarpus*, Stimpson,[1] of which I have examined no specimens, the carapace as in *Cryptosoma* is not broader than long, and the merus of the outer maxillipedes (in the type) bears the next joint at its antero-internal angle. It is distinguished, according to the descriptions, by the extraordinarily developed carpal spine of the chelipede; but in the figures recently published by A. Milne Edwards, this spine is represented as originating from the merus-joint, and is apparently homologous with the smaller meral spine of *Mursia*.

Besides the species referred to below, *Cryptosoma granulosum* (de Haan, *Cyclöes*), Japan, and *Cryptosoma bairdii*, Stimpson, California, belong to this genus.

Cryptosoma cristatum (Leach).

Mursia cristata, Leach, MSS.
    „ „ White (*Mursia*), List Crust. Brit. Mus., p. 45, 1847, not *Mursia cristata*, auctorum.
Cryptosoma cristatum, Brullé, Crust. in Webb et Berthelot, Hist. Nat. des Iles Canaries, pl. i. fig. 2, 1836–1844.
    „ „ Milne Edwards, Hist. Nat. Crust., vol. ii. p. 110, 1837.
    „ „ Lucas, Ann. Soc. Entom. France, ser. 6, vol. ii., Bulletin, p. cxv., 1882.
    „ *dentatum*, Brullé, tom. cit., p. 17, 1836–1844.

Cape Verdes, St. Vincent, July 1873. (Four males, adult and young, and a female.)
The largest male has the following dimensions:—

Adult ♂.		Lines.	Millim.
Length of carapace, about	.	24½	52·5
Breadth of carapace, about	.	25½	54
Length of a chelipede,	.	32½	68·5

The ambulatory legs are imperfect.[2]

Family II. MATUTIDÆ.

Matutidæ, Dana, U. S. Explor. Exped., vol. xiii., Crust. 1, p. 390, 1852.

Characters of Calappidæ except as regards the exterior maxillipedes, whose endognath has the carpal and following joints concealed by the triangular acute merus-joint.

[1] Bull. Mus. Comp. Zoöl., vol. ii. p. 153, 1870.
[2] The locality of Leach's type is "Africa" and not "Indian Ocean," as stated by White, tom. cit., who substituted the latter mentioned for the original locality, both in the Register and Collection of the Museum, for what reason I know not. This species is recorded by Studer (Abhandl. d. k. Akad. d. Wiss. Berlin, p. 15, 1882), from Porto Praya, Cape Verde Islands, in 10 to 30 fathoms.

Subfamily 1. HEPATINÆ.

Hepatinæ, Stimpson, Bull. Mus. Comp. Zoöl., vol. ii. p. 154, 1870.

Carapace somewhat cancroid in form, with the antero-lateral margins arcuated, the dactyli gressorial, not natatorial.

Genera :—*Hepatus*, Latreille; *Osachila*, Stimpson; *Actæomorpha*, Miers.

This subfamily is not represented in the Challenger collection.

Subfamily 2. MATUTINÆ.

Carapace usually suborbiculate; the dactyli of the eight posterior legs natatorial, *i.e.*, with the dactyli laminated and dilated.

Genus :—*Matuta*, Fabricius.

Matuta, Fabricius.

Matuta, Fabricius, Entom. Syst. Suppl., p. 369, 1798.
  ,,   Milne Edwards, Hist. Nat. Crust., vol. ii. p. 113, 1837.
  ,,   Miers, Trans. Linn. Soc. Lond. (Zool.), ser. 2, vol. i. p. 243, 1877.
  ,,   de Man, Notes Leyden Mus., vol. iii. p. 109, 1881.

Carapace much depressed, and usually very slightly broader than long, with the antero-lateral margins slightly arcuated and irregularly dentated or tuberculated, and terminating in a strong acute lateral spine (which in *Matuta inermis* is reduced to a tubercle) placed at the junction of the antero-lateral and postero-lateral margins. Front narrow, about equalling the orbit in width, with a projecting median lobe, which is sometimes entire, sometimes notched. The dorsal surface is usually armed with six tubercles, placed, three in a median transverse series, one, anterior, on each side of the gastric, and one, posterior, on the cardiac region, and there is also usually a tubercle on the postero-lateral margin. The orbits are rather large, with a hiatus, communicating with an excavation on the subhepatic region, below the exterior orbital angle. Antennules nearly longitudinally plicated. Antennæ very small, and placed below the enlarged basal joint of the antennules, with the flagellum obsolete. The exterior maxillipedes cover the whole of the buccal cavity; their ischium-joint is distally truncated, the merus triangulate and distally subacute, and covers the following joints ; the exognath is externally arcuated and reaches but little beyond the distal extremity of the ischium of the endognath. Chelipedes subequal, robust, and closely applicable to the body, with the merus trigonous, carpus externally slightly tuberculated, palm armed with spines or tuberculated ridges, fingers distally acute, the dactyl usually with a tuberculated or striated ridge on its exterior surface. Ambulatory legs of moderate length, with the joints compressed, the penultimate and terminal joints in all laminiform and dilated ; in the fifth legs the dactyl is oval as in the Portunidæ.

The following species, besides those referred to below, are probably well established :—

Section A :—

Matuta lunaris (Herbst) = *Matuta planipes*, Fabricius (*fide* Hilgendorf), and *Matuta rubrolineata*, Miers, *Matuta lenuifera*, Miers, var. Indian and Indo-Malaysian Seas ; Pacific Ocean ; Chefoo, Australia.

Matuta circulifera, Miers. Indo-Malaysian Seas. (Regarded by M. de Man as a variety of *Matuta lunaris*.)

Section B :—

Matuta granulosa, Miers. Oriental Seas ; Amboina.

Matuta maculata, Miers. Chinese Seas ; Panagatan Shoal.

Matuta picta, Hess (= *Matuta distinguenda*, Hoffman, and *Matuta obtusifrons*, Miers, var. ?). Indo-Pacific Seas.

Matuta victrix, Fabricius.

> *Cancer victor*, Fabricius, Species Insect., ii. (Appendix) p. 502, 1781.
> *Matuta victor*, Fabricius, Entom. Syst. Suppl., p. 369, 1798.
> ,, Milne Edwards, Hist. Nat. Crust., vol. ii. p. 115, pl. xx. figs. 3–6, 1837.
> ,, Hilgendorf, Monatsber d. k. preuss. Akad. d. Wiss. Berlin, p. 810, 1878.
> *victrix*, Miers, Trans. Linn. Soc. Lond. (Zool.), ser. 2, vol. i. pt. v. p. 243, pl. xxxix. figs. 1–3, 1877, et synonyma.
> ,, ,, de Man, Notes Leyden Mus., iii. p. 110, 1881.

Philippine Islands, Samboangan, 10 to 20 fathoms. Two males, and an adult and a sterile female.

Adult ♂ .	Lines.	Millims.
Length of carapace, . .	19	40·5
Breadth of carapace, about .	20½	43

Matuta victrix, var. *crebrepunctata*, Miers.

> *Matuta victrix*, var. *crebrepunctata*, Miers, *tom. cit.*, p. 244, pl. xxxix. fig. 4, 1877, not synonym.

Ternate, October 15, 1874 (an adult male).

Adult ♂ .	Lines.	Millims.
Length and breadth of carapace, about	15	31·5

Matuta banksii, Leach.

> *Matuta banksii*, Leach, Zool. Miscell., vol. iii. p. 14, 1817 (?).
> ,, ,, Miers, *tom. cit.*, p. 245, pl. xl. figs. 1, 2, 1877.
> ,, ,, de Man, *tom. cit.*, p. 115, 1881.

Samboangan, Philippine Islands, 10 fathoms (two males and a female, with *Matuta victrix*).

The purple spots which cover the legs are fewer and larger than the spots on the carapace, and the penultimate joints and dactyli of the first, third, and fourth ambulatory legs are marked with a large purple blotch in the two males, but not in the female.

The striations of the dactylus of the mobile finger of the chelipedes are nearly obsolete.

	Lines.	Millims.
Adult ♂.		
Length and breadth of carapace,	12½	26·5

Matuta lævidactyla, Miers.

Matuta lunaris, Miers, tom. cit., p. 247, pl. xl. figs. 10, 11, 1877 ; nec Cancer lunaris, Herbst.
„ lævidactyla, Miers, Ann. and Mag. Nat. Hist., ser. 5, vol. v. p. 316, 1880.

Port Jackson, 3 fathoms (a small male).

Its dimensions are as follows :—

	Lines.	Millims.
Length of carapace,	7	15
Breadth of carapace, nearly	7	14·5
Length of a chelipede,	6	12·5

In this specimen the striated ridge of the outer surface of the dactyl of the chelipede is obsolete, and this joint is externally smooth. The series of tubercles on the outer margin of the palm is parallel (nearly) with the lower margin. The front is distinctly notched, and (the specimen being of small size) all the tubercles of the dorsal surface of the carapace are distinct.

Matuta inermis, Miers.

Matuta inermis, Miers, Crust. in Rep. Zool. Coll. H.M.S. "Alert," p. 256, pl. xxvi. fig. C, 1882.

Torres Strait, 6 fathoms, in lat. 10° 36′ 0″ S., long. 141° 55′ 0″ E. (Station 187). An adult female and three smaller males are in the collection.

The female is of much larger size than the type of the species described in the above report, having the following dimensions :—

	Lines.	Millims.
Adult ♀.		
Length of carapace, nearly	16	33·5
Breadth of carapace, nearly	15	31·5
Length of a chelipede, nearly	14	29·5
Length of second ambulatory leg,	22	46·5

The colour (in spirit) is yellowish-brown, purple on the anterior part of the carapace, sides of the branchial regions, and the dactyli of the fifth ambulatory legs.

Family III. LEUCOSIIDÆ.

Leucosiens, Milne Edwards, Hist. Nat. Crust., vol. ii. p. 118, 1837.
Leucosidæ, Dana, U.S. Explor. Exped., vol. xiii., Crust. 1, p. 390, 1852.

Afferent channels to the branchiæ opening at the antero-lateral angles of the palate and not behind the pterygostomian regions. The carpal and following joints of the endognath of the exterior maxillipedes are wholly concealed by the triangulate merus-joint. (The intromittent sexual appendages in the male are exserted from the sternum.)

The genera of this family are numerous, and vary remarkably in the form of the carapace and chelipedes. No satisfactory classification of them has been proposed. I have thought it advisable to establish only two subfamilies, one of which (Leucosiinæ) is restricted to the single genus *Leucosia*, Fabricius, but it may be found preferable hereafter to separate this genus more definitely under a primary section, and to regard some or all of the sectional divisions of the subfamily Iliinæ as distinct subfamilies.

Subfamily 1. ILIINÆ.

The anterior frontal region of the carapace is not narrowed and produced anteriorly. No thoracic sinus is developed.

Section I. The carapace is laterally produced and expanded, so as to cover in great part the ambulatory legs. The palms of the chelipedes are moderately robust; the fingers compressed, not filiform (Oreophorinæ).

Genera :—*Oreophorus*, Rüppell; *Spelæophorus*, A. Milne Edwards; *Tlos*, Adams and White; *Cryptocnemus*, Stimpson; *Uhlias*, Stimpson.

This section is not represented in the Challenger collection.

Section II. The carapace is not produced over the bases of the ambulatory legs. The palms of the chelipedes are short and turgid; the fingers elongated, very slender or filiform, incurved at the apices, and armed on the interior margins with spinuliform teeth (Myrodinæ).

Genera :—*Myrodes*, Bell; *Nursilia*, Bell; *Iphiculus*, Adams and White.

Myrodes, Bell.

Myrodes, Bell, Trans. Linn. Soc. Lond. (Zool.), vol. xxi. p. 299, 1855.

This genus only differs from *Myra* (with which it is united by A. Milne Edwards and Haswell) in the form of the chelipedes, whose palms are much shorter than the fingers, and turgid, ovoid or subglobose; the fingers are elongated and very slender; strongly incurved at the tips and armed with spinuliform teeth, some of which are more elongated, so that the fingers are rostelliform. This remarkable peculiarity in the structure of the chela, which occurs with more or less distinctness in certain other genera (e.g., *Nursilia, Iphiculus* and *Callidactylus*), is, I think, sufficient to distinguish *Myrodes* from (both) *Myra* and *Persephona*.

From *Nursilia* and *Iphiculus*, *Myrodes* is at once distinguished by a very different form of the carapace, which in those genera has not the three posterior tubercles which exist in Myrodes. *Callidactylus*, Stimpson, is less certainly distinguishable from *Myrodes*; it differs, however, according to Dr. Stimpson, in the absence of an indurated ridge from the basal joint of the antennules (which ridge is not very distinctly developed in *Myrodes*), in the somewhat contorted palms of the chelipedes, and in the character of the dactyli of the ambulatory legs.

The type of this genus, *Myrodes eudactylus*, occurs in the Indo-Malaysian and Australian Seas, and at New Caledonia.

Myrodes eudactylus, Bell.

Myra dilatimanus, White, List Crust. Brit. Mus., p. 46, 1847, *descrip. nullå.*
Myrodes eudactylus, Bell, Trans. Linn. Soc. Lond., tom. cit., p. 299, pl. xxxii. fig. 6, 1855;
    Cat. Leucosiidæ in Brit. Mus., p. 13, 1855.
Myra eudactyla, A. Milne Edwards, Nouv. Archiv. Mus. Hist. Nat., vol. x. p. 46, pl. iii. fig. 3, 1874.
   ,,   ,, Haswell, Cat. Australian Crust., p. 123, 1884.
Myrodes gigas, Haswell, Proc. Linn. Soc. N.S.W., vol. iv. p. 52, pl. v. fig. 5, 1880.

South of New Guinea, 28 fathoms (Station 188) a small, perhaps young, specimen.

In this example the carapace is somewhat broader in proportion to its length, and the lobes of the rostrum less prominent than in the specimens from the Philippines in the collection of the British (Natural History) Museum. The dorsal surface is indistinctly carinated in the median line, as in two out of three specimens in that collection. There is no granulated border on the inner margin of the palm of the chelipede in any specimen I have examined of this species, and Bell's figure is probably in error as regards this particular.

♀.	Lines.	Millims.
Length of carapace, about	6½	14
Breadth of carapace, .	5½	11·5

Section III. The carapace is not produced over the bases of the ambulatory legs. The palms of the chelipedes are very slender and elongated; the fingers very slender and nearly straight, or with the tips slightly incurved. The pterygostomian channels have usually two notches at the distal extremity (Iliinæ, Stimpson).

Genera :—*Ilia*, Fabricius; *Arcania*, Leach (=*Iphis*, Leach); *Ixa*, Leach; *Iliacantha*, Stimpson; *Myropsis*, Stimpson; *Callidactylus*, Stimpson (this genus establishes the transition to the section Myrodinæ).

Arcania, Leach.

Arcania, Leach, Zool. Miscell., vol. iii. p. 19, 1817.
    ,,   Milne Edwards, Hist. Nat. Crust., vol. ii. p. 133, 1837.
    ,,   Dana, U.S. Explor. Exped., vol. xiii., Crust. 1, p. 392, 1852.
    ,,   Bell, Trans. Linn. Soc. Lond. (Zool.), vol. xxi. p. 309, 1855.
Iphis, Leach, Zool. Miscell., vol. iii. p. 19, 1817.
    ,,   Milne Edwards, Hist. Nat. Crust., vol. ii. p. 135, 1837.
    ,,   Dana, tom. cit., p. 392, 1852.
    ,,   Bell, tom. cit., p. 311, 1855.

Carapace convex, subglobose, or somewhat rhomboidal, with the dorsal surface tuberculated, granulated or spinuliferous; the lateral and posterior margins nearly always armed with spines. Front usually rather prominent and bilobed. Orbits (as usual) with three marginal fissures and a rather wide interior hiatus; the interior subocular angle usually spiniform. The post-abdomen in the male is narrow, with two or three of the intermediate segments coalescent. Eyes small. Antennules obliquely plicated. Antennæ with a slender basal joint, which does not fill the interior orbital hiatus; flagellum moderately developed. The merus of the endognath of the exterior maxillipedes is usually much shorter than the ischium, and is often only subacute at the distal extremity; the exterior margin of the exognath is straight. The chelipedes (in the adult males) are slender and somewhat elongated, and, as usual in this subfamily, the merus is subcylindrical, and the palm slender and somewhat swollen at the base; the fingers open in a vertical plane and are armed with minute teeth, some of which are spinuliform. Ambulatory legs slender and somewhat elongated, with the dactyli styliform.

The species occur throughout the Indo-Pacific region in shallow or moderately deep water.

To the species enumerated by Bell are to be added:—

Arcania globata, Stimpson. Seas of China and Japan (16 to 25 fathoms).
Arcania orientalis, Miers. Japan, 30 to 36 fathoms. (This species is intermediate between this genus and *Ebalia*.)
Arcania novem-spinosa, Adams and White, var. *aspera*, Miers. Malaysian Seas.[1]

[1] I have proposed the name *Arcania duodecimspinosa* for a specimen from the Seychelles (4 to 12 fathoms), of whose specific distinctness I am somewhat uncertain, on account of its very small size. *Arcania granulosa*, Miers, from Moreton Bay, is, as I have elsewhere noted, probably identical with *Arcania undecimspinosa*, de Haan, and *Arcania pulcherrima*, Haswell, with *Arcania septemspinosa*, Bell, but as the latter name is preoccupied by Fabricius and Leach for *Arcania (Iphis) septemspinosa*, Haswell's name may be conveniently used to designate Bell's species.

Arcania septemspinosa (Fabricius).

Cancer septemspinosus, Fabricius, Mantissa Insectorum, vol. i. p. 325, 1787.
Iphis septemspinosa, Leach, Zool. Miscell., vol. iii. p. 25, 1817.
    „   „   Milne Edwards, Hist. Nat. Crust., vol. ii. p. 139, 1837 ; Atlas in Règne
                      Animal de Cuvier, ed. 3, Crust., pl. xxv. fig. 4 (after Herbst).
    „   „   Bell, Trans. Linn. Soc. Lond., vol. xxi. p. 311, 1855; Cat. Leucosiidæ in
                      Brit. Mus., p. 22, 1855.

Hong Kong, 10 fathoms (an adult and two younger females) ; Kobé, Japan, 8 to 10 fathoms (a small female); and at Station 233B, in lat. 34° 18′ 0″ N., long. 133° 35′ 0″ E., 15 fathoms (a young female and two rather small males).

The largest male measures as follows :—

♂ .	Lines.	Millims.
Length of carapace, nearly .	7	14·6
Breadth of carapace, nearly .	6½	13·5

The genus *Iphis*, which is retained as distinct from *Arcania*, both by Milne Edwards and Bell, differs merely in its slightly more rhomboidal carapace, and must, I think, be united with that genus.

Ixa, Leach.

Ixa, Leach, Trans. Linn. Soc. Lond., vol. xi. p. 334, 1815.
  „ Milne Edwards, Hist. Nat. Crust., vol. ii. p. 134, 1837.
  „ Dana, U.S. Explor. Exped., vol. xiii., Crust. i, p. 392, 1852.
  „ Bell, Trans. Linn. Soc. Lond. (Zool.), vol. xxi. p. 311, 1855.

Carapace transversely rhomboidal, or somewhat elliptical, and prolonged at the junction of the antero-lateral and postero-lateral margins into a cylindrical lobe, which often equals in length the transverse width of the carapace, and is rounded or tipped with a spinule at the distal extremity ; the carapace is usually longitudinally divided into three parts by two wide fossæ, which border the cardiac, gastric, and hepatic regions, and thence are continued over the pterygostomian regions to the bases of the chelipedes, but these fossæ are sometimes obsolete. The front projects but slightly, and is anteriorly concave ; the orbits have three marginal fissures and a rather wide interior hiatus. The endostomian ridges are strongly defined, and the channels exterior to them, communicating with the branchial cavities, are emarginate at the distal extremity. The post-abdomen (in the male) covers the sternum at base, and is five-jointed, with three of the intermediate segments coalescent. Eyes small. Antennules slightly oblique. Antennæ with the basal joint very slender and not filling the interior hiatus of the orbit. The endognath of the exterior maxillipedes is narrow, with the ischium longitudinally sulcated, the merus triangulate ; the exognath is broad, with the exterior margin straight; it is rounded at the distal extremity, and does not wholly cover the channel which communicates with the branchiæ. The chelipedes (in the male) are cylindrical and very slender, almost filiform, with the merus and palm elongated; fingers slightly incurved at the tips and denticulated on the inner margins. The ambulatory legs are moderately elongated and very slender, with the dactyli styliform.

The four recent species described by authors are probably, as I have stated below, identical with *Ixa cylindrus*, Fabricius, which is probably distributed throughout the Indo-Pacific region from the Mauritius to the Philippines and Borneo, and also occurs in a fossil state in recent alluvial or perhaps quaternary deposits (*vide* A. Milne Edwards, Ann. Soc. Entom. France, p. 156, 1863).

Ixa cylindrus (Fabricius), var. *megaspis*.

Ixa megaspis, Adams and White, Crust. in Zool. H.M.S. "Samarang," p. 55, pl. xii. fig. 1, 1848.

Manila, 1 fathom (an adult male). Presented to Dr. Willemoes-Suhm by Mr. Baer.[1]

Iliacantha, Stimpson.

Iliacantha, Stimpson, Bull. Mus. Comp. Zoöl., vol. ii. p. 155, 1870.

Carapace convex, ovoid or subglobose, with the lateral margins arcuated, with a protuberance or tubercle upon the pterygostomian regions, and with three posterior lobes or spines, as in *Myra* and *Persephona*. The front (in the species I have examined) is narrow and anteriorly slightly concave, and the orbit has more or less distinct indications of three marginal fissures and a wide interior hiatus. Pterygostomian channels distally very strongly defined and bi-emarginate. Post-abdomen (in the young male) distinctly seven-jointed. Eyes small. Antennules (in the species I have examined) slightly oblique. Antennæ with a very slender basal antennal joint, which does not fill the interior orbital hiatus; flagellum of moderate length. Exterior maxillipedes with the ischium-

[1] The late Mr Bell, in his monograph of the Leucosiidæ (*Trans. Linn. Soc. Lond.* (Zool.), vol. xxi. p. 311, 1855), united all the then described species of *Ixa* under the common designation *Ixa cylindrus* (Fabricius). It is probable, indeed, that no characters can be discovered of sufficient value to separate three of these forms specifically, but it may be of service here to indicate the distinctions by which the specimens in the collection of the British (Natural History) Museum may be at present separated.

In that which I think to be the typical *Ixa cylindrus*, Fabricius (*Ixa canaliculata*, Leach) the median portion of the carapace (circumscribed by the deep and wide lateral and postfrontal channels), has its margins sinuated or notched; in the variety *megaspis*, Adams and White, they are entire, and in both the tubercles of the posterior margin are small or obsolete. A third (unnamed) variety, represented by a single female from the Philippines (Cuming) resembles *megaspis* in all particulars except that the later processes of the carapace are without the terminal spinule of that variety and *cylindrus*.

Ixa inermis, of Leach, with which I think *Ixa edwardsii* of Lucas (*Ann. Soc. Entom. France*, vol. vi. p. 179, pl. iv. fig. 3, 1838) to be identical, may be distinguished by the absence of the lateral and postfrontal channels of the carapace (which, however, are represented in the type of *inermis* by an impressed suture on either side of the cardiac region), by the large rounded tubercles of the posterior margin of the carapace, and by the somewhat distally narrowed lateral processes of the carapace (which are without terminal spinules), and may with more probability be regarded as specifically distinct.

joint (in *Iliacantha intermedia*) longitudinally sulcated; merus as usual triangulate; exognath rather narrow, with the exterior margin straight.   Chelipedes slender and rather long, with the merus subcylindrical and granulated; palm slender, narrowing distally and somewhat contorted, so that the fingers open vertically; the fingers (as in *Ilia*) are very slender, and are armed with fine, usually spinuliform, teeth.   Ambulatory legs slender and of moderate length; tarsi styliform.

From *Ilia* this genus differs in having but three posterior spines on the carapace, and from *Myropsis* in that the fingers open in a vertical and not in a horizontal plane.

To the species described by Stimpson, *Iliacantha globosa* and *Iliacantha sparsa*, dredged in the Florida Straits in 30 to 60 fathoms, I have to add a third, *Iliacantha intermedia*, obtained by the Challenger Expedition at Bahia, in shallow water.

Iliacantha intermedia, n. sp. (Pl. XXVI. fig. 3).

This species is distinguished from the West-Indian *Iliacantha subglobosa*, Stimpson, and *Iliacantha sparsa*, Stimpson,[1] by the length of the chelipedes, which in *Iliacantha subglobosa* considerably exceed, but in *Iliacantha intermedia* do not attain the length of the palm, and by the form of the postero-lateral spines or teeth of the carapace, which in *Iliacantha intermedia* are flattened and triangulate; but in *Iliacantha sparsa* are similar in shape to, and more than half as long as, the posterior median spine.

The carapace is moderately convex, longer than broad, and is everywhere very distinctly and evenly granulated; the antero-lateral margins, at the hepatic regions, are bluntly angulated; at some distance behind the hepatic angle there is another slight angular projection, as in *Iliacantha subglobosa*.   The median and posterior spine of the carapace is prominent and acute, and very slightly recurved at the distal extremity; the lateral spines or teeth are flattened and triangulate, and rounded at the apices.   The front (as seen in a dorsal view), projects slightly beyond the eyes; it is concave above, truncated in front, with the anterior margin nearly straight.   The orbits are small, without fissures in the upper margin; the endostome is strongly longitudinally ridged; the ridges define the lateral channels (pterygostomian channels of Stimpson), which terminate distally in three strong spiniform teeth.   The sternum is evenly granulated; the post-abdomen (in the young male), has the segments, except the two first and the last, coalescent, and is granulated at and near the base.   The ischium-joint of the endognath of the outer maxillipedes is longitudinally sulcated on its outer surface, which is nearly smooth; the merus-joint is triangulate, acute, much shorter than the ischium, and strongly granulated on the outer surface; the exognath is externally strongly granulated, with the exterior margin nearly straight; its rounded

[1] *Bull. Mus. Comp. Zoöl.*, vol. ii. pp. 155, 156, 1870.

distal extremity does not attain the apex of the merus-joint of the endognath. The chelipedes are very slender; the merus is nearly as long as the carapace, subcylindrical and granulated; carpus very short, smooth; palm shorter than the merus, smooth, and tapering distally; fingers little more than half as long as the palm, very slender, with the tips incurved and armed with several distant teeth, between which are smaller granuliform teeth. Ambulatory legs slender and moderately elongated, with the dactyli styliform and slightly longer than the penultimate joints. Colour (in spirit) light yellowish-brown.

♂.	Lines.	Millims.
Length of carapace and rostrum,	7½	15
Breadth of carapace, nearly	6½	13·5
Length of a chelipede.	14	30
Length of first ambulatory leg,	11	23·5

Bahia (shallow water), a sterile female.

Stimpson's description of *Iliacantha subglobosa* was also based on a sterile female; it is evident, therefore, that the difference in the length of the dactyli of the chelipedes is not a sexual character, and it would appear from his description that the carpus and palm of the chelipede in *Iliacantha subglobosa* are granulated, and the teeth of the dactyli more acute.

Section IV. The carapace is not produced so as to cover in great part the ambulatory legs. The chelipedes are robust, not slender; the palms and fingers compressed, not slender and elongated (Ebaliinæ, Stimpson, part).

Genera:—*Ebalia*, Leach (= *Bellidilia*, Kinahan, *Phlyxia*, Leach, subgenus); *Persephona*, Leach (= *Guia*, Milne Edwards); ? *Myra*, Leach; *Leucosilia*, Bell; *Randallia*, Stimpson; *Nucia*, Dana; *Lithadia*, Bell; *Carcinaspis*, Stimpson; *Merocryptus*, A. Milne Edwards; *Onychomorpha*, Stimpson; *Nursia*, Leach (in this genus the carapace is slightly produced over the bases of the legs); *Leucisca*, MacLeay; *Pseudophilyra*, Miers; *Philyra*, Leach.

Ebalia, Leach.

Ebalia, Leach, Zool. Miscell., vol. iii. p. 18, 1817.
„ Milne Edwards, Hist. Nat. Crust., vol. ii. p. 128, 1837.
„ Dana, U.S. Explor. Exped., vol. xiii., Crust. 1, p. 392, 1852.
„ Bell, Trans. Linn. Soc. Lond., vol. xxi. p. 303, 1855.
Phlyxia, Bell (subgenus), Trans. Linn. Soc. Lond., *tom. cit.*, p. 303, 1855.
Bellidilia, Kinahan, Journ. Roy. Dublin Soc., vol. i. p. 128, 1858.

Carapace transverse and often rather longer than broad, suborbiculate or subrhomboidal, with the dorsal surface moderately convex or depressed, uneven, tuberculated or obscurely carinated ; the front emarginated or subtruncated or quadridentated, the hepatic regions are usually concave ; the lateral margins are entire or more or less distinctly tuberculated or toothed, but the tubercles are very rarely spiniform. The orbits are very small, circular, with two or three closed marginal fissures and an interior hiatus ; the buccal cavity is not separated from the antennulary fossæ by any distinct epistoma, and the endostomian ridges are very stongly defined. The post-abdomen in the male is narrow and four or five-jointed, with several of the intermediate segments coalescent, and it covers the sternum at base, between the bases of the fifth ambulatory legs. Eyes very small. Antennulary fossæ oblique, or nearly transverse. The antennæ are very small and their basal joint enters the interior orbital hiatus. The exterior maxillipedes cover the whole of the buccal cavity ; the merus of the endognath, as usual, is triangulate, and the exognath has a straight or somewhat curved exterior margin.

The chelipedes (in the male) are subequal and of moderate length, or more rarely, considerably elongated ; with the merus subcylindrical or somewhat trigonous, palm and fingers usually compressed. The ambulatory legs are slender and small, with the joints usually smooth ; dactyli styliform.

The species are small and numerous and inhabit the sublittoral or deeper waters both of the Atlantic and Indo-Pacific regions. The species of the section *Phlyxia*, Bell, are, as far as at present known, restricted to Australia.

From the genus *Nursia* (to which it is nearly allied), *Ebalia* differs in the less produced margins of the carapace, which are not cristated, &c.

The genera *Ebalia* and *Phlyxia* are now connected by so many intermediate species, that not one of the distinctive characters mentioned by Bell can be regarded as constant. I propose, therefore, to unite these genera, but to separate the species under two primary sections or subgenera (for which the names *Ebalia* and *Phlyxia* may conveniently be retained) as follows :—

I. Front slightly concave or truncated, not quadridentated (*Ebalia* [1]) :—

Ebalia tuberosa (Pennant)= *Ebalia insignis*, Lucas, *fide* Heller. European Seas; Adriatic; Mediterranean, to 250 metres (Heller).

Ebalia tumefacta (Mont.)= *Ebalia aspera*, Costa, *fide* Heller. European Seas ; Mediterranean ; Adriatic.

Ebalia cranchii (Leach)= *Ebalia discrepans*, Costa, and *Ebalia deshayesii*, Lucas, *fide* Heller. European Seas ; Mediterranean ; Adriatic. This species and the preceding are mentioned by Heller to occur in 30 to 40 fathoms.

Ebalia granulosa, Milne Edwards. Mediterranean, Corfu (Coll. Brit. Mus.).

Ebalia edwardsii, Costa = *Ebalia algerica*, Lucas. Mediterranean.

[1] The synonyms of the Mediterranean species are as given by Dr. Heller, Crust. südlichen Europa, pp. 124-128, 1863.

Ebalia costæ, Heller. Adriatic. (Is probably a variety of *Ebalia granulosa*.)
Ebalia setubalensis, Capello = *Ebalia elegans*, Capello. Setubal.
Ebalia nux, Norman. Mediterranean, to 300 métres.
Ebalia maderensis, Stimpson. Madeira. (Is perhaps not distinct from *Ebalia tuberosa*.)
Ebalia fragifera, Miers. Canaries.
Ebalia tuberculata, Miers. Senegambia; Goree Island, to 15 fathoms.
Ebalia affinis, Miers. Senegambia; Goree Island, to 15 fathoms.
Ebalia stimpsonii, A. Milne Edwards. Barbados, to 50 fathoms.
Ebalia granulata (Rüppell). Red Sea; Providence Group, to 24 fathoms.
Ebalia orientalis, Kossmann. Red Sea.
Ebalia tuberculosa (A. Milne Edwards). South Africa; Agulhas Bank, to 150 fathoms; East and South Australia; New Zealand, to 150 fathoms.
Ebalia lambriformis (Bell) = *Phlyxia petleyi*, Haswell. North, East, and South Australia; south of New Guinea, to 28 fathoms.
Ebalia lævis (Bell). New Zealand Seas, to 150 fathoms.
Ebalia erosa (A. Milne Edwards). Bass Strait; New Caledonia.
Ebalia quadrata (A. Milne Edwards). Bass Strait. (Perhaps a species of *Nursia*.)
Ebalia granulosa (Haswell). Sydney Heads, "deep water."
Ebalia minor, Miers. Japanese Seas.
Ebalia bituberculata, Miers. Japanese Seas, to 52 fathoms.
Ebalia rhomboidalis, Miers. Japanese Seas.
Ebalia pulchella, A. Milne Edwards. Fiji Islands.
Ebalia miliaris, A. Milne Edwards. Samoan Islands, Upolu. (Perhaps a distinct genus.)

II. Front with four distinct (usually tuberculiform) lobes or teeth, including the tooth at the interior angle of the orbit. (*Phlyxia*.)

Ebalia crassipes (Bell). South and East Australia (to 40 fathoms).
Ebalia quadridentata, Gray, with var. *spinifera*, nov. Australia, Port Jackson.
Ebalia spinosa (Kinahan). Australia, Port Philip.
Ebalia spinosa, var. *orbicularis* (Haswell) = *Bellidilia serratocostis*, Kinahan (?). East and South Australia (to 15 fathoms); Tasmania.
Ebalia ramsayi (Haswell). Port Jackson.
Ebalia dentifrons, n. sp. South Australian Coast.
Ebalia intermedia, n. sp. South Australia, near Port Philip (38 fathoms).[1]

[1] The *Ebalia spinosa*, A. Milne Edwards, from Upolu, is perhaps referable to the genus *Arcania*.

Ebalia lambriformis (Bell).

> *Phlyxia lambriformis*, Bell, Trans. Linn. Soc. Lond., vol. xxi. p. 304, pl. xxxiv. fig. 2, 1855.
> ,,   ,,   Miers, Crust. in Rep. Zool. Coll. H.M.S. "Alert," p. 252, 1884, et synonyma.

South of New Guinea, 28 fathoms, in lat. 9° 59′ 0″ S., long. 139° 42′ 0″ E., Station 188 (an adult female).

Adult ♀.						Lines.	Millims.
Length of carapace,	.	.	.	.	.	5½	11·5
Breadth of carapace, rather over	.	.	.	.	5	11

This species, in the form of the carapace and in other characters, resembles somewhat the genus *Nursia*.

Ebalia lævis (Bell).

> *Phlyxia lævis*, Bell, Trans. Linn. Soc. Lond., vol. xxi. p. 305, pl. xxxiv. fig. 3, 1855; Cat. Leucosiidæ in Brit. Mus., p. 18, 1855.
> ,,   ,,   Miers, Cat. New Zealand Crust, p. 56, 1876.

New Zealand Seas, not far from South Island, lat. 39° 32′ 0″ S., long. 171° 48′ 0″ E., in 150 fathoms, Station 167 (two adult males); Queen Charlotte Sound, near Long Island, 10 fathoms, Station 167A (two adult females).

The deep-water specimens in the Challenger collection do not differ appreciably from the shallow-water examples of the species I have examined, except in the somewhat more elongated merus-joints of the chelipedes. One of them has the following dimensions:—

Adult ♂.			Lines.	Millims.
Length and breadth of carapace, about	.	.	5	10·5
Length of a chelipede, about .	.	.	13	27·5
Length of first ambulatory leg, nearly .	.	7½	15·5

Ebalia tuberculosa (A. Milne Edwards) (Pl. XXV. fig. 1).

> *Persephona tuberculosa*, A. Milne Edwards, Journ. des Mus. Godeffroy, vol. i. Heft iv. p. 86, 1873.
> ,,   ,,   Haswell, Cat. Australian Crust., p. 132, 1882.
> ? *Phlyxia granulosa*, Haswell, Proc. Linn. Soc. N.S.W., vol. iv. p. 54, pl. vi. fig. 3, 1880; Cat. Australian Crust., p. 126, 1882.

Port Jackson, 30 to 35 fathoms (three females); off Twofold Bay, 150 fathoms, Station 163A (two small females); Bass Strait, off East Moncœur Island, 38 fathoms, Station 162 (several specimens, males and females); New Zealand Seas, off South Island, in 150 fathoms, Station 167 (numerous specimens, mostly adult males).

A fully-grown male has the following dimensions :—

Adult ♂.							Lines.	Millims.
Length of carapace,	.	.	.	.	.	.	5	10·5
Breadth of carapace, about	.	.	.	.	.	.	4½	10
Length of a chelipede,	.	.	.	.	.	.	12½	26·5
Length of ambulatory leg,	.	.	.	.	.	.	7	15

The specimens dredged near New Zealand, in deep-water (Station 167), are generally of larger size, and in most, but not all of them, the tubercles of the gastric and branchial regions are less prominent and the chelipedes are more developed than in the specimens from Australian localities.

In small females the two rounded protuberances of the posterior margin are not developed, and the margin is straight.

There are in the collection three small females from the Agulhas Bank, 150 fathoms, lat. 35° 4′ 0″ S., long. 18° 37′ 0″ E. (Station 142), which cannot, I think, be distinguished specifically from *Ebalia tuberculosa*.[1]

Ebalia (*Phlyxia*) crassipes (Bell).

> *Phlyxia crassipes*, Bell, Trans. Linn. Soc. Lond., vol. xxi. p. 304, pl. xxxiv. fig. 2, 1855 ;
> Cat. Leucosiidæ in Brit. Mus., p. 17, 1855.
> „ „ Miers, Crust. in Rep. Zool. Coll. H.M.S. " Alert," p. 252, 1884.

Specimens of this, which is perhaps the commonest species of the subgenus (or genus) *Phlyxia* were obtained at the following localities :—

Port Jackson (Sow and Pig's Bank), 6 fathoms, a good series of specimens; also at Port Jackson, 8 to 15 fathoms (an adult male), and 30 to 35 fathoms (two adult males) ; also several males and females dredged off the South Australian coast in 2 to 10 fathoms, in April 1874 ; and several specimens from East Moncœur Island, Bass Strait, 38 fathoms (Station 162).

In the adult males the chelipedes are often very considerably elongated ; one in the Challenger series from Port Jackson (30 to 35 fathoms) presents the following dimensions :—

Adult ♂.					Lines.	Millims.
Length of carapace,	.	.	.	.	5½	11·5
Breadth of carapace, about	.	.	.	.	5	11
Length of a chelipede, over	.	.	.	.	12	26
Length of first ambulatory leg,	.	.	.	.	7½	16

The carapace is always more or less distinctly tuberculated on the dorsal surface, with greyish or fuscous markings, which exist also on the chelipedes.

[1] *Ebalia tuberculosa* is connected through *Ebalia fragifera*, Miers (*Ann. and Mag. Nat. Hist.*, vol. viii. p. 268, 1881), with *Ebalia tuberculata*, and certain other Atlantic and European forms. *Ebalia fragifera* is, however, distinguished from *Ebalia tuberculosa* by the deeply concave hepatic regions of the carapace, the less prominent front, and the fewer granules of the palms of the chelipedes. The Mascarene (Providence Island) specimen referred by me to *Ebalia granulata*, Rüppell, differs somewhat in the broader carapace, less prominent front, &c.

Ebalia (Phlyxia) intermedia, n. sp. (Pl. XXV. fig. 2).

I thus very doubtfully designate some small specimens (perhaps not fully grown) which are intermediate in their distinctive characters between *Phlyxia crassipes* and *Phlyxia orbicularis*. They are distinguished from the former species by the absence of tubercles from the dorsal surface of the carapace (except sometimes an obscure tubercle in front of the median posterior spine, which is more elongated than in *Phlyxia crassipes*) and from *Phlyxia orbicularis* by the more convex and less orbiculate carapace, more prominent front, and more slender chelipedes. It may be a variety of the latter species; it can hardly, I think, be identified with *Bellidilia serratocostis*, Kinahan, which is very briefly diagnosed.

Carapace moderately convex, longer than broad, granulated but not tuberculated on the dorsal surface, which is somewhat uneven, with a slight protuberance on the intestinal region, in front of the posterior spine, and with three or four small tubercles or granules on the lateral margins, the most prominent being one placed in the middle of the postero-lateral margin; front rather prominent and quadridentated; the median posterior spine of the carapace is prominent and acute, and below it the posterior margin is straight, with the lateral angles not prolonged as lateral spines. There is a small tubercle on each pterygostomian region. The sternum and post-abdomen are finely granulated; the latter is smooth, with all of the segments (except the last) coalescent in the male; in the (young?) female one or two of the basal segments are partially distinct; the terminal segment is narrow and rounded at the distal extremity. The maxillipedes are finely granulated; the exognath has a nearly straight outer margin, and its distal extremity nearly attains the acute apex of the merus of the endognath. The chelipedes are moderately elongated and slender; merus slightly compressed (but not carinated) and granulated, the granules of the posterior surface most prominent; wrist and palm granulated and compressed; the latter has on its exterior margin a small indentation at base close to its articulation with the wrist; fingers thin, compressed, as long as or slightly longer than the palm, and obscurely denticulated on their inner margins near to the apices, which are incurved; ambulatory legs, of which only one or two detached, remain very slender, with the dactyli styliform and considerably longer than the penultimate joints. Colour, in spirit, light yellowish-brown. The most perfect specimen, which is, I think, an immature female, has the following dimensions:—

♀						Lines.	Millims.
Length of carapace, nearly	.	.	.	.	.	4	8
Breadth of carapace,	.	.	.	.	.	3½	7·5
Length of a chelipede, about	.	.	.	.	.	6	13
Length of first ambulatory leg, nearly	.	.	.	.	.	5	10·5

Four small specimens, all more or less imperfect, were dredged in 33 fathoms, off the entrance to Port Philip (Station 161).

Ebalia (Phlyxia) undecimspinosa (Kinahan), var. *orbicularis.*

? *Bellidilia serratorostis,* Kinahan, Journ. Roy. Dublin Soc., vol. i. p. 129, 1858.
?   „   *undecimspinosa,* Kinahan, *tom. cit.,* p. 128, pl. iii. fig. 2, 1858, var.
Phlyxia orbicularis, Haswell, Proc. Linn. Soc. N.S.W., vol. iv. p. 54, pl. vi. fig. 2, 1880; Cat.
Australian Crust., p. 125, 1882.

Port Jackson (Sow and Pig's Bank), 6 fathoms, 8 to 15 fathoms (two adult females),
8 fathoms (an adult male and female); South Australian Coast, 2 to 10 fathoms, April
1874 (numerous specimens).

The largest male presents the following dimensions :—

Adult ♂ .	Lines.	Millims.
Length of carapace, rather over .	11½	24·5
Breadth of carapace, rather less than	11½	24
Length of a chelipede, . .	27½	58
Length of first ambulatory leg, about .	15	32

Although commonly occurring with *Phlyxia crassipes,* this is a perfectly distinct
species, characterised by the less prominent front and the absence of tubercles from the
gastric and branchial regions of the carapace, which has, besides the three prominent
posterior tubercles, only some smaller ones upon the lateral margins, and occasionally two
or three upon the median longitudinal carina ; the fuscous markings, characteristic of
Phlyxia crassipes, do not exist in the specimens of *Phlyxia orbicularis* I have examined.
It also attains a larger size than *Phlyxia crassipes,* and has a more depressed and more
regularly orbiculate carapace.[1]

Ebalia (Phlyxia) quadridentata, Gray, var. *spinifera* (Pl. XXV. fig. 3).

? *Ebalia quadridentata,* Gray, Zool. Miscell., vol. ii. p. 40, 1831.
? *Phlyxia quadridentata,* Stimpson, Proc. Acad. Nat. Sci. Philad., p. 160, 1858.

Phlyxia quadridentata is only known to me by the very short and insufficient
diagnosis of the late Dr. Gray, which, as regards the tubercles of the dorsal surface of
the carapace, does not accurately apply to any species of this genus.

It is not improbable that the Challenger specimen, of which a description follows,
may belong to a distinct species.

Carapace moderately convex, rather longer than broad, covered with small granules
and with larger granules (or small spiniform tubercles) which are disposed as follows :—
three in a triangle on the gastric region, one on the cardiac region, one or two on each

---

[1] I refer to this species under Haswell's name, *Phlyxia orbicularis,* because Kinahan's *Bellidilia serratorostis* is so
briefly characterised, that its identification with *Phlyxia orbicularis* must remain uncertain, and I am unable to discover
the Tasmanian type specimen, which, according to Kinahan, existed in the collection of the British Museum. *Bellidilia
undecimspinosa,* Kinahan, differs from all specimens I have seen in possessing an additional tooth on each postero-lateral
margin, and should (I think) be regarded at least as a distinct variety.

branchial region, one on the hepatic and pterygostomian regions, and three or four on the lateral margins; and with a posterior median spine, on either side of which is a smaller triangulate lobe or tooth. The front is prominent, and is quadridentated, the lateral somewhat stouter but not more prominent than the median teeth, which are separated by a deeper fissure than that between the median and lateral teeth. The orbits have three deep fissures as in other species of the genus. The surface of the sternum and post-abdomen is minutely granulated, the post-abdomen without tubercles, and with all of the segments except the last coalescent, but with indications of the suture defining the basal segment, and with a protuberance at base on each side of the compound segment. The maxillipedes are covered with prominent granules, similar to those of the carapace; their exognathi have a nearly straight outer margin and attain nearly to the distal extremity of the merus of the endognath. The chelipedes are moderately elongated and slender; merus subcylindrical and granulated, the granules of the posterior surface larger than the others; carpus short, palm compressed but scarcely carinated, both carpus and palm are granulated on their exterior margins; fingers about as long as the palm, compressed and crossed at the tips, with scarcely any indications of denticules on the inner margins. Ambulatory legs very slender; dactyli longer than the penultimate joints. Colour (in spirit) light yellowish-brown.

The single specimen has the following dimensions :—

♂.					Lines.	Millims.
Length to base of posterior spine, about	.	.	.	.	3	6·5
Breadth of carapace, rather under	.	.	.	.	3	6
Length of a chelipede, about	.	.	.	.	5	11
Length of first ambulatory leg, about	.	.	.	.	3½	7·5

and was dredged at Port Jackson (Sow and Pig's Bank), in 6 fathoms.

In the arrangement of the tubercles of the dorsal surface of the carapace and in other points this form nearly resembles *Ebalia granulosa*,[1] dredged, according to Haswell, by H.M.S. Challenger, outside of Sidney Heads in deep water; the latter, however, differs in the bidentate front, carinated palms of the chelipedes, &c., and has not the prominent median posterior spine which exists in *Phlyxia quadridentata* (?), var. *spinifera*.

Ebalia (Phlyxia) dentifrons, n. sp. (Pl. XXV. fig. 4).

The carapace is longer than broad, and moderately convex; it is covered, both above and below, with close-set prominent granules; its dorsal surface is very uneven, having several rounded prominences separated by marked intervening depressions; of these, one (very prominent) is situated on each hepatic region, one on each branchial region, and one on each intestinal region. The lateral margins bear, on each side, four small spines

[1] *Phlyxia granulosa*, Haswell, *Proc. Linn. Soc. N.S.W.*, vol. iv. p. 55, pl. vi. fig. 3, 1880; Cat. Australian sessile-eyed Crust., p. 126, 1882.

or teeth, three of which are placed midway between the front and posterior margin, and one near the postero-lateral margin ; the front is prominent and truncated, and four-toothed, the teeth separated by nearly equal notches ; there is also a tooth at the exterior angle of the orbit. The post-abdomen (in the female) covers the whole of the sternal surface of the body between the bases of the legs, and has all of the segments, except the two first and the last, coalescent ; the first and second segments are granulated ; the terminal segment is narrow and deeply encased in a cavity of the sternum, which attains the bases of the outer maxillipedes. The maxillipedes are coarsely granulated, their exognathi robust, with a nearly straight outer margin, and rounded at the distal extremity, which does not quite attain the acute apex of the merus of the endognath. The chelipedes (in the female) are moderately elongated and closely granulated, but not so coarsely as the carapace, and the joints are without spines or teeth ; the merus is sub-cylindrical ; the palm is slightly compressed, but not carinated ; fingers about as long as the palm, compressed, with the tips incurved, obscurely denticulated on their inner margins, and faintly longitudinally-sulcated on the sides. The ambulatory legs are slender, with the antepenultimate joints angulated ; the penultimate joints slightly dilated and carinated on the superior and inferior margins, and shorter than the slender dactyli. Colour of carapace (in spirit) greyish or brownish ; the legs paler.

The best-preserved specimen presents the following dimensions :—

Adult ♀.						Lines.	Millims.
Length of carapace,	.	.	.	.	.	5	10·5
Breadth of carapace, nearly	.	.	.	.	5	10
Length of a chelipede, rather over	.	.	.	5	11
Length of first ambulatory leg,	.	.	.	4½	9·5

Two females were dredged in 2 to 10 fathoms, in April 1874, on the South Australian coast.

The form, granulation, and spinulation of the carapace distinguishes this from any species with which I am acquainted.

Persephona, Leach.

Persephona, Leach, Zool. Miscell., vol. iii. pp. 18, 22, 1817.
,,   Milne Edwards, Hist. Nat. Crust., vol. ii. p. 136, 1857.
,,   Dana, U.S. Explor. Exped., vol. xiii., Crust. 1, p. 399, 1852.
,,   Bell, Trans. Linn. Soc. Lond., vol. xxi. p. 292, 1855.
Guia, Milne Edwards, Hist. Nat. Crust., vol. ii. p. 127, 1837.

The characters separating this genus from Myra are very slight. Persephona is, in fact, scarcely distinguishable from Myra, except by the somewhat more depressed and orbiculate carapace, the more transversely plicated antennules, the much more robust chelipedes and ambulatory legs (the palms and fingers of the chelipedes being dilated and compressed, except in *Persephona lichtensteinei*), and the narrower exognath of the exterior maxillipedes, whose exterior margin is straight, not arcuated, and slightly dilated at base, as in *Myra*. The post-abdomen of the male in *Persephona punctata* (the only species in which I have examined it) is five-jointed, with the penultimate, as well as the first and second and terminal segments distinct.

This genus apparently represents *Myra* on the shores of the American continent and islands adjacent.[1]

Persephona punctata (Browne) (Pl. XXV. fig. 5).

Cancer punctatus, Browne, Civil and Nat. Hist. of Jamaica, vol. ii. p. 422, pl. xlii. fig. 3, 1756.
„   „   Linné (*partim*), Syst. Nat., ed. xii. p. 1045, 1766.
Persephona guaia, Bell, Trans. Linn. Soc. Lond., tom. cit., p. 292, 1855.
„   *punctata*, Stimpson. Ann. Lyc. Nat. Hist. New York, vol. vii. p. 70, 1860.
„   „   Kingsley, Proc. Acad. Nat. Sci. Philad., p. 403, 1879, et synonyma.

Bahia (shallow water), a small male.

♂.						Linea.	Millima.
Length of carapace, about	.	.	.	.	.	11	23·5
Breadth of carapace, about	.	.	.	.	.	10	21

The characteristic coloration is in this specimen almost obliterated.

Myra, Leach.

Myra, Leach, Zool. Miscell., vol. iii. pp. 19, 23, 1817.
„   Milne Edwards, Hist. Nat. Crust., vol. ii. p. 125, 1837.
„   Dana, U.S. Explor. Exped., vol. xiii., Crust. 1, p. 392, 1852.
„   Bell, Trans. Linn. Soc. Lond. (Zool.), vol. xxi. p. 296, 1855.

Carapace dorsally very convex, and more or less distinctly granulated, with the lateral margins regularly arcuated; it has three posterior spines or protuberances (the median one being more elevated and situated on the posterior margin), and there is usually a more or less distinct protuberance upon the pterygostomian regions. The front is concave and does not project beyond the anterior margin of the buccal cavity. The orbits are very small and circular, with usually three deep marginal fissures, and with a rather large inferior hiatus. The post-abdomen (in the male) is usually four-jointed, with all of the segments except the first, second, and last, coalescent. Eyes very small. Antennules somewhat obliquely plicated. Antennæ with a slender basal

[1] Its range may, however, extend over the whole Atlantic region; since there is a specimen, perhaps not distinct from *Persephona punctata*, from South Africa (Sir A. Smith), in the collection of the British (Natural History) Museum, and if the habitat of Herbst's *Cancer mediterraneus* be correctly given, this genus must also occur in the Mediterranean Sea. I have referred the species from Bass Strait, described by Dr. A. Milne Edwards as *Persephona tuberculosa*, to the genus *Ebalia*.

antennal joint, which scarcely attains or does not attain the front. The exterior maxillipedes cover the buccal cavity, and the merus of the endognath, as usual in this family, is triangulate and acute ; the exognath is rather broad, about as broad as the endognath, and its exterior margin is slightly arcuated. The chelipedes (in the adult males) are subequal, and sometimes considerably elongated ; the merus subcylindrical, and sometimes exceeding the carapace in length ; palm subcylindrical or compressed, not dilated ; fingers rather robust, compressed, and distally acute ; the ambulatory legs are relatively small, with the dactyli styliform.

The species are found in the Indo-Malaysian, Japanese and Australian Seas, in littoral or shallow water ; one species at least (*Myra fugax*) occurs commonly throughout the Indo-Pacific region.

Besides the different forms regarded as synonyms of *Myra fugax* and *Myra australis*, Haswell, and *Myra darnleyensis*, Haswell, referred to below, I am not aware that any recent species have been described since the publication of Professor Bell's Monograph.

Myra fugax (Fabricius).

?? *Cancer euphæus*, Linné, Mus. Lud. Ulricæ, p. 440, 1764 ; Syst. Nat., ed. xii. p. 1045, 1766.

? „ *punctatus*, Herbst, Naturgesch. der Krabben u. Krebse, vol. i. p. 89, pl. ii. figs. 15, 16, 1782, nec Linn.

Leucosia fugax, Fabricius, Entom. Syst. Suppl., p. 351. 1798.

Myra fugax, Leach, Zool. Miscell., vol. iii. p. 24, 1817.

„ „ Milne Edwards, Hist. Nat. Crust., vol. ii. p. 126, 1837 ; Crust. in Cuvier, Règne Animal, vol. xxv. fig. 3.

„ „ Bell, Trans. Linn. Soc. Lond., tom. cit., p. 296, 1855 ; Cat. Leucosiidæ in Brit. Mus., p. 12, 1855.

? „ „ de Haan, Crust. in v. Siebold, Fauna Japonica, decas. v. p. 134, pl. xxxiii. fig. 1, 1841, ♂ adult.

„ *carinata*, Bell, Trans. Linn. Soc. Lond., tom. cit., p. 297, pl. xxxii. fig. 3, 1855 ; Cat. Leucosiidæ, in Brit. Mus. p. 13, 1855.

„ „ Miers, Ann. and Mag. Nat. Hist., ser. 5, vol. v. p. 316, 1880 ; Crust. in Rep. Zool. Coll. H.M.S. "Alert," p. 250, 1884, young.

? „ *subgranulata*, Kossmann, Malacostraca in Zool. Ergebn. einer Reise Küstengeb. d. rothen Meeres, p. 65, pl. i. fig. 7, 1877.

„ „ *cf.*, Hilgendorf, Monatsber. d. k. preuss. Akad. d. Wiss. Berlin, p. 811, 1878.

? „ *punctata*, Hilgendorf, tom. cit., p. 811, 1878.

„ *coalita*, Hilgendorf, tom. cit., p. 812, pl. i. figs. 6, 7, 1878, var. (?).

„ *dubia*, Miers, Proc. Zool. Soc. Lond., p. 42, 1879, var. (?).

Arafura Sea, south of New Guinea, 28 fathoms (Station 188), an adult and smaller male, and an adult and two younger females ; also in 49 fathoms (Station 190), in lat. 8° 56′ 0″ S., long. 136° 5′ 0″ E., a young male.

---

[1] The genus *Myrodes*, Bell, which is united by Haswell in his Catalogue of the Australian Crustacea, 1882, with *Myra* should, I think, be retained as distinct.

The largest male of the Challenger series, from Station 138, presents the following dimensions :—

Adult ♂.	Lines.	Millims.
Length of carapace, about .	13	28
Breadth of carapace, about .	11	23·5
Length of a chelipede, nearly	34	72
Length of first ambulatory leg,	15½	33

The smallest specimen (male) from the same locality measures as follows :—

Young ♂.	Lines.	Millims.
Length of carapace, .	8	17
Breadth of carapace, nearly .	7	14·5
Length of a chelipede,	14½	31
Length of first ambulatory leg,	8½	18

The examination of the Challenger series compels me to unite, under the designation *Myra fugax*, several species which I have hitherto supposed to be distinct. In young specimens the carapace is more or less distinctly carinated in the median dorsal line, and the post-abdomen in the male is flat and smooth, without the subbasal prominences and the tubercle which sometimes exists on the penultimate segment in adult examples ; the margins of the carapace also are more distinctly granulated, and the chelipedes relatively shorter. In adult males the chelipedes are sometimes very considerably elongated (as in the figure of de Haan cited above) and—in specimens I have examined in the collection of the British (Natural History) Museum—the three posterior spines of the carapace strongly developed and acute ; perhaps these may be referable to a distinct variety or species ; the chelipedes, however, are usually more slender and more elongated in adult males than in adult females.[1]

[1] I may note in regard to the synonymical citations, that the *Cancer euphaeus* of Linné was founded on a specimen wanting the chelipedes, and cannot therefore be identified with certainty with any species of this genus. *Cancer punctatus*, Herbst, resembles *Myra fugax* in the form of the body and limbs, but differs in coloration from any specimen I have examined. Hilgendorf, who refers (*tom. cit.*) to the type, adds no information regarding it. The form I have described, from Japanese types (*Proc. Zool. Soc. Lond.*, p. 43, 1879), as *Myra dubia*, is probably identical with *Myra coalita*, Hilgendorf, described in the preceding year (*Monatsber. d. k. preuss. Akad. d. Wiss. Berlin*, p. 812, pl. i. figs. 6, 7, 1878), from Zanzibar, and may be a variety of *Myra fugax*, although the types are distinguished from all specimens of that species I have seen by possessing a tubercle on the intestinal region in front of the median spine of the posterior margin. They are probably not fully grown. A similar tubercle was observed by Hilgendorf in specimens referred to *Myra fugax*.

Myra affinis, Bell.

Myra affinis, Bell, Trans. Linn. Soc. Lond., vol. xxi. p. 296, pl. xxxii. fig. 2, 1855; Cat.
Leucosidæ in Brit. Mus., p. 12, 1855.
„   „   Haswell, Cat. Australian Crust., p. 122, 1882.

Torres Strait, August 1874, a young female.

♀.	Lines.	Millims.
Length of carapace,	8	17
Breadth of carapace, nearly	7	14·5
Length of a chelipede, nearly	12	25
Length of first ambulatory leg, nearly	7	14·5

This specimen has the median longitudinal line of granules on the dorsal surface of the carapace which is usually found in young specimens of species of this genus.

This form is very nearly allied to *Myra mammillaris*, Bell, but is apparently distinguishable by the less convex and rounded tubercles of the posterior margin of the carapace, the median one being more elongated and acute, and the two lateral ones more triangulate than in *Myra mammillaris*. Young specimens are to be distinguished from *Myra australis*, Haswell (which they much resemble), by the somewhat narrower carapace and by the less numerous but more prominent granulations of the maxillipedes and the adjacent parts of the body.

Myra australis, Haswell.

Myra mammillaris (yg.), Miers, Trans. Linn. Soc. Lond., ser. 2, tom. cit., p. 239, pl. xxxviii.
figs. 25–27, 1877.
Myra australis, Haswell, Proc. Linn. Soc. N.S.W., tom. cit., p. 50, pl. v. fig. 3, 1880; Cat.
Australian Crust., p. 122, 1882.

Torres Strait, in lat. 10° 36′ 0″ S., long. 141° 55′ 0″ E., 6 fathoms (Station 187), a male.[1]

♂.	Lines.	Millims.
Length of carapace,	8	17
Breadth of carapace,	7	15

Myra darnleyensis, Haswell.

? *Myra darnleyensis*, Haswell, Proc. Linn. Soc. N.S.W., vol. iv. p. 52, pl. v. fig. 4, 1880; Cat.
Australian Crust., p. 122, 1882.

Carapace (as in other species of the genus) convex and rather longer than broad; it is rather closely and finely granulated over the whole of the dorsal surface; there is a

[1] The specimens in the collection of the British (Natural History) Museum, designated " *M. mammillaris*, yg." vary slightly in the prominence of the intestinal region and of the posterior median spine, and the abdomen in the figure should have been represented with the sides slightly concave.

slight prominence upon the hepatic regions, and beneath this the sides of the carapace at the pterygostomian regions are slightly convex but scarcely angulated ; the three tubercles of the posterior margin are placed nearly in the same horizontal line ; they are slightly compressed and nearly semicircularly rounded or slightly triangulate. The front is deeply concave above with the lateral lobes prominent. The three sulci in the upper and latter orbital margins are deep and distinct. The post-abdomen (in both sexes) has some scattered punctulations ; in the male, all of the segments, except the first and last, in the female, the third to sixth segments, are coalescent ; the terminal segment in the male is narrow and elongated, and the penultimate segment bears a small tubercle. The exterior maxillipedes are (as in *Myra mammillaris*) coarsely granulated distally ; the exognath as in other species is arcuate externally, and it nearly attains the distal extremity of the merus of the endognath. The chelipedes (as in other species of the genus) are elongated ; merus finely and closely granulated ; palm compressed and rather short ; fingers slender and elongated, longer than the palm, and meeting along their inner edges, which are denticulated, the tips incurved. The ambulatory legs are slender and short (as in other species of the genus). Colour (in spirit) light yellowish ; the chelipedes (except the fingers) brownish-pink, carapace sometimes with markings of the same colour. The single male presents the following dimensions :—

Adult ♂.				Lines.	Millims.
Length of carapace and rostrum,	.			11½	24
Breadth of carapace, about	.	.	.	9½	20
Length of a chelipede,	.	.	.	25½	54
Length of first ambulatory leg,	.	.		15	32

Celebes Sea, 10 fathoms, in lat. 6° 54′ 0″ N., long. 122° 18′ 0″ E. (Station 212). An adult male and seven adult females.

The Challenger specimens are identified with *Myra darnleyensis* with some uncertainty, because, although agreeing with the description and figure of Haswell in what seem to be the essential characteristics of the species, wherein they also differ from others of the genus, *i.e.*, in the broad and compressed posterior tubercles of the carapace and the greatly elongated fingers of the chelipedes, they differ (it would appear) in the more prominent lobes of the front, the less prominent hepatic tubercle, and the non-granulated male abdomen. A figure is ·therefore given. The Challenger specimens are all larger than Haswell's type from Darnley Island.

Randallia, Stimpson.

Randallia, Stimpson, Journ. Boston Soc. Nat. Hist., vol. vi. p. 471, 1857.

Carapace convex, orbiculate, with the lateral margins regularly arcuated ; the posterior margin armed with two lobes or teeth. The front projects but little, and its anterior margin is concave. Orbits very small, subcircular, with two or three fissures. The post-abdomen in the male (in the species I have examined) is narrowed at the distal extremity, and at base it covers the whole width of the sternum, between the fifth ambulatory legs. Eyes small. Antennules obliquely plicated. The basal joint of the antennæ is well developed, but (in the species I have examined) does not attain the front. The exterior maxillipedes (in *Randallia granulata*) cover the whole of the buccal cavity; the triangulate merus-joint is but little produced at the distal extremity, the exognath is nearly as broad as the endognath, with the distal extremity rounded, the exterior margin straight. The chelipedes (in the males) are well developed; with the merus-joint subcylindrical, palm somewhat compressed, fingers compressed and distally acute; the ambulatory legs are moderately elongated, the joints not dilated, the dactyli styliform.

The type of this genus, *Randallia ornata* (Randall), is from Upper California.

Randallia granulata, n. sp. (Pl. XXVI. fig. 1).

The carapace is convex, as broad as or a little broader than long, very coarsely and evenly granulated; the granules larger upon the dorsal than upon the inferior parts of the body. The sulci defining the regions of the carapace are discernible in some places; the most distinct being the cardiaco-branchial sulci. There is no tooth or prominence upon the hepatic and pterygostomian regions, but the antero-lateral margins, behind the hepatic regions, are slightly indented. There are two small granulated lobes or prominences upon the posterior margin. The front (in a lateral view) does not project beyond the buccal cavity; it is deeply concave above (in a dorsal view), and therefore bilobate, with the lobes rounded. The orbits are very small, with two superior and an inferior fissure, and with a very distinct inner subocular hiatus. The epistoma is transverse and deeper on the sides than in the middle. The sternum is granulated, the segments of the post-abdomen, except the first, second, and last, are partially coalescent, but marked with distinct sulci indicating the sutures of the coalescent segments; the segments in the male are distinctly granulated, except the terminal segment, the sides are straight and converge from the base of the third to the terminal segment; the third segment is longest and bears two lateral rounded prominences; the penultimate segment has a small median tooth at its distal extremity, the terminal segment is narrow, nearly smooth and rounded distally. The antennulary fossettes are rather large and deep, and (as in Stimpson's description of *Randallia ornata*) the large basal joints of the antennules close the aperture which exists between the epistoma and front, the basal joint of the antennæ is larger than the following joints, but does not reach the front. The exterior maxillipedes are coarsely granulated; ischium of the endognath longer than the merus, with a smooth inner margin; merus obliquely truncated at the distal extremity; exognath robust, with a nearly straight outer margin, the distal extremity rounded and not quite attaining the apex of the merus of the endognath. The chelipedes (in the adult) are robust and somewhat elongated; with the joints granulated but not carinated or toothed; merus (in the adult male) subcylindrical and nearly as long as the carapace; carpus a little shorter than the palm, which is more than half the length of the merus; fingers nearly as long as the palm, compressed, finely granulated, and scarcely toothed on the inner margins. Ambulatory legs slender and moderately elongated, with the dactyli styliform, hairy on the superior margins, and shorter than the penultimate joints. Colour (in spirit) light yellowish-brown.

Adult ♂.		Lines.	Millims.
Length and breadth of carapace, about		12½	26
Length of a chelipede, .	.	24	51
Length of first ambulatory leg, about .	.	16	33·5

Off Nukalofa, Tongatabu, 240 fathoms (Station 172A), an adult male; Fiji Islands, lat. 20° 56′ 0″ S., long. 175° 11′ 0″ W., 315 fathoms (Station 173), an adult female and three smaller males.

In the small males the terminal post-abdominal segment is as distinctly granulated as the others.

Lithadia, Bell.

Lithadia, Bell, Trans. Linn. Soc. Lond. (Zool.), vol. xxi. p. 305, 1855.

Of this genus I have examined but three specimens, referable to two species, *Lithadia cariosa*, Stimpson, var. and *Lithadia* (?) *sculpta*, Haswell, and I am unable, from the descriptions only, to indicate any constant characters by which it may be distinguished from *Ebalia*, except such as are derived from the carapace, which is very strongly granulated and more or less pitted, sulcated, and eroded on the dorsal surface, as in the genera *Oreophorus* and *Spelæophorus*; but in these genera the carapace is dilated at the postero-lateral margins, and more or less produced over the bases of the ambulatory legs. From *Actæomorpha*, Miers (perhaps = *Osachila*, Stimpson), it is distinguished by the form of the carapace, which in the typical species of *Actæomorpha* (*Actæomorpha erosa*) is cancroid in shape, with the antero-lateral margins regularly arcuated.

The following is, I believe, a complete list of the described species of this genus :—

> *Lithadia cumingii*, Bell.   Central America; Puerto Portrero (13 fathoms). .
> *Lithadia cariosa*, Stimpson.   North Carolina, southward to Florida, Bahia (?).
> *Lithadia cadaverosa*, Stimpson.   West of Tortugas; Couch Reef (to 40 fathoms).
> *Lithadia pontifera*, Stimpson.   Barbados.
> *Lithadia cubensis*, von Martens.   Cuba.
> *Lithadia brasiliensis*, von Martens.   Brazil, Rio de Janeiro.

Lithadia lacunosa, Kingsley. Florida, Sarasota Bay.
Lithadia rotundata, A. Milne Edwards. Mouth of the Bermejo.
Lithadia granulosa, A. Milne Edwards. Santa Cruz (to 115 fathoms).
Lithadia (?) sculpta, Haswell. North-East Australia ; Fitzroy Islands ; Arafura Sea (to 36 fathoms).[1]

Lithadia cariosa, Stimpson, var. (?) (Pl. XXVI. fig. 2).

cf. Lithadia cariosa, Stimpson, Ann. Lyc. Nat. Hist. New York, vol. vii. p. 238, 1860.

A small male obtained at Bahia (shallow water) is in the collection, to which Stimpson's detailed description will apply in all points except the following :—The ridge connecting the front with the median protuberances of the carapace, and also the hepatic regions, are not very prominent and distinctly granulated, and these median protuberances are evenly granulated like the rest of the dorsal surface. The front is without a median fissure. The outer maxillipedes are evenly granulated, and the merus of the chelipedes has a series of more prominent granules on its posterior margin. The dactyli of the ambulatory limbs are indistinctly granulated. The unique specimen is of very small size. The distinctions above cited may prove to be of specific importance, but other specimens are needed for comparison in order to determine this point.

♂.	Lines.	Millims.
Length and breadth of carapace, about	3½	7·5
Length of a chelipede, .	3½	7
Length of first ambulatory leg,	3	6·5

I cannot identify this form with any of the other described species of this genus.

Merocryptus, A. Milne Edwards.

Merocryptus, A. Milne Edwards, Journ. Mus. Godeffroy, vol. iv. p. 85, 1873.

Carapace dorsally convex and uneven and tuberculated, somewhat rhomboidal, concave on the hepatic regions and on the postero-lateral parts of the branchial regions, and with the branchial regions prolonged at the junction of the antero-lateral and postero-lateral margins, the lobes thus formed somewhat cylindrical as in *Ixa*, but less considerably developed ; posterior margin with two prominences, which are most developed in the males. Front concave or bilobated, and rather prominent. Orbits small, with the marginal fissures indistinctly indicated ; the interior hiatus of moderate

---

[1] *Ebalia mammillosa*, Desbonne and Schramm, Crust. de la Guadeloupe, p. 54, belongs, according to Dr. Stimpson, to this genus ; I have never been able to consult a copy of this work.

width. The channel defined by the endostomian ridges is not emarginated distally. Post-abdomen (in the male) narrow, and at base covering the sternum between the fifth ambulatory legs, with all the segments except the first and last coalescent. Antennules nearly transversely plicated. Antennæ with a rather slender basal joint which does not quite attain the front. The triangulate merus-joint of the endognath of the exterior maxillipedes is shorter than the ischium; the exognath is of moderate width, and its exterior margin is nearly straight. Chelipedes rather robust and short, and shaped nearly as in *Lithadia* and *Oreophorus*, with the merus granulated, palm rather short and turgid, fingers compressed and denticulated on the inner margins, but the denticles are not spinuliform. Ambulatory legs short and granulated, with the dactyli styliform and somewhat uncinated.

But a single species has as yet been recorded of this curious genus; *Merocryptus lambriformis*, A. Milne Edwards, whose types were from Upolu.

Merocryptus lambriformis, A. Milne Edwards.

Merocryptus lambriformis, A. Milne Edwards, Journ. Mus. Godeffroy, vol. i., Heft 4, p. 85, pl. ii. fig. 1, 1873.

Bass Straits, off East Moncœur Island, 38 fathoms (Station 162), several males and females; off Twofold Bay, 150 fathoms (Station 163A), two small males.

In the smaller males the gastric and cardiac prominences, and the tubercles of the posterior margin, are more prominent and acute than in the adult and fully-grown specimens.

The largest male measures as follows :—

	Lines.	Millims.
Adult ♂.		
Length of carapace,	4½	9·5
Breadth of carapace, exclusive of lateral lobes, rather over	4	9

Philyra, Leach.

Philyra, Leach, Zool. Miscell., vol. iii. p. 18, 1817.
   „   Milne Edwards, Hist. Nat. Crust., vol. ii. p. 131, 1837.
   „   Bell, Trans. Linn. Soc. Lond. (Zool.), vol. xxi. p. 299, 1855.

This genus is nearly allied to *Leucosia*, but distinguished by the less convex carapace, which is usually granulated or punctulated, but not polished; by the broader front, which is not at all prominent, and usually slightly concave; by the absence of the *sinus thoracicus*, and by the form of the exognath of the exterior maxillipedes, which is usually much more dilated, and is rounded at the distal extremity, with the exterior margin arcuated.

The species are somewhat numerous and occur commonly in the littoral and shallow waters of the Indo-Pacific region.

To the species enumerated by Professor Bell the following must be added :—

Philyra variegata (Rüppell). Red Sea.
Philyra tuberculosa, Stimpson. Hong-Kong.
Philyra unidentata, Stimpson. Chinese Seas (to 30 fathoms).
Philyra marginata, A. Milne Edwards. Samoan Islands.
Philyra longimana, A. Milne Edwards. New Caledonia.
Philyra cristata, Miers. Goree Island.
Philyra lævidorsalis, Miers. Goree Island.
Philyra rectangularis, Miers. Seychelles.
Philyra rudis, Miers. Penang.

Leucosia orbicularis, Bell, should probably be referred to this genus ; it differs from the typical *Leucosiæ* and resembles *Philyra* in the broader less prominent front, and the absence of the thoracic sinus, but resembles *Leucosia* and differs from *Philyra* in the polished carapace.

Philyra platycheira (?), de Haan.

? *Philyra platycheira*, de Haan, Crust. in v. Siebold, Fauna Japonica, decas 5, p. 132, pl. xxxiii. fig. 6, 1841.
       „   „   Bell, Trans. Linn. Soc. Lond., vol. xxi. p. 300, 1855.

An adult female, obtained on the South Australian Coast in 2 to 10 fathoms, in April 1871, is referred, but rather doubtfully, to this species.

The pterygostomian regions are but very slightly angulated.

Adult ♀.	Lines.	Millims.
Length of carapace, . .	5	10·5
Breadth of carapace, rather over	4½	10
Length of a chelipede, .	7½	16

Subfamily 2. LEUCOSIINÆ.

The frontal region of the carapace is narrowed and produced anteriorly. A thoracic sinus is developed (*i.e.*, a shallow pit in front of and above the bases of the chelipedes).[1]

The subfamily is restricted to the single genus *Leucosia*, Fabricius.

[1] This curious cavity is not apparently connected with the respiratory chamber, as is the opening in front of the chelipedes in the families Calappidæ and Dorippidæ.

Leucosia, Fabricius.

Leucosia, Fabricius, Entom. Syst. Suppl., p. 349, 1798.
,,   Milne Edwards, Hist. Nat. Crust., vol. ii. p. 121, 1837.
,,   Bell, Trans. Linn. Soc. Lond., vol. xxi. p. 281, 1855.

Carapace convex, semiglobose, smooth and usually polished, without tubercles or indications of the regions of the dorsal surface, the anterior part of the cervical region usually prominent, the front either triangulate, truncated, or with a median cusp. The antero-lateral margins defined by a granulated line, which may extend for a short distance along the postero-lateral margins; beneath the lateral margins is an excavated pit, defined in front of the bases of the chelipedes by a series of granules, and continued as a shallow excavation beneath the postero-lateral margins (*sinus thoracicus*, Bell). The post-abdomen in the male is large and covers the sternum at the base between the fifth ambulatory legs; the first and last segments are usually distinct, the remainder either consolidated or divided by a median suture. The eyes and orbits are extremely small; the orbits circular; the antennules are somewhat obliquely plicated; the minute antennæ are placed beneath the antennules; their basal joint does not attain the frontal margin and the small flagellum enters the interior hiatus of the orbit. The exterior maxillipedes completely cover the buccal cavity, the triangulate acute merus of the endognath covers the following joints; the exognath is distally obtuse and its exterior margin is straight or nearly so. Chelipedes in the adult male subequal, with the merus strongly granulated at its base and usually along the margins; carpus and palm usually granulated on the interior margins, palm compressed, fingers distally acute. Ambulatory legs rather small, with the joints unarmed, dactyli styliform and compressed.

The species of this well-known genus are numerous and are often remarkable for the beauty of their coloration; they occur commonly in the littoral and shallower waters of the Indo-Pacific region.

Besides the species described or adverted to by myself in the memoir above alluded to, the following have been described since the publication of Mr. Bell's monograph :—

Leucosia splendida, Haswell.   Port Jackson.
Leucosia leslii, Haswell.   Torres Strait (Darnley Island);

and two new species, *Leucosia australiensis* and *Leucosia haswelli*, are described below.[1]

Leucosia australiensis, n. sp. (Pl. XXVII. fig. 1).

The carapace is convex, smooth, and has a few scattered punctulations, which are absent from the parts near to the posterior and postero-lateral margins. The antero-

[1] The species described by Haswell as *Leucosia cherrti* is probably, as I have shown, identical with *Leucosia whitei*, Bell. I have also described a new variety of the common *Leucosia craniolaris*, *lævimana* (Crust. in Rep. Zool. Coll. H.M.S. "Alert," p. 250, pl. xxvi. fig. A) from the Torres Strait. The variety *viridimaculata*, Haswell, of *Leucosia reticulata*, Miers, is not sustained by Mr. Haswell in his Catalogue of the Australian Crustacea, 1882.

lateral margins are bordered with a line of granules, which are largest above the bases of the chelipedes; this line terminates just behind the chelipedes; the posterior margin is bordered by a minutely granulated line, which is prolonged on each side beneath the postero-lateral margins above the bases of the ambulatory legs. The hepatic region is slightly convex, and in front of it, and behind the front, the sides of the carapace are deeply concave. The front is prominent and somewhat triangulate. The post-abdomen (in the male) is narrow, and is divided in the middle line by a suture (where also its margins are notched), and by two others defining the small basal and terminal segments; in the (sterile) female the sutures defining the second, third, and fourth segments are more or less distinctly indicated. The form of the thoracic sinus will be best understood by a reference to the figure (fig. 1a); the lobe which partially defines it in front is granulated on the margin, and contains a series of small tubercles. The exognath of the outer maxillipedes does not attain the distal extremity of the merus-joint of the endognath, and both it and the merus-joint are distally granulated on the outer surface. The chelipedes (of the male) are of moderate length; the merus is granulated near to the base and along the margins nearly as in *Leucosia ocellata*; the wrist is smooth; the palm is also smooth, somewhat compressed, with the margins neither granulated nor carinated; fingers about as long as the palm, slender, obscurely toothed on the distal half of the inner margins, and with the apices incurved. Ambulatory legs slender and smooth, with the penultimate joints compressed and carinated above; dactyli longer than the penultimate joints. Ground colour greyish or yellowish, with three white spots on either side of the gastric region of the carapace, which are more or less distinctly annulated with orange, and with a pair of circles of the same colour on the back of the branchial regions and an orange-coloured spot on each postero-lateral margin. The bases of the fingers of the chelipedes and the joints of the ambulatory legs are banded with the same colour. The larger specimen (the female) has the following dimensions :—

♀.			Lines.	Millims.
Length of carapace and rostrum,	.	.	9½	20
Breadth of carapace, rather over	.	.	8	17·5
Length of a chelipede, nearly	.	.	9	18·5

Port Jackson, 3 fathoms (a sterile female); South Australian Coast, 2 to 10 fathoms (a male).

This species is most nearly allied to *Leucosia neocaledonica*, A. Milne Edwards,[1] and *Leucosia splendida*, Haswell,[2] which also was discovered at Port Jackson; from both of these species it is distinguished by the coloration and by the absence of a line of granules from the inner margin of the palms of the chelipedes.[3]

[1] *Nouv. Archiv. Mus. Hist. Nat.*, vol. x. p. 49, pl. ii. fig. 1, 1874.

[2] *Proc. Linn. Soc. N.S.W.*, vol. iv. p. 47, pl. v. fig. 1, 1879, and Catalogue, p. 119, 1882.

[3] I may note here, that not only *Leucosia neocaledonica*, A. Milne Edwards, but also *Leucosia longifrons*, de Haan, and *Leucosia pulcherrima*, Miers, are regarded by Dr. J. G. de Man as mere synonyms of *Leucosia urania*, Herbst (*cf.*, *Notes Leyden Mus.*, vol. iii. p. 123, 1881).

Leucosia haswelli, n. sp. (Pl. XXVII. fig. 2).

The carapace is shaped nearly as in *Leucosia reticulata*, Miers, to which this species is nearly allied, *i.e.*, it is rounded above, convex, smooth and polished, somewhat rhomboidal, with a prominent front whose anterior margin is straight or very obscurely sinuated, not dentated or emarginated; its upper surface is covered with scattered punctulations which are most abundant on the anterior part; its sides are bordered by a granulated line which extends along the antero-lateral margins and terminates on the postero-lateral margins, just above the bases of the first ambulatory legs; the posterior margin also is defined by a line of smaller granules, which is prolonged along the sides of the body above the bases of the ambulatory legs, and terminates behind the chelipedes. The thoracic sinus is shaped as in *Leucosia reticulata*, and, as in that species, contains several flattened tubercles. The eyes, antennæ, and maxillipedes present nothing remarkable; the rounded prominence which is observable on the ischium of the outer maxillipedes in *Leucosia reticulata* is not developed, or is very obscurely indicated, in *Leucosia haswelli*. As in *Leucosia reticulata*, the merus of the chelipedes is tuberculated in its proximal half, and the tubercles are, as in that species, crowded at the base and extend along the anterior and posterior margins to the distal extremity; the palm too is compressed, granulated on its inner, and slightly carinated on its exterior margin, as in *Leucosia reticulata*. The ambulatory legs (as usual in the genus) are slender, feeble and compressed, with the penultimate joints carinated above. Coloration, greenish or yellowish, with a large spot of a darker green on the back of each branchial region, and with two white spots on each side of the gastric region; the tubercles of the upper surface of the merus-joints of the chelipedes in one specimen are crimson-red; the ambulatory legs in both are yellowish.

The largest example (a sterile female) is somewhat distorted, and has the following dimensions:—

♀.	Lines.	Millims.
Length of carapace and front, nearly	12	25
Breadth of carapace (allowing for its distortion), nearly	11	23
Length of a chelipede, nearly	11½	24
Length of first ambulatory leg, .	9½	20

Arafura Sea, south of New Guinea, in 28 fathoms, lat. 9° 59′ 0″ S., long. 139° 42′ 0″ E. (Station 188), an adult female; Celebes Sea, in 10 fathoms, lat. 6° 54′ 0″ N., long. 122° 18′ 0″ E. (Station 212), a sterile female.

This species in all of its characters nearly resembles *Leucosia reticulata*, Miers,[1] from West Australia, from which it is distinguished not only by the different disposition of the markings of the carapace, which more nearly resemble those of *Leucosia pallida*,

[1] *Trans. Linn. Soc. Lond.* (Zool.) ser. 2, vol. i. p. 237, pl. xxxviii. figs. 13-15, 1877.

Bell, but also by the absence of the group of hepatic tubercles, which form so conspicuous an ornament of *Leucosia reticulata*.

Leucosia ocellata, Bell.

> *Leucosia ocellata*, Bell, Trans. Linn. Soc. Lond., vol. xxi. p. 289, pl. xxxi. fig. 5, 1855; Cat.
> Leucosiidæ in Brit. Mus., p. 8, 1855.
> „   „   Haswell, Cat. Australian Crust., p. 118, 1882.

Arafura Sea, south of New Guinea, 28 fathoms (Station 188), two adult males and a female.

In addition to the four ocellated red spots of the gastric region mentioned by Bell as characteristic of this species, there are, I may add, two obscure yellowish-brown spots on the back of each branchial region. These are always very faintly indicated.

An adult male measures as follows :—

Adult ♂.	Lines.	Millims.
Length of carapace, nearly	9½	19·5
Breadth of carapace, .	8	17

Leucosia whitei, Bell.

> *Leucosia whitei*, Bell, Trans. Linn. Soc. Lond., vol. xxi. p. 289, pl. xxxi. fig. 2, 1855.
> „   „   Haswell, Cat. Australian Crust., p. 118, 1882.

Arafura Sea, south of New Guinea, 28 fathoms, lat. 9° 59′ 0″ S., long. 139° 42′ 0″ E., (Station 188), an adult male.

Adult ♂.	Lines.	Millims.
Length of carapace, nearly	6	12·5
Breadth of carapace, .	5½	11·5

Leucosia craniolaris (Linné) (Pl. XXVII. fig. 3).

> ? *Cancer craniolaris*, Linné, Mus. Lud. Ulricæ, p. 431, 1764 ; Syst. Nat., ed. xii. p. 1041, 1766.
> „   „   Herbst, Naturgesch. der Krabben u. Krebse, vol. i. p. 90, pl. ii. fig. 17,
> 1782.
> *Leucosia craniolaris*, Fabricius, Entom. Syst. Suppl., p. 350, 1798.
> „   „   Milne Edwards, Hist. Nat. Crust., vol. ii. p. 122, 1837.
> „   „   Bell, Trans. Linn. Soc. Lond., tom. cit., p. 283, 1855.
> „   „   var. *larinsana*, Miers, Crust. in Zool. Coll. H.M.S. "Alert," p. 230,
> pl. xxvi. fig. A, 1884.

Specimens are in the collection from the following localities :—

Arafura Sea, south of New Guinea, 28 fathoms, lat. 9° 59′ 0″ S., long. 139° 42′ 0″ E. (Station 188), two small females ; and 49 fathoms, lat. 8° 56′ 0″ S., long. 136° 5' 0" E. (Station 190), a small female; also Kobé, Japan, 50 fathoms, two males, and Hong Kong, 10 fathoms, a fully grown female and smaller male. This, which is the largest male, measures as follows :—

	Lines.	Millims.
Adult ♂.		
Length of carapace, about	7½	15·5
Breadth of carapace, .	6	13

In the smaller examples of this species the lateral lobes of the front are sometimes obsolete, as in the form described by de Haan as *Leucosia rhomboidalis*, which may be a variety of *Leucosia craniolaris*, though perhaps distinguishable by the form of the thoracic sinus.

Family IV. DORIPPIDÆ.

Dorippiens, Milne Edwards (pt.), Hist. Nat. Crust., vol. ii. p. 151, 1837.
Dorippidæ, Dana, U.S. Explor. Exped., vol. xiii., Crust. I, p. 390, 1852.

The afferent channels to the branchiæ open (normally) behind the pterygostomian regions and in front of the chelipedes; the carpal and following joints of the endognath of the exterior maxillipedes are not concealed by the merus-joint. The two to four posterior legs are short and feeble, and raised on the dorsal surface of the carapace, as in many *Anomura*. (The sexual appendages in the male are exserted from the sternum.)

Genera :—*Dorippe*, Fabricius; *Ethusa*, Roux (subgenus *Ethusina*, Smith); ? *Cymopolia*, Roux; *Corycodus*, A. Milne Edwards; ? *Cyclodorippe*, A. Milne Edwards; *Cymonomus*, A. Milne Edwards; *Cymopolus*, A. Milne Edwards; ? *Tymolus*, Stimpson.

This family is not very extensively represented in the Challenger collection, and as I have examined no specimens of the genus *Tymolus*, or of the new genera recently characterised by A. Milne Edwards, I will not attempt to separate the genera under subfamilies or sectional headings.

The genus *Cymopolia*, as I have noted below, is related in many points to the Catometopa. *Cyclodorippe*, A. Milne Edwards, should probably be regarded as the type of a distinct subfamily, since there are no openings communicating with the branchiæ in front of the chelipedes (in this character this genus establishes the transition to the Leucosiidæ).

Dorippe, Fabricius.

Dorippe, Fabricius, Entom. Syst. Suppl., p. 361, 1798.
„ Milne Edwards, Hist. Nat. Crust., vol. ii. p. 154, 1837.

Carapace very much depressed on the dorsal surface and usually broader than long; narrowest in front, and widening to the postero-lateral margins of the branchial regions; the front is narrow and concave anteriorly; the interior supraocular angles of the large orbits are produced as a lobe or tooth, and there is a stronger spine at the exterior orbital angle; they are somewhat incompletely defined below. There is no epistoma, the buccal cavity being narrowed anteriorly and produced between the antennulary fossæ to the front. The post-abdomen (in the male) is distinctly seven-jointed. Eyes well developed. The long and nearly vertically plicated antennules are not capable of being retracted within the antennulary fossettes. The basal antennal joint is short and moderately robust, and occupies the interior hiatus of the orbit; the flagellum is well developed. The exterior maxillipedes do not cover the anterior part of the buccal cavity; the ischium of the endognath is produced at its antero-internal angle; the merus is narrow, shorter than the ischium, and bears the next joint at its antero-internal angle; the exognath is rather narrow and shorter than the endognath. The chelipedes (in the male) are moderately developed and rather large, subequal or unequal, with the merus trigonous; palm moderately compressed (sometimes the palm of one chelipede is considerably enlarged); fingers with regular obtuse denticles on the interior margins, and distally acute. The ambulatory legs of the first and second pairs are very considerably elongated, with the dactyli slender, elongated, and slightly falcated; those of the third and fourth pairs are very slender, short; the last pair raised above the preceding, and both subprehensile, i.e., terminating in a short, slightly arcuate dactylus, which is reflexible against the short penultimate joint.

The species occur both in the Mediterranean, the West Atlantic, and the Indo-Pacific region, in moderately deep water.

The following are species which have been described since the publication of Milne Edwards's work:—

Dorippe japonica, v. Siebold.   Japan (8 to 50 fathoms); Shanghai.
Dorippe granulata, de Haan.   Japan (to 30 fathoms); Hong Kong.
Dorippe sexdentata, Stimpson.   Japan (to 30 fathoms).
Dorippe armata, White (ined.), Miers.   West Africa, Senegambia (to 15 fathoms).
Dorippe australiensis, Miers.   North-East and East Australia.[1]

[1] As I have already noted (Crust. in Rep. Zool. Coll. H.M.S. "Alert," p. 257, 1884), the common Indo-Pacific species designated by Milne Edwards *Dorippe quadridentata*, is probably the *Cancer doripes* of Linné.

Dorippe facchino (Herbst).

> *Cancer facchino*, Herbst (pt.), Naturgesch. der Krabben u. Krebse, vol. i. p. 190, pl. xi. fig. 68, 1782.
> *Dorippe sima*, Milne Edwards, Hist. Nat. Crust., vol. ii. p. 157, pl. xx. fig. 11, 1837.
> „   *facchino*, de Haan, Crust. in v. Siebold, Fauna Japonica, p. 123, 1841.
> „   „   Stimpson, Proc. Acad. Nat. Sci. Philad., p. 163, 1858.

Hong Kong, 10 fathoms (an adult male).

Adult ♂.							Lines.	Millims.
Length of carapace,	.	.	.	.	.	. .	11	23·5
Breadth of carapace,	.	.	.	.	.	.	13	27·5

The chelipedes in this specimen are small, subequal, and not robust.[1]

Dorippe japonica, v. Siebold.

> *Dorippe japonica*, v. Siebold, Spicilegia, Fauna Japonica, p. 14, 1824.
> „   „   de Haan, Crust. in v. Siebold, Fauna Japonica, p. 122, pl. xxxi. fig. 1 (*Dorippe callida*, Fabricius, on plate), 1841.

Japan, Kobé (8 to 15 fathoms), an adult male.

This specimen has the following dimensions:—

					Lines.	Millims.
Length of carapace, nearly	.	.	.	.	8	16·5
Breadth of carapace,	.	.	.	.	8	16·5
Length of a chelipede,	.	.	.	.	8	17
Length of first ambulatory leg,	.	.	.	25½	54

The branchial regions are convex, the right and left chelipedes are similar in form and development.

Ethusa, Roux.

> *Ethusa*, Roux, Crust. de la Méditerranée, pl. xviii., 1828, text not paged.
> „   Milne Edwards, Hist. Nat. Crust., vol. ii. p. 161, 1837.
> *Ethusina*, Smith, subgenus (?), Ann. Rep. Com. Fish and Fisheries, 1882, p. 349, 1844.

Carapace depressed, subquadrilateral, and usually much longer than broad ; the cervical and cardiaco-branchial sutures distinctly defined. Front bilobated, the lobes divided by a deep median sinus, and each terminating in two spines, one of which is the interior orbital spine ; there is also a spine at the exterior orbital angle. Orbits rather large and shallow, incompletely defined, with the superior margin deeply sinuated or emarginated. The buccal cavity is rather abruptly narrowed and triangulate towards the distal extremity. The post-abdomen (in the male) is usually distinctly seven-jointed.

---

[1] *Dorippe sima*, Milne Edwards, is, as de Haan has pointed out (Crust. in v. Siebold, Fauna Japonica, p. 123), almost certainly identical with the crab figured by Herbst as *Cancer facchino*. But Herbst says in his description of this species, wherein he correctly distinguishes between it and the Mediterranean *Dorippe lanata*, " Man findet sie sowohl am mittelländische Meere, als an den Ost-Indischen Küsten," and there is a specimen purporting, though upon no reliable authority, to have been obtained in the Mediterranean, in the collection of the Museum.

The eye-peduncles are sometimes short, but in the typical species they extend beyond the level of the orbits. The exterior maxillipedes do not cover the anterior part of the buccal cavity; the ischium of the endognath is slightly produced at its antero-internal angle, the merus is much shorter than the ischium and distally truncated or subtruncated, with the antero-external angle rounded, and bears the next joint at its antero-internal angle; the endognath is slender. Chelipedes (in the male) either equal or unequal, and of moderate size; with the merus-joint trigonous or subcylindrical, the palm compressed, and sometimes one or the other dilated, the fingers distally acute, and scarcely toothed on the inner margins. The ambulatory legs of the first two pairs are elongated and rather slender; the dactyli styliform and slightly arcuated; those of the last two pairs, as in *Dorippe*, are short and feeble, and the last pair is raised above the preceding, they are subprehensile and terminate in a very short curved claw.

The species of this genus are not numerous and are the forms which evince the greatest degree of degradation from the Brachyuran type; there are also among them forms which inhabit the deepest abysses of the ocean. The species known to me are the following:—

> *Ethusa mascarone*, Roux. Mediterranean (to 445 metres, A. Milne Edwards). Canaries; Senegambia.
>
> *Ethusa americana*, A. Milne Edwards. West Florida (to 20 fathoms).
>
> *Ethusa microphthalma*, Smith. South Coast of New England (to 156 fathoms); Azores (1000 fathoms).
>
> *Ethusa orientalis*, n. sp. Fiji Islands (310 fathoms).
>
> *Ethusa* (*Ethusina*) *abyssicola*, Smith. East Coast of United States (to 1735 fathoms).
>
> *Ethusa* (*Ethusina*) *sinuatifrons*, n. sp. Japan Seas (to 1875 fathoms).
>
> *Ethusa* (*Ethusina*) *gracilipes*, n. sp. Philippines (to 700 fathoms); Arafura Sea (to 800 fathoms), var. *robusta*, nov.[1]

Ethusa microphthalma, Smith.

Ethusa microphthalma, Smith, Proc. U.S. Nat. Mus., vol. iii. p. 416, 1881; vol. vi. p. 22, 1883, published 1884.

Azores 1000 fathoms, in lat. 38° 30′ 0″ N., long. 31° 14′ 0″ W. (Station 73). A small female.

♀		Lines.	Millims.
Length of carapace,		$3\frac{1}{2}$	7·5
Breadth of carapace, about		3	6·5

[1] Professor S. I. Smith (*Proc. U.S. Nat. Mus.*, vol. iii. p. 416, 1881) refers to a species, *Ethusa serdentata*, from Japan. I am unacquainted with this form unless by it be intended the species briefly described by Stimpson as *Dorippe serdentata*. *Ethusa granulata*, Norman, has been recently referred by A. Milne Edwards to a distinct genus (*Cyrnomorus*).

The carapace in this specimen is a little longer than broad; the single chelipede remaining is very slender. In all particulars this specimen agrees very closely with the description of S. I. Smith.

Ethusa orientalis, n. sp. (Pl. XXVIII. fig. 1).

Carapace slightly longer than broad, depressed, everywhere granulated, on the dorsal and inferior surface, with the cervical and cardiaco-branchial sutures strongly defined. The front is divided by a triangulate median sinus into two lobes, each of which is tipped with two spines, so that the front is quadrispinose, and there is a strong triangulate tooth or lobe at the exterior orbital angle, which projects forwards to a level with the frontal teeth; the lateral margins of the carapace are straight and converge slightly to the front as in other species of the genus. There is no distinct epistoma. The post-abdomen (in the male) is distinctly seven-jointed; the terminal segment triangulate and subacute. The eye-peduncles are subcylindrical and small. The bases of the antennules are moderately dilated, and are not armed with distal spinules. The basal joint of the antennæ is slender, and although more elongated than in some other species does not attain the front. The exterior maxillipedes are granulated and are shaped as in other species of the genus, i.e., the ischium of the endognath is slightly produced at its antero-internal angle and the merus is rounded at the antero-internal angle and obliquely truncated along the antero-internal margin; the exognath is very slender. The chelipedes (in the male) are either subequal or unequal, smooth, with the merus very obscurely trigonous, carpus short, the larger palm somewhat dilated and compressed, and rather longer than the fingers, which are distally acute and not denticulated on the inner margins; the palm of the smaller chelipede (or of both chelipedes in one specimen) are very slender, not thicker than the wrist. The legs of the second and third pairs are moderately robust and elongated, but less slender than in the species of the subgenus *Ethusina*, the dactyli are but very little longer than the penultimate joints; those of the fourth and fifth pairs are short and moderately robust, with very small dactyli.

Both the specimens are unfortunately very imperfect.

Two male specimens were collected at the Fiji Islands, in 310 fathoms, Station 173A, in lat. 19° 9′ 32″ S., long. 179° 41′ 55″ E.

The specimen with unequal chelipedes, which is slightly smaller than the other, measures as follows :—

Adult ♂.			Lines.	Millims.
Length of carapace,	.	.	7½	16
Breadth of carapace, about	.	.	7	15
Length of larger chelipede,	.	.	11½	24
Length of third leg,	.	.	23½	49·5
Length of fourth leg,	.	.	8½	18

The larger male differs not only in the small subequal chelipedes, but also in the form of the post-abdomen, which is slightly broader in proportion to its length.

Ethusina, Smith.

Ethusina, Smith, Ann. Rep. Com. of Fish and Fisheries, 1882, p. 349, 1884.

This genus (or subgenus as I prefer to regard it) is, according to Smith, closely allied to Ethusa, but is distinguished by the form of the antennules, whose basal segments are very large and swollen, occupy the whole width of the front, and crowd back the eyes and antennæ into an almost transverse position, nearly beneath the exterior orbital angles, which are reduced to small lateral teeth, far back from the front. The eye-stalks are very small and immovably imbedded in the orbits, which closely inclose them to near the tips, except for a narrow space beneath.

In the typical species Ethusina abyssicola, Smith, dredged off the east coast of the United States (1497 to 1735 fathoms), there are, according to Professor Smith, no podo-branchiæ at the bases of the first gnathopods, so that there are only six branchiæ on each side; two arthrobranchiæ each at the base of the second gnathopod and first perciopod, and one pleurobranchia each for the second and third perciopods.

There are in the collection of H.M.S. Challenger two species, which on account of the structure of the antennules and eye-peduncles I assign to this genus. One of these species is unfortunately represented only by a single mutilated example.

Ethusa (Ethusina) challengeri, n. sp. (Pl. XXVIII. fig. 2).

Carapace about as broad as long, depressed above, with the cervical and cardinco-branchial sutures very indistinctly defined; the lateral margins nearly straight, and converging to the front, so that the body, as in Ethusina abyssicola, is much narrower in front than posteriorly. The front is not quadridentated, but sinuated, and concave in the middle line, where it is prolonged downwards (as in Ethusina abyssicola) between the bases of the antennules, and is in contact with the narrow median process of the epistoma. The orbits are very incompletely defined, and the exterior orbital spine or tooth, which is developed on one side only, is very short. The post-abdomen (in the female) is dis-tinctly seven-jointed. The eye-peduncles are short, and taper from the bases to the distal extremity; the eyes are small, and terminal. The bases of the antennules are very large and swollen, subglobose. The basal joint of the antennæ is short and slender, and does not nearly attain the front (the flagellum is broken in the specimen examined). The exterior maxillipedes do not cover the anterior part of the buccal cavity, which is narrowed very abruptly (as in Ethusina abyssicola); the ischium is produced and rounded at its antero-internal angle. The merus is distally somewhat rounded, and is obliquely truncated along its antero-internal margin; the exognath is very slender and does not attain the distal extremity of the merus of the endognath. The chelipedes (in the female) are subequal and slender; with the merus-joint but very obscurely trigonous; carpus very small, unarmed; palm short, slightly compressed, rounded above and below; fingers compressed, acute, and not denticulated on the inner margins. The legs of the second and third pairs are slender, with the joints subcylindrical; the dactylus (a single one only remains loosely attached to the third left ambulatory leg) is slightly curved, and rather longer than the penultimate joint. The legs of the fourth and fifth pairs are very slender and subcylindrical, much shorter than the preceding, and terminate in a small dactylus which is shorter than the penultimate joint. The body and limbs are scantily clothed with a very short close greyish pubescence.

♀.				Lines.	Millims.
Length and breadth of carapace, about	.	.	.	6	12·5
Length of a chelipede, nearly	.	.	.	9	18·5
Length of third left ambulatory leg, .	.	.	.	20½	43·5
Length of fourth ambulatory leg, about	.	.	.	7½	16

The single specimen, which is a female, probably adult, was taken at Station 237, in 1875 fathoms, lat. 34° 37′ 0″ N., long. 140° 32′ 0″ E.

This is the greatest depth at which any Brachyurous Crustacean was taken by the Expedition, and also, I believe, the greatest depth hitherto recorded for any species of Crab.

Ethusa (Ethusina) gracilipes, n. sp. (Pl. XXVIII. fig. 3).

Carapace depressed, finely granulated, longer than broad, and narrowed anteriorly; with the cervical and cardiaco-branchial sutures distinctly defined; the front is armed with four spines, the two median of which are separated from one another by a somewhat wider and deeper interspace than that which intervenes between them and the two exterior spines; the spine at the exterior orbital angle is strongly developed; the orbits are incompletely defined. The epistoma is very narrow, transverse. The post-abdomen (in the male) is narrow and five-jointed, with the third to fifth segments coalescent. The eyes are small and taper but very slightly, if at all, from the base. The bases of the antennules are considerably dilated, as in *Ethusa sinuatifrons*, and usually bear a small distal spine or tubercle. The basal joint of the antennæ is short and slender and does not nearly attain the front; the flagellum is considerably elongated, reaching, when retracted, to the posterior margin of the carapace. The exterior maxillipedes are shaped nearly as in *Ethusa sinuatifrons*, but the ischium of the endognath is narrower and less robust. The chelipedes (in the male) are subequal and very slender, with the merus subcylindrical; carpus very short; palm but little longer than the carpus, slightly compressed, and shorter than the fingers, which are slender, scarcely denticulated on the inner margins and are slightly decussate at the acute apices. The legs of the second and third pairs (in the typical form) are greatly elongated and very slender, with the joints smooth, the dactyli slightly curved, and exceeding the penultimate joints in length. Those of the fourth and fifth pairs are (in the typical variety) nearly filiform; the dactyli very short. The body and limbs are covered with an extremely short brownish or whitish pubescence.

Of this species two very distinct varieties were collected; the first represented by two adult males and two females, at Station 207, near the Philippines, in 700 fathoms, lat. 12° 21′ 0″ N., long. 122° 15′ 0″ E., the second by three adult females, dredged in 800 fathoms, in the Arafura Sea, Station 191, lat. 5° 41′ 0″ S., long. 134° 4′ 30″ E., and by an adult and larger female obtained in the Banda Sea, in 1425 fathoms, lat. 4° 21′ 0″ S., long. 129° 7′ 0″ E. (Station 195).

The first-mentioned variety was selected for description because there are males of it; the second variety is distinguished by the more dilated bases of the antennules, the slightly tapering, not cylindrical, eye-peduncles, and the more robust chelipedes and ambulatory legs. I propose to designate it var. *robusta*.

Adult ♂.	Lines.	Millims.
Length and breadth of carapace,	4½	9·5
Length of a chelipede, nearly	8	16·5
Length of third leg, .	19	40
Length of fourth leg, nearly .	6	12

The adult female specimen of the variety *robusta*, from 1425 fathoms (Station 195), which only differs from the specimens taken in 800 fathoms (Station 191) in the slightly more convex carapace, with more deeply accentuated sutures and the somewhat shorter exterior orbital spines, measures as follows :—

Adult ♀.	Lines.	Millims.
Length and breadth of carapace, nearly	7½	15·5
Length of a chelipede,	11	23
Length of third ambulatory leg, nearly	25½	53·5
Length of fourth ambulatory leg, about	11	23

Cymopolia, Roux.

Cymopolia, Roux, Crust. de la Méditerranée, pl. xxi., 1828.
    „   Milne Edwards, Hist. Nat. Crust, vol. ii. p. 158, 1837.

Of this genus I have only examined three species, *Cymopolia caronii*, Roux, *Cymopolia jukesii*, White, and *Cymopolia whitei*, Miers, the numerous species recently described by A. Milne Edwards being unrepresented in the collections of H.M.S.

Challenger and of the Museum, as is also *Cymopolia gracilis*, Smith, it is therefore impossible for me to draw up a satisfactory description of the genus, the typical species of which are, however, sufficiently distinguished from *Dorippe* and *Ethusa* by their much more broadly tranverse carapace, with dentated (not spinose) front, and dentated antero-lateral margins, and by the nearly quadrate, not triangulate, buccal cavity, in which characters they more nearly resemble certain Catometopa than the Oxystomata. The afferent channel to the branchiæ opens immediately at the bases of the chelipedes, and is not separated from them, as in the species of *Dorippe*.

This genus is not very nearly allied either to *Dorippe* or *Ethusa*, and should not perhaps be referred to the same family, but it is retained in the vicinity of *Ethusa* by A. Milne Edwards and other authors, and here, accordingly, I retain it for the present.

The following species have been described :—

> *Cymopolia caronii*, Roux. Mediterranean; Cape Verde Islands.
> *Cymopolia jukesii*, White. North and North-East Australia; Sir C. Hardy
>   Island ; Celebes Sea (to 10 fathoms).
> *Cymopolia obesa*, A. Milne Edwards.
> *Cymopolia dilatata*, A. Milne Edwards.
> *Cymopolia dentata*, A. Milne Edwards.
> *Cymopolia cristatipes*, A. Milne Edwards.   Gulf of Mexico and Florida Straits
> *Cymopolia cursor*, A. Milne Edwards.   (to 298 fathoms).
> *Cymopolia gracilipes*, A. Milne Edwards.
> *Cymopolia sica*, A. Milne Edwards.
> *Cymopolia acutifrons*, A. Milne Edwards.
> *Cymopolia gracilis*, Smith. New England (to 142 fathoms).
> *Cymopolia whitei*, Miers. Seychelles (4 to 12 fathoms).

Cymopolia caronii, Roux.

> *Cymopolia caronii*, Roux, Crust. de la Méditerranée, pl. xxi. figs. 1–7.
>   „   „   Milne Edwards, Hist. Nat. Crust, vol. ii. p. 159, 1837.
>   „   „   Lucas, Animaux articulés in Explor. Sci. de l'Algérie, Crust., p. 25, pl.
>   iii. fig. 1, 1849.
>   „   „   Heller, Crust. des südlichen Europa, p. 140, pl. iv. fig. 8, 1863.

St. Vincent, Cape Verde Islands, July 1873. An adult female, bearing ova.

♀ .	Lines.	Millims.
Length of carapace, . . .	4½	9·5
Breadth of carapace, rather over . .	5	11

The specimen agrees closely with the figure of Roux, and with specimens referred to this species from the Canary Islands, in the collection of the Museum, but the merus-joints are perhaps slightly more dilated than is usual.

Cymopolia jukesii, White.

Cymopolia jukesii, White, App. to Jukes' Voy. H.M.S. "Fly," p. 338, pl. ii. fig. 1.
    ,,   ,,   Miers, Crust. in Zool. of H.M.SS. "Erebus" and "Terror," No. xx. p. 3, pl. iii. fig. 4, 1874.

Celebes Sea, lat. 6° 54′ 0″ N., long. 122° 18′ 0″ E., in 10 fathoms (Station 212). An adult female, bearing ova.

Adult ♀.	Lines.	Millims.
Length of carapace,	$4\frac{1}{2}$	9·5
Breadth of carapace,	$5\frac{1}{2}$	11·5

The specimen is imperfect, having lost two of the legs, but one of these, the first of the ambulatory legs on the right hand side, is replaced by a new and imperfectly developed limb, measuring only 5 lines (10·5 mm.).

APPENDIX.

This long list of Errata must be attributed to the cause adverted to in the Introduction, which rendered it impossible for me to correct the sheets as they were passing through the press.

Page 282, line 21, *insert* the following definition of the group Oxystomata or Leucosiidea, which was unfortunately wanting in the MSS. when sent to the editor for press, I being uncertain whether the group should be sustained:—

OXYSTOMATA or LEUCOSIIDEA.

Carapace convex or depressed, transverse, with the antero-lateral margins arcuated or orbiculate, or even subglobose, or more or less oblong, with subparallel or slightly convergent margins (Dorippidæ). Epistome very much reduced or rudimentary. Buccal cavity more or less triangulate, nearly always produced and narrowed in front, with the margins anteriorly convergent. The afferent channels to the branchiæ enter either behind the pterygostomian regions and in front of the chelipedes, or, more rarely, at the antero-lateral angles of the palate (Leucosiidæ). Branchiæ six to nine (Claus). Antennules longitudinally or obliquely plicated. The carpal joint of the exterior maxillipedes is articulated either at the antero-internal angle or at the antero-external angle or at the distal extremity of the merus, and is frequently concealed beneath it. The verges of the male are exserted either from the sternal surface or more usually from the bases of the fifth pair of legs, which are either gressorial, natatorial, or feeble and raised upon the dorsal surface of the carapace.

The Oxystomata constitute a large but somewhat heterogeneous group, characterised generally by the triangulate or narrowed buccal cavity and the position of the afferent branchial channels, and related on the one hand to the Oxyrhyncha through the Leucosiidæ, and with the Anomura through the Dorippidæ. This group includes among the highly-specialised Leucosiidæ some of the most beautiful of the littoral species, and others (Calappidæ, Leucosiidæ) no less remarkable for peculiarity of form and structure. *Matuta*, in which genus all the legs are natatorial, is one of the best adapted for swimming of all the genera of Brachyura, and among the remarkable genera of the group Dorippidæ are found the forms which inhabit the deepest ocean depths, and those which most nearly approach the Anomura in the structure of the buccal organs and of the ambulatory legs.

Page 19, line 2 from bottom, for "*Menæthiun*" read "*Menæthius.*"

Page 40, line 8 from bottom, for "*Halmius*" read "*Halimus.*"

Page 40, line 3 from bottom, for "*Peltina*" read "*Peltinia.*"

Page 53, line 7 from bottom, for "*Aretopisis*" read "*Aretopsis.*"

Page 56, lines 27–29. This short paragraph, beginning with the words, "The name *Hyastenus,*" should be placed in a footnote, and the following paragraph, which contains the enumeration of species of Section 2 of the genus *Hyastenus*, should run on after the words, "Targioni-Tozzetti" in that section.

Page 83, line 11, for "*tenuidus*" read "*tumidus.*"

Page 87, line 14, for "*rubei*" read "*ruber.*"

Page 92, line 10, for "*Parthenopoides*, Miers" read "*Parthenolambrus*, A. Milne Edwards."

Page 99, line 10, *for* "crowded" *read* "eroded."

Page 108 (footnote). In this footnote, which exhibits in a tabulated form the parallelism existing between the genera comprised in the subsections *a* and *b* of the typical Cancridæ, the genera with acute finger-tips are placed in the left hand column and those with excavated finger-tips in the right hand column. The genus *Lophoxorynus*, in which the fingers are not excavated, should be transferred from the right hand column and placed after *Lophactæa*, A. Milne Edwards, and in like manner the genera *Panopeus*, Milne Edwards, and *Micropanope*, Stimpson, in which the finger-tips are acute, should be placed in the left hand column, after *Xanthodes*, Dana.

Page 113, line 8 from bottom, *for* " A. Milne Edwards" *read* " Milne Edwards."

Page 118, lines 18 and 24, for " *Malæus granulatus* " read " *Medæus hanselli.* "

Page 118, line 3 from bottom, for " *Banareia* " read " *Banarcia.* "

Page 135, line 14, *for* " ranged " *read* " ranges."

Page 136, line 10, *for* " plain " *read* " plane."

Page 139, line 20, for " *angulatus* " read " *ungulatus.* "

Page 146, line 23, *for* " Kepler " *read* " Kessler."

Page 147, line 16, for " *melanacanthus* " read " *melanacanthus.* "

Page 149, line 13, for " *vanquelinii* " read " *vanquelinii.* "

Page 172, line 21, *for* " T. J. Smith " *read* " S. I. Smith." (See also p. 219, line 7, p. 220, line 14, p. 222, line 29, p. 267, line 21, where *for* " S. J. Smith" *read* " S. I. Smith.")

Page 180, line 3 from bottom, for " *Amphitrite* " read " *Achelous.* "

Page 184, line 10 from bottom, for " *Goniosoma* " read " *Goniosoma.* "

Page 189, line 5 from bottom, for " *Goniosoma bispinosum* " read " *Cronius bispinosus* "

Page 190, line 13 from bottom, *for* " Hoffman " *read* " Hoffmann."

Page 208, line 12 from bottom, for " *Plagusetis* " read " *Plagusetes.* "

Page 210, line 11, after " *Hypopeltarium* " erase the words " n. gen."

Page 220, line 6, for " *crassum* " read " *crassum.* "

Page 226, last line, *for* " Limoda " *read* " Simoda."

Page 255, line 6, *for* " fossetta " *read* " fossettes," and at line 23, for " *trigosus* " read " *strigosus.* "

Page 260, line 11 from bottom, *insert* a semicolon after the word " distinct."

Page 269, last line, *insert* a comma after the word " typæs."

Page 274, lines 17, 18, 19 ; page 275, line 17 ; page 278, lines 2, 23, 25, 27, 28 ; and at page 279, line 7, for " *Myctiris* " read " *Mycteris.* "

Page 275, line 13, and at page 278, line 1, *for* " Myctirinæ " *read* " Mycterinæ."

Page 282, line 20, *for* " 1825 " *read* " 1853."

Page 295, line 4, for " *lineifera* " read " *lineifera.* "

Page 301, line 8 from bottom, *for* " later " *read* " lateral."

Page 302, line 7 from bottom, *for* " young male " *read* " sterile female."

Page 303, line 10 from bottom, for " *Leucisca* " read " *Leucisca.* "

Page 304, line 29, *for* " Heller " *read* " Haller."

Page 305, lines 6, 7 from bottom, for " *spinosa* " read " *undecimspinosa.* "

Page 308, line 10 from bottom, *insert* a comma, and the word " are " *after* " remain."

Page 310, line 3 from bottom, *for* " each intestinal region " *read* " the intestinal region."

Page 323, line 12, *before* the word " contains " *insert* " it."

Page 329, line 14 from bottom, for " *sinuatifrons* " read " *challengeri.* "

Page 332, line 22, *for* " Pl. XXVIII. fig. 3 " *read* " Pl. XXIX. fig. 1."

Page 333, line 17, after the words " var. *robusta* " insert a reference to the figure "(Pl. XXIX. fig. 2)."

In the Explanation to the Plates :—

Plate I. fig. 3a is magnified 5 diameters.	Plate VIII. fig. 1b is magnified 3 diameters.
Plate IV. fig. 1c is magnified 3 diameters.	Plate X. fig. 3b is magnified 4 diameters.
Plate IV. fig. 2, for " *mosleyi* " read " *moseleyi.* "	Plate XI. fig. 2c is magnified 6 diameters.
Plate IV. fig. 2c is magnified 3 diameters.	Plate XVI. fig. 1d is magnified 5 diameters.

INDEX.

Note.—Synonyms are printed in *italics*; the more important pages are indicated by darker type.

PLATE I.

(ZOOL. CHALL. EXP.—PART XLIX.—1886.)—Coc.

PLATE I.

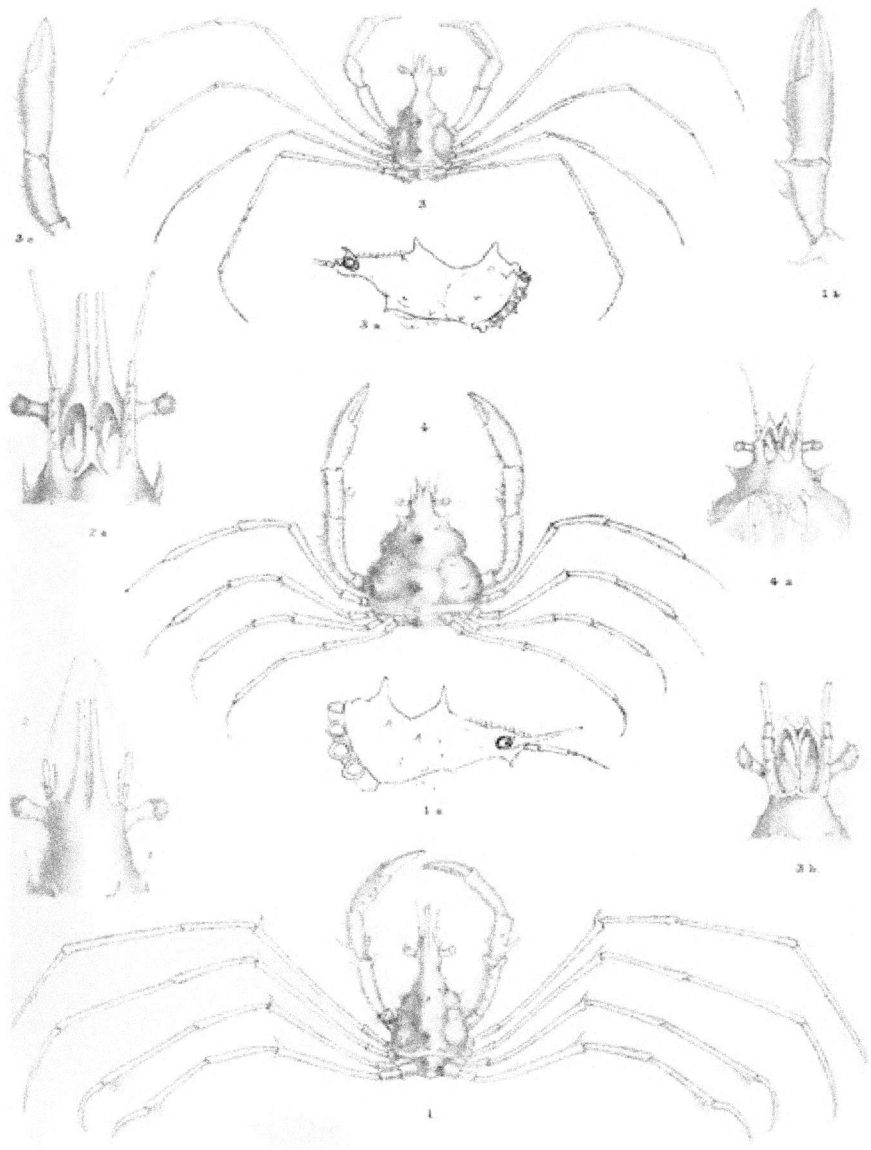

3 c

3

3 a

1 b

2 a

4

4 a

2

1 a

3 b

1

PLATE II.

PLATE II.

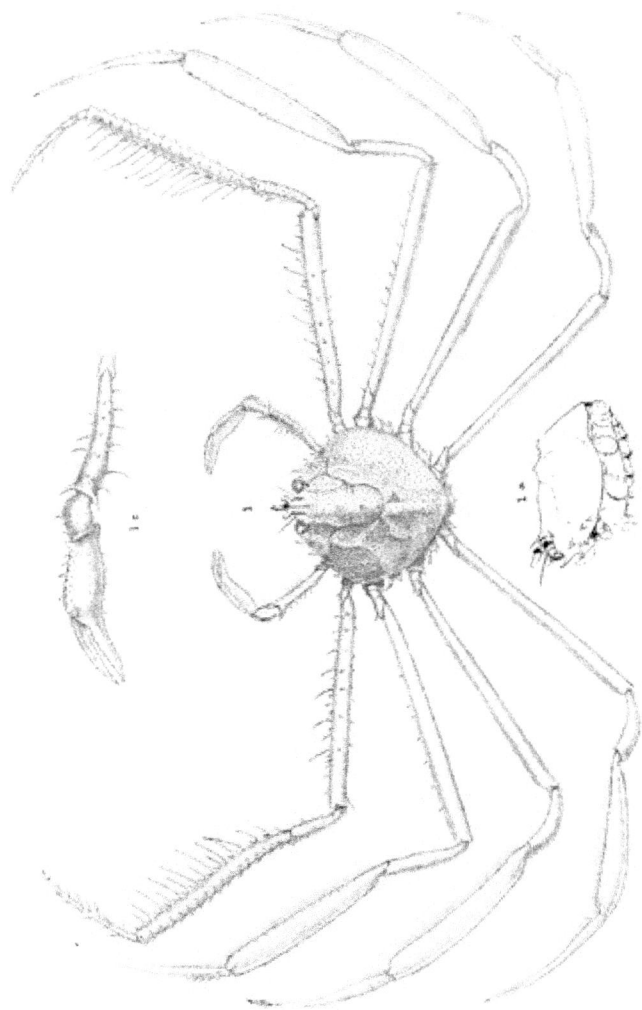

PLATE III.

PLATE III.

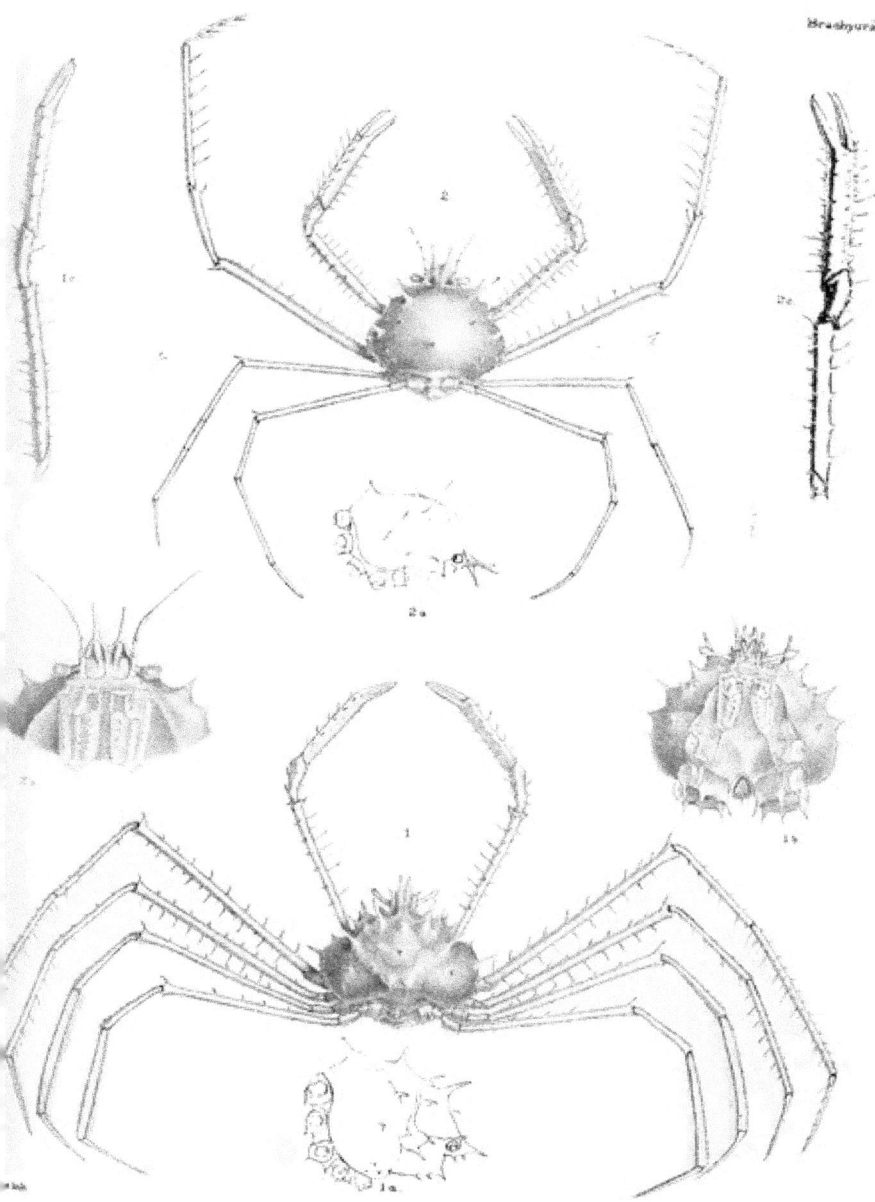

CYRTOMAIA

PLATE IV.

PLATE IV.

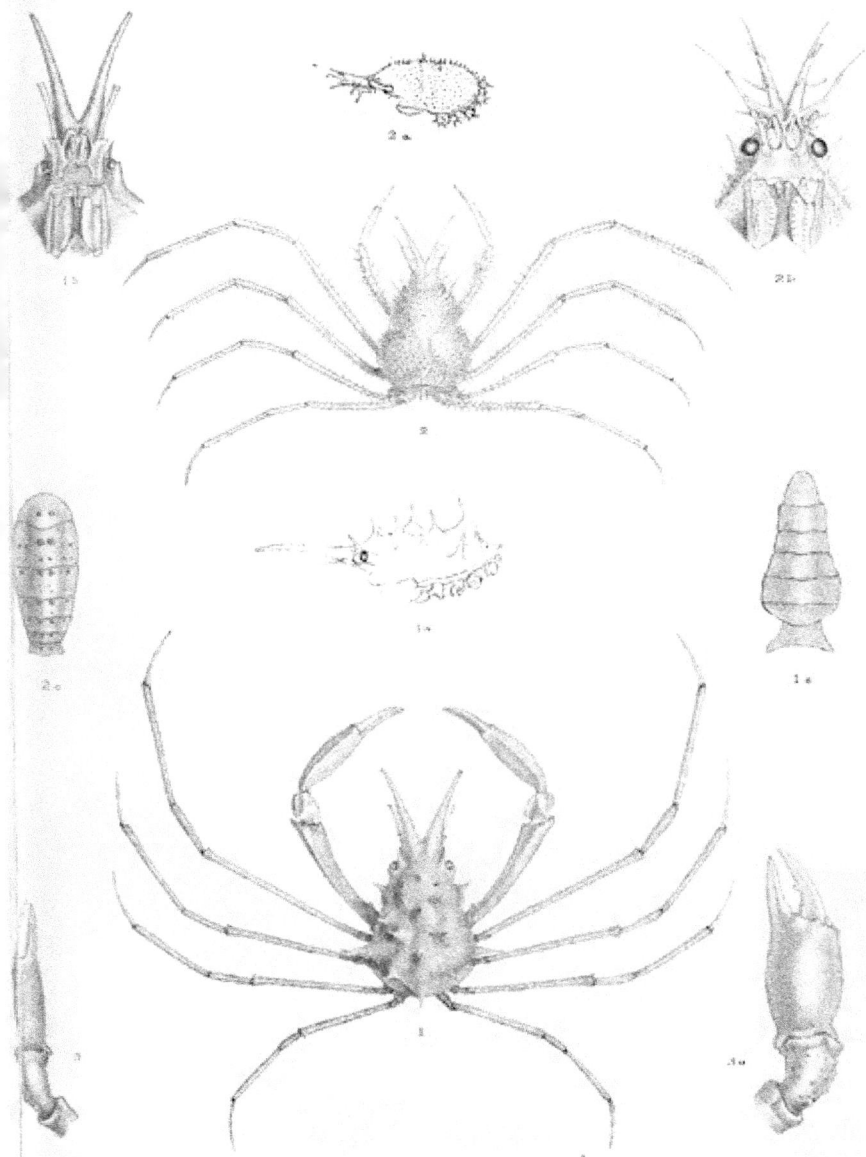

AMATHIA — ECHINOPLAX — EURYPODIUS.

PLATE V.

(ZOOL. CHALL. EXP.—PART XLIX.—1886.)—Ccc.

PLATE V.

2 a.

2.

1 a.

3 a.

2 c.

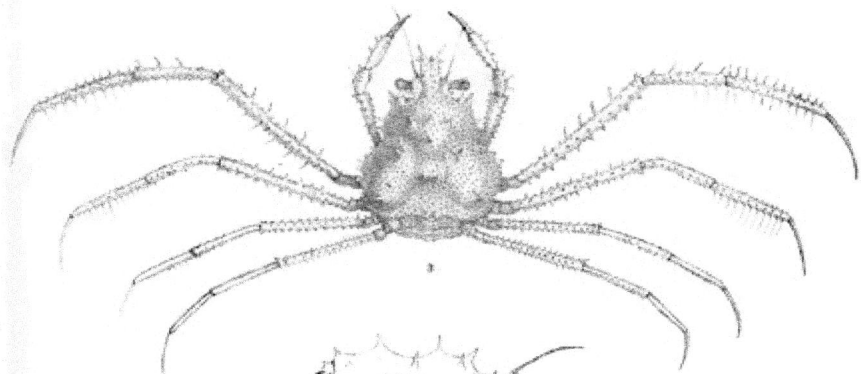

3.

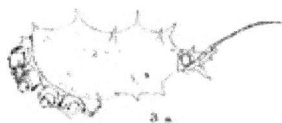

3 a.

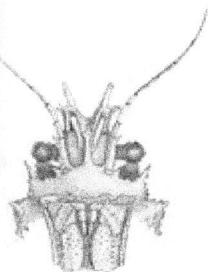

3 b.

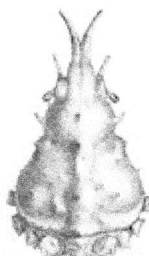

1.

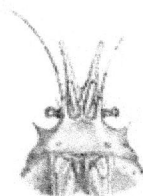

2 b.

PLATE VI.

PLATE VI.

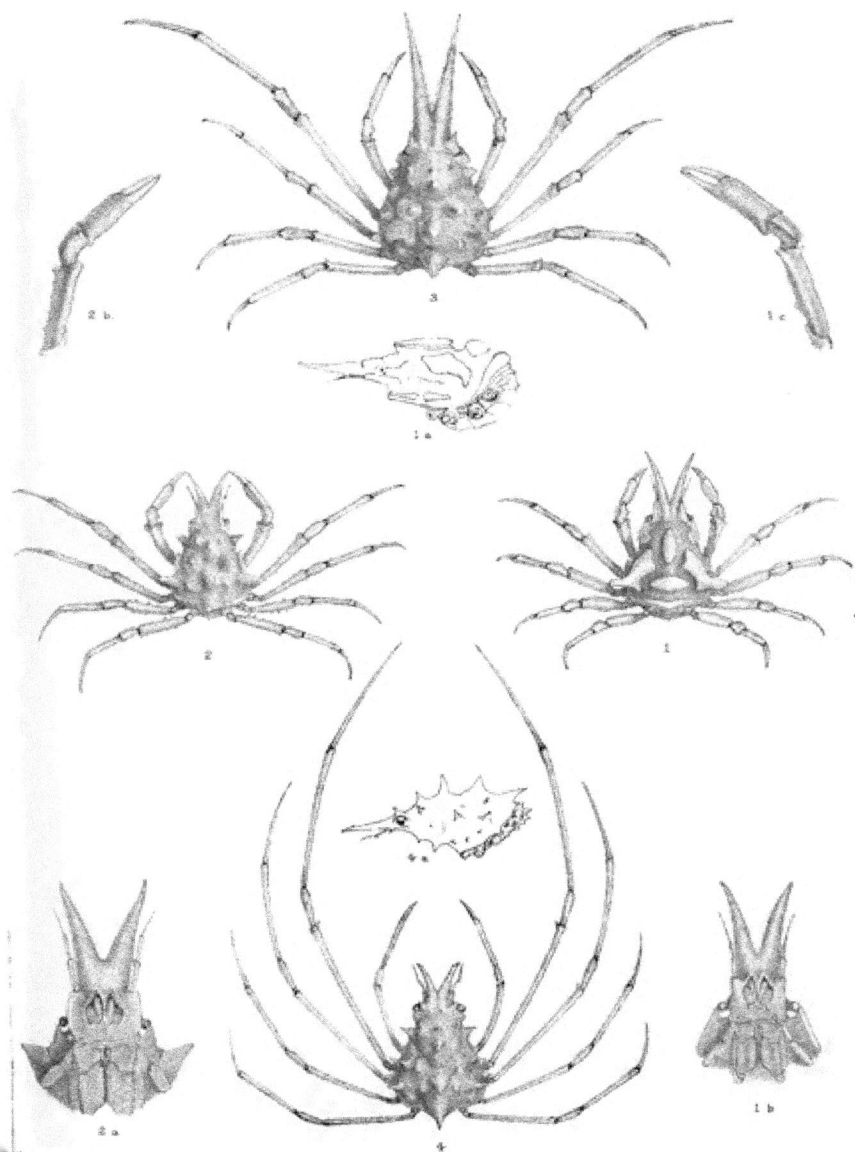

PLATE VII.

(ZOOL. CHALL. EXP.—PART XLIX.—1886.)—Ccc.

PLATE VII.

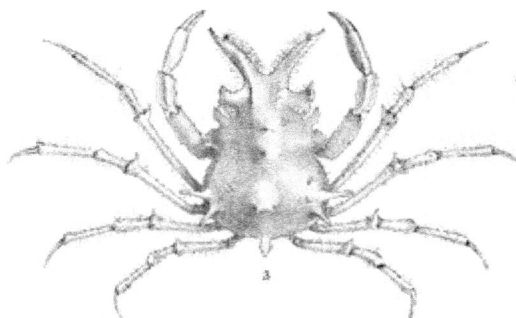

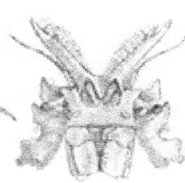

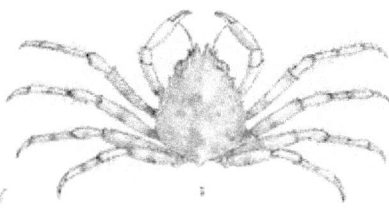

PLATE VIII.

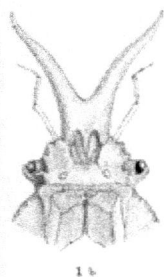

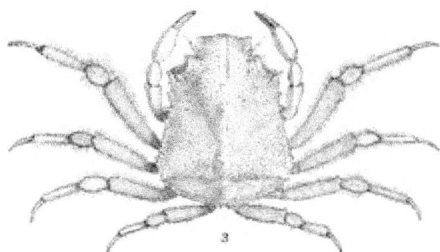

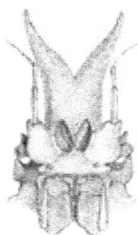

1 b

3

3 a

1 c

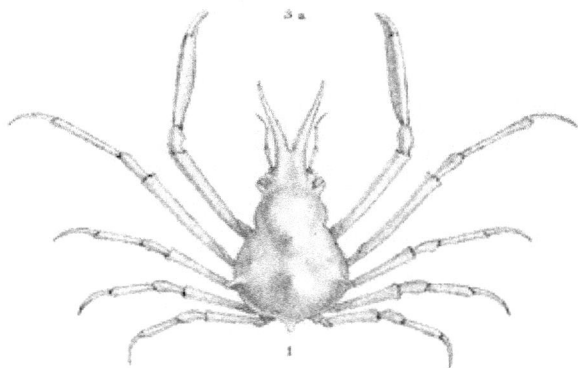

3 b

1

2 b

2 a

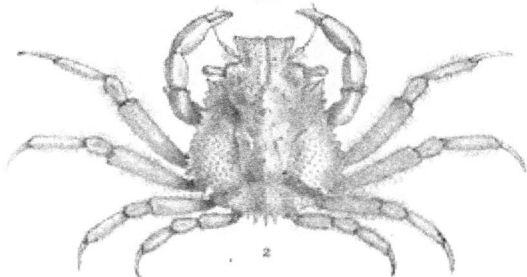

1 a

2

3 c

NOTOLOPAS — MICIPPA.

PLATE IX.

(ZOOL. CHALL. EXP.—PART XLIX.—1886.)—Coc.

PLATE IX.

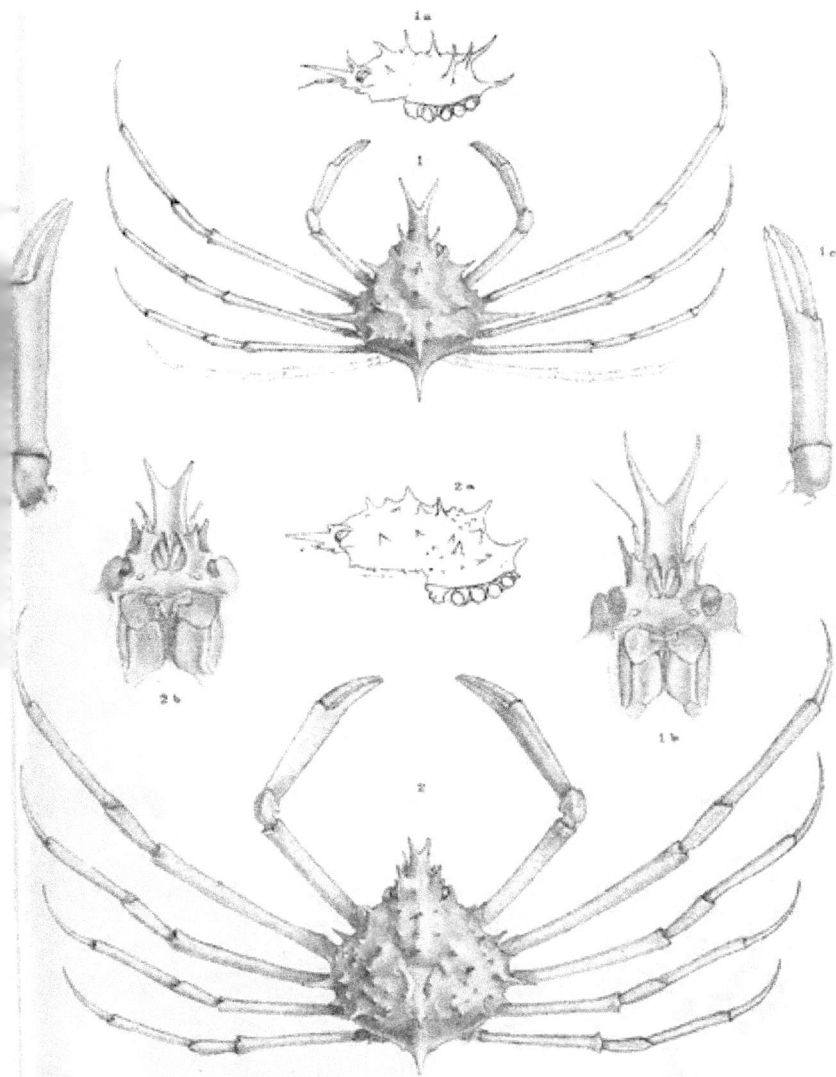

1a

1

1c

2a

2b

1b

2

LIBINIA.

West, Newman & C.º imp

PLATE X.

PLATE X.

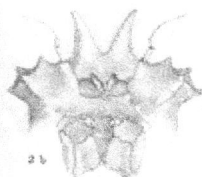
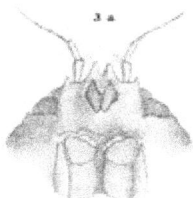
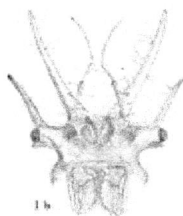

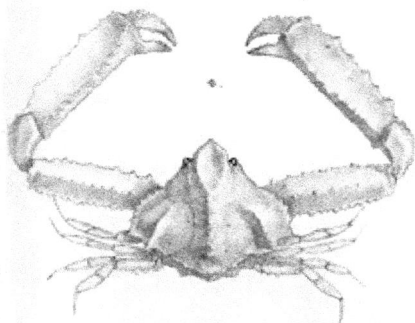
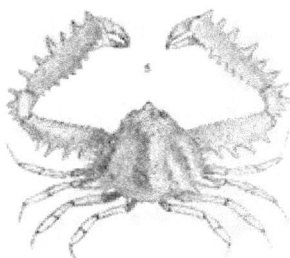

PICROCEROIDES — MACROCŒLOMA · MITHRAX · LAMBRUS

PLATE XI.

(ZOOL. CHALL. EXP.—PART XLIX.—1886.)—Ccc.

PLATE XI.

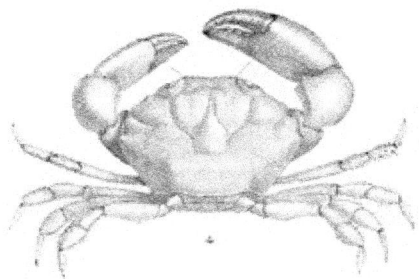

1 a

4

4 b

2 a

2 b

4 a

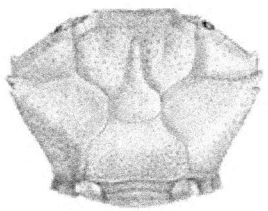

2

3 a

1 a

3 b

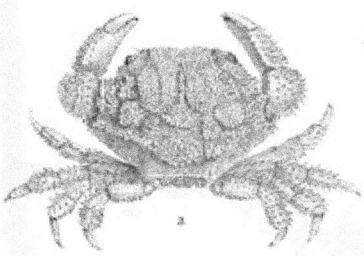

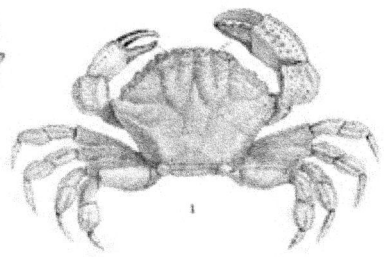

3

2 c

1

Morgan del. et lith.   West, Newman & Co imp.

LOPHOZOZYMUS – MEDŒUS – ACTÆA – XANTHO

PLATE XII.

PLATE XII.

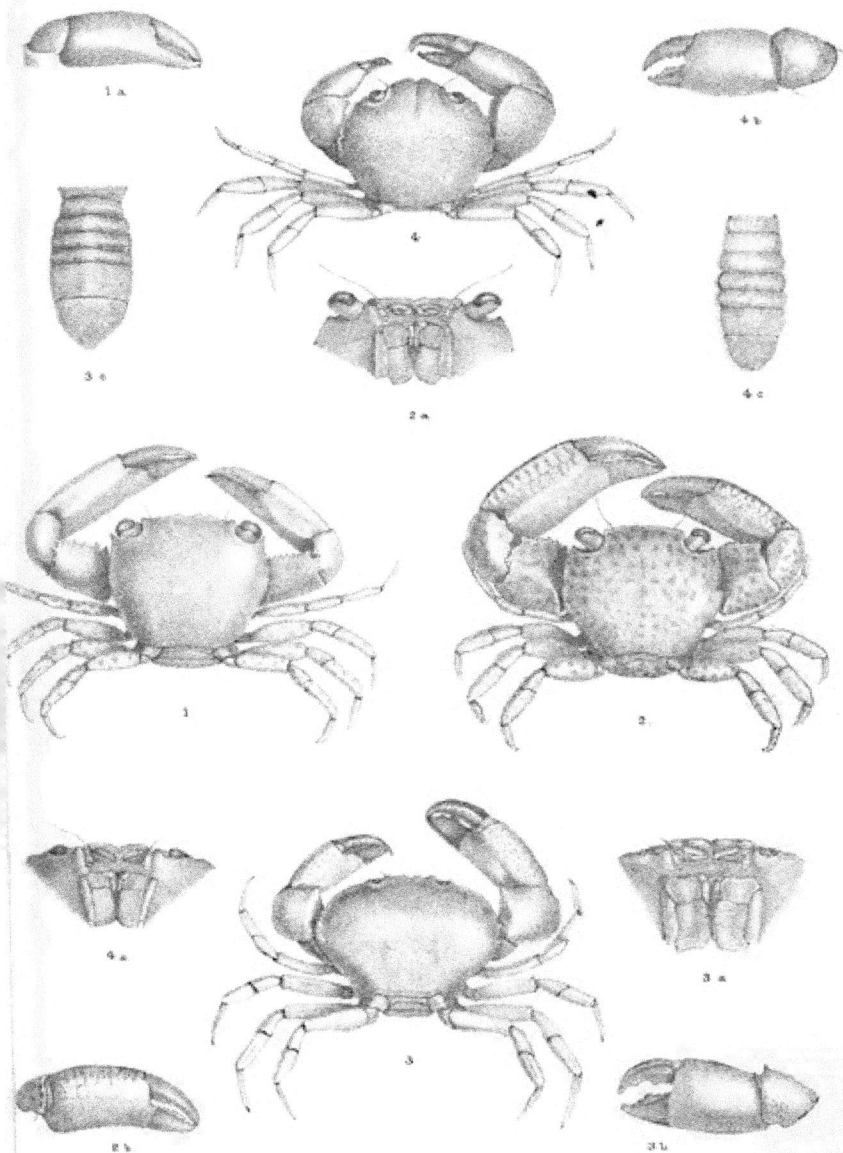

PLATE XIII.

(ZOOL. CHALL. EXP.—PART XLIX.—1886.)—Ccc.

PLATE XIII.

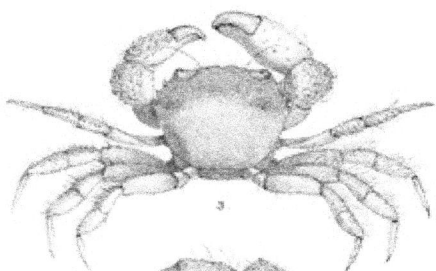

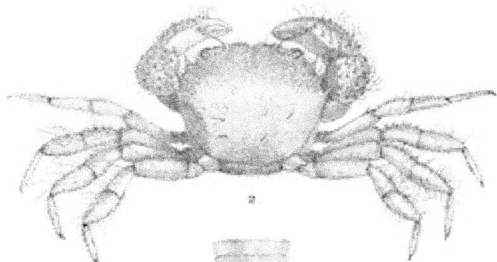

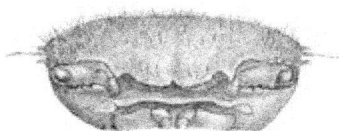

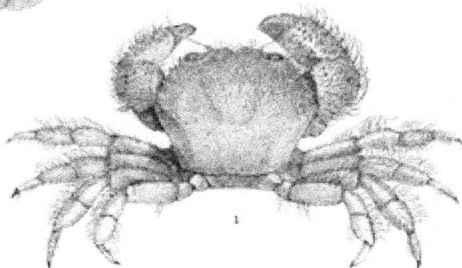

PILUMNUS.

PLATE XIV.

R. Mintern del. et lith. West, Newman & C.o imp.

PILUMNUS.

PLATE XV.

PLATE XV.

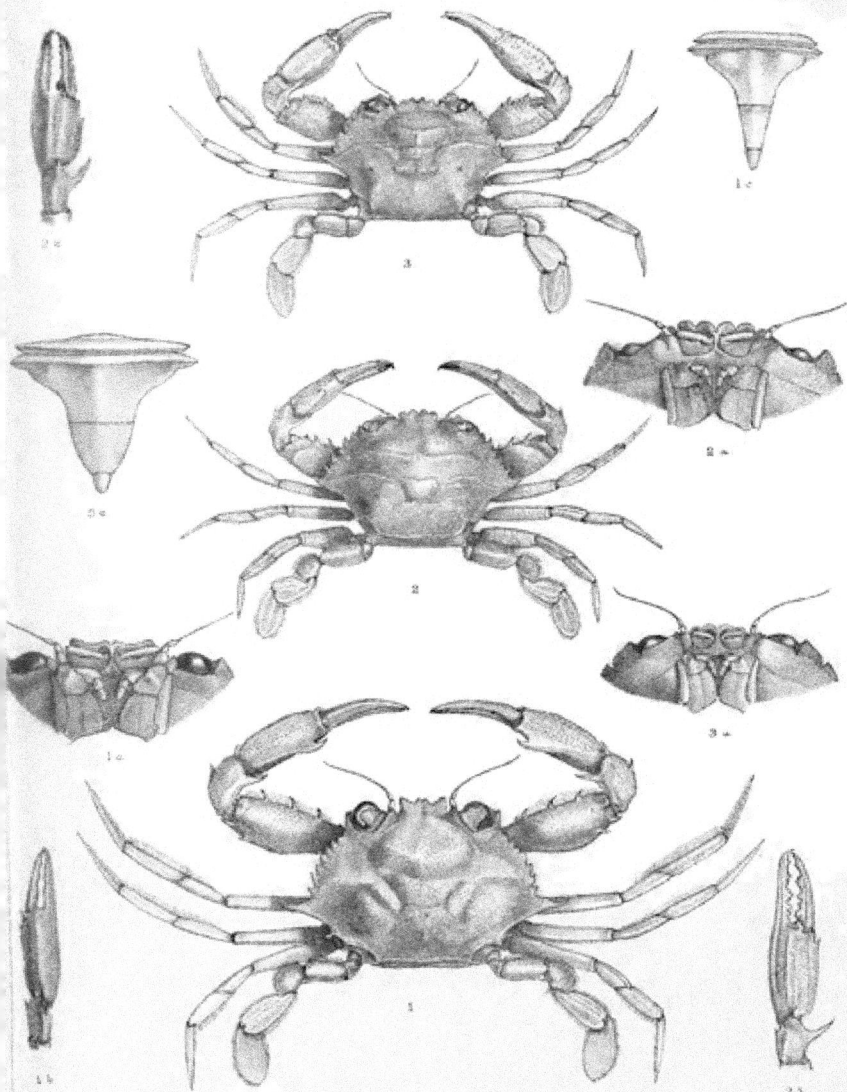

NEPTUNUS — CRONIUS — GONIOSOMA.

PLATE XVI.

PLATE XVI.

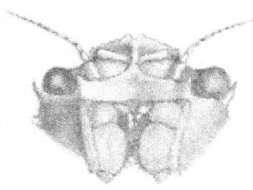

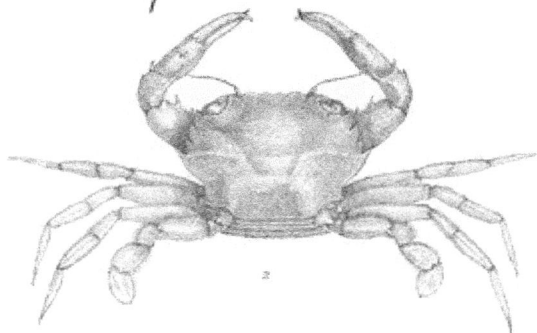

PLATE XVII.

PLATE XVII.

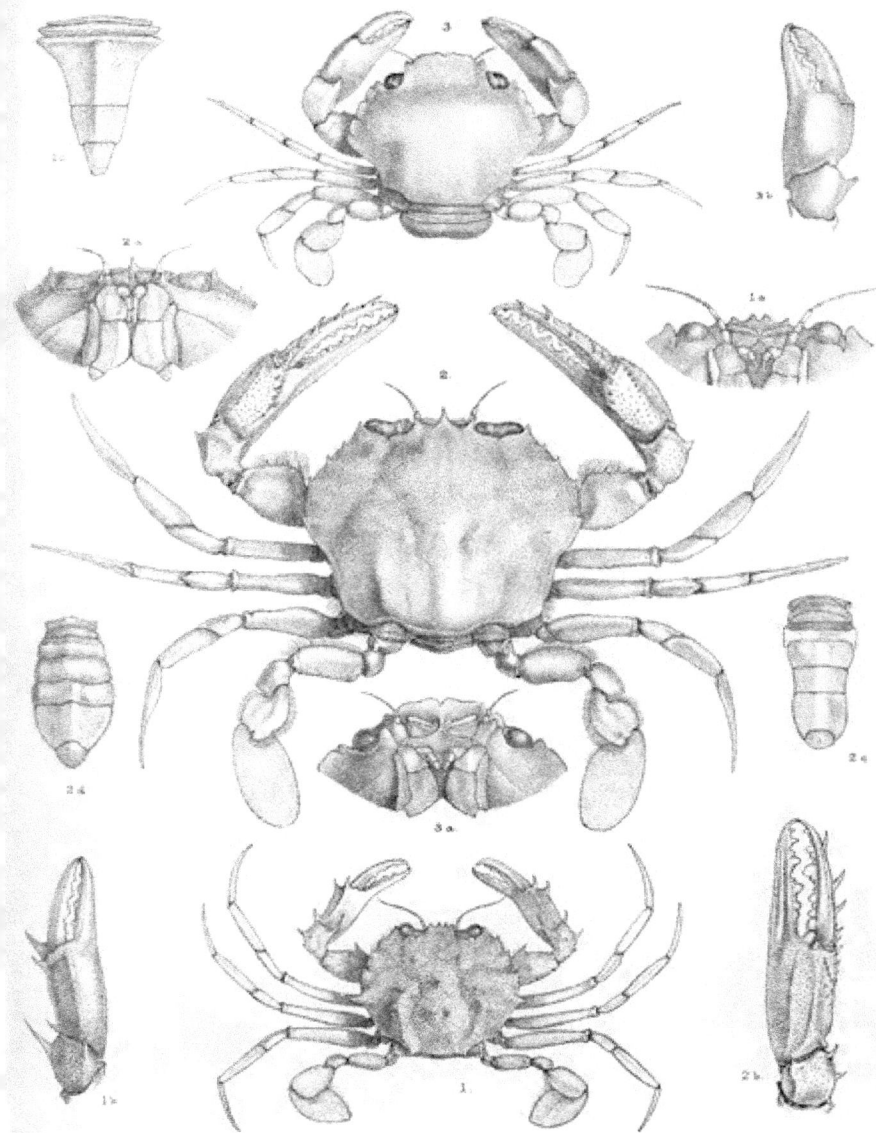

LUPOCYCLUS — PLATYONYCHUS — LISSOCARCINUS.

PLATE XVIII.

PLATE XVIII.

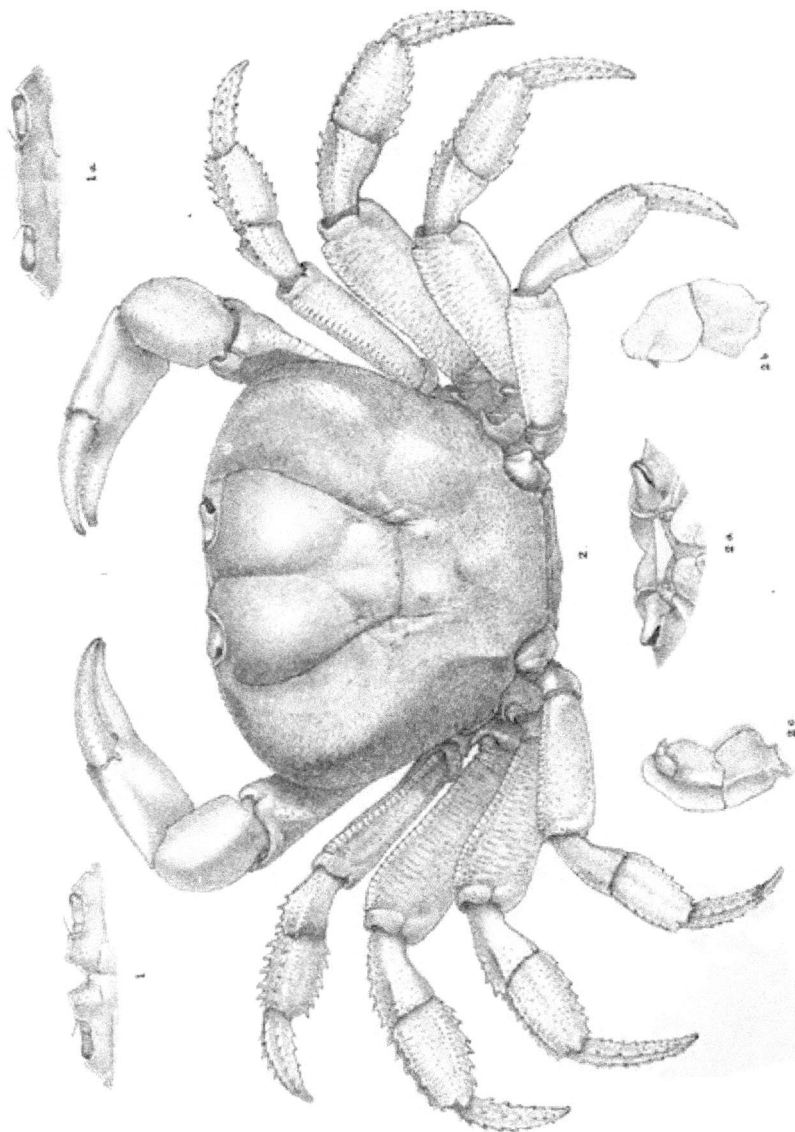

PLATE XIX.

(ZOOL. CHALL. EXP.—PART XLIX.—1886.)—Cc.

PLATE XIX.

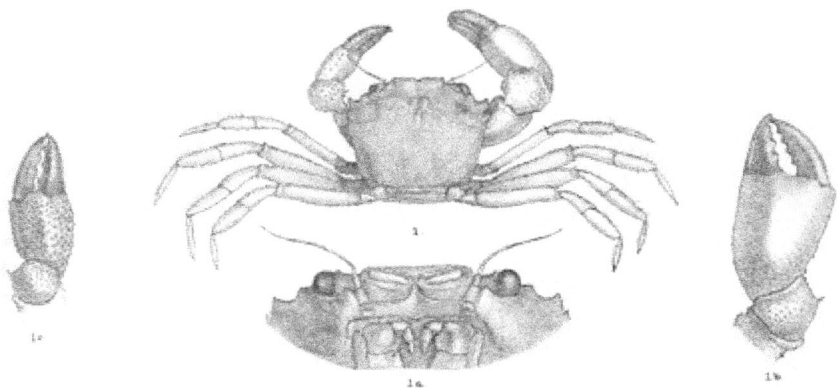

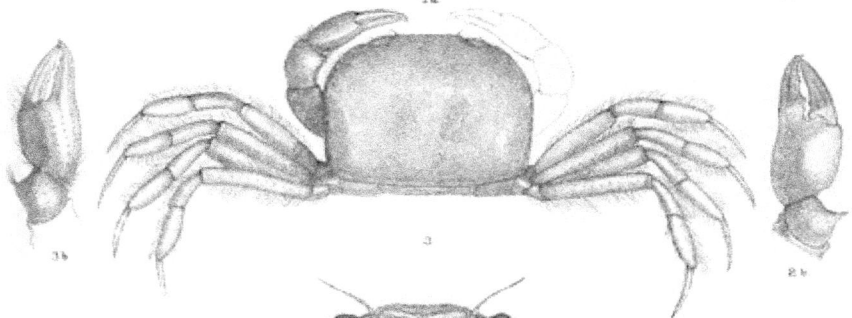

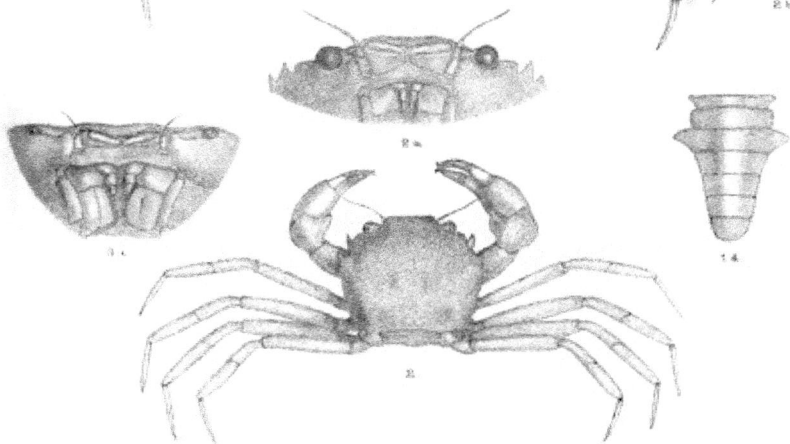

PLATE XX.

PLATE XX.

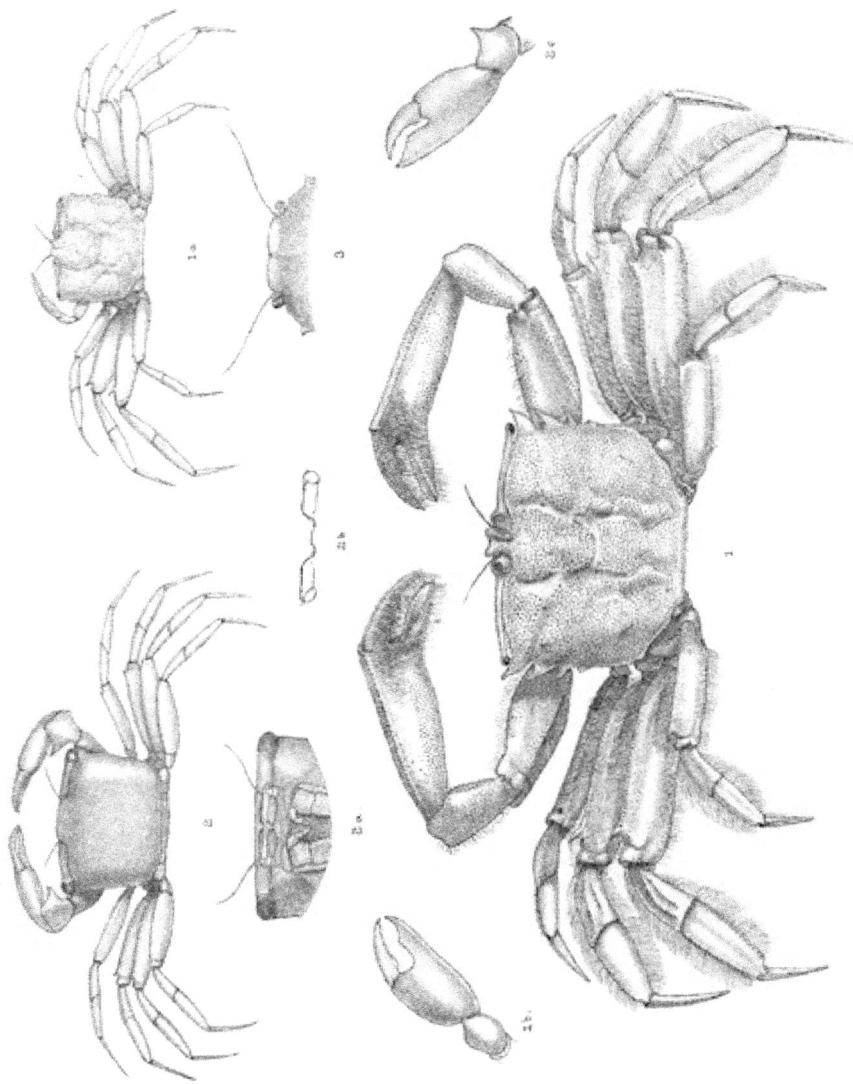

PLATE XXI.

(ZOOL. CHALL. EXP.—PART XLIX.—1886.)—Ccc.

PLATE XXI.

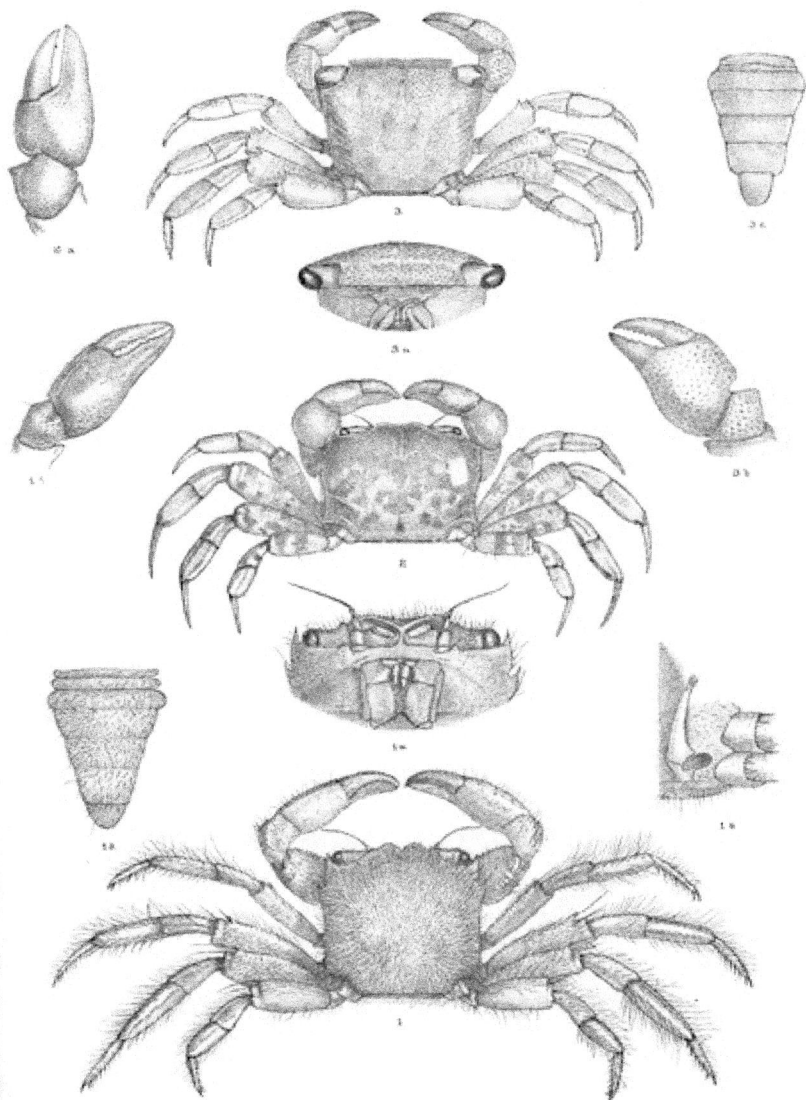

PLATE XXII.

PLATE XXII.

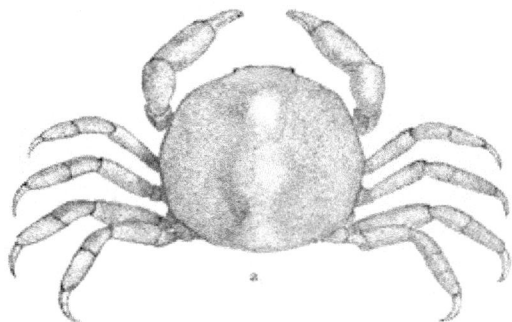

2 c.

2

2 b.

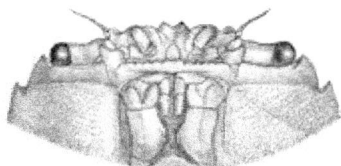

1 a.

1 c.

2 a.

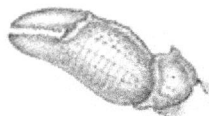

1 d.

1 b.

1.

PLATE XXIII.

(ZOOL. CHALL. EXP. —PART XLIX.—1886.)—Ccc.

PLATE XXIII.

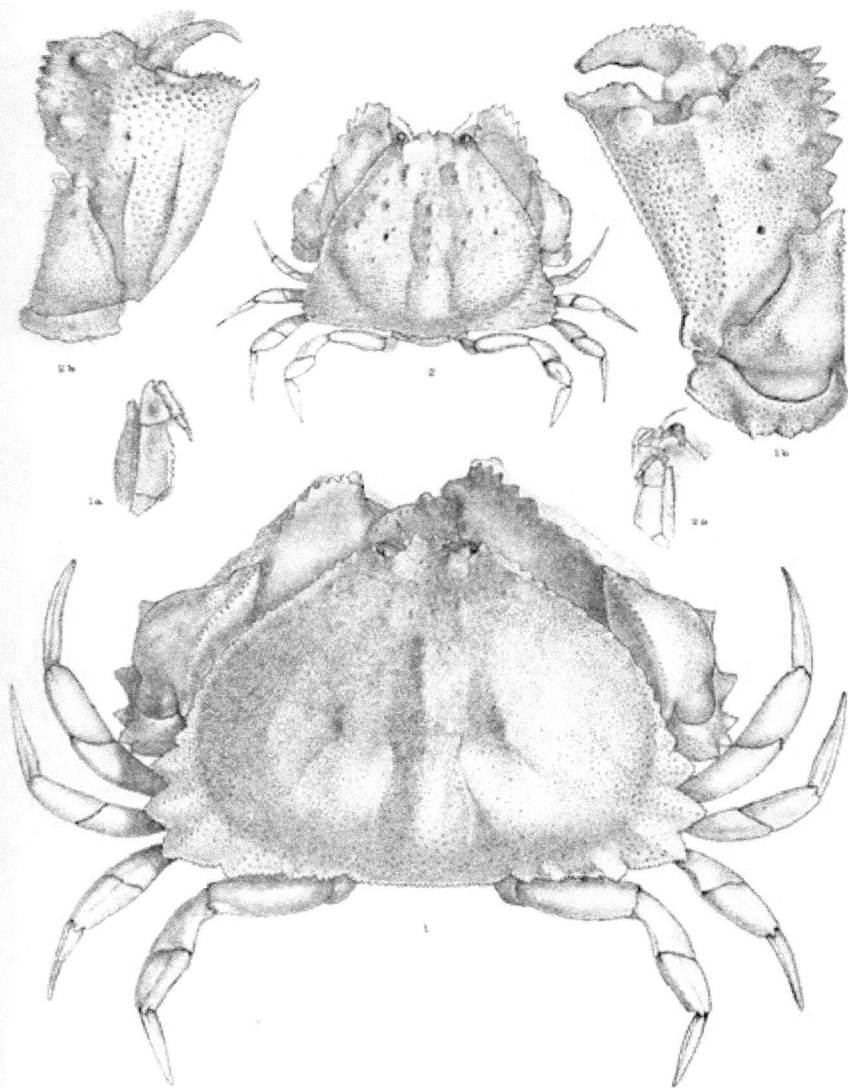

PLATE XXIV.

PLATE XXIV.

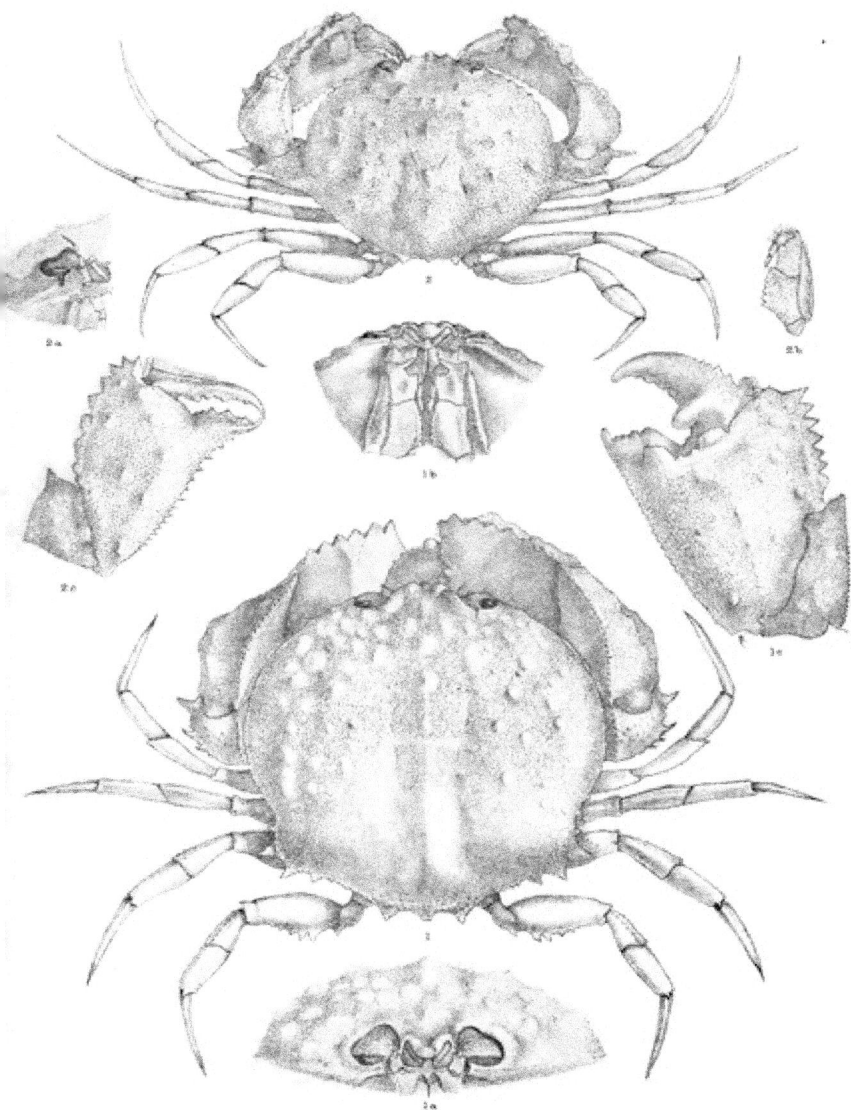

PARACYCLOIS · MURSIA

PLATE XXV.

PLATE XXV.

PLATE XXV.

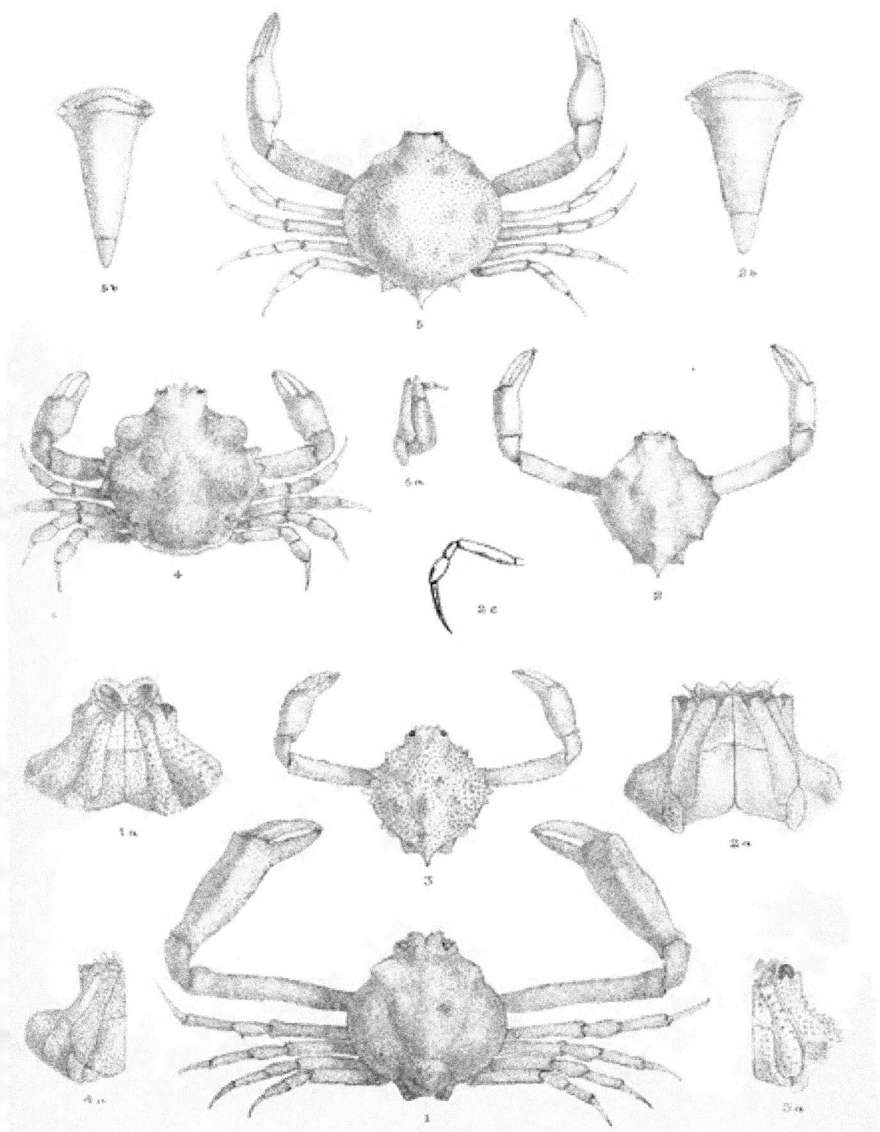

PLATE XXVI.

PLATE XXVI.

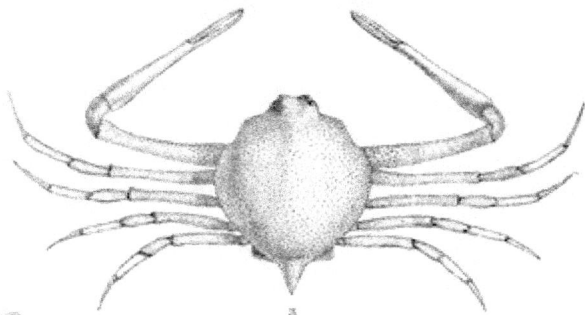

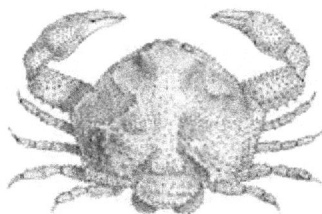

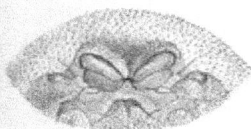

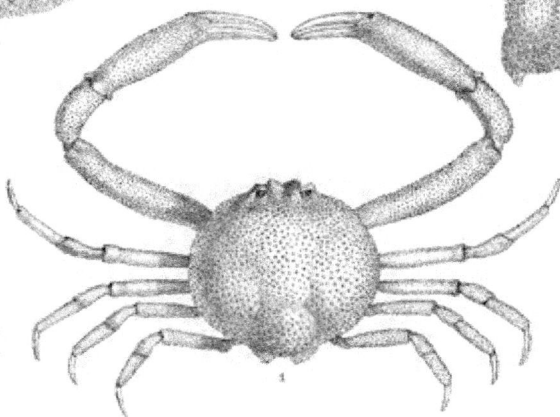

Morgan del. et lith.   West, Newman & Co. imp.

RANDALLIA · LITHADIA · ILIACANTHA

PLATE XXVII.

(ZOOL. CHALL. EXP.—PART XLIX.—1886.)—Ccc.

PLATE XXVII.

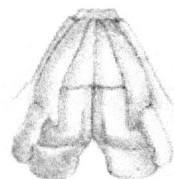

1 b.

1 a.

2 b.

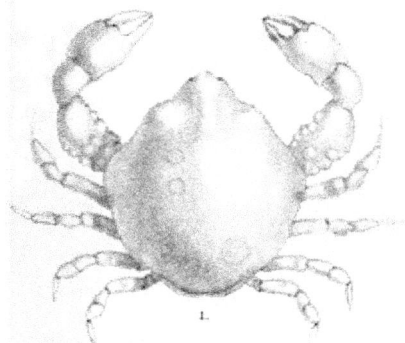

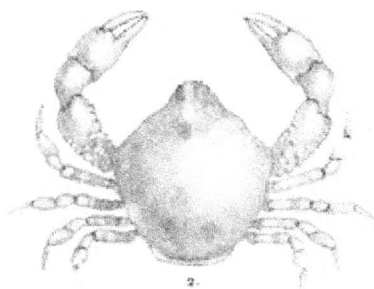

1.

2.

3 a.

2 a.

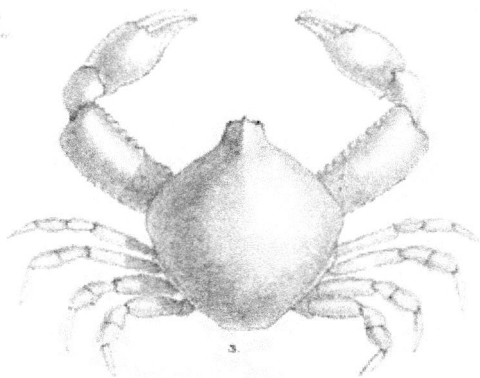

1 c.

3.

3 b.

PLATE XXVIII.

PLATE XXVIII.

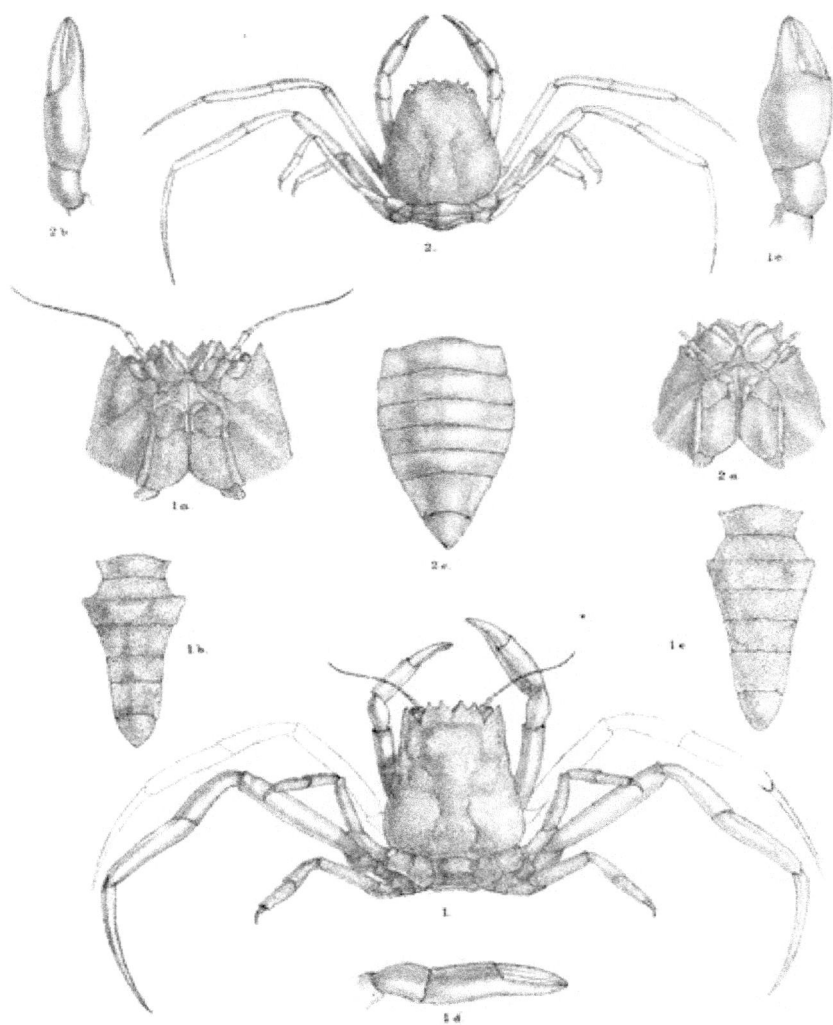

2 b.

2.

1 c.

1 a.

2 e.

2 a.

1 b.

2 c.

1 e.

1.

1 d.

PLATE XXIX.

(ZOOL. CHALL. EXP.—PART XLIX.—1886.)—Ccc.

PLATE XXIX.

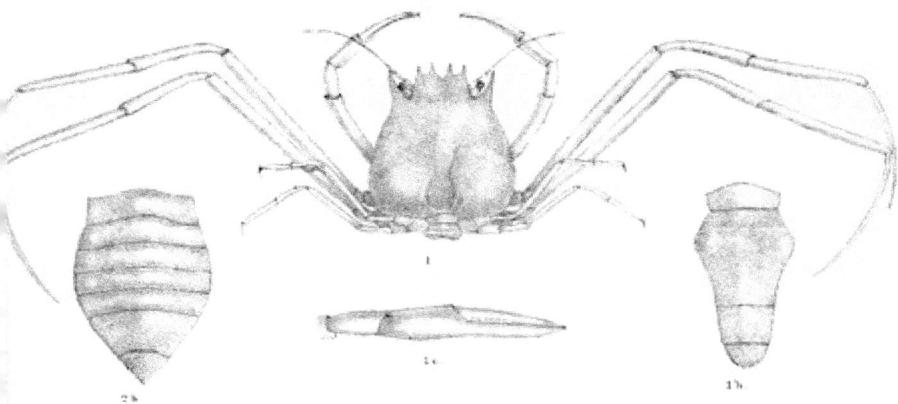

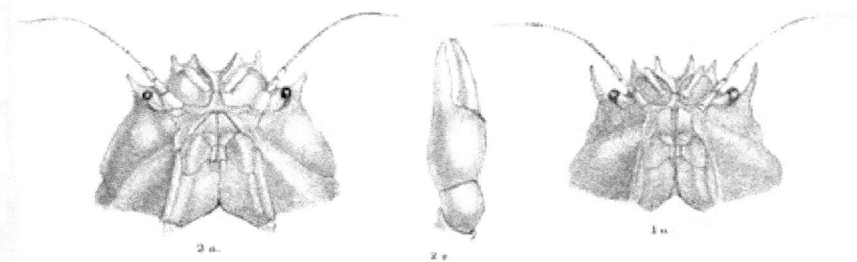

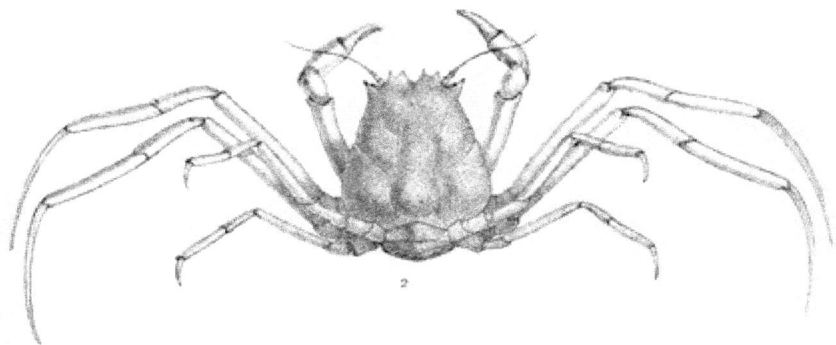

ETHUSA

VOYAGE OF H.M.S. CHALLENGER.

ZOOLOGY.

REPORT on the POLYZOA collected by H.M.S. Challenger during the years 1873–76. By GEORGE BUSK, F.R.S., V.P.L.S., &c.

PART II.—The CYCLOSTOMATA, CTENOSTOMATA, and PEDICELLINEA.

INTRODUCTION.

So much has been written of late years concerning the Polyzoa, that it is not possible for me here to name all those to whom I have been indebted for assistance in the preparation of this Report, but among the most important recent contributions to our knowledge of all the three orders now under consideration, must be mentioned the valuable works of Mr. Hincks[1] and Professor Smitt.[2] To M. L. Joliet[3] we owe papers, both anatomical and descriptive, on the Polyzoa of the French Coast. Mr. P. H. Macgillivray,[4] has described and figured many new species of CYCLOSTOMATA belonging to the Australian region. Mr. Waters[5] has given descriptions of recent species from the Bay of Naples, as well as of fossil Australian species; and Dr. J. Jullien[6] of species collected in the

---

[1] Hincks, Brit. Mar. Polyz., 1880, and various papers in *Ann. and Mag. Nat. Hist.*, &c.

[2] Smitt, Haŝ-Bryoz. Uteckl. och Fettkr., 1864; Bryoz. Mar., 1867; Kritisk Fertockn., 1865, 1871; Floröl Bryoz., pt. i., 1872.

[3] Joliet, Les Bryozoaires des Côtes de France, *Archives d. Zool. Expér.*, t. vi., p. 193, 1877.

[4] Macgillivray, P. H., *Nat. Hist. Vict.*, decades iv. and vii.; *Proc. Roy. Soc. Vict.*, Polyzoa, parts iv. vi., vii., viii., ix.

[5] Waters, *Ann. and Mag. Nat. Hist.*, ser. 5, vol. iii.; *Quart. Journ. Geol. Soc.*, vol. xl. p. 674, 1884.

[6] Jullien, *Bull. Soc. Zool. de France*, t. vii., 1882.

(ZOOL. CHALL. EXP.—PART L.—1886.) Ddd a

Atlantic by the "Travailleur." With regard to the PEDICELLINEA I must express my obligations to Drs. Nitsche,[1] Hatschek,[2] Salensky,[3] &c., for the information derived from their researches.

The total number of species enumerated in this part of my Report is forty-six, of which number thirteen still appear to me to be new, and three or four others have only been described and named since the Challenger collection was made; if I have overlooked the descriptions of any others I can only express my regret that it should be so, as I have made every endeavour to keep pace with the recent additions to our knowledge on the subject.

Of the number of species above mentioned by far the greater part, viz., thirty-three, belong to the sub-order CYCLOSTOMATA, of which, however, only five are new. Of the CTENOSTOMATA there are only eleven species in all, but of these seven are new, though none of them present any new generic type; this very much larger proportion of new species can be accounted for probably by the fact that this sub-order has been much less fully studied hitherto than either the CHEILOSTOMATA or CYCLOSTOMATA. Belonging to the PEDICELLINEA, only two species have come under my notice, of which but one is now described for the first time, the other having been named by Mr. Hincks in 1884.

CLASSIFICATION.

1. The arrangement of the sub-order CYCLOSTOMATA followed in this Report, as exemplified in the accompanying Table (see p. viii.), is nearly the same as that adopted in my Monograph of the Crag Polyzoa, 1857, and in the British Museum Catalogue, pt. iii., 1875; the number of species procured by H.M.S. Challenger belonging to this sub-order not having been sufficiently large to lead to any material change in it.

2. In the sub-order CTENOSTOMATA, again, the number of species in the collection was too small to justify the attempt at forming any different general scheme, and therefore I have followed, as nearly as may be, the arrangement adopted by Mr. Hincks in his British Marine Polyzoa, only differing from him in the definition of the two principal divisions.

3. With respect to the order PEDICELLINEA all that I need say is, that though the total number of species collected on the voyage only amounted to two, they have appeared to me to deserve recognition as a new generic type, to which I have given the name of *Ascopodaria*, but it does not seem necessary for me to enter into any disquisition as to a scheme of classification beyond what has been already written by Mr. Hincks, Dr. Nitsche, Professor Smitt and others.

[1] Nitsche, Zeitschr. f. wiss. Zool., Bd. xx. p. 343, 1870.
[2] Hatschek, Zeitschr. f. wiss. Zool., Bd. xxix. p. 502, 1877.
[3] Salensky, Ann. d. Sci. Nat., sér. 6, t. v. p. 27, 1877.

DISTRIBUTION.

1. *Geographical.*

I have thought it convenient for the sake of uniformity, and to facilitate reference, to arrange the species enumerated in this second part of the Report into the following seven divisions of the ocean, in the same manner that the CHEILOSTOMATA were divided in the first part.

A. North Atlantic Region, between the parallels of 70° W. and 20° E.
B. South Atlantic Region, from 70° W. to 20° E.
C. South Indian or Kerguelen Region, from 20° E. to 110° E.
D. Australian Region, from 110° E. to 160° W. and S.
E. Philippine or Japanese Region, from 110° E. to 160° W. and N.
F. North Pacific Region, from 160° W. to the coast of North America.
G. South Pacific Region, from 160° W. to 70° W.; but from this region no species belonging to any of the orders referred to in this part of the Report were procured.

In the following List, to the names of the species procured at each Station a reference is added, by corresponding letters, to the other regions in which it was found, so that its geographical distribution may be seen at a glance.

2. *Bathymetrical.*

The Stations in each geographical region, in the List, are arranged in bathymetrical order, beginning with those of the greatest depth; it will be seen that only two species of CYCLOSTOMATA occur at depths greater than 1000 fathoms, viz., one at 1600 and one at 1450, the former, however, also being found at various depths, from 50 fathoms downwards; four or five others were found at 450 to 600 fathoms, but by far the larger number were procured at between 50 and 150 fathoms, and ten in shallow water. Of the CTENOSTOMATA only three occurred at depths as great as 150 fathoms, the remaining eight having all come from depths less than 40 fathoms. The only two species belonging to the PEDICELLINEA group both came from 150 fathoms.

3. *Geological.*

To the sub-order CYCLOSTOMATA belong most of the oldest fossil Polyzoa that have been found up to this time, whilst "as yet we have no clear evidence that Cheilostomatous types existed in Palæozoic times;"[1] although in the Mesozoic and Tertiary strata fossil CHEILOSTOMATA are numerous. The palæontological evidence as to the antiquity of the CYCLOSTOMATA is fully confirmed and strengthened by the embryological researches that have recently been so carefully and accurately made by various authors; for instance M. Barrois[2] says that the study of the structure of the larva and of the formation of the cell coincides with palæontology in furnishing us with perfectly concordant results, which are conclusive as to the antiquity of the CYCLOSTOMATA. Many of the species even have a wide distribution in time, for out of the thirty-three included in this collection, fourteen have already been identified in the fossil state.

No fossil forms belonging either to the CTENOSTOMATA or to the Entoproctan Polyzoa have hitherto been identified, but Mr. Vine has thrown out a hint, in a paper on *Ascodictyon*,[3] that perhaps *Ascodictyon filiforme* may be a primitive representative of the stoloniferous Vesiculariadæ, or possibly of the Entoprocta. That this latter order is of great antiquity is also confirmed by its embryonic history, for the same eminent authority, above quoted, M. Barrois,[4] after the most careful and elaborate comparison of the larvæ of the various Ectoproctan and Entoproctan groups, comes to the conclusion that "the larvæ of Entoprocta represent the primitive type from which all the others are derived."

A.—NORTH ATLANTIC REGION.

STATION 75, lat. 38° 38′ N., long. 28° 28′ 30″ W.; 450 fathoms; volcanic mud.

Idmonea milneana, C.	*Idmonea irregularis*.

CAPE VERDE ISLANDS, 100 to 120 fathoms.

Hornera frondiculata.

STATION 36, lat. 32° 7′ 25″ N., long. 65° 4′ W., off Bermuda; 30 fathoms; coral.

Crisia denticulata, var. *patagonica*.	*Amathia lendigera*.

CAPE VERDE ISLANDS, 10 fathoms.

Crisia conferta, D.

STATION 109, 0° 55′ 38″ N., long. 29° 22′ 35″ W., off St. Paul's Rocks; shallow water.

Crisia denticulata, B, D.

[1] Vine, *Quart. Journ. Geol. Soc.*, vol. xl. p. 332, 1884.
[2] Barrois, *Ann. and Mag. Nat. Hist.*, ser. 5, vol. x. p. 391 (footnote).
[3] Vine, *Ann. and Mag. Nat. Hist.*, ser. 5, vol. xiv. p. 87.
[4] Barrois, *loc. cit.*, p. 401.

B.—SOUTH ATLANTIC REGION.

STATION 320, lat. 37° 17′ S., long 53° 52′ W.; 600 fathoms ; green sand.

Crisia acuminata.	*Idmonea fissurata.*
Idmonea marionensis, C.	*Hornera lichenoides.*

STATION 135, lat. 37° 1′ 50″ S.; long. 12° 19′ 10″ W., off Tristan da Cunha Islands ; 60 to 360 fathoms.

Crisia biciliata.	*Alecto granulata.*
Crisia denticulata, A, D.	*Diastopora patina.*
Crisia acuminata.	*Lichenopora fimbriata.*
Crisia cylindrica.	*Lichenopora hispida.*
Idmonea atlantica.	*Fasciculipora ramosa.*

Ascopodaria discreta.

STATION 142, lat. 35° 4′ S., long. 18° 37′ E.; 150 fathoms ; green sand.

Alcyonidium flustroides.

OFF BAHIA, 10 to 40 fathoms.

Amathia distans.	*Amathia brasiliensis.*

Farrella atlantica.

SIMON'S BAY, Cape of Good Hope ; 18 fathoms.

Idmonea atlantica.

STATION 315, lat. 51° 40′ S., long. 57° 50′ W.; 12 fathoms ; sand and gravel.

Tubulipora flabellaris.	*Tubulipora fimbria.*

C.—SOUTH INDIAN OR KERGUELEN REGION.

STATION 147, lat. 46° 16′ S., long. 48° 27′ E.; 1600 fathoms ; Diatom ooze.

Idmonea marionensis, B.

PRINCE EDWARD ISLAND, 80 to 150 fathoms.

Crisia holdsworthii.	*Idmonea milneana,* A.
Idmonea marionensis, B.	*Pustulopora proboscidea.*

STATION 149, off Kerguelen Islands ; 28 to 105 fathoms.

Crisia eburnea, var. laxa.	*Idmonea atlantica.*

STATION 151, off Heard Island; 75 fathoms; volcanic mud.

Idmonea marionensis, B.   *Pustulopora proboscidea.*
Hornera violacea.   *Pustulopora deflexa.*
                    Supercytis tubigera.

STATION 144A, lat. 46° 48′ S., long. 37° 49′ 30″ E., Marion Islands; 69 fathoms; volcanic mud.

Crisia holdsworthii.   *Idmonea australis*, D.
Idmonea marionensis, B.   *Pustulopora proboscidioides.*

D.—AUSTRALIAN REGION.

STATION 176, lat. 18° 30′ S., long. 173° 52′ E.; 1450 fathoms; Globigerina ooze.

Crisia elongata.

STATION 163A, lat. 36° 59′ S., long. 150° 20′ E., off Twofold Bay; 150 fathoms; green mud.

Crisia conferta, A.   *Amathia tortuosa.*
Amathia spiralis.   *Ascopodaria fruticosa.*

STATION 167, lat. 39° 32′ S., long. 171° 48′ E.; 150 fathoms; blue mud.

Supercytis digitata.

STATION 162, lat. 39° 10′ 30″ S., long. 146° 37′ E., Bass Strait; 38 fathoms; sand and shells.

Pustulopora regularis.   *Amathia spiralis.*

STATION 163B, lat. 33° 51′ 15″ S., long. 151 22′ 15″ E., off Port Jackson; 35 fathoms; hard ground.

Idmonea australis, C.   *Hornera foliacea.*

STATION 161, lat. 38° 22′ 30″ S., long. 144° 36′ 30″ E., off Port Philip; 33 fathoms; sand.

Crisia acropora.   *Hornera foliacea.*
                    Amathia spiralis.

STATION 188, lat. 9° 59′ S., long. 139° 42′ E.; 28 fathoms; green mud.

Amathia semispiralis.   *Vesicularia papuensis.*
                    Cylindræcium papuense.

STATION 172, lat. 20° 58′ S., long. 175° 9′ W., off Nukalofa, Tongatabu ; 18 fathoms ; coral mud.

Idmonea radians, F.

STATION 186, lat. 10° 30′ S., long. 142° 18′ E., off Cape York ; 8 fathoms ; coral mud.

Crisia denticulata, A, B. | *Idmonea eboracensis.*

Amathia connexa.

E.—PHILIPPINE OR JAPANESE REGION.

OFF ZEBU, Philippine Islands.

Crisia denticulata, var. *gracilis.*

F.—NORTH PACIFIC REGION.

OFF HONOLULU, Sandwich Islands ; 20 to 40 fathoms.

Idmonea radians, D.

G.—SOUTH PACIFIC REGION.

None.

MEASUREMENTS.

The metrical system is now so generally in use for scientific purposes that I have adopted it for all the measurements of the magnified figures, and they are accordingly given in millimetres or tenths of a millimetre. The figures are nearly all enlarged by 25 or 50 diameters, and scales are appended to all the plates.

Table of Classification.

DESCRIPTION OF GENERA AND SPECIES.

GROUP A. **ECTOPROCTA.**

SUB-ORDER II. **CYCLOSTOMATA**, Busk.

Cyclostomata, Bk., Smitt, Hincks, Auctt.
Tubuliporina, M.-Edwards, Johnst., &c.
Auloporina and *Myrioporina* (pars), Ehrenberg.
Cerioporina (pars), Bronn.
Centrifuginea (pars), d'Orbigny.
Milléporés à cellules (pars), Blainville.

Character.—Zoœcia tubular, calcareous, partially free or wholly connate or immersed; aperture terminal, inoperculate.

DIVISION I.—ARTICULATA seu RADICATA.

Radicellata, d'Orbigny, Hincks, Smitt, &c.
Articulatæ s. Radicatæ, Bk., Crag Polyz.

Character.—Zoarium branched, divided into distinct internodes by flexible chitinous joints; affixed by chitinous, or calcareous jointed stoloniform radical tubes. Zoœcia disposed in single or double longitudinal series, all facing in one direction.

Family I. CRISIADÆ.

Les Crisies, M.-Edwards.
Crisiadæ, Johnst., Busk, &c.
Crisiidæ, Busk, Hincks, &c.
Crisiæ, Smitt.
Crisidea, Reuss.
Crisidæ, d'Orbigny.

Character.—The only family.

(ZOOL. CHALL. EXP.—PART L.—1886.)

Ddd 1

The Family here contains : —

> 1. *Crisia.*
>
>> (1) *Crisia biciliata*, Macgilliv. (Pl. I. figs. 1, 2).
>> (2) *Crisia eburnea*, var. *laxa* (Pl. II. fig. 1).
>> (3) *Crisia denticulata*, Lamk. (Pl. II. fig. 3).
>> *Crisia denticulata*, var. *a*, *gracilis* (Pl. I. fig. 4).
>> *Crisia denticulata*, var. *β*, *patagonica*.
>> (4) *Crisia elongata*, M.-Edw. (Pl. I. fig. 3).
>> (5) *Crisia acuminata*, n. sp. (Pl. III. fig. 1).
>> (6) *Crisia acropora*, Bk.
>> (7) *Crisia holdsworthii*, Bk. (Pl. III. fig. 2).
>> (8) *Crisia conferta*, Bk. (Pl. II. fig. 5).
>> (9) *Crisia cylindrica*, n. sp. (Pl. II. figs. 2, 4).

1. *Crisia.*

Sertularia (pars), Linné.
Cellularia (pars), Pallas.
Cellaria (pars), Solander, Lamk.
Tibiana (pars), Lamk., Schweigger, Blainville, Lister.
Falcaria (pars), Oken, Gray.
Eucratea (pars), Esper, Lamx., Fleming, &c.
Crisia (pars), Lamx., Johnst., Gray, Auctt.
Crisidia (pars), M.-Edwards, Busk, d'Orbigny (1852).
Filicrisia, d'Orbigny (1851).

Character.—Zoœcia disposed in a single or double longitudinal series. Oœcia modified zoœcia, with a tubular aperture, walls punctate.

> § *a.* Uniserial.
>
>> *Crisidia*, M.-Edwards, Busk, Auctt.
>> *Unicellaria* (pars), Blainville.
>> *Falcaria*, Oken, Gray.
>> *Tibiana*, Lister (nec Blainville, Lamx.).
>> *Unicrisia*, d'Orb., Vine.

No species belonging to this section occurs in the Challenger collection.

> § *β.* Zoœcia disposed in a double series, opposite or alternate.
>
>> §§ *a.* Zoœcia opposite in pairs, from two to four (rarely six) in an internode.
>>
>>> *Bicrisia*, d'Orbigny.
>>> *Crisia* (pars), Auctt.

(1) *Crisia biciliata*, Macgilliv. (Pl. I. figs. 1, 2).

Crisia biciliata, Macgilliv., Nat. Hist. Vict., Dec. iv. p. 37, pl. xxxix. fig. 2.

Character.—An opposite pair of perfect zooecia in each internode, with a third intermediate aborted one between them, from which the succeeding pair or a branch arises. A pair of long, jointed spines articulated to the outer part of each zooecium, excepting those bearing the secondary branches. Ooecia small, much elongated, pyriform, situated at the angle of a bifurcation. Zooecia 0·07 mm. in diameter.

Habitat.—Station 135, Tristan da Cunha, 60 to 1100 fathoms, rock and shells; [Williamstown, Mapleston; Warrnambool, Watts].

As Mr. Macgillivray remarks (p. 38) "The aspect and general arrangement of the cells are the same as in *C. edwardsiana*, d'Orb. There are two cells in each internode, except in those from which the branches originate. The cells, as he observes, are not so long as those represented in M. d'Orbigny's figure of *C. edwardsiana*, and are wider superiorly, and the free part is not so long, and is much more abruptly curved forwards: the ooecial cell is smaller, more elongated, and each lateral cell supports usually two, but sometimes only one long-jointed spine."

Mr. Macgillivray further remarks that he is doubtful to what species my description and figures of *Crisia edwardsiana* (Brit. Mus. Cat., pt. iii. p. 5, pl. ii. figs. 5–8) refer, and I am compelled to say that subsequent examination of the specimens from which that description was chiefly drawn, some of which were from Tierra del Fuego, collected by Mr. Darwin, and others from New Zealand, procured by Dr. Sinclair, has left considerable doubt in my mind as to the identity of these two forms; the latter, it is highly probable, is distinct from the Patagonian *Crisia edwardsiana*, in which the zooecia, as represented by M. d'Orbigny, are very long and erect, whilst in the New Zealand species they are short and curved forwards. There is also another form or variety closely approaching the New Zealand species, but in some measure intermediate between that and the South American one, which may turn out to be distinct from either, and to form an intermediate variety, characterised by a tendency to have two or more pairs of cells in some of the internodes, and less exactly opposite. All three, however, are furnished with only a single articulated spine, which arises close below the mouth, instead of low down the back as in *Crisia biciliata*. One character is observable in the New Zealand form which I have not noticed in the others, viz., that the dorsal aspect of the pair of zooecia is entire, and faintly striated transversely.

§§ *b.* Zooecia numerous (more than six) in each internode, alternate on the two sides. Ooecia irregularly disposed, often absent.

Crisia (pars), Ancit.

(2) *Crisia eburnea*, var. *laxa* (Pl. II. fig. 1).

Sertularia eburnea, Linné.
La Sertolara d'avorio, Cavolini, Sprengel.
Cellularia eburnea, Pallas, Bruguières.
Cellaria eburnea, Soland., Bosc, Lamarck, Johnst., Transact. Newc. Soc., vol. ii. p. 262, pl. xi. fig. 5.
Crisia eburnea, Lamx., M.-Edwards, Fleming, Johnst., Templeton, Blainville, Risso, Couch,
    Hincks, Norman, Alder, Hassall, Van Beneden, Rech. s. l'Anat. des Bryoz., Mém. Brux.,
    t. xviii. p. 52, pl. vi. figs. 12–16; Busk, d'Orbigny, Pal. Franç., p. 598; Joliet, Arch. de
    Zool. Expér. et Gén., vol. vi., 1877.

Character.—Zoarium small, tufted. Zoœcia usually three to seven in each internode (rarely nine to eleven) loosely aggregated. Branches usually arising from the first or lowest zoœcium in an internode, sometimes from the second or third. Aperture circular, sometimes slightly pointed on one side. Ocœcia pyriform, surface even. Branches 0·15 mm., zoœcia about 0·07 mm. in diameter.

Habitat.—Station 149, D. and I., off Kerguelen Island, 28 to 105 fathoms, volcanic mud.

[Seas of Europe; Spitzbergen, Smitt; Madeira; Adriatic; Roscoff, Joliet.]

Except in its rather more loose mode of growth this form does not differ in any essential character from the common *Crisia eburnea*.

(3) *Crisia denticulata*, Lamarck, sp. (Pl. II. fig. 3).

Cellaria denticulata, Lamk.
Crisia lunata, Fleming, Blainville, Johnston (ed. 1), Couch.
Crisia denticulata, M.-Edw., Johnst., ed. 2, p. 284, pl. l. figs. 5–6; Gray, Sars, d'Orb., Gosse,
    Alder, Busk, Hincks, Norman, Smitt, Krit. Fört. öfver Skand. Hafs-Bryoz., p. 117 (*nec*
    Floridan Bryoz.); Waters, Ann. and Mag. Nat. Hist., ser. 5, vol. iii. p. 269; Vine, Joliet.
Cellaria arctica, Sars (teste Smitt).

Character.—Zoarium 1 to 3 inches high, of straggling growth. Zoœcia almost straight, connate nearly throughout their entire length. Aperture elliptical, usually pointed on one side. Oœcial cells truncate, often with three or more transverse annulations. Branches usually arising in an internode from or above the fourth cell (sometimes from the second in the lower parts of the zoarium). Branches 0·23 mm., zoœcia 0·07 mm. wide.

Habitat.—Off St. Paul's Rocks, shallow water. Station 186, lat. 10° 30′ S., long. 142° 18′ E., 8 fathoms, coral mud. Off Inaccessible Island, Tristan da Cunha, 60 to 90 fathoms.

[Coasts of Britain and Ireland; Norway; Spitzbergen? Grand Manan? Stimps.; Bay of Naples; Roscoff.]

Var. α. *gracilis* (Pl. I. fig. 4).

Character.—Closely resembles *Crisia denticulata* but of far slenderer habit, rarely if ever presenting any longitudinal interspaces between the series of zoœcia; branches not more than 0·2 mm. wide; zoœcia about 0·06 in diameter.

Habitat.—Off Zebu, Philippine Islands.

Var. β. *patagonica*, d'Orbigny (?)

Crisia patagonica, d'Orb., Voy. Amér. Mérid., Polyp., p. 7, pl. i. figs. 1–3.

"Cells from nine to nineteen, straight, very distinct; branches arising from second or third cell; sometimes two from an internode, when the second arises from the sixth cell. Joints black." Diameter of branches about 0·23 mm., and of zoœcia 0·08 mm.

Habitat.—Station 36, off Bermudas, 30 fathoms, coral.

[Patagonia.]

(4) *Crisia elongata*, M.-Edw. (Pl. I. fig. 3).

Crisia elongata (?), M.-Edwards, Réch. sur les Crisies, p. 10, pl. vii. fig. 2; Busk, Brit. Mus. Cat., pl. iii. p. 3, pl. iv. figs. 5–6; Waters.

Character.—Zoarium composed of long straight branches. Zoœcia, twelve to twenty-one or more in each internode; often much produced and curved forwards. Aperture circular, even; branches arising from the fifth to the seventh zoœcium. Oœcial cells unknown. Surface finely granular. Branches 0·3 mm., zoœcia 0·07 mm. wide.

Habitat.—Station 176, lat. 18° 30′ S., long. 173° 52′ E., 1450 fathoms, Globigerina ooze.

[Red Sea or Mediterranean? M.-Edw.; Algoa Bay.]

Whether the specimen (the only one in the Challenger collection) here described and figured really be the form described by M. Milne-Edwards I am by no means now convinced, but it is the same as that to which I have given the same appellation in the British Museum Catalogue. One reason for the doubt is that M. Milne-Edwards describes his *Crisia elongata* as narrower than *Crisia denticulata*, while that I have to name is certainly quite as wide, if not wider, than the usual form of *Crisia denticulata*.

(5) *Crisia acuminata*, n. sp. (Pl. III. fig. 1).

Character.—Zoarium 1 to 2 inches high, composed of long, straggling, flexuose branches dividing once or twice dichotomously and terminating in two short bifurcations. One of the terminal zoœcia (usually the penultimate), is often produced into a long, acute, tapering spine. The internodes comprise from seven to eleven zoœcia, usually seven or nine, and the branches always arise from the second. Zoœcia about half connate, produced above and curved abruptly forwards; aperture circular, even, border thin. Surface sparsely punctate; dorsal aspect finely striated, with a row of punctures down each interspace. Branches convex before and behind, and without any intermediate longitudinal space. Joints white or pale brown. Oœcial cells? Branches 0·2 to 0·25 mm., zoœcia 0·1 mm. wide.

Habitat.—Station 320, lat. 37° 17′ S., long. 53° 52′ W., 600 fathoms, green sand.

Somewhat like *Crisia denticulata*, but differing in the general habit which is characterised by the very long, straggling, wavy, or flexuose, sparingly forked branches, terminating in two or three short forks. Many of the ultimate segments exhibit a longer or shorter, acutely tapering, pointed spine, formed of a metamorphosed zoœcium. A similar disposition may be occasionally seen in a species of *Crisia*, to which I have given the name of *Crisia sinclairensis* (Brit. Mus. Cat., pt. iii. p. 6, pl. iv. figs. 7–11), but in this species the spinous process thus formed is much more obtuse, and there are other differences which prevent their being considered the same.

(6) *Crisia acropora*, Busk.

Crisia acropora, Bk., Voy. of Rattlea., vol. i. p. 351; Brit. Mus. Cat., pt. iii. p. 6, pl. v. figs. 3–4; Macgilliv., *loc. cit.*, Dec. iv. p. 38, pl. xxxix. fig. 3.

Character.—Cells nine to thirteen in each internode; a conical tooth (sometimes bifid) behind the orifice. Zoœcia slightly compressed; surface closely punctured, brilliant, sometimes porcellanous. Branches arising from the second to the fourth zoœcium. Oœcial cells large, pyriform, frequently annulated. Branches 0·25 mm., and zoœcia about 0·06 mm. wide.

Habitat.—Station 161, off Port Philip, 33 fathoms, sand.

[Bass Strait, R.; Williamstown and Queenscliff, Macgilliv.]

In this species the radical tubes are much curled, always arising from the bottom of the lowest cell in an internode, behind.

(7) *Crisia holdsworthii*, Busk (Pl. III. fig. 2, oœcium).

Crisia holdsworthii, Busk, Brit. Mus. Cat., pt. iii. p. 7, pl. vi.ᴮ fig. 2.

Character.—Zoœcia nine to eleven in each internode, connate throughout, with a short, tubular, cylindrical prolongation projecting directly forwards; walls very delicate, sparsely punctured; branches arising usually from the third, but in the lower internodes not unfrequently from the second cell. Ooecial cell infundibuliform, rounded at the top. Branches 0·2 mm., and zooecia 0·08 mm. wide.

Habitat.—Off Marion Island, 50 to 75 fathoms; off Prince Edward Island, 80 to 150 fathoms.

[Ceylon, Holdsworth.]

Habit very delicate and slender. May be allied to *Crisia tenuis*, Macgilliv. (*loc. cit.*, p. 39, pl. xxxix. fig. 5), in which, however, the zooecia appear to be less closely connate. In the specimens brought by Mr. Holdsworth from Ceylon there were no ooecia; on which account I have given a figure of that organ from the Challenger collection.

(8) *Crisia conferta*, Busk (Pl. II. fig. 5).

Crisia conferta, Bk., Brit. Mus. Cat., pt. iii. p. 7, pl. vi.a fig. 5.

Character.—Zoarium tufted, composed of short, thick, curved branches radiating as it were from a short central stem. Zooecia thirteen to twenty-one in an internode; nearly the upper half free, cylindrical, curved abruptly forwards; orifice orbicular or subelliptical, of the same diameter as the tube; branches one to four from each internode, not opposite. Ooecial cells closely adnate, median or axillary; usually broadly truncate. Branches 0·35 mm., and zooecia 0·07 mm. wide.

Habitat.—Off St. Vincent, Cape de Verde Islands, depth 10 fathoms. Station 163A, off Twofold Bay, 150 fathoms.

[Cape de Verde Islands, H.M.S. "Herald."]

A well-marked species, growing usually in dense tufts, and peculiar for the number of branches springing from an internode. The curved free portion of the cell is not, as is most usually the case, a mere production of the peristome marked with annular lines of growth, since the wall of that part is punctured like the rest of the zooecium.

(9) *Crisia cylindrica*, n. sp. (Pl. II. figs. 2, 4).

Character.—Zoarium about ⅓ an inch high, furcately branched; ten to thirty zooecia in an internode; usually two branches given off from the longer segments, the lower from about the seventh zooecium, and the upper near the summit of the internode. Zooecia about half immersed, the upper part curved forwards, exceedingly delicate and thin-walled, without puncta; orifice circular, margin even. Branches perfectly cylindrical, with an even shiny surface, distinctly punctate; dorsal aspect obliquely striated, but quite even. Ooecial cells pyriform, usually axillary, with a wide tubular orifice. Diameter of branches about 0·15 mm., and of zooecia 0·06 to 0·08 mm.

Habitat.—Off Nightingale Island, Tristan da Cunha, 100 to 150 fathoms.

A beautifully delicate form, distinguishable by its very slender habit and the perfectly cylindrical aspect of the branches, with the projecting cylindrical zoœcia, the projecting portion wholly oral. Its nearest ally would be *Crisia holdsworthii*.

Division II.—INARTICULATA.

Centrifuginés empâtés à cellules non operculées, d'Orb., Palæont. Franç., p. 605 (pars).
Inarticulatæ seu affixæ, Bk., Crag Polyzoa, p. 93.
Incrustata, d'Orbigny, Smitt.

Character.—Zoarium continuous, not divided into distinct internodes, fixed by a contracted calcareous base, either erect and free, or immediately adnate upon foreign bodies, and recumbent in whole or in part.

Subdivision A. ERECTA.

Family II. Idmoneidæ, Busk.

Tubigeridæ (pars), d'Orbigny, *loc. cit.*, p. 698.
Tubuliporidæ (pars), Johnst., Smitt, Hincks.
Les Tubuliporiens (pars), Milne-Edwards.
Idmoneidæ, Bk., Crag Polyzoa, p. 94; Brit. Mus. Cat., pt. iii. p. 10; Macgilliv.
Idmoneadæ, Bk., Engl. Cyclopedia, art. Polyzoa.
Horneridæ, Hincks.

Character.—Zoarium usually erect and rarely adnate, simple or branched; branches cylindrical, subcylindrical, or triangular, free or anastomosing.

The Family here contains the following genera :—

1. *Idmonea*, Lamx.

§ α. The zoœcia all disposed in alternate series on each side of the front of the branches; the innermost the longest.

(1) *Idmonea atlantica*, E. Forbes.
(2) *Idmonea radians*, Lamk.
(3) *Idmonea marionensis*, Busk.
(4) *Idmonea australis*, Macgilliv. (Pl. III. fig. 3).
(5) *Idmonea eboracensis*, n. sp. (Pl. III. fig. 4).

§ β. The outermost zoœcia in the lateral series the longest ; isolated zoœcia opening in the space on the front between the lateral series.

(6) *Idmonea milneana*, d'Orb.
(7) *Idmonea irregularis*, Meneghini.
(8) *Idmonea fissurata*, n. sp. (Pl. III. fig. 5).

2. *Hornera*, Lamx.

§ α. Zoarium branched, branches free or rarely inosculating.

§§ *a.* Oœcia dorsal ; anterior surface of branches reticulato-fibrillate.

(1) *Hornera frondiculata*, Lamx.
(2) *Hornera lichenoides*, Linn.

§§ *b.* Oœcia anterior, either wholly or in part ; surface in front not fibrillated or sulcate.

(3) *Hornera violacea*, Sars.

§ β. Zoarium foliaceous, branches connected by transverse tubules.

(4) *Hornera foliacea*, Macgilliv.

3. *Pustulopora*, Blainv.

(1) *Pustulopora proboscidea* (Pl. IV. fig. 1).
(2) *Pustulopora proboscidioides*, (Pl. IV. fig. 4).
(3) *Pustulopora deflexa*, Smitt (Pl. IV. fig. 3).
(4) *Pustulopora regularis*, Macgilliv. (Pl. IV. fig. 2).

1. *Idmonea*, Lamouroux.

Idmonea, Lamx., Exp. Méth., p. 80 ; Defrance, Blainville, M.-Edwards, Johnst., Lonsdale, Reuss, Michelin, Hagenow, d'Orb, 1852 (pars) ; Busk, Van Beneden, Hincks, Smitt (subgenus), &c.
Crisia (pars), d'Orbigny, Stoliczka.
Retepora (pars), Goldfuss, Lamk.
Diastopora (pars) Michelin.
Tubulipora (pars), Lamk., Smitt.
Crisina (pars), d'Orbigny, Smitt.

Character.—Zoarium erect, free or very partially adnate, branched dichotomously or irregularly ; springing from a single tubular cell, having a constricted, basal, discoid expansion. Branches free or anastomosing ; orifices of zoœcia disposed in parallel or subparallel, transverse or oblique, usually alternate series on the sides of the front of the branches, which are usually flattened behind, and either angular or rounded on the anterior aspect.

§ α. The zoœcia all disposed in alternate series on each side of the front of the branches ; the innermost the longest.

(1) *Idmonea atlantica*, E. Forbes.

Idmonea radians, Van Beneden, Bryoz. de la Mer du Nord, Bull. Brux., xvi. pt. ii. p. 646, pl. i. figs. 4, 6.

? *Idmonea coronopus*, Defrance, Dict. d. Sci. Nat., vol. xxii. p. 565 ; d'Orbigny, Milne-Edwards, Réch. sur les Crisies, p. 23, pl. x. fig. 3.

Idmonea atlantica, E. Forbes, MSS., Smitt, Johnst., Gray, Sars, Busk, Ann. and Mag. Nat. Hist., ser. 2, vol. xviii. p. 34, pl. i. figs. 6a-e ; Quart. Journ. Micr. Sci., vol. vi. p. 128, pl. xviii. fig. 5 ; Rep. Brit. Assoc., 1859 (Trans. Sect.) p. 146 ; (var. *tenuis*) Brit. Mus. Cat., pt. iii. p. 11, pl. ix.; Smitt, Florid. Bryoz., p. 6, pl. ii. figs. 7, 8 ; Hincks, Waters, &c.

? *Idmonea augustata*, d'Orb., Palæont. Franç., p. 731.

Character.—Zoarium irregularly branched and usually more or less in one plane ; branches triangular, one to four or five cells in each series, the innermost the longest ; dorsal surface very minutely punctate ; peristome entire, even. Oœcium anterior, subpyriform.

Habitat.—Off Nightingale Island, 100 to 150 fathoms. Station 149E, off Cape Maclear, Kerguelen, 30 fathoms. Simon's Bay, Cape of Good Hope, 18 fathoms.

[Arctic Seas ; coast of Norway and Finmark ; Shetland ; var. *tenuis*, North Atlantic ; Gulf of Florida, Smitt; Madeira (?) ; Bay of Naples; fossil in Italian Miocene and Canadian Post Pliocene (?).]

(2) *Idmonea radians*, Lamarck (sp.).

Retepora radians, Lamk., d'Orbigny.

Idmonea radians, Busk., Brit. Mus. Cat., pt. iii. p. 11, pl. vii. figs. 1-4 ; Macgilliv., Nat. Hist. Vict., Dec. vii. p. 30, pl. lxviii. fig. 3; Waters; Haswell, Proc. Linn. Soc. N. S. Wales, vol. iv. p. 350, vol. v. p. 35.

? *Hornera radiata*, Blainv., Man. d'Actin., p. 419.

Idmonea rayonnante, M.-Edw., loc. cit., p. 25, pl. xii. fig. 4.

Character.—Zoarium usually procumbent, radiate in a more or less regular circle, stipitate, sometimes suberect, with elongated, straight, subparallel bifurcating branches. Branches keeled in front, rounded behind. Dorsal aspect longitudinally sulcate, with a series of long perforations or alveoli along the sulci ; the sides and front pitted, sometimes almost reticulate. Zoœcia produced, gently curving forwards, somewhat tapering, often with a bilabiate orifice, about 0·06 mm. Branches 0·3 mm., series about 0·4 mm. apart. Usually only one or two zoœcia on the sides of the front, alternate, when more than one the inner one the longer. Oœcial chambers subglobular on the anterior aspect, close below, but not at a bifurcation ; surface coarsely pitted or foveolate.

Habitat.—Station 172, off Tongatabu, 18 fathoms, coral mud. Off Honolulu, 20 to 40 fathoms.

[Australian Seas *ubique*, New Zealand, &c.]

This species appears to occur under two rather distinct forms. In one (the typical) it constitutes a beautiful circular expansion, about one inch or more in diameter, composed of short bifurcating branches radiating from a central short stem (*vide* Brit. Mus. Cat., pt. iii. pl. vii. fig. 1*a*), or a more straggling growth in which the branches, though still springing from a more or less excentric point, are much longer, forming elongated forks (*loc. cit.*, fig. 1*b*), constituting a variety to which the term *stricta* might be applied. Very fine specimens of this form occur in the Challenger collection, from Honolulu, whilst a beautiful specimen of the radiate type was procured at Station 172, off Nukalofa, Tongatabu.

In all the Challenger specimens there is rarely more than one zoœcium to represent the lateral series, but in others there are occasionally as many as four (*vide* Brit. Mus. Cat., pt. iii. pl. vii. fig. 4), as described by Mr. Macgillivray. The great peculiarity of the species, however, is seen in the coarsely pitted or foveolate surface on the sides of the branches, and the series of large pores or pits along the dorsal sulci; the oœcial chambers also have peculiar sculpture.

(3) *Idmonea marionensis*, Busk.

> *Idmonea marionensis*, Busk, Brit. Mus. Cat., pt. iii. p. 13, pl. xiii. figs. 3, 5; pl. vii. figs. 7, 8 (young state); Waters, Haswell.
> ? *Crisia hochstetteriana*, Stoliczka, "Novara" Exp., Geol. Theil, Bd. i. p. 113, Taf. xviii. fig. 3; Smitt, Florid. Bryoz., p. 6, pl. ii. figs. 11–13.

Character.—Zoarium slender, elongate, very sparingly branched; stem and branches cylindrical or subcylindrical. Zoœcia two to four in each series (most usually two) about 0·15 to 0·20 mm., series wide apart, 0·7 to 0·8 mm., when entire the innermost the longest. Surface finely but sparsely punctured; dorsal surface convex, with a fine longitudinal striation.

Habitat.—Off Marion Islands, 50 to 75 fathoms; off Prince Edward Island, 80 to 150 fathoms. Station 147, lat. 46° 16′ S., long. 48° 27′ E., 1600 fathoms, Diatom ooze. Station 151, off Heard Island, 75 fathoms, volcanic mud. Station 320, lat. 37° 17′ S., long. 53° 52′ W., 600 fathoms, green sand.

[? Auckland, New Zealand, fossil, Stoliczka; Bay of Naples and Marion Island, Waters; Queensland, Haswell.]

As I have remarked in Brit. Mus. Cat., this species seems to mark a transition between *Pustulopora* and *Idmonea*. The cells, however, are always disposed in rows or series on each side of the anterior aspect of the branch, and are for the most part deeply immersed;

the exserted portion being of a very thin and delicate texture rarely shows the orifice in a perfect condition. It may perhaps be identical with M. d'Orbigny's *Idmonea canariensis* (Palæont. Franç., p. 732); but as neither figure nor sufficient description of that species are given, and it is merely stated to be "slender as a thread and almost round, with very few cells," it is impossible to be certain.

(4) *Idmonea australis*, Macgillivray (Pl. III. fig. 3).

Idmonea australis, Macgilliv., loc. cit., Dec. vii. p. 30, pl. lxviii. fig. 2.

Character.—Zoarium of small size ($\frac{1}{2}$ to $\frac{3}{4}$ inch), irregularly branched once, each short branch terminating in a single fork; branches contorted and sometimes twisted; four to six zoœcia in each series, the inner the longest; no intermediate longitudinal space in front between the series. Zoœcia very slender (0·12 mm.), connate below, but when perfect much produced or free for one-half or two-thirds of their length, slightly tapering, some nearly straight and ascending obliquely, but towards the upper part of the branches curved forwards, not flattened in front; series 0·7 to 1·0 mm. apart. Branches compressed, rounded both in front and behind, about 0·6 mm. wide, everywhere minutely dotted, up to the border of the aperture; dorsal surface very finely striated longitudinally, intermediate spaces with very minute dots in irregular longitudinal series.

Habitat.—Station 163B, off Port Jackson, 30 to 35 fathoms, rock. Off Marion Island, 50 to 75 fathoms.

[Port Philip Heads, 10 to 15 fathoms, Macgilliv.]

A very distinct and well-marked species, easily recognisable by the compressed form of the branches, which on section are oval, as well as by the extremely fine punctation, or rather white dotting of the surface, and the fine or close longitudinal striation of the dorsal aspect. It appears to vary very much in the length of the exserted portion of the zoœcia, which, in the specimens from Marion Island, forms more than half the length of the cell (Pl. III. fig. 3). The exserted part is very slightly tapering and no part of it appears to be peristomal, as the wall exhibits the minute punctation quite up to the orifice, and there is very rarely any appearance of annular lines of growth. None of the specimens present any oœcial chamber.

Mr. Macgillivray suggests that this species may prove to be a form of *Idmonea atlantica*, but for this view I can see no grounds whatever.

(5) *Idmonea eboracensis*, n. sp. (Pl. III. fig. 4).

Character.—Zoarium very small, not more than $\frac{1}{4}$ inch high; branches very short and irregular, once furcate, much compressed, 0·6 mm. wide; dorsal aspect rounded, longitudinally striated but not grooved. Striæ (fig. 4c), about 0·01 mm. apart, a single irregular row of dots down each interspace. Zoœcia (0·1 mm.), usually four in each series, of uniform length, except the innermost, which is the longest ; series 0·4 to 0·5 mm. apart.

Habitat.—Station 186, off Cape York, 8 fathoms, coral mud.

The collection affords only a single such specimen, but apparently mature, inasmuch as two of the branches are widely dilated at the second bifurcation into an elongated, deeply immersed oœcial chamber.

§ *β.* (subgenus *Tervia*).—The outermost zoœcium in each lateral series the longest ; scattered zoœcia opening irregularly in the space between the lateral series.

Tervia, Jullien, Bull. Soc. Zool. de France, vol. vii. p. 500, 1882.

(6) *Idmonea milneana,* d'Orbigny.

Idmonea milneana, d'Orb., Voy. Amér. Mérid., "Polypiers" p. 20, pl. ix. figs. 17–21; Palæont. Franç., p. 732 ; Smitt, Florid. Bryoz., p. 8, pl. iii. figs. 14–17 ; Macgilliv., *loc. cit.,* Dec. vii. p. 29, pl. lxviii. fig. 1 ; Busk, Brit. Mus. Cat., pt. iii. p. 12, pl. xi. ; Waters, Haswell, Ridley.

? *Idmonea transversa,* Milne-Edw., *loc. cit.,* p. 26, pl. ix. fig. 3.

Character.—Zoarium spreading from a central peduncle, branching dichotomously. Branches depressed, broad, flattened or slightly rounded behind, 0·8 to 1·5 mm. wide ; surface thickly punctate ; on dorsal aspect irregularly striated longitudinally, and, except in the younger part, transversely wrinkled. Zoœcia about 0·2 mm. in diameter, usually four or more in a series, the outer the longer ; a few intermediate zoœcia opening in the space in front between the lateral series ; series 0·6 to 1 mm. apart. Oœcial chamber ?

Habitat.—Station 75, lat. 38° 38′ N., long. 28° 28′ 30″ W., 450 fathoms, volcanic mud. Station 151, off Heard Island, 75 fathoms, volcanic mud. Off Prince Edward Island, 80 to 150 fathoms.

[Port Philip Heads, 10 to 15 fathoms, Macgilliv.; Falkland Islands, d'Orbigny ; coast of Tierra del Fuego, and Patagonia, 30 fathoms ; Chonos Archipelago, Darwin; Port Jackson and Queensland, Haswell.]

The cells, as Mr. Macgillivray observes, are usually four in series, the inner the least prominent, the others gradually increasing in length to the outer, which projects very much. They are united side to side throughout almost their whole length, so as to form regular walls, rising up and projecting far beyond the sides of the branches. As in several other species of *Idmonea* numerous radical tubes are given off from the back of the branches by which the growth is attached. The anterior median single zoœcia are few in number and usually nearly level with the surface. *Idmonea milneana* belongs to the group for which M. Jullien has proposed the name of *Tervia,* characterised by the possession of an intermediate or azygos set of zoœcia on the front, and, it may be added, by the circumstance that the outermost zoœcium in the lateral series is the longest.

(7) *Idmonea irregularis*, Meneghini.

Idmonea irregularis, Meneghini, Mem. sui Polypi della Famiglia Tubuliporiani, p. 12 (*teste* Heller); Heller, Adriat., p. 121; Busk, Brit. Mus. Cat., pt. iii. p. 13, pl. xii.; Waters, Haswell. ? *Terria folini*, Jullien, Bull. Soc. Zool. de France, vol. vii. p. 501, pl. xiii. figs. 8–9, 1882.

Character.—Zoarium composed of irregularly dichotomous, slender, rounded branches, 0·5 to 0·6 mm. wide. Zoœcia four to six in each lateral series, and the intermediate space in front occupied by numerous others irregularly scattered ; series 0·6 to 0·8 mm. apart. Diameter of zoœcia 0·1 to 0·15 mm.; surface uniformly and thickly punctate ; dorsum with longitudinal striæ wide apart. Ooœcial chamber ?

Habitat.—Station 75, lat. 38° 38′ N., long. 28° 28′ 30″ W., 450 fathoms, volcanic mud.

[Adriatic, Meneghini, Heller ; Mediterranean, H.M.S. "Porcupine"; Bay of Biscay, Jullien ; Bay of Naples, Waters ; Queensland, Haswell.]

(8) *Idmonea fissurata*, n. sp. (Pl. III. fig. 5).

Character.—Zoarium procumbent, composed of branches radiating irregularly from a central peduncle, dividing about twice dichotomously and appearing strongly serrated on the borders ; branches 0·7 to 1 mm. wide. Zoœcia 0·15 to 0·2 mm. in diameter, with an elliptical aperture, closely adnate, the outermost the longest ; a few isolated median zoœcia whose apertures are usually level with the surface. Lateral series 0·9 to 1·1 mm. apart. Surface of branches in front and behind deeply and irregularly sulcate and pitted.

Habitat.—Station 320, lat. 37° 17′ S., long. 53° 52′ W., 600 fathoms, green sand.

This species is at once recognisable by the peculiar fissured aspect of the surface, both in front and behind, and the numerous irregular pits with which it is studded. Its habit in some specimens is something like that of *Idmonea radians*, but much stronger. Like the others of the group to which it belongs, the branches appear serrated on each side.

2. *Hornera*, Lamouroux.

Hornera, Lamx., Expos., p. 41 (1821); Milne-Edwards (pars), Reuss (pars), Blainville (pars), Defrance, Michelin, Hagenow, d'Orbigny, Smitt, Hincks, Busk, Sars, &c. *Millepora* (pars), Linné, Pallas, Esper, Solander. *Retepora* (pars), Lamarck, Goldfuss. *Syphodictyum* (pars), Lonsdale.

Character.—Zoarium ramose or foliaceous; zoœcia opening only on one side; oœcia dorsal or anterior.

§ *a. Ramosa.*—Zoarium branched, branches free or rarely inosculating, cylindrical or subcompressed.

§§ *a.* Oœcia dorsal; anterior surface of branches reticulato-fibrillate.

(1) *Hornera frondiculata,* Lamouroux.

Hornera frondiculata, Lamx., Exp., p. 41, pl. lxxiv. figs. 7–9; Milne-Edwards, *loc. cit.*, p. 17, pl. ix. figs. 1–1c; Blainville, Man. d'Actin., p. 419; Heller, Adriat., p. 134; Busk, Brit. Mus. Cat., pt. iii. p. 17, pl. xx. figs. 1, 2, 3, 6; Manzoni, Waters.

Millepora tubipora, Ell. and Soll, p. 139, pl. xxxi. fig. 1.

Millepora lichenoides, Linné, Pallas, Esper.

? *Hornera affinis,* M.-Edw., *loc. cit.*, p. 9, pl. x. fig. 1.

? *Hornera undegrcensis,* Mich., Icon. Zooph., p. 318, pl. lxxvi. fig. 8.

Hornera serrata et tubulosa (?), Meneghini, *loc. cit.*, p. 10.

Character.—Branches tapering, ramifying more or less in one plane, cylindrical or subcompressed; anterior surface strongly fibro-reticulate, divided into rhomboidal spaces in which are situated the orifices of the zoœcia, surrounded with numerous small pores. Orifice exserted, usually bifid; dorsal surface coarsely reticulate, granular or nearly smooth, with small elongated pores in the sulci. Oœcia oblong, carinate, ribbed; aperture tubular, superior. Diameter of branches about 0·6 mm., and of orifice 0·08 mm.

Habitat.—Porto Praya, St. Iago, Cape de Verde, 100 to 120 fathoms.

[Mediterranean, Adriatic (very abundant). Fossil in the Crag and Upper Tertiaries of Sicily, &c.]

(2) *Hornera lichenoïdes,* Linné (sp.).

Corallium, Pontoppidan, Norg. Naturl., vol. i. p. 258, pl. xiv. figs. D,E.

Millepora lichenoides, Linné, Müller, Prodr., p. 252, No. 3046; Ström, Act. Hafn., vol. xii. p. 309, pl. iii. fig. 12; Fabricius, Zool. Samnl. (*teste* Smitt), and Fauna Groenl., p. 432 (*nec* Pallas).

Hornera frondiculata, Sars, Reise Lof. og Finm., Nyt Magazin f. Nat.-Vid., t. vi. p. 146; Busk, Ann. and Mag. Nat. Hist., ser. 2, vol. xviii. p. 34, pl. i. fig. 7a.

Hornera borealis, Busk, Crag Polyzoa, pp. 95–103; Alder, New British Polyzoa, Quart. Journ. Micr. Sci., N.S., vol. iv. p. 108, pl. v. figs. 1–6.

Hornera lichenoides, Smitt, Kritisk Förteckn., pp. 404, 469, pl. vi. fig. 10, pl. vii. figs. 1–14; Busk, Brit. Mus. Cat., pt. iii. p. 17, pl. xviii. figs. 5–6; Hincks, Brit. Mar. Polyz., p. 468, pl. lxvii. figs. 1–5.

Character.—Zoarium rising from a rather thick furrowed base, branched irregularly; branches dichotomous, crowded and straggling, often spreading out in a flabelliform manner. Anterior surface obscurely fibro-reticulate, sparsely furnished with subtubular pores. Zoœcia on the front of the branches, immersed or very slightly prominent, those of the outer sides of the branches tubular, slightly expanded towards the orifice, which is more or less elliptical, with an entire border often much produced on one side; dorsal surface finely sulcate with minute pores in the sulci. Oœcia dorsal, subglobular or gibbous, the surface reticulate or coarsely punctured. Aperture tubular, lateral (from which a very slender rib stretches across to the opposite side, Hincks). Diameter of branches 0·5 to 0·6 mm., and of orifice about 0·1 mm.

Habitat.—Station 320, lat. 37° 17′ S., long. 53° 52′ W., 600 fathoms, green sand.

[Arctic Seas, Lovén; coasts of Scandinavia, Pantoppidan, Sars, Macandrew; Shetland, Barlee; Hebrides, Norman; Nova Zembla; Greenland, Lütken; St. Georges Banks, Smitt and Harger.]

§§ *b.* Oœcia anterior, either wholly or in part; dorsal surface granular or finely striate.

(3) *Hornera violacea*, Sars.

> *Hornera violacea (forma violacea),* Smitt, *loc. cit.*, p. 404, Tab. vi. fig. 6–9; Sars, Geol. og zool. Jagtt. Reise Trondhj., 1862, Nyt Mag. Nat.-Vid., vol. xii. p. 282; Busk, Brit. Mus. Cat., pt. iii. p. 18, pl. xviii. figs. 1–4; Norman, Shetl. Dredg., Rep. Brit. Assoc., 1867, p. 310; Hincks, Brit. Mar. Polyz., i. p. 469, pl. lxvii. figs. 6–8, pl. lxii. figs. 2, 3.
> *Pustulopora orcadensis,* Bk., Quart. Journ. Micr. Sci., 1860, vol. viii. p. 214, pl. xxxix. figs. 1, 2.

Character.—Zoarium more or less of a violet colour, branched, straggling; branches short, truncate, irregularly dichotomous. Zoœcia very irregularly disposed, sometimes crowded into small fasciculi, usually elongated; surface punctate (not reticulate); dorsal surface granular, or very finely striated with minute pores, often a rib-like elevation down the centre of the branch. Oœcia elongated, situated in the axils of the branches, usually more towards the dorsal aspect; surface smooth, finely punctate, with a thin median costa; orifice bilabiate near the top. Diameter of branches 0·4 to 0·7 mm., and of orifice 0·12 mm.

Habitat.—Station 151, off Heard Island, 75 fathoms, volcanic mud.

[Arctic and Northern Seas, generally distributed; Shetland.]

§ *β. fenestrata* (subgenus *Retihornera*).—Branches sub-parallel, connected by transverse tubules, so as to form an expanded frond with quadrangular fenestræ.

> *Retihornera* (pars), Kirchenpauer, Cat. iv. of the Museum Godeffroy, 1869; Busk, Brit. Mus. Cat., pt. iii. p. 19.

(4) *Hornera foliacea*, Macgillivray.

Hornera foliacea, Macgilliv., Proc. Roy. Soc. Vict., vol. ix., p. 142, 1868.
Retihornera foliacea, Busk, Brit. Mus. Cat.,'pt. iii., p. 19, pl. xiii. figs. 1, 2, and pl. xix.; Haswell.
? *Retihornera dentata*, ? *Retihornera plicata*, Kirchenpauer.

Character.—Zoarium expanded, foliaceous, irregularly plicate or convoluted, rising from a short stem with a discoid base; main branches straight, parallel, connected by numerous transverse celliferous branches or trabeculæ, forming quadrangular fenestræ of pretty uniform size, from 0·7 to 2 mm. in length by 0·3 to 0·4 mm. wide, or about the width of the branches. Zoœcia in the young state exserted, with a usually bifid or toothed orifice, about 0·05 mm. in diameter. In the older condition more immersed, with an irregularly bifid or toothed, thickened, somewhat expanded orifice. Anterior surface fibro-reticulate, obscurely punctate and uneven; posterior sulcate, granular, obscurely punctured or pitted. Oœcia subglobose, dorsal; usually three zoœcia in the width of a branch, and one in a trabecula.

Habitat.—Station 161, off Port Philip, 33 fathoms, sand.   Station 163a, off Port Jackson, 35 to 38 fathoms, rock.

[South Australia, Gould, Macgillivray; Queensland, Haswell.]

Although at one time inclined to regard the fenestrate form of *Hornera* as entitled to the rank of a distinct genus or subgenus, I no longer regard it as forming more than a subgenus, as in all essential characters it perfectly agrees with such forms as *Hornera lichenoides*, *Hornera frondiculata*, and *Hornera cæspitosa*, mihi., differing as do those species from *Hornera violacea* in having the anterior aspect fibro-reticulate, and the oœcia dorsal.   In the Brit. Mus. Cat., pt. iii., p. 19, I have described *Hornera foliacea* as being furnished with delicate spiculæ projecting from the sides of the fenestræ, but it is highly probable that these are merely the spiculæ of some parasitic encrusting Sponge; in all other respects the form brought by Mr. Gould from South Australia in my collection, from which the account in the Brit. Mus. Cat. was drawn up, exactly agrees with the specimens in the Challenger collection, which again are undoubtedly the *Hornera foliacea* of Mr. Macgillivray.   In one of the specimens is a shallow, circular, cup-shaped depression on the dorsal aspect, doubtless the remnant of an oœcium, but these organs would appear to be very rare.

3. *Pustulopora*, Blainville.

Pustulopora, Blainville (text), Man. d'Actin., p. 418, 1834 ; Milne-Edwards, Hagenow (*nec* Geinitz), Reuss, Michelin (?), Grube, Meneghini ; Busk, Brit. Mus. Cat., pt. iii. p. 20, &c.; Macgilliv., Proc. R. Soc. Vict., December 1880, p. 6.
Pustulipora, Blainv. (index), Johnston, Gray, Sars, Joliet.
Tubulipora (pars), Couch.
Entalophora (pars), d'Orbigny (*nec* Lamouroux), Hincks, Brit. Mar. Polyz., p. 455 ; Smitt, Florid. Bryoz., vol. i. p. 11 ; Stoliczka, Waters, &c.

Character.—Zoarium erect, simple or branched, cylindrical ; branches irregular, composed of tubular zooecia partially or wholly connate or immersed ; opening on all sides of the branch, and disposed quincuncially or irregularly, sometimes in more or less annular or subspiral order.

Although most recent writers, including such high authorities as Professor Smitt and Mr. Hincks, have adopted the name *Entalophora* for the genus here intended, I am inclined, with the greatest deference, to prefer M. de Blainville's and M. Milne-Edwards' name, for the reason that the species named *Entalophora* by Lamouroux appears to me to differ in at least one most important respect, it may be said, from all the other known Cyclostomata, and most certainly from all with which I am acquainted, either recent or fossil, viz., in the appendages, as he terms them, being trumpet-shaped, or gradually increasing in diameter as they increase in length. Whether this arises from an error of observation on the part of Lamouroux or of his draughtsman, or is the true condition, may perhaps admit of doubt ; with the exception of M. Michelin (Iconog., pl. lvi. fig. 4), whose figure very strongly resembles that of Lamouroux, no one seems to have recorded any other form with trumpet-shaped tubes, and as even his figure does not represent them as having that form, I am much inclined to assume that Lamouroux's specimen is unique in that respect, and if correctly figured and described, that it must on that account alone be referred to a distinct generic type from all other known Pustuloporidæ, and in fact, as above observed, from all other Cyclostomata. (May it not be a coralline?).

On the other hand, M. de Blainville's definition of *Pustulopora*, as distinguished from Lamouroux's *Entalophora*, is so clear and precise, and his genus has met with the acceptance of M. Milne-Edwards, Hagenow, Reuss, and numerous others, and in fact may be said, until quite recently, to have been in full possession of the field, that I feel no hesitation in retaining it for all forms with cylindrical tubes of the same diameter throughout ; and in relegating those forms, if there really be any, with trumpet-shaped tubes, to at least a distinct genus.

With respect to the spelling of the name there can be no doubt that *Pustulopora* is the correct way, *Pustulipora* being apparently merely a printer's error in the index to the Manuel d'Actinologie. In the text (*loc. cit.*) M. de Blainville has it *Pustulopora*. Mr. Macgillivray has passed over a similar misprint (*loc. cit.*) the name being spelled *Pustulopera* in the text and *Pustulopora* in the description of the plates.

(1) *Pustulopora proboscidea*, Milne-Edwards (Pl. IV. fig. 1).

Pustulopora proboscidea, M.-Edw., Sur les Crisies, p. 27, pl. xii. fig. 2 ; Heller, Adriat., p. 125 ;
Grube, Die Insel Lussin, p. 68 ; Busk, Brit. Mus. Cat., pt. iii. p. 21 (*see* pl. xviii.x) ; Haswell.
Pustulipora proboscidea, Johnst., p. 279, pl. xlviii. fig. 4 ; Gray.
Entalophora proboscidea, Waters, Bryoz. of the Bay of Naples, Ann. and Mag. Nat. Hist., ser. 5,
        vol. iii. p. 274 (1879) ; Vine.
To these Mr. Waters, Fossil Cyclostomatous Polyzoa from Australia, Quart. Journ. Geol. Soc., vol. xl.,
        1884, p. 686, adds the following fossil forms :—
Entalophora raripora, d'Orb., Prod. Pal. Strat., p. 267 ; Palæont. Franç., p. 787, pl. dexxi.
        figs. 1–3, pl. dexxiii. figs. 15–17 ; Deissel, Bryoz. Aach. Kreid., p. 82, pl. x. figs. 120–128.
Pustulopora cingula, Hagenow, Maast. Kreid., p. 17, pl. i. fig. 3.
Entalophora immersa, d'Orb., Palæont. Franç., p. 781, pl. dexvi. figs. 12–14.
Entalophora attenuata, Stol., Bry. von Lemberg., p. 77, pl. i. fig. 1 ; Reuss (?), Bry. Cretacé,
        p. 74, pl. xxxvi. figs. 1, 2.
Entalophora anomala, Manzoni, Bry. Mioc. Aust. ed Ungh., p. 10, pl. ix. fig. 33.
Entalophora lusitanica, Stol., Bry. Ora. Bay., p. 102, pl. xvii. figs. 4, 5.

Character.—Zoarium composed of long, straight, alternate, furcate, cylindrical branches, 0·4 to 0·9 mm. in diameter, constituted usually of four longitudinal alternate series of zoœcia which are exserted for about one-third of their length, the exserted tube curving outwards nearly at a right angle and constituted mainly of a peristomal production, ringed, thin, not punctate, with an entire circular orifice about 0·15 mm. in diameter; surface elsewhere finely punctate. Oœcia consisting apparently of a terminal expansion, or of one situated close to a bifurcation.

Habitat.—Station 151, off Heard Island, 75 fathoms, volcanic mud. Off Prince Edward Island, 80 to 150 fathoms.

[Shetland (?), E. Forbes; Mediterranean, Milne-Edwards; Adriatic, Heller; Naples, Waters; Teneriffe and Canaries, d'Orbigny; Madeira, I. G. L; Gulf of Florida, Smitt ; Port Jackson and Queensland, Haswell.]

With respect to the Shetland locality assigned to this species in the Brit. Mus. Cat. on the authority of E. Forbes, as its occurrence in that locality is not noticed either by Mr. Norman or by Mr. Hincks as a British form, and, moreover, is omitted by Professor Smitt in Lists of Scandinavian or Arctic Polyzoa, it is highly probable that some mistake has arisen. It would appear to be a Mediterranean and Atlantic form, extending as far south as the Kerguelen region.

(2) *Pustulopora proboscidioides*, Smitt (sp.) (Pl. IV. fig. 4).

Entalophora proboscidioides, Smitt, Florid. Bryoz., vol. i. p. 11, pl. iv. figs. 26, 27.

Character.—Zoarium composed of elongated, forked, cylindrical branches, 0·7 to 1 mm. in diameter, constituted of about six longitudinal series of zoœcia, almost entirely immersed, except a short, cylindrical, exserted portion with a circular orifice, about

0·17 to 0·2 mm., which are disposed irregularly in circular whorls, about 1 mm. apart ; surface rough, punctate, even, with very faint indication of longitudinal striation.

Habitat.—Off Marion Island, 50 to 75 fathoms.

(3) *Pustulopora deflexa*, Smitt (sp.) (Pl. IV. fig. 3).

> *Entalophora deflexa*, Smitt, Florid. Bryoz., vol. i. p. 11, pl. v. figs. 28-30 ; Waters, Ann. and Mag. Nat. Hist., ser. 5, vol. iii. p. 274.
>
> *Pustulipora deflexa*, Johnst., p. 279, pl. xlviii. fig. 5 ; Norman, Rep. Brit. Assoc., 1868, p. 310 ; Marion, Ann. d. Sci. Nat., sér. 6, t. viii. p. 1 ; Joliet.
>
> *Pustulopora deflexa*, Heller, Adriat., p. 125.
>
> ? *Tubulipora deflexa*, Couch, Corn. Fauna, vol. iii. p. 107, pl. xix. fig. 4.
>
> ? *Stomatopora deflexa*, Hincks, Brit. Mar. Polyz., p. 437, pl. lvii. fig. 4.
>
> ? *Pustulopora clavata*, Busk, Crag Polyzoa, p. 107, pl. xvii. fig. 1.

Character.—Zoarium composed of very irregular furcate branches, 1 to 1·2 mm. in diameter, constituted of very long cylindrical or very slightly terete ascending zoœcia, often crowded together in fasciculate bundles and varying greatly in length ; slightly produced orally, the produced portion curving slightly outwards ; surface granular or uneven, sometimes transversely rugose, about 0·2 mm. in diameter. Oœcia ?

Habitat.—Station 151, off Heard Island, 75 fathoms, volcanic mud.

[South coasts of Britain ; Shetland, Norman ; Gulf of Florida, Smitt ; Bay of Naples, Waters ; Adriatic, Heller ; Marseilles, Marion ; Roscoff, Joliet.]

A peculiar feature of this species is the great disposition of the zoœcia to be collected into fasciculate bundles, somewhat in the same way apparently as they are described by Mr. Hincks in his *Stomatopora fasciculata*, from which, however, it differs in other more important particulars, such as the perfectly free and erect habit, and to judge from Mr. Hincks's figure, the less diameter of the zoœcial tubes, and their less entire immersion or connation. In *Stomatopora fasciculata*, moreover, the zoarium is described as having a dense and smooth surface and a dark brown colour.

Mr. Couch's description of *Tubulipora deflexa* is far too incomplete to afford any assistance towards its determination, and his figure is still less reliable. All he says is that the zoarium is erect, cylindrical, with waved tubes projecting from all parts. Mr. Hincks, however, states, with respect to his *Stomatopora deflexa*, that the "zoarium is in great part adherent ; with linear branches expanding very slightly upwards, the extremities free, erect, subclavate. The zoœcia slender, disposed in pairs along the creeping portion, and semialternate or alternate, the oral extremity free, bent upwards, and projecting considerably." To this it may be added that Mr. Hincks rejects Professor Smitt's *Entalophora deflexa* as a synonym of his *Stomatopora deflexa*. So that on the whole it seems extremely doubtful what name should be assigned to the form here described, with respect to which all that appears to me to be certain is that the specimens (mere fragments) in the Challenger collection are identical with the form described by Professor Smitt as *Entalopora deflexa*. It is scarcely possible that Mr. Hincks's *Stomatopora* should be the same, but to indicate the possibility that it may be a recent variety, I have retained the specific appelation for a decided *Pustulopora*, for which otherwise the name "*fasicularis*" would have been very appropriate.

(4) *Pustulopora regularis*, Macgillivray (Pl. IV. fig. 2).

Pustulopora (sic) regularis, Macgilliv., Proc. Roy. Soc. Vict., vol. xix., 1882, p. 292, pl. i. fig. 3.
? *Pustulopora subcerticillata*, Busk, Crag Polyzoa, p. 108, pl. xviii. fig. 1d.

Character.—Zoarium branched, branches of uniform thickness (about 1 to 2 mm.), furcate and ending in a short fork. Entire surface studded with the exserted extremities of the immersed zooecia, disposed quincuncially or in nearly regular subspiral series, and about eight in the width of the zoarium. Orifice 0·12 to 0·15 mm. in diameter. Surface of zooecia punctate, except the peristomal production which is clear, vitreous, shining and ringed. Occia?

Habitat.—Station 162, off East Moncœur Island, Bass Strait, 38 fathoms, sand and shells.

[Port Philip Heads, Macgilliv.]

Probably, as suggested by Mr Macgillivray, allied to *Entalophora subregularis*, d'Orbigny.

SUBDIVISION B. ADNATA seu DECUMBENTIA.

Family III. TUBULIPORIDÆ, Busk.

Tubuliporidæ, Bk., Crag Polyzoa ; Brit. Mus. Cat. pt. iii. p. 23.
Tubuliporidæ (pars), Johnst., Blainville, Milne-Edwards (" *Tubulipores* "), Smitt, Alder, Gray, Hincks, Vine, &c.
Sporsidæ (pars) d'Orbigny.

Character.—Zoarium entirely adnate, partially erect from an expanded base, in shape linear, reniform, flabelliform, or claviform ; simple or divided into branching lobes. Zooecia distinct, more or less free and much elongated ; irregularly disposed or ranged in more or less regular series, diverging from a mesial line. Occia formed by an inflation of the branch.

The Family here contains :—

    1. *Alecto*, Lamx.

        (1) *Alecto granulata*, M.-Edw.

    2. *Tubulipora*, Lamk.

        (1) *Tubulipora flabellaris*, Fab. (Pl. V. fig. 1).
        (2) *Tubulipora fimbria*, Lk. (Pl. V. fig. 2).

1. *Alecto*, Lamouroux.

Alecto, Lamx., 1821 ; Blainville, Johnston, M.-Edw., Busk., Michelin, Gray, Norman, Heller, &c.
Tubulipora (pars), Lamk., Smitt.
Stomatopora, Bronn, d'Orb., Hincks (pars), &c.
Aulopora (pars), Goldfuss, Reuss.
Diastopora, (pars), Smitt.
† *Proboscina* (pars), Audouin, d'Orb., Smitt (subgenus).

Character.—Zoarium closely adnate throughout ; simple or branched ; linear or ligulate. Zoœcia uniserial, or disposed in distant, more or less regular, transverse series of two to four.

(1) *Alecto granulata*, Milne-Edwards.

Alecto granulata, M.-Edw., Réch. sur les Crisies, p. 13, pl. xvi. fig. 3, 3a ; Johnston (pars),
    Busk, Brit. Mus. Cat., pt. iii. p. 24, pl. xxxii. fig. 1 ; Joliet.
† *Alecto parasitica*, Heller, Manzoni.
Stomatopora granulata, d'Orb., loc. cit., p. 836, pl. 628, figs. 5–8 ; Hincks, Brit. Mar. Polyz.,
    p. 425, pl. lvii. figs. 1, 2.

Character.—Zoarium linear, branched, the branches often anastomosing. Zoœcia uniserial, decumbent, with the oral extremities raised, subventricose below ; surface granular or coarsely ringed.

Habitat.—Off Inaccessible Island, Tristan da Cunha, 60 to 90 fathoms (on dead shell). [British and Irish coasts ; Norway and Sweden, Roscoff, &c.]

2. *Tubulipora*, Busk.

Tubulipora, Bk., Engl. Cyclop., art. Polyzoa, col. xv.; Crag Polyzoa, p. 110 ; Lamouroux,
    Hagenow, &c., Bk., Brit. Mus. Cat., pt. iii. p. 24.
Tubulipora (pars), Lamarck, Blainville, M.-Edwards, Johnston, Lonsdale, Michelin, Reuss,
    d'Orbigny, Gray, Smitt (1867), Hincks, &c.
Ceriopora (pars), Hagenow.
Phalangella sp., Gray, Smitt (subgenus).
Proboscina sp., d'Orbigny.

Characters.—Zoarium springing from a minute subglobular or discoid cell, and expanding as it grows, into an irregularly lobate, or entire, reniform or fan-shaped, adnate growth, from which spring the elongated, cylindrical, tubular zoœcia, free or partially connate and ascending ; disposed more or less regularly in series diverging from a mesial line or irregularly.

(1) *Tubulipora flabellaris*, Fabricius (sp.) (Pl. V. fig. 1).

Tubipora flabellaris, Fab., Faun. Græal., p. 430, 1780.

Tubulipora flabellaris, Manzoni, Hincks, Brit. Mar. Polyz., vol. i. p. 446, pl. lxiv. figs. 1–3.

Tubulipora (subgenus *Phalangella*) *flabellaris*, Smitt, 1866 ; Kritisk Förteckn., pp. 401, 455, pl. ix. figs. 6, 8.

Tubulipora phalangea, Couch, Corn. Fauna, vol. iii. p. 106, pl. xix. fig. 7 ; Johnston, Busk, Crag Polyzoa, p. 111, pl. xviii. fig. 6 ; Hincks, Devon. Cat., Ann. and Mag. Nat. Hist., ser. 3, vol. ix. p. 308 ; Busk, Brit. Mus. Cat., pt. iii. p. 25, pl. xxiii.; Waters, &c.

Tubulipora verrucaria, M.-Edw. (pars), Rech. sur les Tubulipores, p. 3, pl. xii. fig. 1 ; Heller.

Phalangella pholangea, Gray.

Character.—Zoarium wholly adnate, suborbicular or reniform, or obsoletely lobate ; tubular cells long, slender, 0·15 mm. in diameter, disposed in more or less regular, uniserial rows radiating from a mesial line. Walls of free portion of zoœcia ringed, not punctate. Basal expansion thickly punctate.

Habitat.—Station 315, lat. 51° 40′ S., long. 57° 50′ W., 12 fathoms, sand and gravel.

[British and Irish Seas ; Arctic Sea ; coast of Norway ; South Labrador ; Adriatic, Bay of Naples.]

Professor Smitt and Mr. Hincks appear to be so convinced that this is the form intended by Fabricius that I have thought it better to adhere to their determination and to adopt his appellation instead of Mr. Couch's.

One peculiarity as distinguishing this form from the very closely allied *Tubulipora fimbria*, Lamk., consists in the absence, so far as I have observed, of punctation of the walls of the tubular or free portion of the zoœcia, which in the latter species are sparsely punctate up to the border of the orifice, also the punctation of the basal expansion in *Tubulipora fimbria* is rather more sparse, and the spots or pores larger.

(2) *Tubulipora fimbria*, Lamarck (Pl. V. fig. 2).

? *Tubipora extusa*, Fab.

Tubulipora fimbria, Lamk., Hist. Anim. sans Vert., ed. 1, vol. ii. p. 163, ed. 2, vol. ii. p. 243 ; Smitt, (subgenus *Phalangella*), loc. cit., p. 401, 452, pl. ix. fig. 5 ; Hincks, Brit. Mar. Polyz., p. 448, pl. lx. fig. 3.

Tubulipora fimbriata, M.-Edw., loc. cit., p. 10, pl. xiv. fig. 2 ; Michelin, Iconog., p. 321, pl. lxxvii. fig. 7.

Tubulipora flabellaris, Johnst., p. 274, pl. xlvi. figs. 5, 6 ; Landsb., Pop. Hist. Brit. Zooph., p. 274, pl. xv. fig. 50 ; Busk, Crag Polyzoa, p. 111, pl. xviii. fig. 3, pl. xx. fig. 9 ; Brit. Mus. Cat., pt. iii. p. 25, pls. xxiv., xxv.; Hincks, Ann. and Mag. Nat. Hist., ser. 4, vol. xix. p. 109 ; Haswell, Joliet.

Character.—Zoarium adnate, flabelliform, often recurved on the sides. Zoœcia decumbent, irregularly disposed or very obscurely serial. Surface of basal expansion and tubes punctate, often transversely rugose. Zoœcia about 0·15 to 0·17 mm.

Habitat.—Station 315, lat. 51° 40′ S., long. 57° 50′ W., 12 fathoms, sand and gravel.

[Northern coast of Britain ; Shetland ; Ireland ; Greenland, Fabricius ; Davis Strait,

100 fathoms; Gulf of St. Lawrence; Spitzbergen; West of Nova Zembla, lat. 72° 30′ N., long. 52° 45′ E., 5 to 20 fathoms, Stuxberg and Théel; Roscoff, Joliet; Port Jackson, Haswell.]

Distinguished from the preceding species, as above observed, by the tubular portions of the zoœcia being punctate up to the border of the orifice, and their rather larger size. There can, I now think, be no doubt as to the distinctness of the two species.

Family IV. DIASTOPORIDÆ, Busk.

Diastoporidæ (pars), Bk., Crag Polyzoa, p. 113, Smitt.
Tubuliporidæ (pars), Hincks, &c.

Character.—Zoarium crustaceous or foliaceous, discoid or of indefinite outline; adnate and sessile, or pedunculate and erect; no interstitial cancelli.

The Family here contains :—

1. *Diastopora*, Johnston.

(1) *Diastopora patina.*

1. *Diastopora*, Johnston.

"*Diastopora simplex,*" M.-Edwards, Réch. sur les Crisies, p. 39.
Diastopora, Johnst., Bk., Crag Polyzoa, p. 113; Brit. Mus. Cat., pt. iii., p. 28; Hincks, Brit. Mar. Polyz., p. 457.
Diastopora (pars), Lamx., M.-Edwards, Plainville, Reuss, Hagenow, Michelin, d'Orbigny, Smitt.
Tubulipora sp., Johnst., Auctt.
Berenicea (pars), Lamx., d'Orbigny.
Patinella, Gray, Hincks, Zooph. S. Devon, Ann. and Mag. Nat. Hist., ser. 3, vol. ix. p. 468.

Character.—Zoarium adnate, discoid or flabelliform, centric or excentric; outline lobed or entire. Zoœcia towards the centre wholly immersed, usually sub-erect and partially free towards the periphery; orifice orbicular or elliptical; horizontal or oblique.

(1) *Diastopora patina*, Lamarck (sp.).

Tubulipora patina, Lamk., Johnst., Gosse, Mar. Zool., pt. iii., p. 8, fig. 1 (*nec* Milne-Edwards); Joliet.
? *Patinella verrucaria*, Gray.
Patinella patina, Hincks, Zooph. S. Devon, Ann. and Mag. Nat. Hist., ser. 3, vol. ix. p. 468.
Diastopora patina, Smitt., Busk, Brit. Mus. Cat., pt. iii., p. 28, pl. xxix. figs. 1, 2, pl. xxx. fig. 1; Hincks, Brit. Mar. Polyz., vol. i. p. 458, pl. lxvi. figs. 1–6; Waters.
? *Discosparsa marginata* (pars), d'Orbigny.
Discosparsa patina, Heller, Marion.

Character.—Zoarium when mature, discoid, circular, cupped, and bordered by a thin expansion. Central zoœcia immersed and usually occluded; marginal ones erect and open, usually disposed in irregular wavy lines radiating from the centre. (Sometimes gemmiparous, Hincks).

Habitat.—Off Nightingale Island, Tristan da Cunha, 100 to 150 fathoms. [British and Irish Coasts, northern and southern ; North Sea, Arctic Ocean, 5 to 10 fathoms, on Fucus, Flustra, &c., and on shells and coral from 50 to 100 fathoms, Smitt ; coast of Norway, Lovén ; Shetland, Barlee, 170 fathoms ; Adriatic, Heller ; South Labrador ; Marseilles, Marion ; Roscoff, Joliet.]

Family V. LICHENOPORIDÆ.

Lichenoporidæ, Smitt, Hincks, &c.
Discoporadæ, Bk., Engl. Cyclop.
Cascidæ (pars), d'Orbigny.
Tubigeridæ (pars), d'Orbigny.
Discoporellidæ, Bk., Brit. Mus. Cat., pt. iii. p. 30, 1875.

Character.—Zoarium discoid, simple or confluent ; adnate or substipitate, interzooecial spaces cancellate (cancelli sometimes obsolete). Zooecia erect or suberect, disposed more or less regularly in series diverging from an open central area.

The Family here contains :—

    1. *Lichenopora.*

        (1) *Lichenopora fimbriata,* Bk.
        (2) *Lichenopora hispida,* Flem.

1. *Lichenopora,* Defrance.

Madrepora (pars), Fabr., Exper.
Lichenopora, Defrance (1823), Blainville, Michelin, Smitt (1878), Hincks, Brit. Mar. Polyz., p. 471.
Discoporella, Gray, Brit. Mar. Rad. ; Busk, Crag Polyzoa, and Brit. Mus. Cat., pt. iii. p. 30; Smitt, Kritisk Förteckn., p. 405 (1865).
Discopora (pars), Flem. (non Lamk.), Busk, Engl. Cyclop. Polyz.
Tubulipora (pars), Johnst., M.-Edw., &c.
Defrancia (pars), *Actinopora, Discocavea, Unicavea,* &c., d'Orb.
Heteroporella sp., Hincks.

Character.—Zoarium sessile, usually closely adnate, with a thin calcareous border ; discoid, raised in the centre (hemispherical, conical, or subconical). Zooecia partly free, disposed irregularly or in lines radiating from the centre. Mouth acuminate or toothed.

(1) *Lichenopora fimbriata*, Busk.

Discoporella fimbriata, Bk., Brit. Mus. Cat., pt. iii. p. 32, pl. xxvii.

Character.—Zoarium subconical or hemispherical; zoœcia indistinctly serial, distant; interstitial cancelli or pores small, circular, often more or less obsolete; orifice somewhat expanding; peristome fimbriate, with a variable number of pointed teeth.

Habitat.—Off Nightingale Island, Tristan da Cunha, 100 to 150 fathoms.

[Chonos Archipelago, 13 fathoms; Tierra del Fuego, Cape Horn, 40 fathoms; Chiloe, 96 fathoms, Darwin; Tasmania, Mr. Smith.]

(2) *Lichenopora hispida*, Fleming (sp.).

Discopora hispida, Fleming, Blainville, Couch.
Tubulipora hispida, Johnst.
Discoporella hispida, Gray; Busk, Crag Polyzoa, p. 115, pl. xviii. fig. 5; Brit. Mus. Cat., pt. iii. p. 30, pl. xxx. fig. 3; Smitt, Sars, Alder, &c.
Heteroporella radiata, Bk., Crag Polyzoa, p. 127, pl. xix. fig. 2.
Heteroporella hispida, Hincks, Ann. and Mag. Nat. Hist., ser. 3, vol. ix. p. 469.
Lichenopora hispida, Hincks, Brit. Mar. Polyz., p. 473, pl. lxviii. figs. 1–8.

Character.—Zoarium suborbicular, convex, with or without a narrow marginal lamina; surface uniformly covered with circular openings level with the surface, of tolerably uniform size; towards the border some of the orifices raised, subtubular, and bi- or tridenticulate, disposed in obscure irregular series.

Habitat.—Stations 135 to 135a, Tristan da Cunha, 100 to 1100 fathoms.

[British Coasts, north and south; Northern Seas, Greenland, Labrador, &c. (fossil, Coral Crag; Post Pliocene, Canada).]

Family VI. FRONDIPORIDÆ, Smitt.

Fasciculinea (pars), d'Orbigny, Smitt, 1866.
Fasciporidæ (pars), d'Orbigny.
Frondiporidæ, Smitt, Kritisk Förteckn., pp. 407, 408 (1866); Busk, Brit. Mus. Cat., pt. iii. p. 37.
Cerioporidæ (pars), Busk, Crag Polyzoa, p. 118.
Cerioporinæ (pars), Hagenow.

Character.—Zoarium massive, stipitate, simple or lobate, or ramose. Zoœcia connate throughout, aggregated into fasciculi; lumen of tubes angular; walls finely porous; sides of lobes or fasciculi faintly striated or subporcellanous; no intermediate pores or cancelli.

The Family here contains :—

    1. *Fasciculipora*, d'Orb.

        (1) *Fasciculipora ramosa*, d'Orb.

    2. *Supercytis*, d'Orb.

        (1) *Supercytis digitata*, d'Orb. (Pl. V. fig. 3).

        (2) *Supercytis tubigera*, n. sp. (Pl. V. fig. 4).

        1. *Fasciculipora*, d'Orb.

Fasciculipora, d'Orb. (1839), Busk, Brit. Mus. Cat., pt. iii. p. 37 (pars).
Frondipora, Michelin (pars) ; Hagenow.
Corymbopora (pars), Michelin.
Corymbosa, sp., d'Orbigny.
Fungella, Hagenow, Bk., Crag Polyzoa, p. 118.

Character.—Zoarium stipitate ; capitulum lobate. Zooecia opening only at the ends of the fasciculi.

    (1) *Fasciculipora ramosa*, d'Orbigny.

Fasciculipora ramosa, d'Orb., Voy. Amér. Mérid., Polypiers, p. 20, pl. ix. figs. 22–24.
? *Frondipora ramosa*, Hagenow.
Corymbosa ramosa, d'Orb., Cours Élém. de Pal., tom. ii. p. 109, 1851.
? *Fungella prolifer*, Hagenow, Maast. Kreid., p. 37, pl. iii. figs. 6, 7 (?).

Character.—Zoarium fungiform ; capitulum composed of numerous obtuse, rounded lobes (usually in pairs); each lobe constituted of a thick fasciculus of tubular cells of large calibre and very thin walls, with a few intermediate tubes of less diameter interspersed ; outer surface smooth, dotted obscurely, showing the outline of the elongated zooecia, or thickened and porcellanous.

Habitat.—Off Inaccessible and Nightingale Islands, Tristan da Cunha, 60 to 150 fathoms.

[South Patagonia, 48 fathoms, Darwin, d'Orbigny.]

Fasciculipora ramosa bears a close resemblance to *Fungella multifida*, mihi, of the Crag (pl. xvii. fig. 4), but in that species, which probably corresponds with *Frondipora marsilli* of Michelin (Iconog., p. 68, pl. xiv. fig. 4). The whole growth appears more squat or depressed, and the lobes shorter and not in pairs, whilst the outer surface towards the base is marked with hexagonal areolæ, an appearance not seen in *Fasciculipora ramosa*. Otherwise the two forms appear to be closely allied.

2. *Supercytis*, d'Orbigny.

Supercytis, d'Orbigny, Palæont. Franç., p. 1060 ; Waters, Quart. Journ. Geol. Soc., vol. xl.,
       1884, p. 692.
Fasciculipora (pars), d'Orbigny ; Busk, Brit. Mus. Cat., pt. iii. p. 37.

Character.—Zoarium stipitate ; capitulum expanded, flat or cupped, with numerous
furcate or trifid fasciculi projecting round the border.   Fasciculi compressed, constituted
of coalesced, almost completely immersed zoœcia of varying lengths, all of which open on
the upper flattened side of the fasciculus or at the extremity.   Dorsal surface rounded,
even, longitudinally striated and minutely punctate.   Oœcia (when present) hemispherical, at the base of the fasciculi, and usually on the upper surface.

It is not easy to assign its proper family place to this peculiar type, but on the
whole it would perhaps be more at home among the Fasciculinæ or Frondiporidæ, than
elsewhere, the main difference between it and the other members of the group consisting
in the openings of the zoœcia not being altogether terminal but partly on the upper side
of the lobes or lateral fasciculi, or more rarely on the central area of the capitulum,
which in one of the forms here described, in the perfect and perhaps more or less
immature state, is covered with an even, calcareous, minutely punctate lamina, marked
out into very regular hexagonal areolæ, from some of which, towards the border, may be
seen the slightly projecting orifices of zoœcia.   In the second species the hexagonal areolation is apparently wanting, and in this form a few long tubular zoœcia project at the base
of some of the fasciculate lobes.

In the British Museum Catalogue I have described and figured, under the name of
Fasciculipora digitata, a species, which as pointed out by Mr. Waters (*loc. cit.*, p. 692), is
in all probability identical with M. d'Orbigny's *Supercytis digitata*, but in that specimen,
which was a good deal worn, the hexagonally areolated, calcareous lamina of the central
area is absent, and nothing is seen but the open orifices of what might be taken for the
interstitial cancelli characteristic of the Lichenoporidan group.   There can, however, I
think, be no doubt that they represent the orifices of stunted or undeveloped zoœcia,
because, firstly, towards the base of the digitiform lateral fasciculi many of the areolæ
are actually developed into short zoœcial tubes ; and secondly, in none can be traced
a vestige of the internal ciliary processes which are seen almost universally in true
interstitial cancelli.   Besides these marginal stunted zoœcia, there may be seen in all
parts of the central area similar projecting orifices, which are described by Mr. Waters
as the ends of central zoœcia slightly exserted, and which, as he remarks, give this portion
the aspect of a *Diastopora*, such as *Diastopora sarniensis*.

(1) *Supercytis digitata*, d'Orbigny (Pl. V. fig. 3).

Supercytis digitata, d'Orb., Palæont. Franç., p. 1061, pl. deexcviii. figs. 6–9; Waters, *loc. cit.*, p. 692, pl. xxxi. figs. 22, 26, 27.

Fasciculipora digitata, Bk., *loc. cit.*, p. 37, pl. xxxiii. fig. 1.

Character.—Zoarium oblong, 0·12 × 0·8 mm.; the stipitate capitulum flattened above, presents a large central area covered with a hexagonally areolated lamina, and from the sides project twelve digitate, forked, or sometimes trifid compressed lobes, composed of longer or shorter tubular zoœcia, about 0·2 mm. in diameter, almost completely immersed or sometimes slightly projecting, and opening throughout the whole length of the lobe on its upper flattened aspect, and some from the areolæ of the central area. Dorsal aspect of the lobes rounded, even, longitudinally striated and minutely punctate. Oœcia ?

Habitat.—Station 167, lat. 39° 32′ S., long. 171° 48′ E., 150 fathoms, blue mud.

[Cape Capricorn, Australia, H.M.S. "Rattlesnake." Fossil, Cretaceous, Meudon, &c., d'Orbigny ; South Australia, Waters.]

(2) *Supercytis tubigera*, n. sp. (?) (Pl. V. fig. 4).

Character.—Zoarium stipitate, capitulate; capitulum irregular or inequilateral; central area small, not areolated, but covered with a thickish calcareous lamina, with concentric rugæ, giving it a conchoidal aspect ; ten or twelve marginal fasciculate or digitate bi- or trifurcate compressed processes, in which the zoœcia are disposed more or less regularly in series of connate tubes, opening either at the extremity of the fasciculus or on its upper flattened aspect ; at the base of some of the fasciculi a few much elongated tubular zoœcia arise nearly vertically, with punctate walls, and about 0·25 mm. in diameter. Dorsal surface of fasciculi and capitulum striated and minutely punctate. Oœcia in the form of hemispherical projections at the base of the lateral fasciculi and usually on the upper aspect.

Habitat.—Station 151, off Heard Island, 75 fathoms, volcanic mud.

As the collection affords only a single specimen, which conveys the impression of a somewhat distorted growth, it may, perhaps, be merely a variety of the preceding. But the absence of areolation of the central area of the capitulum and the presence of the much elongated tubular zoœcia, together with the occurrence of the hemispherical oœcia, appeared to me to justify its being considered specifically distinct.

SUB-ORDER III. **CTENOSTOMATA**, Busk.

Ctenostomata, Busk, Hincks, Smitt, Auctt.
Halcyonellea and *Vesicularina*, Johnst.

Character.—Zoœcial orifice simply circular, bilabiate (?) or quadrangular; retractile; border contractile, furnished with a setose or membranous fringe or velum. Zoarium corneous or membranaceous, or carnose; never calcified. No marsupial or appendicular organs.

DIVISION I.—HALCYONELLEA.

Halcyonellea, Ehrenberg, Hincks.
Polyzoa carnosa, Gray.
Alcyonidulæ, Johnst.
Halcyonelleæ, Smitt.

Character.—Zoarium membranous or carnose, or semigelatinous, developed by continuous gemmation of the zoœcia from each other.

This division, embracing in Mr. Hincks' classification the genera

Alcyonidium,
Flustrella,
Arachnidium,

is represented in the Challenger collection by a single species referable to *Alcyonidium*.

Family I. ALCYONIDULÆ.

Alcyonidulæ, Couch.
Halcyonelleæ, Smitt.
Alcyonidiidæ, Hincks.

Character.—Zoœcia more or less closely united, immersed in an expanded and adherent gelatinous crust, or forming an erect, cylindrical or compressed zoarium; orifice closed by the invagination of the tentacular sheath.

The Family here contains:—

    1. *Alcyonidium*, Lamouroux.

       (1) *Alcyonidium flustroides*, n. sp. (Pl. X. figs. 13, 14).

1. *Alcyonidium.*

Alcyonidium, Lamx., Johnst., Couch, Busk, Engl. Cyclop., art. Polyz.; Hincks, &c.
Alcyonium (pars), Linné, Pallas, Müller, Fleming, &c.
Halodactylus, Farre, v. Benedan.
Cycloum and *Sarcochitum,* Hassall.

Character.—Zoœcia immersed or subimmersed. Orifice usually papillæform, more or less exsertile. Zoarium erect and lobate or crustaceous or repent.

(1) *Alcyonidium flustroides,* n. sp. (Pl. X. figs. 13, 14).

Character.—Zoarium erect and foliaceous, much branched, extending to 4 or 5 inches; bilaminate, compressed and flustroid. Zoœcia polygonal, arranged in irregular longitudinal series, the septa between which are raised and strongly marked. The substance of the walls semigelatinous, irregularly dotted with small black granules. Orifice minute, papillæform, superior. Polypide with about sixteen tentacles. Ova scattered, usually singly, in the zoœcia. Width of branches about 4 mm.; zoœcia irregular in size, from about 0·8 × 0·4 mm. to 1·6 × 0·6 mm.

Habitat.—Station 142, lat. 35° 4′ S., long. 18° 37′ E., 150 fathoms, green sand.

This species forms straggling tufts of loosely entwined and sometimes anastomosing branches, which are quite soft and flexible in the upper part, though the stem and lower branches become hard and firm near the base. Sometimes the branches embrace and adhere firmly to some foreign substance, such as worm tubes, &c. The structure is at first sight very obscure, as the substance is very thick and opaque; immersion for a short time in acid, however, renders it much more transparent and enables the nature of the zoœcia to be seen. Many of these contain polypides alone, others polypides and ova together, and others again either "brown bodies" or scattered ova only. The orifices are very small and often quite obscure. The walls seem to be partly membranous and partly of a semigelatinous nature, irregularly dotted with small black granules which are possibly argillaceous. In the form of the cell and the raised septa this species resembles *Alcyonidium mytili,* as described by Mr. Hincks,[1] but entirely differs from that form in its erect bilaminate mode of growth.

[1] Brit. Mar. Polyz., p. 498, pl. lxx. figs. 2, 3.

Division II.—VESICULARINA.

Vesicularina (pars), Johnst.
Stolonifera, Ehlers, Hincks.
Les centrifuginés radicellés (pars), d'Orb.
Vesiculariea, Smitt.

Character.—Zoarium corneous, developed by the continuous segmentation of a branching stem or stolon, having a transverse diaphragm at each node. Zoœcia budding directly from the internodes and not from each other.

Family II. VESICULARIDÆ.

Vesicular iadæ, Johnst., Alder, &c.
Vesicular iidæ and *Valkeriidæ*, Hincks.
Vesiculariea, Smitt.

Character.—Zoarium erect, free and ramose or radicate, repent or stoloniferous. Zoœcia deciduous or readily detached, leaving a circular area filled in by a perforated diaphragm. Wall entire all round, without any membranous area.

The Family here contains the following genera :—

   1. *Amathia*, Lamouroux.

      (1) *Amathia lendigera*, Linn.
      (2) *Amathia distans*, n. sp. (Pl. VII. fig. 1).
      (3) *Amathia brasiliensis*, n. sp. (Pl. VII. fig. 2).
      (4) *Amathia spiralis*, Lamx. (Pl. VI. fig. 2).
      (5) *Amathia tortuosa*, Woods (Pl. VI. fig. 1).
      (6) *Amathia connexa*, n. sp. (Pl. VI. fig. 3).
      (7) *Amathia semispiralis*, Kirchenpauer (Pl. VIII. fig. 3).

   2. *Vesicularia*, J. V. Thompson.

      (1) *Vesicularia papuensis*, n. sp. (Pl. VIII. fig. 1).
      (2) *Vesicularia trichotoma*, n. sp. (Pl. VIII. fig. 4).

   3. *Farrella*, Ehrenburg.

      (1) *Farrella brasiliensis*, n. sp. (Pl. VII. fig. 3).

1. *Amathia*, Lamouroux.

Sertularia (pars), Linn.
Amathia, Lamx., d'Orbigny (pars); Kraus, Heller, Hincks, &c.
Serialaria, Lamk., Fleming, Johnst., Blainville, Kirchenpauer, d'Orbigny (pars), Busk, Engl.
   Cyclop., art. Polyz.; Joliet, Barrois, &c.
Valkeria (pars), Dalyell.

Character.—Zoarium radicate, erect or creeping, with free dichotomous branches. Zoœcia cylindrical or subovoid, with a broad adherent base, and more or less connate laterally; arranged alternately in a double series, disposed spirally, entirely or partially surrounding the internodes, or in a straight line parallel with the axis, or in short distant groups at the upper ends of the internodes.

(1) *Amathia lendigera*, Linné (sp.).

" Nit Coralline," Ellis.
Sertularia lendigera, Linné, Pallas, Ellis and Sol., Lister.
La Sertolara lendinosa, Cavolini.
Amathia lendigera, Lamx., Pol. flex., p. 159; Heller, Hincks, Brit. Mar. Polyz., p. 516, pl. lxxiv.
   figs. 7–10, &c.
Serialaria lendigera, Lamk., Auctt.
Valkeria lendigera, Dalyell.

Character.—Zoarium much branched, tangled, filamentous; branches dichotomous, about 0·5 × 0·15 mm. Zoœcia subcylindrical, ovate, subcompressed, tapering gradually in a long neck, disposed obliquely in series of four to five pairs close below each joint or bifurcation, and gradually diminishing in length from below upwards. Cells sub-distinct, and scarcely connate.

Habitat.—Station 36, off Bermuda, 30 fathoms, coral.

[British and European Seas, *ubique.*]

(2) *Amathia distans*, n. sp. (Pl. VII. fig. 1).

Character.—Zoarium very slender and delicate, straggling, filamentous, very regularly dichotomous. Internodes long, straight, rigid, thick-walled, about 0·15 mm. in diameter, of pretty uniform length, and each completely encircled with a spiral coil, occupying usually not more than the upper half of the internode, the lower portion of which is left bare. Zoœcia ovoid or oblong, about 0·4 or 0·5 mm. by 0·1 to 0·15 mm., distinct or not closely connate, with a short conical neck.

Habitat.—Off Bahia, 10 to 20 fathoms, mud.

This form is at once recognisable by its delicate filamentous growth, and the great distance between the spiral coils, giving it somewhat the aspect of *Amathia lendigera*. Another character is the comparative shortness and distinctness of the zoœcia.

(3) *Amathia brasiliensis*, n. sp. (Pl. VII. fig. 2).

Character.—Zoarium several inches high (?), irregularly branched; branches frequently terminating in two long jointed filaments, which occasionally throw out one or two isolated zocecia or stunted branches at the nodes. The branches sometimes also commence with two or three barren internodes. Stems about 0·3 mm. in diameter, wall white and delicate. Zocecia disposed in a more or less complete spiral, which in the younger internodes occupies only the upper part, but in the older nearly their entire length; very distinct, subcylindrical, tapering from the base upwards, about 0·6 × 0·1 mm.; neck long and slender.

Habitat.—Off Bahia, 10 to 20 fathoms, mud.

The striking peculiarity of this form is the tendency of the branches to terminate in two long jointed tags (terminals), usually barren, but sometimes giving off one or two scattered isolated cells in the manner of some of the Vesiculariæ, &c. The comparative distinctness of the zocecia in the spiral series shows a tendency in the same direction.

(4) *Amathia spiralis*, Lamouroux (Pl. VI. fig. 2).

Amathia spiralis, Lamx., Polyp. Flex., p. 161, pl. iv. fig. 2.; Expos., p. 10, pl. lxv. figs. 16–17; Encyclop., p. 44.

Serialaria convoluta, Lamarck, Schweigger.

? *Serialaria spiralis*, Woods, Proc. Roy. Soc. N.S.W., vol. xi. p. 84, 1877.

? *Amathia bicornis*, Woods, Trans. Roy. Soc. Vict., vol. xvi. p. 102, 1880.

Character.—Zoarium very thick, branched subdichotomously or irregularly, several inches high. Zocecia subcylindrical, of uniform diameter, very closely connate, 1·0 mm. long, by 0·2 mm. in diameter (at base); the exserted neck thick and very strongly wrinkled; transversely disposed in an apparently continuous spiral from one internode to another, though in reality probably discontinuous at each internode.

Habitat.—Station 161, off entrance to Port Philip, 33 fathoms, sand. Station 162, off East Moncœur Island, Bass Strait, 38 fathoms, sand and shells. Station 163a, off Twofold Bay, 150 fathoms, green mud.

[Bass Strait and Australia, *ubique* (?), Lamx., &c.]

(5) *Amathia tortuosa*, Woods, (Pl. VI. fig. 1).

? *Amathia tortuosa*, Woods, Proc. Roy. Soc. Vict., vol. xvi. p. 89, fig. 6.

Character.—Zoarium 3, 4 or 5 inches high, in thick tufts, when spread out; divaricate, with long subalternate branches; rooted by radical fibres. Dull olive green colour. Stem 0·5 to 0·6 mm. in diameter. Zocecia slender, about 1 mm. × 0·15 to 0·2 mm., disposed spirally round the internodes, but not always forming quite one turn.

Habitat.—Station 163A, off Twofold Bay, 150 fathoms, green mud.
[Australia, J. T. Woods.]

It appears to me very doubtful whether this is really the form so named by Mr. Woods, who may probably not have distinguished it from the next species, which in its general habit seems to resemble the figure of his *Amathia tortuosa* more than the present. However, I am led to suppose that he had this one in view from his remark respecting the great length of the cells, which in my *Amathia connexa* are rather short. What Mr. Woods intends by "a crescentic mouth, without setæ or spines," I do not clearly understand ; and it should moreover be remarked, that in his figure of *Amathia tortuosa* the cells are not represented by any means as unusually long.

(6) *Amathia connexa*, n. sp. (Pl. VI. fig. 3).

Character.—Zoarium 3 to 4 inches high, very irregularly branched, straggling, forming dense tufts. Stem and branches from 0·5 to 0·6 mm. in diameter, transparent as glass, each internode encircled with a spiral series of zoœcia not extending its entire length, but leaving a space at each end clear. Branches here and there connected by transverse barren tubes. Zoœcia oblong, 0·5 × 0·13 mm., abruptly rounded (the neck projecting about 0·2 mm.), connivent, very delicate walls, so that the outlines towards the summits are very indistinct.

Habitat.—Station 186, off Cape York, 8 fathoms, coral mud.

The main characteristics of this form consist in—

1. The comparatively large diameter of the segmented stems and the beautiful glassy transparency of their walls, upon which the encircling series of zoœcia appear to stand out in strong relief, so as at first sight to seem as if they were disposed on one side only of the segment ; but examination shows that in reality they form nearly or wholly complete circles round the stem.

2. A second very striking feature, that I have not noticed in any other species of *Amathia*, is the occasional connection of the branches by transverse, barren, segmented tubes, resembling a similar arrangement in some of the Cheilostomata. As observed in the description of the preceding species, the general habit of *Amathia connexa*

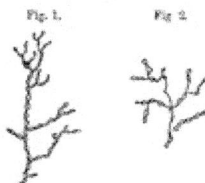

Fig. 1.   Fig. 2.

FIG. 1.—*Amathia connexa.*
FIG. 2.—*Amathia tortuosa*, Woods.

closely resembles that of Mr. Woods' *Amathia tortuosa*, as shown in his figure, which is copied in the accompanying woodcut. But that *Amathia connexa* should be the species intended by him under the name *tortuosa* is contradicted, as has been remarked, by the comparative shortness of the zoœcia.

(7) *Amathia semispiralis*, Kirchenpauer (sp.) (Pl. VIII. fig. 3).

Serialaria semispiralis, Kirchenpauer, Cat. Mus. Godeffroy, vol. iv. p. xxxiv., 1869.

Character.—Zoarium filamentous, dichotomous; stem brownish, about 0·3 mm. in diameter. Zoœcia disposed in short series, composed of four to eight pairs of cells; series distant, placed obliquely, three or four in each internode; zoœcia slender, cylindrical or somewhat quadrangular in form, about 0·6 to 0·7 × 0·15 mm., obtuse and strongly wrinkled transversely at the closed end; neck very short, and setæ short.

Habitat.—Station 188, lat. 9° 59′ S., long. 139° 42′ E., 28 fathoms, green mud.

[Samoa, Kirchenpauer.]

It may be doubtful whether this form is really that intended by Kirchenpauer, who (*loc. cit.*) describes the zoarium, as "dichotomously branched. The cells dispersed in biserial groups, at certain distances apart, and spirally round the stem; each spiral, however, extending only about one-half round the axis. From *Serialaria semiconvoluta*, Lamk., it differs in the character that in that species the tubular cells form much longer rows, whilst it also has an entirely different habit." How M. Kirchenpauer can bring it at all in comparison with the *Serialaria continhii*, which in habit more resembles a Mimosella, is more than I can imagine. No two things would appear to be more distinct.

2. *Vesicularia*, Thompson.

Vesicularia, J. V. Thompson (pars); Farre, Johnst. Busk, Engl. Cyclop.; Hincks, Brit. Mar. Polyz., p. 512, &c.
Sertularia (pars), Linné, Pallas, &c.
Laomedea (pars), Lamx., Blainville.
Valkeria (pars), Fleming.

Character.—Zoarium erect, radicate, or rooted by a fibrous base. Zoœcia distinct, usually distant, disposed in a single row or alternately in two rows on one face only of the stems. (Polypides with a gizzard, Hincks).

(1) *Vesicularia papuensis*, n. sp. (Pl. VIII. fig. 1).

Character.—Zoarium about ¾ inch high, very delicate, branching dichotomously in one plane, at an acute angle, rooted by short radical fibres. Branches composed of three or four internodes, about 0·15 mm. in diameter, and each supporting on one face a double row of zoœcia disposed alternately on each side. Zoœcia ovate, 0·5 × 0·15 mm. (with the neck retracted); neck thick and rather bulbous; surface generally smooth. No gizzard.

Habitat.—Station 188, lat. 9° 59′ S., long. 139° 42′ E., 28 fathoms, green mud.

The distinction between this form and *Vesicularia spinosa* is too obvious to require remark. But there is a second species, which occurred in the "Rattlesnake" collection, and was procured between Cumberland Island and Point Slade, which appears to be very closely allied, and I have, therefore, thought it might be useful to give a description of it, though not strictly belonging to the Challenger Expedition.

(2) *Vesicularia trichotoma* (Pl. VIII. fig. 4).

Character.—Zoarium of irregular straggling growth, main stems or primary branches about 0·2 in diameter, and usually barren, thick-walled. Primary branching, trichotomous, secondary usually furcate. Secondary and tertiary branches much slenderer, thin-walled. The internodes support eight to ten zoœcia, disposed alternately in a double series on only one side of the branch; in the youngest segments there is only a single row. Zoœcia very readily detached, ovoid, and about 0·4 × 0·2 mm. The surface is smooth and the neck short.

Habitat.—Bass Strait, between Cumberland Island and Point Slade, Voyage of H.M.S. "Rattlesnake."[1]

A peculiarity at once distinctive of this form is the trichotomous division of the primary branches; the great difference in diameter of the branches is also characteristic, as distinguishing it from the preceding, with which it agrees in the alternate arrangement of the zoœcia on the internodes.

2. *Farrella*, Ehrenburg.

Farrella, Ehrenb., Johnst., Busk, Engl. Cyclop.; Hincks, Brit. Mar. Polyz., p. 528, &c.
Lagenella, Farre, W. Thomson, Hassall.
Laguncula, Van Beneden.

Character.—Zoarium stolonate, free, or creeping and adherent; branching irregularly or at definite intervals. Zoœcia cylindrical or subventricose below, pedunculate. Orifice, when the neck (goulot) is retracted, bilabiate or quadrangular. Neck long, tapering, with or without a crown of setæ.[2] (No gizzard).

(1) *Farrella atlantica*, n. sp. (Pl. VII. fig. 3).

Character.—Zoarium stolonate, filamentous, jointed at regular intervals, free or creeping and adnate. Stolon 0·02 to 0·04 mm. in diameter, throwing out three or four short branches or zoœcia close below each joint at regular intervals. Zoœcia with the

[1] This species does not occur in the Challenger collection.
[2] It does not seem to have hitherto been remarked that the genus *Laguncula* (V. B.), as exemplified in *Laguncula repens* and *Laguncula elongata*, has no setæ.

neck retracted, about 0·4 mm. long by 0·07 to 0·09 in diameter, cylindrical and of uniform diameter throughout, quadrangular at the orifice when the neck is retracted. Supported on a long peduncle. Surface of peduncle and lower part of cells very finely wrinkled.

Habitat.—Off Bahia, 10 to 20 fathoms, mud. On *Amathia brasiliensis* and *Amathia distans.*

Family III. CYLINDRŒCIDÆ.

Cylindrœcidæ, Hincks, Brit. Mar. Polyz., p. 534.
Vesicularidæ (pars), Bk., Alder, Hincks.
Vesiculariœ (pars), Smitt.

Character.—Zoœcia not deciduous, arising from, and apparently continuous with, the stolon or segment from which they spring. Walls earthy or argillaceous.

The Family here contains the following :—

1. *Cylindrœcium*, Hincks.

(1) *Cylindrœcium papuense* (Pl. VIII. fig. 2).

1. *Cylindrœcium*, Hincks.

Farrella (pars), Bk.
Avenella, Hincks, Gosse (nec Dalyell).
Cylindrœcium, Hincks, Brit. Mar. Polyz., p. 535.
Vesicularia (subg. *Avenella*, pars), Smitt.

Character.—Zoœcia uniformly cylindrical or slightly ventricose below, sometimes dilated at the base. Stolon slender, creeping, with occasional enlargements. Zoœcia wide apart or sometimes crowded.

(1) *Cylindrœcium papuense*, n. sp. (Pl. VIII. fig. 2).

Character.—Zoarium a creeping adherent stolon with occasional bulbous thickenings. Zoœcia cylindrical, of uniform diameter and not dilated at the base, 1·3 mm. long by 0·1 to 0·11 mm. in diameter, springing either singly and widely apart from the stolon, or four or five together from a bulbous thickening.

Habitat.—Station 188, lat. 9° 59′ S., long. 139° 42′ E., 28 fathoms, green mud. Parasitic on *Amathia semispiralis*, Kirchenp.

The forms known to me as referable to the genus *Cylindrœcium* as above defined
are :—

    (1) *Cylindrœcium giganteum*, Busk.

        Farrella gigantea, Bk., Quart. Journ. Micr. Sci., vol. iv. p. 93, pl. v. figs. 1, 2 ; Gosse,
          Mar. Zool., pt. ii. p. 22, fig. 40.
        Avenella gigantea, Hincks, Ann. and Mag. Nat. Hist., ser. 3, vol. ix. p. 473.
        Avenella fusca (forma producta), Smitt, Kritisk Förteckn., p. 503.
        Cylindrœcium giganteum, Hincks, Brit. Mar. Polyz., p. 535, pl. lxxvii. figs. 3, 4.

    (2) *Cylindrœcium dilatatum*, Hincks.

        Farrella dilatata, Hincks, Quart. Journ. Micr. Sci., vol. viii. p. 279, pl. xxx. fig. 7.
        Avenella dilatata, Hincks, Ann. and Mag. Nat. Hist., ser. 3, vol. ix. p. 473.
        Cylindrœcium dilatatum (pars), Hincks, Brit. Mar. Polyz., p. 536, pl. lxxix. figs. 1–3.

    (3) *Cylindrœcium fuscum*, Busk.

        Farrella fusca, Bk., Quart. Journ. Micr. Sci., vol. iv. p. 94, pl. vi. fig. 3.
        Avenella fusca, Alder, North. Cat., p. 69 ; ? Norman, Shetland Polyz., Rep. Brit. Assoc.,
          1868, p. 311 ; Smitt (*forma simplex*), Kritisk Förteckn., p. 503.
        Cylindrœcium dilatatum (pars), Hincks, Brit. Mar. Polyz., p. 537, pl. lxxvii. figs. 1, 2.

    (4) *Cylindrœcium pusillum*, Hincks.

        Cylindrœcium pusillum, Hincks, *loc. cit.*, p. 537, pl. lxxx. fig. 8.

    (5) *Cylindrœcium pusillum*, var. " *dwarf*," Hincks.

        Cylindrœcium pusillum, var. " *dwarf*," Hincks, *loc. cit.*, p. 538, pl. lxxx. fig. 9.

    (6) *Cylindrœcium papuense*, n. sp.

The respective dimensions of the zoœcia in these species, so far as they can be made
out (mostly from Mr. Hincks's careful figures), are as follows, given in millimetres:—

	Length.		Breadth.
1. *Cylindrœcium giganteum*,	3·5	×	0·20–0·25
2. *Cylindrœcium dilatatum*,	0·9	×	0·13
3. *Cylindrœcium fuscum*, .	1·7	×	0·2–0·3
4. *Cylindrœcium pusillum*,	1·3	×	0·10–0·13
5. *Cylindrœcium pusillum*, var.,	0·7	×	0·1
6. *Cylindrœcium papuense*,	1·3	×	0·10–·11

From these dimensions it would seem that the form most nearly approaching that in
the Challenger collection is the one named *Cylindrœcium pusillum* by Mr. Hincks ; from
this, however, it differs in the uniform diameter of the zoœcia, which in the latter
become wider or subventricose below.

I may take this opportunity of remarking that the form named *Farella fusca* by me,
from specimens collected at Newhaven, Firth of Forth, by Mr. W. Thompson in 1849,
and which is considered by Mr. Hincks synonymous with his *Cylindrœcium dilatatum*,

seems to differ so much in its comparative dimensions as perhaps to deserve recognition as a distinct species, which might be named *Cylindrœcium fuscum*, as being the first of the genus to which that appellative was given.

GROUP B. ENTOPROCTA, Nitsche.

Entoprocta, Nitsche, Zeitschr. f. wiss. Zool., Bd. xx. p. 34 ; Hincks, Brit. Mar. Polyz., p. 563.

Character.—Both oral and anal orifices lying within the crown of tentacles ; no tentacular sheath. Tentacles contractile but not retractile, arranged bilaterally and symmetrically.

Order PEDICELLINEA.

The only order.

Family I. PEDICELLINIDÆ, Hincks.

Polypiaria pedicellinea, Gervais, 1837.
Pedicellina, Johnst.
Pedicellinidæ, Smitt, 1867 ; Hincks, 1880 ; Jullien, 1885

Character.—Polypides deciduous, borne on a more or less muscular, rigid or contractile peduncle ; united into colonies by a chitinous ramified stem or stolon.

The general characters of the family Pedicellinidæ are so well and succinctly given by Mr. Hincks[1] as scarcely to require further observation. The chief points to be noticed, as he remarks, besides the Entoproctous anal orifice are—

1. That there is no invagination of the anterior region and therefore no tentacular sheath, and consequently an absence of the retractor muscular fibres by which it is retracted in the Ectoproctæ.

2. That the integument is soft and never calcified, and is closely applied to its contents ; *i.e.*, there is no perivisceral cavity containing a fluid as in most other Polyzoa, such small space as there is between the inner wall of the calyx and the contained organs is occupied by a more or less delicate parenchymatous tissue. The integument is composed of a very delicate outer membrane lined by a layer of flattened polygonal cells. The outer membrane or ectocyst is prolonged beyond the visceral mass and forms the side of the vestibular cup-like chamber, within the transparent walls of which the tentacles are usually seen coiled. The tentacles arise from the upper edge of the inner surface of this cup, and their outer surface is formed by a prolongation of the transparent ectocyst, whilst the inner is covered by a more opaque layer of ciliated cells. The vestibular chamber is separated from the visceral part of the polypide by a thin lamina

[1] Brit. Mar. Polyz., vol. i. p. 563.

(intra-tentakuläre Leibeswand, Nitsche), through which passes on one side the œsophagus and on the other the rectum.

3. All the Pedicellinidæ are furnished with a more or less mobile stem, which is either flexible and contractile throughout, or as in *Pedicellina gracilis*, partially flexible and partially rigid, or as in *Ascopodaria*, wholly rigid and chitinous, its motions being effected by a peculiar muscular apparatus at the lower end.

The only forms belonging to this Family that I have noticed in the Challenger collection belong to an apparently distinct genus, to which in 1878 I had given the MS. name of "*Ascopodaria*," as stated in Professor Allman's Linnæan address of 1879. In 1880, however, Mr. Hincks, in a description of some Arctic Polyzoa, described and figured under the name of *Barentsia bulbosa*, a pedicelline species, which, though apparently quite distinct specifically from either of the two Challenger forms, evidently belongs to the genus I had already proposed to establish. In strict right of priority of publication, therefore, his name should have precedence over that merely provisionally given by me, and I should without hesitation have adopted it, but since then he has described and figured a second species, referable, from my point of view, to the same genus, under the name of *Pedicellinopsis fruticosa*,[1] thus giving two distinct names to the same generic type. I have, therefore, felt justified in retaining my original appellation, and in regarding *Barentsia* and *Pedicellinopsis* as synonyms. As an additional argument, though one of less weight, in favour of the course I have pursued, I might cite the appropriateness of the title I chose, indicative as it is of the main peculiarity of the genus.

The Family here contains :—

    1. *Ascopodaria*.

        (1) *Ascopodaria fruticosa*, Hincks, sp. (Pls. IX., X. figs. 1–5).
        (2) *Ascopodaria discreta*, n. sp. (Pl. X. figs. 6–12).

1. *Ascopodaria*, Busk.[2]

Ascopodaria, Bk. (MS.), Add. by Prof. Allman, Journ. Linn. Soc. Lond. (Zool.), vol. xv. p. 2.
Barentsia, Hincks, Ann. and Mag. Nat. Hist., ser. 5, vol. vi. p. 285 ; Vigelius, Bijd. Dierk. Genoot.,
    Nat. Art. Mag., Amsterdam, Afev. ii. pt. 2, p. 85.
Pedicellinopsis, Hincks, loc. cit., vol. xiii. p. 363.
Pedicellina (pars), Sars, Lealy.

Character.—Polypide budding from and supported at the extremity of a chitinous, tubular, perforated stem, which expands below into a cylindrical, barrel-shaped dilatation, lined internally by a layer of longitudinal muscular tissue.

---

[1] This species is the same as one of the two Challenger forms to which I had given the name of *Ascopodaria socialis*, but I have now as a matter of course adopted Mr. Hincks specific name.

[2] From ἀσκός, a wine-skin, πούς, a foot.

The structure of the peduncle is the character by which this genus is distinguished from *Pedicellina*. The pedicel is rigid and chitinous throughout, and depends for its motion on the muscular fibres which line the barrel-shaped expansion at the base; the central cavity of this expansion as well as of the rest of the stem being filled with an extremely delicate parenchymatous tissue.

The anatomy of the polypides appears to agree almost entirely, as far as I have been able to observe it in the spirit specimens, with the very careful descriptions given by Dr. Nitsche in his paper on *Pedicellina echinata*.[1] The whole polypide or calyx is enveloped in a delicate transparent membrane or ectocyst, lined with a more or less distinct tesselated epithelium. The alimentary canal consists of an œsophagus, stomach, intestine and rectum (Pl. IX. fig. 6); the liver cells extending along the upper side of the stomach present the usual deep yellow colour. In all the specimens that I have examined the rectum lies in a horizontal position forming an angle with the rest of the intestine; whether this is its normal position, as it appears to be in the closely allied genus *Urnatella*[2] or whether it merely is the case during a young stage of growth, as mentioned by Dr. Nitsche, I am unable to decide. I have not been able to observe with any certainty the reproductive organs; but in nearly all the polypides of one species, *Ascopodaria fruticosa*, between the stomach and the base of the vestibular cavity, there are two large, round, ovarian masses (Pl. IX. figs. 6, 8, 9), which are separated from one another by a thin lamina (Pl. IX. fig. 9). In the other species these masses are not apparently always present.

Mr. Hincks has suggested[3] that his genus *Pedicellinopsis* would properly include the *Pedicellina gracilis* of Sars; in this I am disposed fully to agree with him and should therefore propose to include it in my genus *Ascopodaria*. Professor Leidy[4] refers to a species of *Pedicellina* found by him in 1859, which, from the short description given, if not identical with *Pedicellina gracilis*, ought also to be placed in this genus. The known species therefore would be four or five, as follows :—

(1) *Ascopodaria gracilis*, Sars.

(2) *Ascopodaria bulbosa*, Hincks.

(3) *Ascopodaria fruticosa*, Hincks = *socialis*, Bk., MS.

(4) *Ascopodaria discreta*, Bk.

(5) *Ascopodaria* (?), Leidy.

(1) *Ascopodaria fruticosa*, Hincks, sp. (Pls. IX., X. figs. 1–5).

Pedicellinopsis fruticosa, Hincks, Ann. and Mag. Nat. Hist, ser. 5, vol. xiii p. 364, pl. xiv. fig. 3.

Character.—Zoarium arborescent, constituted of thick, erect, chitinous, jointed, branching stems, arising from tubular stoloniform fibres. The deciduous polypides (or

[1] *Zeitschr. f. wiss. Zool.*, Bd. xx. p. 13.   [2] Leidy, *Journ. Acad. Nat. Sci. Philad.*, 1884, vol. ix. p. 12.
[3] *Loc. cit.*, vol. xiii. p. 364.   [4] *Loc. cit.*, p. 14.

calices, Auctt.) are seated on the upper end of slender tubes or pedicels, which are produced into a single or double point on one side at the top; at its base the pedicel dilates into a thick barrel-shaped cylinder (Pl. IX. fig. 7), which is covered by a transparent, ringed, chitinous envelope (Pl. X. fig. 1), lined with a strong muscular layer, the cavity being occupied by a very delicate fibro-cellular tissue (Pl. IX. fig. 14). The chitinous pedicels have four more or less regular longitudinal series of funnel-shaped perforations. These polypiferous peduncles are seated in a cup-shaped hollow, and attached by a much restricted termination in a spiral direction around the upright stems, communication with the interior of which is maintained through a fine funnel-shaped orifice (Pl. XI. figs. 12, 13). The polypides are of the usual pedicelline character, and have a very short flexible stalk, which is attached just within the upper edge of the chitinous pedicels, and when young is continuous with the inner cellular tissue; when mature the polypides appear to be quite cut off from the pedicels on which they are placed, and from which they bud in succession (Pl. IX. fig. 5). The tentacles vary in number from twenty in a bud to twenty-six or twenty-eight in an adult, and are arranged more or less bilaterally and symmetrically. The pedicels and stems are of a bright light brown colour usually; the stems turning nearly black when old. The polypides are white and the barrels white or nearly so, the transparent chitinous envelope being so thin that the white inner layer shows through.

The total length of the calyx and peduncle is 3·5 to 3·8 mm. The polypide measuring about 0·65 × 0·5 mm., the pedicel 2·3 × 0·07 mm., and the barrel 0·75 × 0·5 mm.

Habitat.—Station 163A, off Twofold Bay, 150 fathoms.

[Port Philip Heads.]

The arborescent growth of this beautiful species distinguishes it at once from all other known Pedicelline forms, but the rest of its structure leaves no doubt as to its belonging to that order.

At first sight it is difficult to observe that the tentacles are not arranged in a perfectly regular and continuous circle, but here and there indications may be noticed that a wider space does occur between two at opposite sides of the circumference, viz., at the two ends of the symmetrical plane of the animal; the bilateral arrangement is most clearly seen in a young budding Polypide (Pl. X. fig. 2) which appears closely to resemble the figures given by Hatschek[1] in his paper on *Pedicellina echinata*, and also the figure and description by Salensky.[2] The buds arise in succession spirally and somewhat in pairs (Pl. X. fig. 1) round the growing ends of the chitinous stems and branches. Fresh polypides also bud from the ends of the pedicels after others have died and dropped off; that this also occurs in *Pedicellina* has been noticed and described by Salensky,[3] and

[1] *Zeitschr. f. wiss. Zool.*, Bd. xxix. pl. xxx. figs. 39, 40, 45.   [2] *Ann. d. Sci. Nat.*, sér. 6, t. v. p. 36, pl. iv. fig. 35.
[3] *Loc. cit.*, pp. 30, 31.

is mentioned by Professor Leidy[1] as occurring in *Urnatella*, but I have not found it referred to by any other writers on the Pedicellinea.

Pl. X. fig. 1, represents a group of buds at the end of one of the branches, and also shows the barrel-shaped expansion at the base of one of the peduncles, from which the transparent ringed covering has been partially loosened and torn off by the process of boiling. Figs. 3 to 5 on the same plate are taken from sketches made by the late Sir C. Wyville Thomson when the specimens were fresh and alive.

(2) *Ascopodaria discreta*, n. sp. (Pl. X. figs. 6–12).

Character.—The zoarium consists of a creeping stoloniferous stem, jointed at intervals where the branches are given off or where the polypides arise. The deciduous polypides are seated at the upper end of slender chitinous pedicels, which are dilated below into barrel-shaped cylinders that have a thick, ringed, chitinous envelope, and exactly resemble those of the preceding species. The polypiferous peduncles are seated by a broad base on the stoloniform stems; usually singly on the somewhat expanded jointed bifurcation of four branches (fig. 11), but sometimes scattered along the stolons (fig. 12). The chitinous pedicels are irregularly punctured by minute funnel-shaped pores. The polypides are united to the pedicels by a spirally ringed flexible joint (fig. 12). The tentacles are from sixteen to twenty in number. The pedicels and stolons are of a bright brown, horny colour, the polypides white, and the barrels also white or very light brown, appearing darkest when quite young, the chitinous envelope becoming thinner and more transparent as the animal grows older.

The total length varies considerably, apparently according to age; the majority of the older ones measure as much as from 4·25 to 4·4 mm. The polypide being about 0·5 × 0·4 mm., the pedicel 3·0 × 0·6 mm., and the barrel 0·7 × 0·24 mm. This species is, therefore, on the whole, taller and more slender than the preceding one.

Habitat.—Station 135, off Nightingale Island, Tristan da Cunha, 100 to 150 fathoms.

There were very few specimens in all of this species in the collection, and, therefore, it has not been possible to enter into a full and minute examination of the polypide, but it appears to present all the usual Pedicelline characters.

[1] *Journ. Acad. Nat. Sci. Philad.*, vol. ix. pt. i. p. 12.

INDEX.

GENERA, SPECIES, AND VARIETIES.

(Synonyms printed in *italics*.)

	PAGE
CRISIEÆ,	1
CRISIES, Lea,	1
CRISIIDÆ,	1
CRISINA,	9
hochstetteriana,	11
CTENOSTOMATA,	30
CYCLOSTOMATA,	1
CYCLOUM,	31
CYLINDRŒCIDÆ,	38
CYLINDRŒCIUM,	38
papuense,	38
DEFRANCIA,	25
DIASTOPORA,	9, 22
DIASTOPORA,	24
patina,	24
DIASTOPORES simples,	24
DIASTOPORIDÆ,	24
DISCOCAVEA,	25
DISCOPORA,	25
hispida,	26
DISCOPORADÆ,	25
DISCOPORELLA,	25
fimbriata,	26
hispida,	26
DISCOPORELLIDÆ,	25
DISCOSPARSA marginata,	24
patina,	24
ENTALOPHORA,	18
anomale,	19
attenuata,	19
deflexa,	20
haastiana,	19
icaunensis,	19
proboscidea,	19
raripora,	19
ENTOPROCTA,	40
ERECTA,	8
EUCRATEA,	2
FALCARIA,	2
FARRELLA,	38
FARRELLA,	37
atlantica,	37
FASCICULINEA,	26
FASCICULIPORA,	28
FASCICULIPORA,	27
digitata,	29
ramosa,	27
FASCIGERIDÆ,	26
FILICRISIA,	2

	PAGE
FRONDIPORA,	27
ramosa,	27
FRONDIPORIDÆ,	26
FUNGELLA,	27
prolifer,	27
HALCYONELLEA,	30
HALCYONELLEÆ,	30
HALODACTYLUS,	31
HETEROPORELLA,	25
radiata,	26
hispida,	26
HORNERA,	14
affinis,	15
andegavensis,	15
borealis,	15
foliacea,	17
frondiculata,	15
lichenoides,	15
radiata,	10
serrata,	15
tubulosa,	15
violacea,	16
HORNERIDÆ,	8
IDMONEA,	9
angustata,	10
atlantica,	10
australis,	12
coronopus,	10
eboracensis,	12
fissurata,	14
marionensis,	11
milneana,	13
irregularis,	14
radians,	10
transversa,	13
IDMONEADÆ,	8
IDMONEIDÆ,	8
INARTICULATA,	8
INCRUSTATA,	8
LAGENELLA,	37
LAGONCULA,	37
LAOMEDEA,	36
LICHENOPORA,	25
fimbriata,	26
hispida,	26
LICHENOPORIDÆ,	25
MADREPORA,	25
MILLEPORA,	14
lichenoides,	15

PLATE I.

(ZOOL. CHALL. EXP.—PART L.—1886.)—Mm1.

PLATE I.[1]

CRISIA.

[1] A scale in millimetres is marked near the lower right-hand corner of Plates I.-VIII., which applies to all the figures on each, except where a different scale is marked.

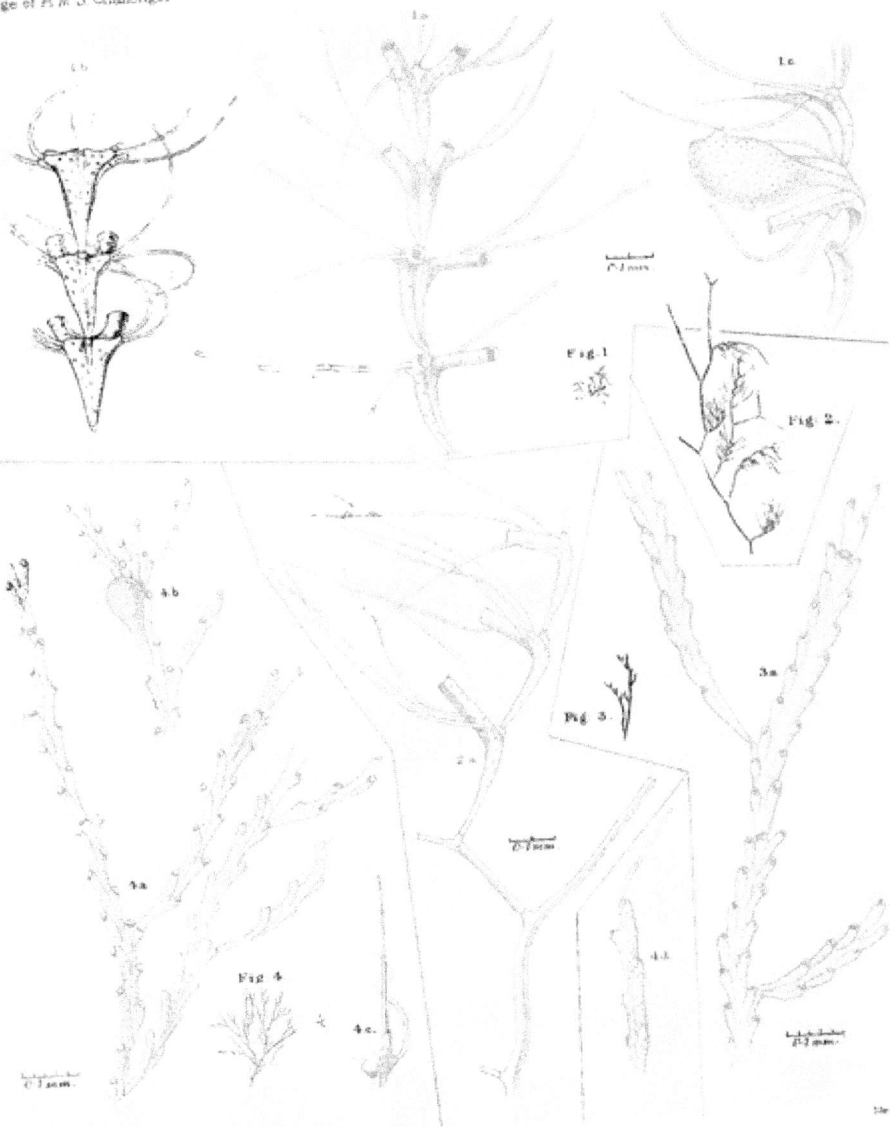

PLATE II.

PLATE II.

CRISIA.

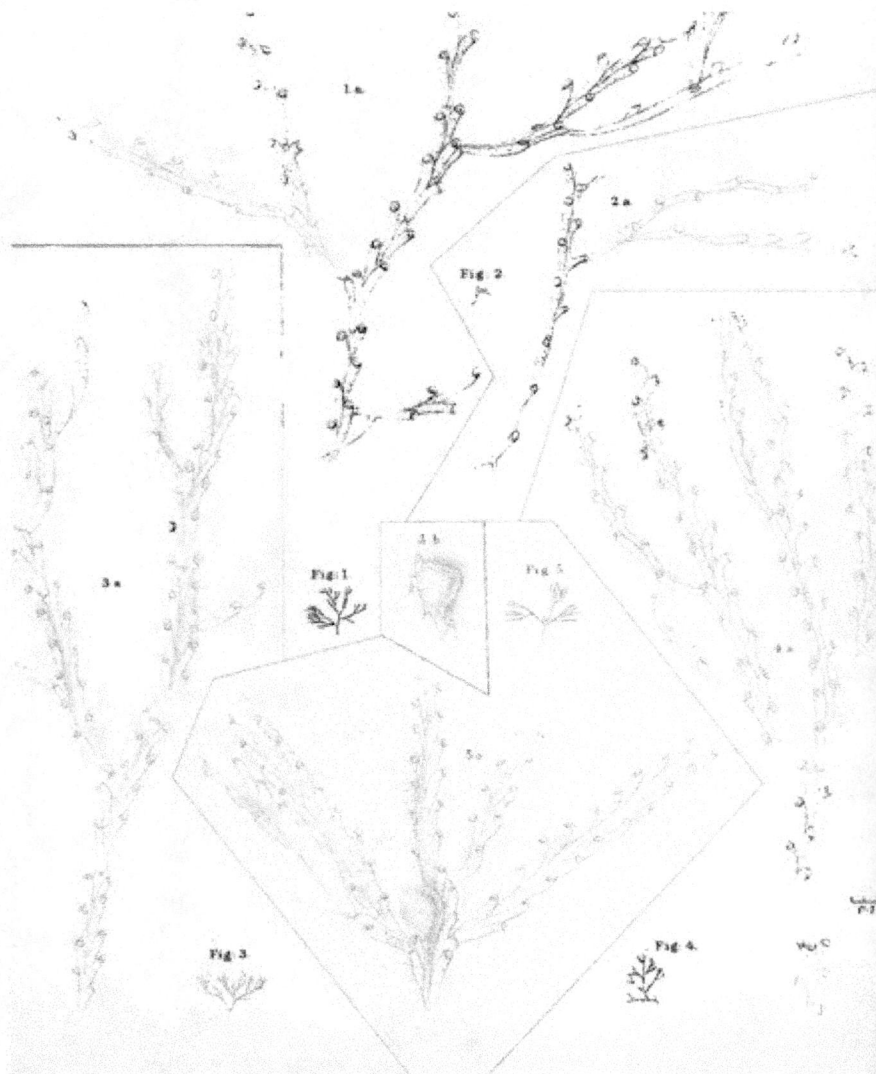

Fig. 2

Fig. 1

1 b

Fig. 5

Fig. 3

Fig. 4

CRISIA.

PLATE III.

PLATE III.

CRISIA.—IDMONEA.

PLATE IV.

PLATE IV.

PUSTULOPORA.

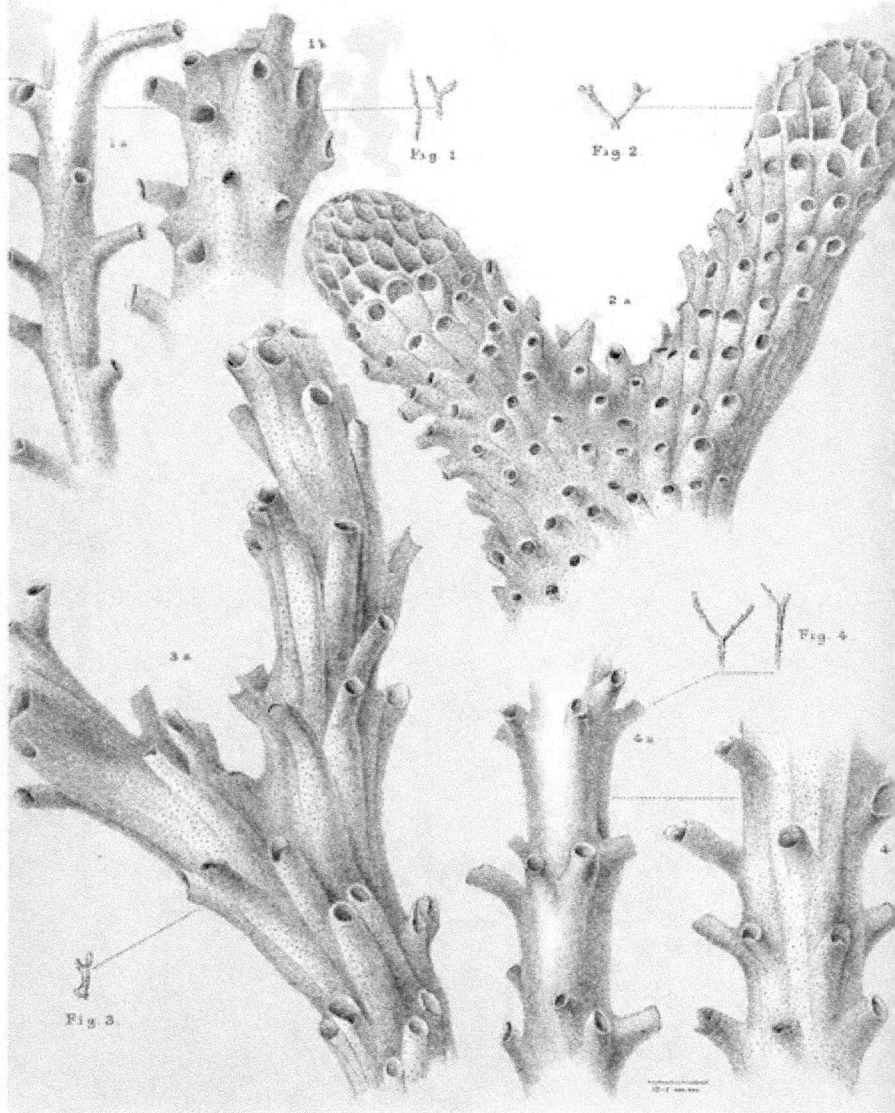

PLATE V.

(ZOOL. CHALL. EXP.—PART L.—1886.)—Ddd.

PLATE V.

TUBULIPORA.—SUPERCYTIS.

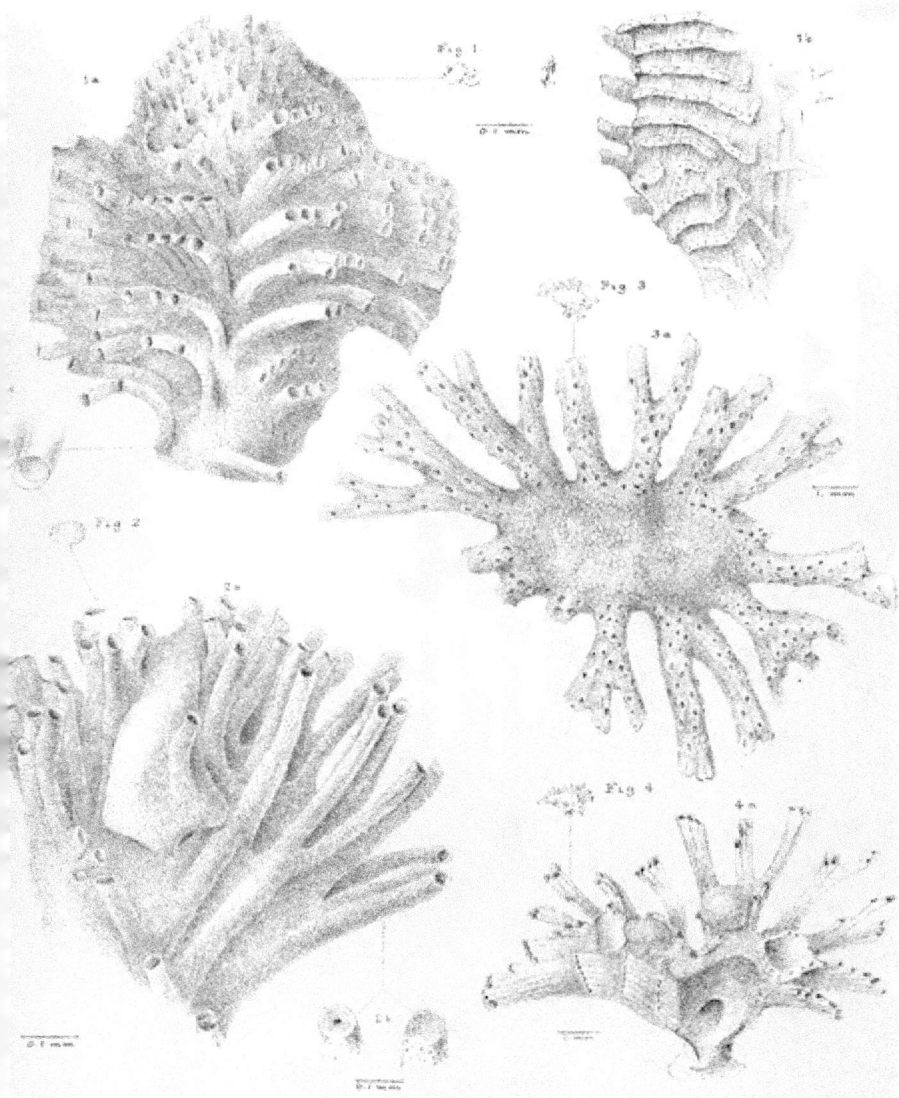

TUBULIPORA, SUPERCYTIS.

PLATE VI.

PLATE VI.

AMATHIA.

Fig. 1

Fig. 2

Fig. 3

AMATHIA.

PLATE VII.

PLATE VII.

AMATHIA.—FARRELLA.

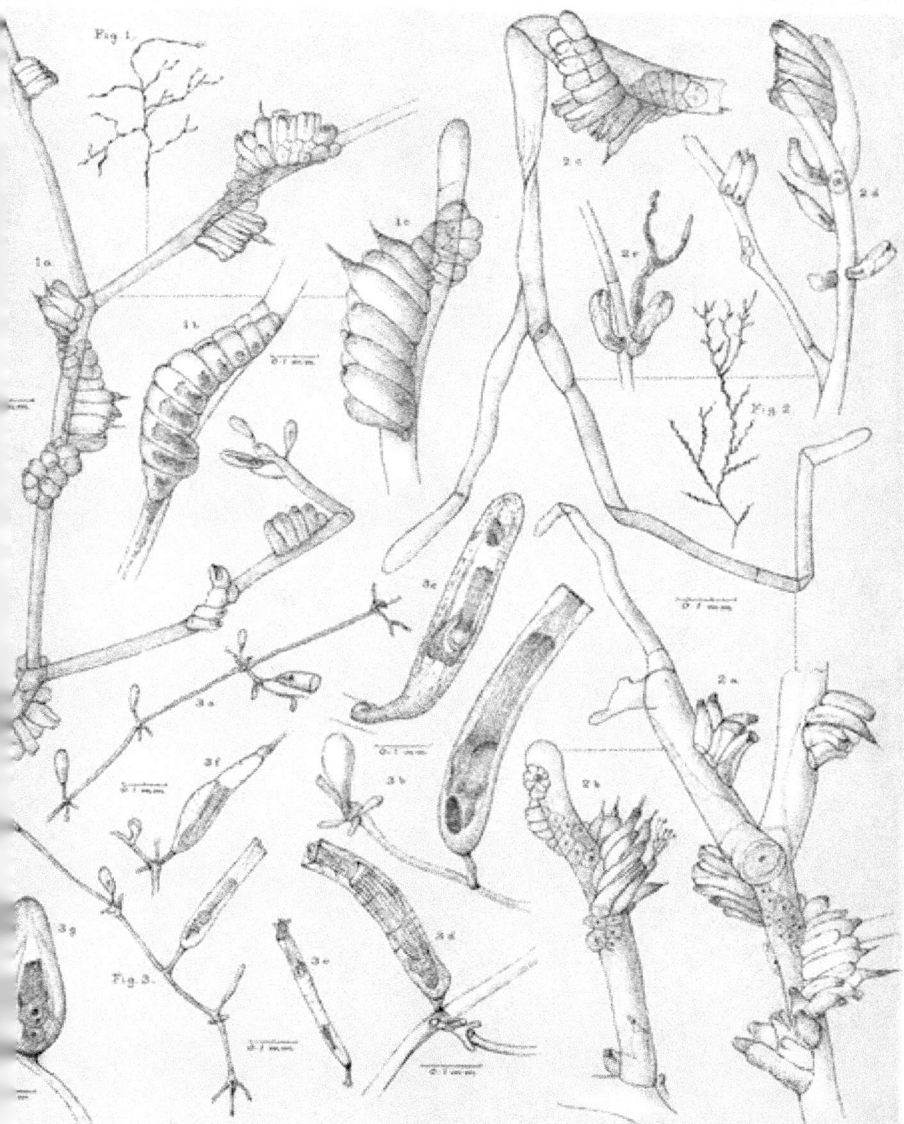

AMATHIA, FARRELLA.

PLATE VIII.

PLATE VIII.

AMATHIA.—VESICULARIA.—CYLINDRŒCIUM.

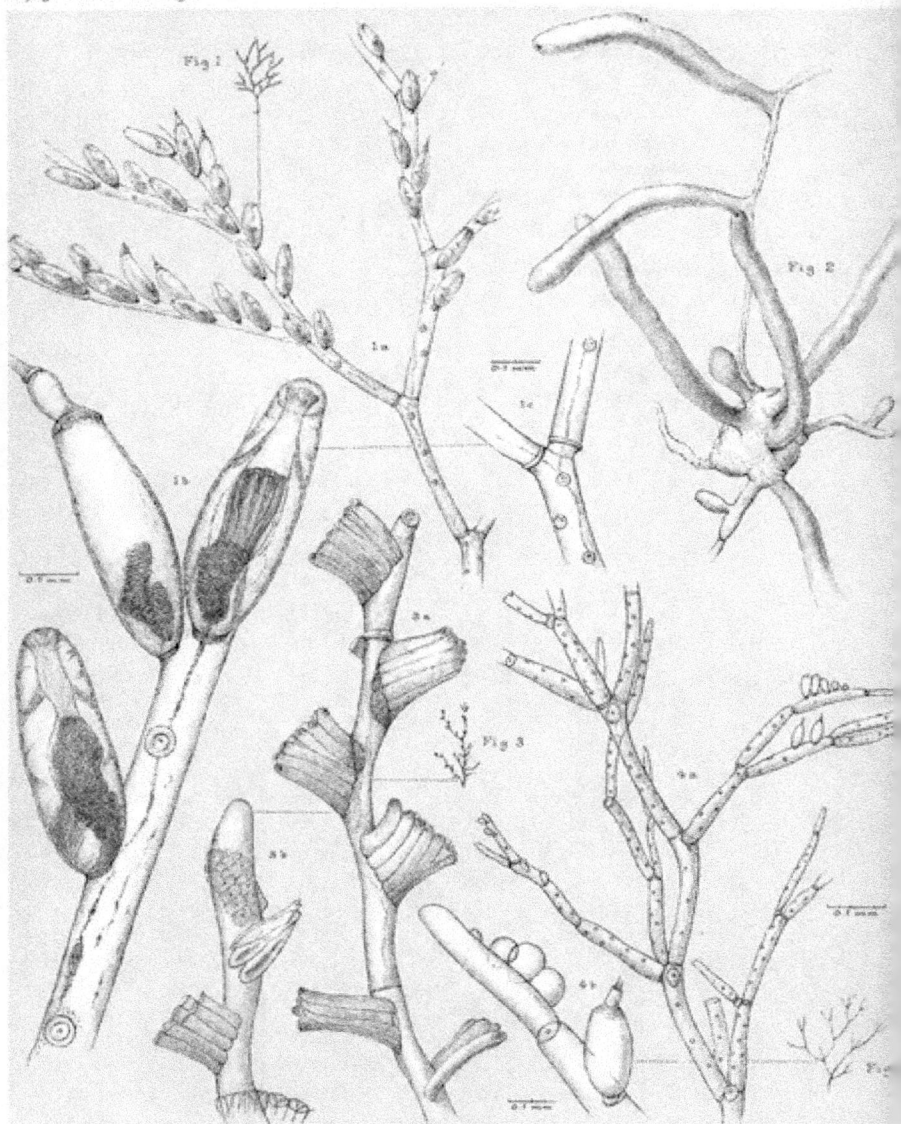

AMÁTHIA, VESICULARIA, CYLINDRŒCIUM.

PLATE IX.

(ZOOL. CHALL. EXP.—PART I.—1886.)—Ikkk.

PLATE IX.[1]

Ascopodaria.

		Diam.	Page
Figure	1.—*Ascopodaria fruticosa*, natural size, ...		42
,,	2.—A single zoœcium and polypide, . .	× 25	
,,	3.—A single polypide, with the tentacles coiled within the open edge of the cup, . . .	× 50	
,,	4.—Another polypide, with the edge of the cup contracted, ...	× 50	
,,	5.—A young polypide budding from the end of an old pedicel, ...	× 50	
,,	6.—A single polypide, showing the visceral organs, .	× 110	
,,	7.—Barrel-shaped expansion at the base of a pedicel, .	× 50	
,,	8, 9.—Longitudinal sections through a polypide, . .	× 110	
,,	10.—Portion of stem, with bases of peduncles, . .	× 25	
,,	11.—Longitudinal section of stem, . .	× 50	
,,	12, 13.—Transverse sections of stem, .	× 50	
,,	14.—Portions of the parenchymatous tissue, from the centre of the stem, . . .	× 360	

[1] Several references used on this plate:—*i*, intestine; *l*, liver cells; *œ*, œsophagus; *om*, ovarian masses; *r*, rectum; *s*, stomach; *t*, tentacles.

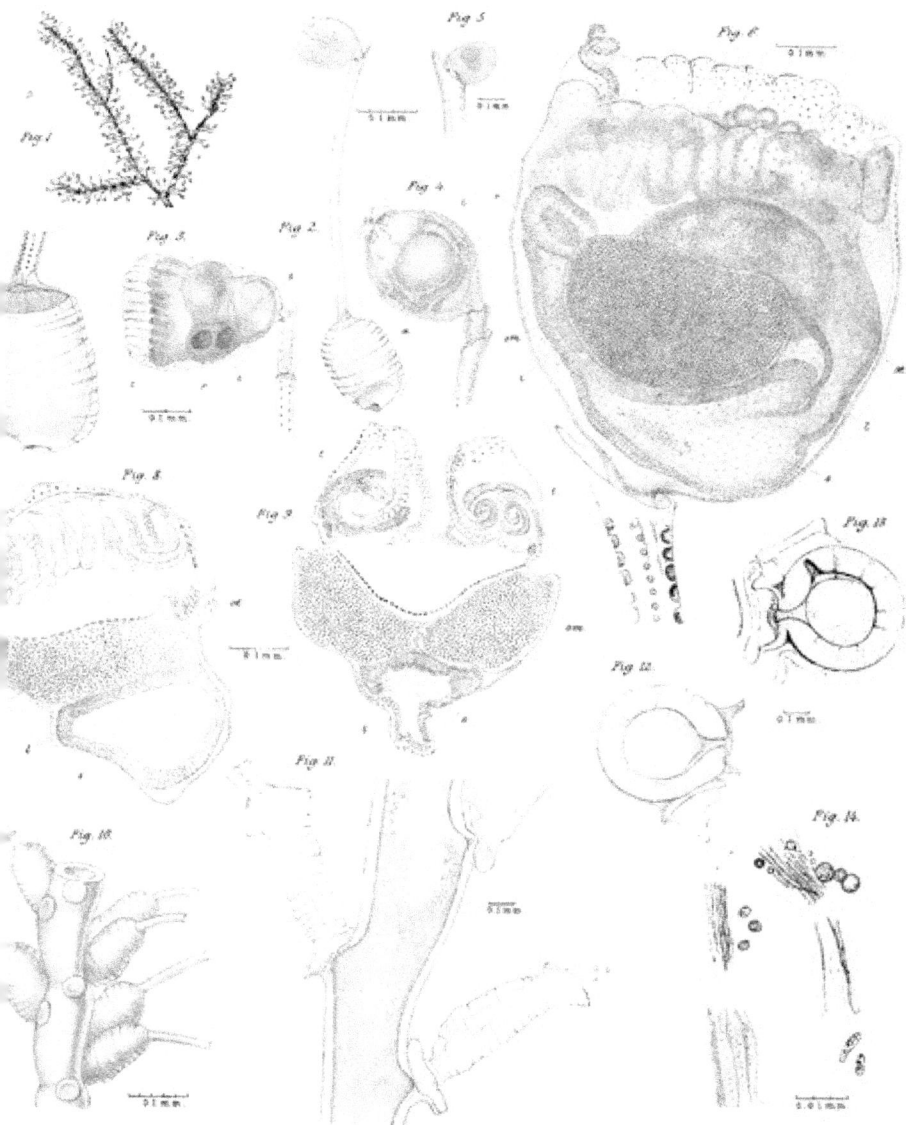

Fig. 1.

Fig. 2.

Fig. 3.

Fig. 4.

Fig. 5.

Fig. 6.

Fig. 7.

Fig. 8.

Fig. 9.

Fig. 10.

Fig. 11.

Fig. 12.

Fig. 13.

Fig. 14.

ASCOPODARIA.

PLATE X.

PLATE X.[1]

ASCOPODARIA.—ALCYONIDIUM.

[1] Several references used on this plate:—*i*, intestine; *l*, liver cells; *œ*, œsophagus; *om*, ovarian masses; *r*, rectum; *s*, stomach; *t*, tentacles.